Environment and Biodiversity

Environment and Biodiversity

Edited by **Neil Griffin**

SYRAWOOD
PUBLISHING HOUSE
New York

Published by Syrawood Publishing House,
750 Third Avenue, 9th Floor,
New York, NY 10017, USA
www.syrawoodpublishinghouse.com

Environment and Biodiversity
Edited by Neil Griffin

International Standard Book Number: 978-1-68286-279-7 (Hardback)

Printed in China.

Contents

Preface

I am honored to present to you this unique book which encompasses the most up-to-date data in the field. I was extremely pleased to get this opportunity of editing the work of experts from across the globe. I have also written papers in this field and researched the various aspects revolving around the progress of the discipline. I have tried to unify my knowledge along with that of stalwarts from every corner of the world, to produce a text which not only benefits the readers but also facilitates the growth of the field.

Earth is home to a wide variety of species and all of them play a crucial part in maintaining equilibrium in their surrounding environment and respective biomes i.e., ecosystems. This book traces the progress of this field and highlights some of its key aspects such as environmental conservation, management and planning, environmental health and biodiversity, etc. It strives to provide a fair idea about this discipline and to help develop a better understanding of the latest advances within this field. For someone with an interest and eye for detail, this book covers the most significant topics in the field of environment and biodiversity.

Finally, I would like to thank all the contributing authors for their valuable time and contributions. This book would not have been possible without their efforts. I would also like to thank my friends and family for their constant support.

Editor

Population dynamics of *Chrysoperla externa* (Hagen) (Neuroptera: Chrysopidae) in a silvopastoral system

Simone Alves de Oliveira[1], Alexander Machado Auad[2*], Brígida Souza[3], Marcy das Graças Fonseca[2] and Tiago Teixeira Resende[2]

[1]Federal University of Juiz de Fora, MG, Brazil.
[2]Embrapa Dairy Cattle Research Center, Cep. 36038-330. Juiz de Fora - MG. Brazil.
[3]Federal University of Lavras, Brazil, Caixa Postal 3037, CEP 37200-000, Brazil.

In this study we investigated lacewings (Neuroptera: Chrysopidae) in a silvopastoral system. The study was conducted between July, 2006 and June, 2008 in a pasture with *Brachiaria decumbens* and trees. Adult specimens were caught twice a month by using a malaise trap, while the larvae were caught in the shoots of fodder and by visual observation. Flowering panicles were also observed for the presence of immature specimens. All adult lacewings (1186) belonging to one species *Chrysoperla externa* (Hagen, 1861). The adult population peaked between April and June, but this occurrence was significantly affected by temperature and rainfall. We only captured 8 adult lacewings by using the malaise traps. There was a significant correlation between the number of immature specimens and relative humidity and the number of captured adults. A higher population density of larval and adult lacewings was associated with the flowering of signalgrass, leading to the inference that pollen was used as a source of protein. This strategy ensures that these predators remain on crops during periods of prey scarcity.

Key words: Lacewing, pasture, predator, seasonality, survey.

INTRODUCTION

Insects represent the most abundant animals in almost all ecosystems and can be used to evaluate the impact of environmental changes (Rocha et al., 2010). It is important to study insects to understand their close relationship with the environment. It is of great importance to know the population dynamics of predators such as green lacewings (Neuroptera: Chrysopidae) among insects. Green lacewings are found in many natural environments or environments modified by humans (Costa et al., 2003; Miliczky and Horton, 2005; Silva et al., 2006), where they feed on various pest insects. We know very little about the population dynamics of Neuroptera in most of these environments, including silvopastoral systems. These systems can accommodate a great diversity of arthropods because, as an exploitation model, they closely resemble a natural forest and have trees, grasses, and cattle in the same area, showing several ecological interactions. Among the different species of grasses, those in the genus *Brachiaria* (Poaceae), such as *Brachiaria decumbens* Stapf have shown good acceptance (Boddey et al., 2004).

Lacewing may be present in almost all agroecosystems (McEwen et al., 2001). It is necessary to explore the relationship of these predators with different food sources present in these environments. The ecological study of green lacewings in the fields is very important for conservative biological control. Thus, the maintenance of such habitats, which provide sites for development and permanence of these predators in crops, is of prime importance for the success of these insects as biological control agents. Knowledge of the population dynamics of a species is of paramount importance for their effective management. The presence of an insect in a given ecosystem may represent only a casual introduction, making it necessary to confirm its interaction with the crop by observing its developmental stages or

*Corresponding author. E-mail: amauad@cnpgl.embrapa.br.

seasonality. For most Chrysopidae, there is more or less of an established regional fluctuation in population every year (Souza and Carvalho, 2002). However, factors such as climate change and decreased food supply may alter the synchrony of a population (Szentkirályi, 2001).

According to Rao et al. (2000), research on understanding the influence of ecological factors that promote the population dynamics of natural enemies in agroforestry systems is necessary for sustainable pest management. Therefore, knowledge of the population dynamics of lacewings and interactions with their habitat is important for their use as biocontrol agents. Thus, the aim of our study was to evaluate the population dynamics of green lacewings in a silvopastoral system.

MATERIALS AND METHODS

Tests were conducted on a silvopastoral system in Coronel Pacheco MG, Brazil (21º 33′ 22″ S and 43º 06′ 15″ W; altitude, 410 m) between July, 2006 and June, 2008. The climatic of the region is cwa (mesothermal), according to Koppen's classification, with a dry season from May to October and a rainy season from November to April. The temperature ranges from 16°C in winter to ≥ 23°C in summer (Antunes, 1986). The 4 ha sampling area consisted of *B. decumbens* maintained in a 30 m-wide band alternating with 10 m bands comprising *Eucalyptus grandis* W. Hill ex Maiden (Myrtaceae), *Acacia mangium* Wild (Fabaceae), *Acacia angustissima* Wild (Fabaceae), and *Mimosa artemisiana*, Heringer and Paula (Fabaceae).

Adult green lacewings were sampled biweekly for a period of 2 years for 2 consecutive hours in the morning. The sampling was performed in the entire area in a zigzag-type pattern, and the green lacewings were captured using a hand net. A malaise trap (Townes model) remained installed throughout all sampling periods, permitting uninterrupted collection of the insects. The specimens were preserved in a vial containing 70% ethyl alcohol and formaldehyde drops and removed every 15 days. The lacewings were identified by Dr. Renildo Ismael Félix Costa of the Federal Institute of Northern Minas. Larvae were sampled every fortnight by using a sweep net to reach the grass shoots and through observations of plants in 1 m^2 areas that were randomly selected between April, 2007 and March, 2008. During the flowering periods, approximately 500 inflorescences were collected and inspected for immature lacewings in the laboratory.

Data on temperature, rainfall, and relative humidity were obtained from the meteorological station of Embrapa Dairy Cattle located 100 m from the sampling area. The results were analyzed for correlation between climatic factors and the occurrence of insects by using the Spearman test (Saeg, 2007).

RESULTS AND DISCUSSION

Over 24 months of active collection, peaks in the populations of adult chrysopids were observed in April, May, and June 2007, when 421, 447, and 310 insects, respectively, were collected (Figure 1A). *Chrysoperla externa* (Hagen, 1861), a very common and well-studied species in the neotropical region (Souza and Carvalho, 2002) were the only species collected in our study. The high population density in the months mentioned earlier suggests a possible migration of adults from surrounding areas, since in the preceding months, the insect was either absent or in insufficient numbers to cause these population peaks (Figure 1A). It is interesting to note that, *B. decumbens* was in the flowering period in the months of the highest abundance of *C. externa*; this may be one of the factors that stimulated the migration of the insects. The population of *Chrysoperla sinica* that migrates via long-distance flights was observed by Liu et al., (2011).

Abiotic factors influenced the population dynamics of *C. externa*, but the number of *C. externa* adult specimens was not influenced by relative humidity (r = 0.006, z = 0.293, P = 0.384). There was a significant correlation between the number of adults and temperature (r = - 0.552, z = - 2.648, P = 0.004) and rainfall (r = - 0.508, z = - 2.438, P = 0.007). The highest number of insects was observed during the months in which temperature and rainfall showed the lowest indices (Figure 1B). Similar findings were obtained by Gitirana et al. (2001) and Souza and Carvalho (2002), who recorded the highest populations of *Ceraeochrysa* spp. and *C. externa*, respectively, in citrus orchards during the months with low temperature and rainfall. Silva et al. (2006), who studied population fluctuations in *C. externa* in coffee crops in Brazil, recorded a reduction in the number of insects with increase in rainfall and temperature. Thus, these climatic factors limit the occurrence as well as the survival of these species under natural conditions. Souza and Carvalho (2002) reported that rainfall, for example, can influence many aspects of the population behavior of these insects, through direct mechanical action or indirect effect on their habitats.

In contrast with the present study, Cardoso et al. (2003) and Costa (2006) verified positive correlations between temperature and abundance of green lacewings in the forest and observed a higher number of insects in hot periods. Accordingly, the absence of significant correlations between rainfall and lacewing populations in that environment could be attributable to the collection of samples near a creek, where water is available throughout the year. Although, Penny (2002) used the malaise trap to catch Neuropterans, this collecting method was not effective in our study. During the 2 sampling years, we captured only 8 *C. externa* adults. We caught 1 adult in December 2006, 1 in May 2007, 3 in June, 2007 and 3 in October, 2007. This record is in agreement with the findings of Carvalho and Souza (2000), who reported that hand nets were more effective than any other method for collection of lacewings. Ábrahám et al. (2003) investigated the Neuroptera community by using different capture methods and observed that suction traps were more efficient than malaise traps.

We collected 31 larvae over the year (Figure 1A) in the flowers of *Brachiaria*. The number of larvae did not correlate with temperature (r = - 0.054, z = - 0.179, P =0.428) or rainfall (r = 0.054, z = 0.179, P = 0.428). However, there was a significant correlation between the number of immature specimens and relative humidity(r =+0.545, z = 1.807, P = 0.035) and the number of

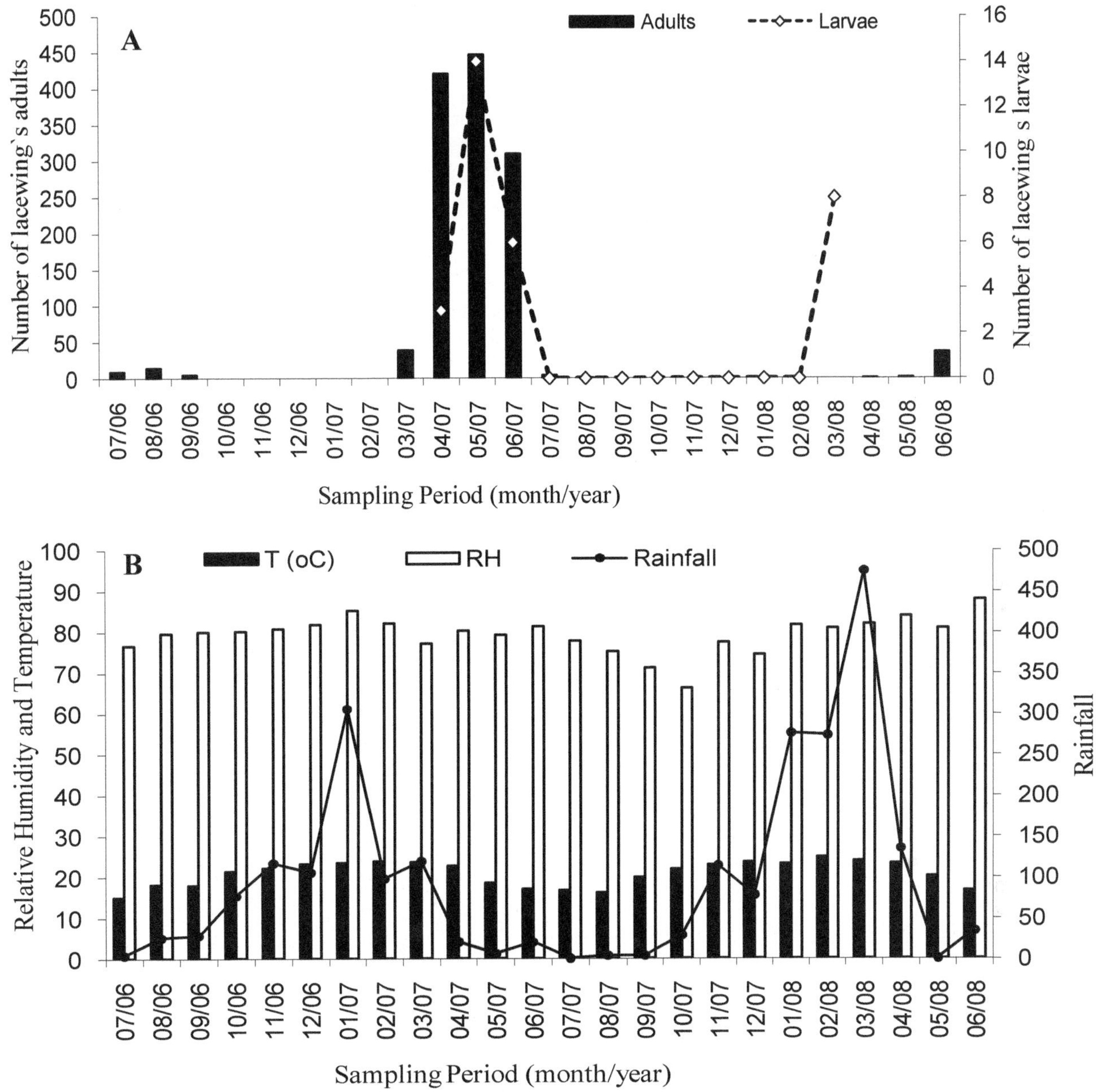

Figure 1. A - Seasonal distribution of adults and lacewings larvae collected in the silvopastoral environment, Coronel Pacheco, MG. B - Average monthly temperature (T°C), relative humidity (RH%) and rainfall (mm) during the sampling period in silvopastoral environment, Coronel Pacheco, MG. July 2006 to June 2008.

captured adults (r = 0.598, z = 1.984, P = 0.023). We considered the *C. externa* population as a function of plant phenology and observed that the population peak observed in May coincided with the time of flowering of *Brachiaria*, which usually occurs from May to July, according to Lascano et al. (2002). Silva et al. (2006) observed that the presence of grasses in an agroforest increased the occurrence of insects who consumed the pollen as food.

Thus, we believe that the population density of *C. externa* in silvopastoral systems is probably related to the availability of pollen, which serves as the prime food source for the adults. Such findings were reinforced by the absence of any other arthropod species that could serve as a diet for this predator. Thus, pollen, which is known as a food source for adult Chrysopidae alone, may also allow larval development being used as an important protein source at the times when preys are insufficient.

These results were in accordance with those obtained by Oliveira et al. (2009) who studied the biological response of *C. externa* larvae fed on grass pollen diets and observed that wet pollen can be an alternative protein source for chrysopids. This justifies the positive correlation between relative humidity and the number of larvae sampled in this study. These observations may imply an interesting and important strategy used by larvae that guarantees the permanence of these predators on crops in times of prey shortage. Moreover, the presence of higher plant diversity in the silvopastoral system provides appropriate conditions for ensuring the permanence of these predators on the crops.

According to McEwen et al. (2001), green lacewing may be present in almost all agroecosystems. It is necessary to explore the relationship of these predators with the different food sources (pests) present in these environments. The ecological study of green lacewings in the fields is of fundamental importance for conservative biological control. Thus, the maintenance of such habitats that provide sites for development and permanence of these predators in crops is of prime importance for the success of these insects as biological control agents.

Conclusions

1. *C. externa* was the only species of lacewing captured in a silvopastoral system, with maximum populations recorded during the months with low rainfall and temperature.
2. *C. externa* larvae were associated with *B. decumbens* inflorescences during periods of prey scarcity and they fed on pollen, the best source of protein.

ACKNOWLEDGMENTS

We thank the Conselho Nacional de Desenvolvimento Científico e Tecnológico (CNPq, Brazil) and Fundação de Amparo à Pesquisa do Estado de Minas Gerais (FAPEMIG, Brazil) for supporting our research.

REFERENCES

Ábrahám L, Markó V, Vas J (2003). Investigations on a Neuropteroid community by using different methods. Acta Phyt. Ent. Hungarica, Budapest, 38: 199–207.

Antunes FZ (1986). Characterization the climate of the state of Minas Gerais. Inf. Agropec., 12: 9-13.

Boddey RM, Macedo R, Tarré R, Ferreira E, Oliveira OC, Rezende CP, Cantarutti RB, Pereira JM, Alves BJR, Urquiaga S (2004). Nitrogen cycling in *Brachiaria* pastures: the key to understanding the process of pasture decline. Agric. Ecosyst. Environ., 103: 389-403.

Cardoso JT, Lázarri SMN, De Freitas S, Iede ET (2003). Occurrence and population fluctuation of *Chrysopidae* (Neuroptera) on *Pinus taeda* (L.) (Pinaceae) in Southern Paraná. Rev. Brasil. Entomol., 47: 473-475.

Carvalho CF, Souza B (2000). Methods for rearing and production of chrysopids. Biological control of pests: mass production and quality; Lavras, pp. 91-110.

Costa RIF (2006). Taxocenosis of green lacewings (Neuroptera: Chrysopidae) in natural and agricultural ecosystems. (Thesis Doctoral in Agronomy/Entomology). Federal University of Lavras, Lavras, p. 124.

Costa RIF, Carvalho CF, Souza B, Loreti J (2003). Influence of individuals density on rearing of Chrysoperla externa (Hagen, 1861) (Neuroptera: Chrysopidae) larvae. Ciên. Agrotec, 27: 1539-1545.

Gitirana Neto J, Carvalho CF, Souza B, Santa-Cecília LVC (2001). Population dynamics ofCeraeochrysa Adams, 1982 (Neuroptera: Chrysopidae) species on citrus trees at the Lavras – MG region. Ciên. Agrotec., 25: 550-559.

Lascano C, Pérez R, Plazas C, Medrano J, Pérez O, Argel PJ (2002) Toledo grass pasture (Brachiaria brizantha CIAT 26110): it is a grass that grows fast and can intensify the Colombian cattle production. Villavicencio: Corporación Colombiana de Investigación Agropecuaria; Cali: Centro Internacional de Agricultura Tropical, p. 18.

Liu Z, Wyckhuys KAG, Wu K (2011) Migratory Adaptations in *Chrysoperla sinica* (Neuroptera: Chrisopidae). Environ. Entomol., 40: 449-454

McEwen PK, New TR, Whittington AE (2001). Lacewings in the crop environment. Cambridge, Cambridge University.

Miliczky ER, Horton DR (2005). Densities of beneficial arthropods within pear and apple orchards affected by distance from adjacent native habitat and association on natural enemies with extra-orchard host plants. Biol. Control, 33: 249-259.

Oliveira SA, Auad AM, Souza B, Carvalho CA, Souza LS, Amaral RL, Silva DM (2009). Benefits of honey and pollen of forrage on the biological parameters of *Chrysoperla* externa (Hagen, 1861) (Neuroptera: Chrysopidae). Arq. Inst. Biol., 76: 583-588.

Penny ND (2002). A guide to the lacewings (Neuroptera) of Costa Rica. Proc. Calif. Acad. Sci., 53: 161-457.

Rao MR, Singh M p, Day R (2000). Insect pest problems in tropical agroforestry systems: Contribuitory factors and strategies for management. Agrofor. Syst., 50: 243-277.

Rocha JRM, Almeida JR, Lins GA, Durval A. (2010). Insects as indicators of environmental changing and pollution: A review of appropriate species and their monitoring. Holos Environ., 10: 250-262.

SAEG (2007). System for statistical analysis, version 9.1. Federal University of Viçosa, Brazil.

Silva RA, Reis PR, Souza B, Carvalho CF, Carvalho GA, Cosme LV. (2006). Population dynamics of adults of Chrysoperla externa (Hagen, 1861) (Neuroptera: Chrysopidae) in coffee crops conducted in organic and conventional systems in Santo Antônio do Amparo County, MG. Man. Integr. Plagas Agroecol., 77: 44-49.

Souza B, Carvalho CF (2002). Population dynamics and seasonal occurrence of adults of *Chrysoperla externa* (Hagen, 1861) (Neuroptera: Chrysopidae) in a citrus orchard in southern Brazil. Acta Zool. Acad. Sci Hung., 48: 301-310.

Szentkirályi F (2001). Lacewing infruit and nut crops. In: Mcewen PM, New TR, Whittington AE. Lacewing inthe Crop Environment, Cambridge, pp. 172-220.

Distribution, conservation and folk uses of Vaibidang (*Embelia ribes* Burm. f.)

Rama Shankar[1], G. S. Lavekar[2], Sourabh Deb[3], B. K. Sharma[4] and M.S.Rawat[5]

[1]Ayurveda Regional Research Institute, Itanagar, India-791111.
[2]Central Council for Research in Ayurvedic Science, New Delhi- 110058, India.
[3]Department of Forestry and Biodiversity, Tripura University, Suryamaninagar, Tripura, India.
[4]Department of Microbiology, Tripura University, Suryamaninagar, Tripura, India.
[5]National Medicinal Plants Board, Chandralok Building, Janpath, New Delhi, 110001, India.

Vaibidang (*Embelia ribes* Burm. f.) is a red listed climbing shrub found in the semi-evergreen to evergreen forests of India, Sri Lanka, Malaysia and China, whose fruits are used in a large number of Ayurvedic formulations. It is traditionally used as analgesic, antipyretic, antibacterial, antifertility and anthelmintic in different parts of India. The dosage of traditional medicine prepared from this herb varies from region to region. Ripening and timely harvesting are also an important factor for the potency of this herb. Introducing fully mature seeds for raising planting materials is only an accessible way of its propagation.

Key words: Traditional, Myrsinaceae, ayurvedic, embelin, Arunachal Pradesh.

INTRODUCTION

Vaibidang is a Sanskrit name of *Embelia ribes* Burm. f. (Myrsinaceae), which is an important medicinal plants belonging to family Myrsinaceae. The plant is a scandent shrub, whose fruits are used in a large number of Ayurvedic formulations. This shrub is slender branched with elliptic-lanceolate and gland-dotted leaves. The fruit is globular and wrinkled, varying in colour from dull red to nearly black; a short pedicel is often present; the pericarp is brittle enclosing a single seed covered with a membrane. Flowers are dull white in colour with violet red fruits. Mature fruits turn brown or black. Root is deeply penetrating in the soil and takes help of trees for their climbing habit. In North East region, it is commonly distributed in Arunachal Pradesh, Meghalaya and Mizoram in the altitudinal zone of 500 to 2500 msl. It is a red listed climbing shrub found in the semi-evergreen to evergreen forests of India, Sri Lanka, Malaysia and China (Guhabakshi et al., 2001). The species is also reported to be vulnerable in the Western Ghats of Tamil Nadu and Karnataka states of India and at lower risk in Kerala state (Ravikumar and Ved, 2000). The plant is locally called as Biakol-lata (Assamese), Baberang (Hindi), Biranga (Bengali), and is recorded from Itanagar, Kimin, Hunli, Dibang Valley, Nechephu and Dirang areas (West Kameng district), Joram, and Hija areas near Hapoli in Lower Subansiri district of Arunachal Pradesh. The species was also reported from upper Assam and from Cherapunji and Jowai areas of Khasi/Jaintia hills, and Mao areas of Manipur, Kawnpui - Aizawl, Serchhip areas of Mizoram in both high and low altitudes of north east India. It was also found that there are some morphological variations in leaf characters among the species collected from different altitudes.

*Corresponding author. E-mail: rshankar58@gmail.com.

MATERIALS AND METHODS

The study is purely based on exploration of different forest areas in North eastern India as well as interaction with traditional healers to know about different uses of *E. ribes* in their traditional healing practices. Trials on germination of seeds in different conditions were undertaken for examining the earlier statements regarding the cultivation of *E. ribes*. Pharmaceutical use of fruits and other part of the plant was explored from literature and from various pharmaceutical products where fruits of *E. ribes* are used in compound formulation. During exploration, status and systemic identity of the plant was established which in turn was followed by the characterization and distribution of the highly demanded medicinal plant for pharmaceutics. The methodology for use of plant drug against particular disease was recorded. To study the possible means for cultivation, both vegetative and seed germination trials were made. For vegetative propagation, stem cuttings of different age groups were taken in first instance. After getting only shoot development without root initiation by treating the stem cuttings with gibberelic acid-3, stem cuttings of basal region of stem with attachment of few rootlets was taken. For seed germination, both immature red coloured and mature brown coloured seeds were taken under the study. Counseling for the conservation of plant of *E. ribes* was made by involving local people by providing approximate possible methodology in field for conservation and sustainable use of drug parts. Voucher specimens were prepared for authentic records by matching with the authentic samples kept in the Herbarium of Botanical Survey of India, Eastern Circle, Shillong. Voucher specimens were deposited in the Herbarium of Ayurveda Regional Research Institute Itanagar (Acronym- arri).

RESULTS AND DISCUSSION

Medicinal uses

The plant is an ingredient of about 75 traditional ayurvedic drug formulations. Dried fruits of the plant are the main source of an ayurvedic drug. The dosage of traditional medicine prepared from this herb varies from region to region. Ripening and timely harvesting are also an important factor for the potency of this herb.

The plant contains embelin, quercitol and fatty ingredients, an alkaloid, christembine, tannins and minute quantities of a volatile oil (Krishna and Verma, 1941). Embelin occurs in golden yellow needles insoluble in water and soluble in alcohol, chloroform and benzene. It is reported to be effective against tapeworm but not against roundworm or hookworm. Embelin dyes silk and wool from an alcoholic solution. The dark colored fatty oil is reported to be similar to linseed and rapeseed oil in its properties. The plant is reported as analgesic, antipyretic (Atal et al., 1984), antibacterial (Narang et al., 1961), antifertility (Seshadri and Venkataraghavan, 1981) and anthelmintic (Chopra et al., 1966) in different parts of India. In North East India, *E. ribes* is used for liver disorder and for removal of worms from subtropical region of Arunachal Pradesh. Sinha (1996) and Mahanti (1994) reported this plant from Manipur and Mizoram as anthelmintic, astringent, carminative and stimulant used in constipation as purgative, fever and skin diseases. Root bark is used for cough and diarrhoea, and young leaves are used for abdominal pain (Anonymous, CCRAS, 1999). Mahanti (1994) added the use of *E. ribes* for scorpion sting and snake bite.

The pharmacological and clinical investigations by various workers gave promising results about its antifertility activity without any side effects (Mitra, 1995; Anon, 2002). Different medicines prepared from this plant product are Vaibidang Churna (classical ayurveda for worm repellent), Abana (cardiac disorders), Diak of (cough care), Gasex (gastri care), Geriforte (geri care/old aged management), Herbolax (laxative), Koflet (cough care), Mentat (nervine management), Mentat syrup, Purim (blood purification), blood purifier capsules and syrup, dental cream, anxocare, Appetonic forte Vet, Appetonic Vet, Geriforte Aqua, Geriforte Vet, etc.

Identity and cultivation

It was reported that the cultivation of *E. ribes* is difficult, hence the conducting of different experiments on various economic mode of cultivation, that is, seed germination, vegetative cultivation by stem cutting, basal stem cutting with some root pieces in medicinal plant garden of Ayurveda Regional Research Institute , Itanagar. During the period of experiments made on various economic aspects of cultivation accessible for farmer's field for a period of five years, it was observed that maturation of fruits in one bunch takes place at an interval of certain period. Accordingly, it was found that the seed germination in case of fully mature seed has about 90% viability, and introducing fully mature seeds for raising planting materials is the only accessible way of propagation (Table 1 & 2). Generally, all seeds of the plants do not mature at the same time, and maturation starts from the apical towards bottom region. It was observed that rooting does not take place in the stem cuttings of different age groups. Whereas stem cuttings of base of stem where few rootlets are intact, develops more roots. In case of seed germination trials, only fully mature seeds having brown black colour initiate germination. Results of vegetative trials and seed germination are given in Table 2. The confusion regarding its cultivation and regenerations starts due to its differentiation and segregation of mature and immature seeds. The species can also be cultivated through vegetative means, and for that, single root of the plant is sufficient for its multiplication but without root its propagation is just impossible. Hence, it is not worth reporting that the regeneration of this plant is poor or not easily cultivated. This plant can give 80 to 90% viability if it can be cultivated in a proper technique and only with matured seeds. Raghu et al. (2006) mentioned that embryos are small when present and most of the seeds are abortive. Therefore, specific habitat conditions are required for its survival and growth. Identity of the plant was confirmed as some market samples were not genuine drug material. To strengthen the cultivation part of the study various experiments were made in the agroclimatic condition of the state of Arunachal Pradesh where plant naturally occurs (Figure 1).

Wrongly collected samples are also found in the markets, as many of the market samples examined and do not contain Embelin, either due to wrong identity or due to collection of immature fruit drugs. Such types of

Table 1. Characteristics of *E. ribes* found in North East India.

S/N	Parameter	Observation
1	Type	Dioecious woody climber
2	Habitat	Liana
3	Distribution	Itanagar, Kimin, Hunli, Dibang Valley, Nechephu, Dirang, West Kameng, Joram and Hija areas near Hapoli in Lower Subansiri district of Arunachal Pradesh. The species was also reported from upper Assam and Khasi/Jaintia hills
4	Flowering/fruiting season	March - May fruiting initiates during July and mature during November to December.
5	Immature fruits	
6	Mature fruits:	
7	Local Name	Biakol-lata (Assamese); Baberang (Hindi); Biranga (Bengali)
8	Trade name	Kala vaivding
9	Price	Rs. 110 - 160/kg
10	Embelin content (%)	2.3 - 3.1%
11	Occupancy (km^2)	10-500
12	Density	1-5/100 ha
13	Polpulation decline	(%): >50
14	Status	Rare
15	Threat	Habitat loss, immature harvest time
16	Cultivation practice	Seed germination, vegetative cultivation by stem cutting, basal stem cutting
17	Seed germination	80 - 90% with mature seeds
18	Phyto-constituents	Embelin, embolic acid and rapanone
19	Medicinal products	Vaibidang Churna, Abana, Diak, Gasex, Geriforte, Herbolax, Koflet, Mentat, Mentat syrup, Purim, Anxocare, Appetonic forte Vet
20	Major activity	Anti-helminthic, anti-tumour, bronchitis, mental disorders, jaundice, anti-fertility, analgesic, antibacterial, anti-inflammatory and antioxidant
21	Annual requirement in Indian markets	Approx. 5 - 8 tons

anomalies are not only with *E. ribes* but also with several other drugs which are under screening. External morphology of the plant growing in different altitudinal range varies in certain features without any scientific terminology. While conducting experiments on propagation of the plants collected from temperate and alpine regions in institute's medicinal garden established in tropical region (450 msl), it was observed that when plants from higher altitude are cultivated in lower altitudinal range it grows well but with gradual change in morphological features and forms the features of intermediary stage of the two climatic zone plants. Leaf found in low altitudinal range is broader and shorter than the plants growing on high altitude. Fruit characters in lower hills up to 800 msl are greenish and turn brown on maturity whereas the fruits in Alpine areas are red in colour which turns brown on maturity. On cultivation of higher altitude plants in the range of 500 msl, leaves gradually become broader.

Conservation

E. ribes plays an important role in Ayurvedic medicine as well as traditional medicine preparation in North East India. Nowadays, several medicinal pharmaceuticals trade its fruits from North East India for preparing different products. Poor people of remote villages in this region collect and exploit this plant from forests area for monetary benefits. The main threat of this plant is its unsustainable and indiscriminate harvesting for commercial purposes. Habitat loss, Jhum cultivation, forest fire and agriculture expansions are also some factors for its decrease in population. On the other hand, the

Table 2. Cultivation experiment on vegetative propagation and seed germination practices of *E. ribes.*

S. No.	Type of Experiment	No. of samples under study	Results	
			Period of observation (1 month)	Period of observation (2 months and above)
1	Stem cutting with full mature stem	100	No root initiation as well as foliage development takes place at different interval of time	stem cuttings get dried
2	Shoot cuttings with young leafy branches	100	No root initiation as well as foliage development takes place at different interval of time	Shoots get dried
3	Mature stem cuttings after treatment of Gibberellic Acid- 3	100	Foliage develops No rooting was observed	Foliage and single young shoot was initiated and dries later on No rooting was observed
4	Young shoot with treatment of Gibberellic Acid- 3	50	Foliage persists and new foliage initiates Rooting does not takes place	Only foliage persists later dries Rooting does not take place
4	Mature brown Seed selected for germination	1000	< 875	Seedlings developed
5	Random collection of seeds (Red and brown)	1000	>250	Seedlings developed

Figure 1. (A) Seedling of *Embelia ribes;* (B) Flowering of *E. ribes*; (C) fruiting of *E. ribes.*

regeneration from seedling of this plant is also very poor. While conducting market survey, fruit of this plant was found to be traded from Arunachal Pradesh via Assam to other parts of India. Hence, it is a matter of concern how this plant can be preserved and conserved. So, the government should take initiative for conserving this prioritized species, and emphasize more research and survey of wild population for evaluating the status of this plant in North East India. Development of *in vitro* propagation may play an important role in conserving this species.

ACKNOWLEDGEMENTS

Authors are thankful to the Chief Executive Officer, National Medicinal Plants Board, Department of AYUSH, Government of India for financial assistance. Thanks are also due to Dr. S. K. Sharma, Advisor (Ayurveda), Department of AYUSH, Ministry of Health and Family

Welfare, Government of India for different technical support.

REFERENCES

Atal CK, Siddique MA, Zutshi VA (1984). Non-narcotic, orally effective, centrally acting analgesic from an ayurvedic drug. J. Ethnol. Pharmacol. 11(3):309-317.

Chopra RN, Nayar SL, Chopra IC (1966). Glossary of Indian Medicines. CSIR Publication, New Delhi p.106.

Guhabakshi DN, Sensarma P, Pal DC (2001). A lexicon Medicinal Plants of India. Calcutta, India, Naya Prakashan, Calcutta pp.135-136.

Krishna S, Verma KS (1941). Embelia ribes. Indian Forest Bulletin: N. S. No. 102.

Mahanti N (1994). Tribal ethnobotany of Mizoram. Tribal studies of India series T 171. Inter India Publication, New Delhi p.64.

Narang GD, Garg LC, Mehta RC (1961). Preliminary studies on antibacterial activity of *E. ribes* (Myrisinaceae). J. Vet. Anim. Husband. Res. 5(1):73-79.

Raghu AV, Geetha SP, Martin G, Balachandran I, Ravindran PN (2006). Direct shoot organogenesis from leaf explants of *Embelia ribes* Burm. f.: A vulnerable medicinal plants. J. For. Res. 11:57-60.

Seshadri C, Venkataraghavan S (1981). Effect of aqueous and alcoholic extract of *E. ribes* on liver plasma and reproductive organs in male rats, a preliminary study. Indian J. Pharmacol. 13(1):88.

Sinha SC (1996). Medicinal Plants of Manipur, Mass and Sinha, Manipur Cultural Integration Conference, Imphal p.69.

Rice landraces of Kerala State of India: A documentation

Latha M.[3]*, Abdul Nizar M.[1], Abraham Z.[3], Joseph John K.[3], R. Asokan Nair[3], Mani S.[3] and M. Dutta[2]

[1]National Bureau of Plant Genetic Resources Regional station, PKV Campus, Akola, Maharashtra 444 104, India.
[2]National Bureau of Plant Genetic Resources, Pusa Campus, New Delhi 110012, India.
[3]National Bureau of Plant Genetic Resources Regional station, KAU P.O., Thrissur, Kerala 680 656 India.

Rice landraces of Kerala State of India are named based on cropping season, growing conditions, crop duration, morphological features, color of seed and other specific traits, if any. The collections of landraces of Kerala are classified as rare, very rare and common. A part of these collected accessions have been characterized. DIVA-GIS was used to map the collection site, which showed that the collections made are well distributed throughout Kerala but very meager collections are represented from Thiruvananthapuram and Pathanamthitta Districts. Till about 20 years back, all the paddy fields in Kerala were being cultivated with various landraces of rice during the southwest and northeast monsoon seasons. Currently, these paddy fields are being progressively filled with soil transported from other places and converted into building sites or sites for growing other upland crops. As a result of this, a large number of primitive cultivars and local landraces of rice have been lost. Genetic erosion can be assessed and target areas for collection in diversity rich area can be predicted based on resurvey of the collection localities and mapping the collection localities using DIVA-GIS tool. The present paper describes the rice landraces of Kerala with their specific location from where they have been collected.

Key words: Rice, landraces, Kerala, DIVA GIS.

INTRODUCTION

The narrow western part of peninsular India, including the coast, spread north south between the river Tapti and Kanyakumari, and comprises the erstwhile Malabar province. Within this narrow geographical area, and arising abruptly from the coast, is the more or less continuous chain of hills, called the Western Ghats. This region of India is unique in its flora and fauna. Throughout its length, of nearly 1600 km it passes through six states viz., Gujarat, Maharashtra, Goa, Karnataka, Kerala and Tamil Nadu. Topographically, the southern half of the Western Ghats is more varied and biologically enriched. The widest gap of Western Ghats is at Palakkad in Kerala adjoining Tamil Nadu. Kerala lies in the south-western corner of the Indian peninsular region.

Kerala is a narrow strip of land, situated in the southern part of the Western Ghats adjoining Tamil Nadu and Karnataka in the east and north-east, and joined by the Arabian Sea in the west. Kerala State is divided into 14 districts viz., Alappuzha (ALP), Ernakulam (EKM), Idukki (IDK), Kannur (KNR), Kasargod (KSD), Kottayam (KTM), Kollam (KLM), Kozhikode (KKD), Palakkad (PKD), Pathanamthitta (PTA), Malappuram (MPM), Thrissur (TSR), Thiruvananthapuram (TVM) and Wynad (WYD). Except Idukki, Kottayam, Palakkad, Pathanamthitta and Wyanad and all other districts have coastal areas. Wyanad (Lakkidi) receives the highest rainfall, whereas Palakkad receives the lowest rainfall among the districts. Rice cultivation in Kerala dates back to 3000 B.C. (Manilal, 1991). Vast diversity of germplasm is available both in wild and cultivated rice. Hence the objective of the present study was to analyze the landraces of Kerala collected so far by this station and map the collection

*Corresponding author. E-mail: lajay5@rediffmail.com.

Table 1. Cultivated rice collected from Kerala.

District	Number of accessions
Alappuzha	53
Ernakulam	113
Idukki	15
Kannur	83
Kasaragod	40
Kollam	10
Kottayam	4
Kozhikode	30
Malappuram	144
Palakkad	42
Pathanamthitta	1
Thiruvananthapuram	6
Thrissur	63
Wayanad	106

localities using DIVA Geographic Information System (GIS). Seven explorations and collection missions were conducted during the period 1990 to 2010 exclusively for rice in all the districts of Kerala. The sampling strategy followed was bulking of seed samples either in farmers' field or store. The information gathered from the farmers and tribals during various exploration missions on landraces of rice were recorded together with germplasm and compiled.

MATERIALS AND METHODS

Seven explorations and collection missions were conducted during the period of 1990 to 2010 exclusively for rice in all the districts of Kerala. The sampling strategy followed was bulking of seed samples either in farmer's field or store. The information gathered from the farmers and tribals during various exploration missions on landraces of rice were recorded together with germplasm and compiled. DIVA GIS is a Geographic Information System designed to assist Plant Genetic Resources and biodiversity communities to map the range of distribution of species (Hijmans et al., 2002). DIVA-GIS Software (Version 7.2, www.diva-gis.org) was used to map the landraces collected from Kerala and analyzing the gap.

RESULTS AND DISCUSSION

NBPGR Regional Station at Thrissur, Kerala has collected a total of 855 accessions belonging to rice and wild rice during various exploration and collection missions carried out up to 2010. This included 710 accessions of cultivated (Table 1) and 145 of wild rice (Table 2).

The majority of germplasm is from Malappuram (144), Ernakulam (113), Wyanad (106), Kannur (83), Thrissur (63) and Alapuzha (53) districts. The first documented account of association of tribals with the conservation of landraces in Kerala is that of Manilal (1991). The tribal people of Kerala practice subsistence farming but ten distinct tribes of Malabar region are agriculturists who cultivate 22 primitive named landraces in North Kerala (Manilal, 1991). Joseph and Nizar (1998) documented the collection of 60 landraces of rice in Malabar and adjoining areas, classifying them into 4 categories.

Landraces of Kerala are named based on cropping season, growing condition of the crop, crop duration, morphological features, color of seed and specific uses as detailed below.

Categories of landraces

Among the cereals, rice is a major crop of this region with considerable variability in duration, adapting to various agro-climatic and cultural situations and seasons, seed coat colour, shape, size, pigmentation of plant parts, plant height, tillering, lodging, grain, cooking quality and aroma.

Names associated with season

In Kerala, rice has been grown in three seasons namely *Virippu* (April-May to September-October), *Mundakan* (September-October to December-January) and *Puncha* (December-January to March-April).

Names associated with growing condition

Rice is grown in widely different regions from deeply submerged areas which are below sea level extended to high attitudes. The rice varieties adapted for sowing in submerged areas with 1 to 5 m water are called shallow water or deep water or floating varieties and also locally called *Pokkali* rice. They are long duration, photo-period sensitive and have rapidly elongating internodes. The varieties which are adapted to high altitudes are normally referred to as high altitude varieties or locally as *Malanellu* or *Malabarian* rice. Varieties grown in dry situations are called upland varieties or *Modon* (Figure2 and 3. *Kaima* is the most popular red bold rice of irrigated transplanted belt (Joseph John and Abdul Nizar, 1998).

Names associated with crop duration/maturity

Rice varieties can be classified as early, medium and late according to the length of time required to reach maturity. The duration of early maturing varieties is from 90 to 105 and that of late maturing is more than 140 days. *Thonnuran* means variety maturing in 90 days (early duration).

Names linked to plant height

Rice varieties may be classified into dwarf, semi-dwarf and

Table 2. Wild rice collected from Kerala.

Taxa	District	No. of accessions
	Alappuzha	7
	Ernakulam	2
	Kannur	5
	Kasaragod	7
O. sativa f. *spontanea*	Kollam	4
	Kozhikode	10
	Malappuram	8
	Palakkad	2
	Thrissur	12
	Kollam	3
	Malappuram	12
	Palakkad	12
O. meyeriana subsp. *granulata*	Pathanamthitta	3
	Trivandrum	3
	Thrissur	1
	Wyanad	1
	Kozhikode	1
O. nivara	Malappuram	2
	Palakkad	5
	Thrissur	2
	Ernakulam	1
O. officinalis ssp. *malampuzhaensis*	Malappuram	15
	Palakkad	6
	Thrissur	5
	Ernakulam	1
	Kozhikode	2
	Malappuram	4
O. rufipogon	Palakkad	2
	Thrissur	3
	Trivandrum	3
	Kollam	1

and tall based on their height. Dwarf and semi-dwarf are short stature, lodging resistant and named as *Kutta* or *Katta*. Thus *Kattamoda* all denotes dwarf and semi-dwarf varieties. Likewise, tall varieties are named as *Kodi*or *Kodiyan* as they tend to lodge. Thus *Kodivellai* or *Kodiyan* describe tall *indica* forms.

Names associated with morphotypical features

Awn

Awn is the outgrowth of lemma. Most *indica* varieties are awnless, hence easy to handle during harvesting, threshing and cleaning where these operations are done manually as in India. *Mullanchenna* is an awned landrace from Wyanad.

Grain shape

Rice grains of different varieties can be classified into long, medium and short, according to the length of grain; bold, medium and slender according to the ratio of length and width. Long slender varieties with a uniform translucency in the endosperm and bright lusture are locally referred to as *Channa*, medium bold grains as Matta and long bold as *Thondi*.

Names associated with specific use

Various kinds of special purpose rice have been in use in India from time past. The rice flavour of certain glutinuous types is used for various kinds of pastry products. Similarly, specific types of rice are used for puffed rice, popped rice and avul. These varieties are locally called by various names. For example *Varukunnanellu* is used for preparing popped rice. Njavara is cultivated for its medicinal properties and is an important ingredient of *navarakkizhi* in Ayurvedic treatment. It is also used as weaning food (Joseph John and Abdul Nizar, 1998). *Chennellu*, another landrace of medicinal value grown by Mavilan tribe who administer popped rice soaked in water to diarrhea patients (Joseph and Nizar, 1998).

Names associated with colour of milled rice

Ancient literatures show that rice was classified by the Malabar ayurvedics into two broad groups *Saali* which are white rices and *Vreethi*, which are red rices (Manilal, 1991). Varinellu, Chennellu and Kalamappari are white rices and Njavara, Karimkuruva, Karinellu, Chitteni and Modan are red rice.

Names associated with colour of husks

The colour of husk varies from straw to black through brown to purple like *Karuthanjavara* and *veluthanjavara* are referred to the husk colour (Figure 4 and 5).

Names associated with aroma

Very fine seed size and scent add to the quality of rice used for delicious preparations like pulav and pudding. Scent or aroma is a special genetic trait, restricted to certain varieties only. Such scented rice varieties fetch premium price in the market and are known in trade as *Jeeragasala*, *Gandagasala*, *Briyaniari* (suitable for preparing briyanni), *Kitchadi samba* (suitable for preparing *kitchadi*). Table 3 represents the different landraces meant for their characteristic features

During the 7 specific collection missions, 623 different named landraces of Kerala were collected, out of these 256 landraces are classified as very rare (Table 4), which were collected from one district only; 46 as rare (Table 5) which were collected from two or three districts and landraces namely *Arikarayi, Aryan, Chambavu, Jeerakasala, Kaima, Kuruva, Mundakan, Njavara, Karuthanjavara* and *Chittenias* very common occurring in 4 to 6 districts which are spread throughout with the exception of Thiruvananthapuram, Pathanamthitta and Kottayam districts. Out of the named landraces, 532 accessions have been characterized and conserved in the seed bank of National Gene Bank, National Bureau of Plant Genetic Resources, New Delhi. The qualitative characters showed majority of landraces with certain characteristic features – leaf pubescence either intermediate (51.66%) or pubescent (47.24%) indicating less chances of pest and disease attack; absence of awning (78.45%) and white apiculous color (93.65%); well exerted panicle (62.71%); red seed coat color (92.27%); and straw colored hull (39.23%). The quantita-tive characters showed wide variation in leaf length, culm length, number of effective tillers and grain yield per plant which indicated that morphological variation among the landraces is immense. The variability in grain length and grain width was low indicating that all have the same type of grain shape.

Utility of DIVA-GIS tool

DIVA GIS a Geographic Information System designed to assist Plant Genetic Resources and biodiversity communities to map the range of distribution of species (Hijmans et al., 2002). DIVA-GIS Software (Version 7.2, www.diva-gis.org) was used to map the landraces collected from Kerala and analyzing the gap. It is an excellent and simple mapping tool for documentation, diversity analysis and location of collection sites. Hence this tool was used to map the collection localities of rice from 339 sites in Kerala. This showed that the collections made are distributed throughout Kerala and very meagre collections are represented from Thiruvananthapuram and Pathanamthitta districts (Figure 1). Now the situation of occurrence of named landraces in their known localities is alarming. Till 20 years back, all the paddy fields in Kerala were being cultivated with various land-races of rice during the southwest and northeast monsoon seasons. In uplands, different upland varieties of rice were grown during the southwest monsoon season and sesame, horsegram, greengram and blackgram were grown in the northeast monsoon seasons. But today in most of the places both in the plains and uplands, these crops are not being cultivated as it used to be. Instead, paddy fields are being progressively filled with soil transported from other places and converted in to building sites or sites for growing other crops like banana ("*nenthran*" variety), coconut, arecanut, cassava, betel leaves, ginger and in some places aquaculture of prawns and exotic fishes are being carried out. As a result of this, a large number of primitive cultivars and local landraces of rice, sesame, horsegram, greengram and blackgram have been lost. With the conversion of paddy fields for the cultivation of other crops or for housing there is threat of genetic erosion to most of the *Oryza* species and landraces. Moreover, human population pressure due to building of dams for hydroelectric power projects are great in Kerala and the resultant habitat destruction has undoubtedly reduced the range of distribution of *Oryza* species. Genetic erosion can be assessed and target areas for collection in diversity rich areas can be predicted based on resurvey of the collection localities and mapping of the collection localities using DIVA-GIS tool.

Conclusion

Ecologically, Kerala is bestowed with suitable features that can promote variability. In addition, dispersed village settlements, and ethnic and cultural diversity have demanded varying requirements at different times and these might have incorporated variability. The factors, which might have operated to create intra-varietal differences in the cultivated rice of Kerala are: 1. Long history of cultivation of rice in Kerala, under widely different ecological conditions, during which period, multitude of forms and varieties would have been developed, leading to conscious or unconscious selection by farmers/breeders to suit their specific needs, thus culminating in the present day numerous local forms; 2. Slope and terrace cultivation in high altitude areas which are dry or semi-dry, requiring upland varieties; in irrigated conditions of the plains a different set of suitable varieties are needed; in the coastal areas where the principal soil type is acid, saline or acid sulphate, *pokkali* cultivation is being practiced with a set of suitable varieties; in the interior areas of Palakkad and Thiruvananthapuram districts, the soil is neutral requiring varieties suitable for fresh water wetland conditions; in the cole areas (post monsoon cultivation in the submerged areas) of Thrissur district, another set of submergent tolerant varieties are needed. Considering all these specific agro-ecological conditions within the humid tropical situations, specific varieties adapted to each of these varied conditions are required. Varietal requirement for different seasons, areas, water regimes, input supply, pests and pathogens, farmers/consumers preferences is distinct in different regions. Hence, no single variety can suit all the needs and solve the problems associated with rice cultivation in Kerala which has resulted in the large number of landraces occurring or reported to occur in Kerala.

Table 3. Landraces and their characteristic feature.

S/N	Characteristic feature	Name of landraces	Districts
1	Aromatic	Gandhakasala, Jeerakachampav, Mullanchanna	Wyanad, Idukki
2	Awned	Komban	Idukki
3	Cold tolerant	*Chettuveliyan, Koduveliyan, Raajani*	*Wyanad*
4	Colour	*Chomala, Choman, Chuvannachoman, Chuvannakuruka, Chuvannaponmani, Chuvannaitti, Chuvanthutti, Chuvappan, Karimutty, Karuthachoman, Karuthakuruka, Karuthakutti, Karuthavattan, Karuthavelliyan, Karuthavithu, Karuthacheera, Karuthadukkan, Karuthamundakan, Karutharikannan, Vellachettadi, Vellachitteni, Vellakaravala, Vellakattan, Vellakkoli, Vellakuttadan*	Alappuzha, Ernakulam, Idukki Kannur, Kasargod, Kottayam, Kozhikode, Malappuram, Thrissur, Thiruvananthapuram, Wynad
5	Height	*Kalakodiyan, Kodian, Pookilakodiyan, Thondi, Thonnuranthondi, Veliyan, Vellakkodiyan*	Thrissur, Wynad, Malappuram
6	Medicinal properties	*Njavara (Brown), Njavara (Red),Njavara (Unden), Njavara (White), Njayaruvithu*	Kannur, Kozhikode, Malappuram, Thrissur
7	Suitable for saline deep water ecosystem	*Anakodan, Ayyampillypokkali, Cherayipokkali, CheriyaOrpandy, Chootupokkali, Elamkulampokkali, Kadamakudipokkali, Karunagapallipokkali, Khuzhippallypokkali, Kozhippillipokali, Kulapandi, Kuthirunellu, Kuzhippulipokkali, Nedungodupokkali, Odachan, Oorpandy, Orkyma, Orumundakan (black), Pallipurampokkali, Vadanakkudipokkali, Vellapokkali, Vettakkalchettivirippu*	Alapuzha, Kannur, Kasarakod, Kozhikode, Ernakulam, Thrissur
8	Seasonal specific	*Cherumundakan, Cheruvirippu, CheruvirippuVettakkal, Kamanalichettivirippu, Karamundakan, Karimundakan, Kochumundon, Koorimundakan, Koottumundakan, Malmoodan, Mundakacheera, Mundakan (white), Mundakavella, Mundodan, Puncha, Punchanellu, Punchaparuthu, Punjakayama, Vellamundakan, Veluthamundodan, Virippu (Koothumundakan), Virippu –Kottuvalli, Vrischikapandi*	Alappuzha, Ernakulam, Kannur, Kasargod, Kollam, Kottayam, Kozhikode, Malappuram, Wynad
9	Special purpose	*Nagari, Varukkunnanellu*	Kasarakod
10	Upland ecosystem	*Adimodan, Ambaladan, Buluginellu, Chingappadan, Erinellu, Kallarikkoyala, Karakkozhivalan, Karanellu, Karavala, Karavalakochuvithu, Karimodan, Keeripallan, Keeriputhada, Kurumkaima, Kururkaima, Metariyan, Modan, Mullankkoyala, Palkaima, Peruvayan, Peruvazha, Poothakali, Thalanellu, Thalavirichankaruthothu, Thuvinellu, Vellakoyala, Vellamodan, Vellaperuvalan, Vellapperuvasan, Veluthakoyyala*	Alappuzha, Idukki, Kannur, Kasargod, Kollam, Malappuram, Palakkad, Thrissur

Table 4. District wise collections of very rare rice landraces of Kerala, India.

Cultivar/Landrace name	Number of accessions collected	District(s) of collection
Adhiyan	3	TSR
Adimodan	1	PKD
Adukkan	3	WYD
Aiyirammeni	2	MPM
Amankari	1	MPM
Ambaladan	1	KNR
Arupathamkuruva	1	KKD
Aryankali	1	MPM
Ayyampillypokkali	2	EKM
Bolangatti Kayama	1	KNR
Buluginellu	1	PKD
Cheera	4	TSR
Cheerachamban	1	MPM
Chendamani	1	MPM
Chennavella	1	TVM
Chennelthondi	1	WYD
Chenthadi	5	WYD
Cherady	1	KSD
Cherayipokkali	1	EKM
Cheriyaaryan	1	MPM
CheriyaOrpandy	1	EKM
Cheriyachitteni	1	PKD
Cheruvellari	2	MPM
Cheruchitteni	1	KKD
Cherukumbalam	1	PKD
Cherumallaran	1	KLM
Cherumundakan	1	ALP
Cheruvatta	1	WYD
Cheruvirippu	1	EKM
CheruvirippuVettakkal	1	EKM
Chettivirippu Kannamali	1	EKM
Chettuveliyan	1	WYD
Chindinellu	1	TSR
Chingappadan	1	KNR
Chomala	6	WYD
Choman	1	KSD
Choorappandy	1	WYD
Chootupokkali	1	EKM
Chowri	1	WYD
Chunkayama	1	MPM
Chuvannachittadan	1	MPM
Chuvannachoman	1	KSD
Chuvannakuruka	1	MPM
Chuvannaponmani	1	EKM
Chuvannaitti	1	WYD
Chuvanthutti	1	MPM
Chuvappan	1	MPM
Elamkulampokkali	2	EKM
Ennappatta	3	MPM

Table 4. Contd.

Erinellu	1	PKD
Eruvakkali	1	MPM
Gandhakasala	13	WYD
Ittikkandappan	1	WYD
Jeerakachampav	1	IDK
Kadamakudipokkali	2	EKM
Kalakodiyan	1	TSR
Kalathekkan	1	KKD
Kallarikkoyala	1	KNR
Kalluruli	1	MPM
Kamanalichettivirippu	1	EKM
Kanali	1	WYD
Karakkozhivalan	1	KNR
Karamundakan	2	ALP
Karanellu	1	PKD
Karavala	2	ALP
Karavalakochuvithu	1	KLM
Karikagga	1	EKM
Karimodan	1	PKD
Karimundakan	1	KTM
Karimutty	1	WYD
Karinjan	1	KNR
Karivalan	1	KNR
Karuka	1	EKM
Karunagapallipokkali	1	EKM
Karuthachoman	1	KSD
Karuthakuruka	2	EKM
Karuthakutti	1	MPM
Karuthavattan	1	MPM
Karuthavelliyan	1	MPM
Karuthavithu	1	TVM
Karuthacheera	1	KTM
Karuthadukkan	1	KNR
Karuthamundakan	1	ALP
Karutharikannan	1	KSD
Keeripallan	1	KSD
Keeriputhada	1	KNR
Khuzhippallypokkali	1	EKM
Kochithonnuran	1	MPM
Kochumundon	1	WYD
Kodian	1	TSR
Koduveliyan	1	WYD
Kolathari	1	PKD
Komban	1	IDK
Konna	2	MPM
Kookavalan	1	TVM
Koorimundakan	1	ALP
Koottumundakan	1	KKD
Kothampalarikayama	1	PKD
Kottakkadan	1	TSR
Kotti	1	MPM
Kozhippillipokali	1	EKM

Table 4. Contd.

Kulapandi	1	TSR
Kumbalavan	3	MPM
Kunjikannan	1	EKM
kunjikuruvi	1	EKM
Kunjuvithu	1	KNR
Kuroorai	1	KNR
Kurumkaima	1	KNR
Kuruppan	1	PKD
Kururai	1	KNR
Kururkaima	1	KSD
Kutchadi	1	MPM
Kuthalam	1	IDK
Kuthirunellu	1	KSD
Kuthur	2	KNR
Kuttikandappan	1	KNR
Kuttiveliyan	1	WYD
Kuzhippulipokkali	2	EKM
Malakkariadukkan	1	KNR
Malamutti	1	MPM
Malanji	1	MPM
Malappuram	1	MPM
Malayidumban	1	KSD
Mallikuruva	2	WYD
Malmoodan	1	EKM
Mangalapuram	1	KKD
Manjaparivithu	1	TSR
Mannuvelian	1	WYD
Mattakuruva	1	TVM
Mattathondi	1	WYD
Mattikkaran	1	KNR
Mekkancheera	1	MPM
Metariyan	1	TSR
Modan	1	MPM
Mulaki	1	IDK
Mullanchanna	2	WYD
Mullankkoyala	1	KNR
Mundakacheera	2	MPM
Mundakan (white)	1	KLM
Mundakavella	2	MPM
Mundodan	2	KSD
Mundonkutti	1	PKD
Munduvella	1	KLM
Muthu	1	KNR
Mysori	1	MPM
Nagari	1	KSD
Nambiaparamban	1	ALP
Nayarambalam	1	EKM
Nayaruvella	1	MPM
Nayiruvithu	1	KKD
Nedungodupokkali	1	EKM
Neycheera	1	KKD
Nityakalyani	1	ALP

Table 4. Contd.

Njavara (Brown)	1	TSR
Njavara (Red)	1	MPM
Njavara (Unden)	1	TSR
Njavara (White)	1	KNR
Njayaruvithu	1	KKD
Nooranvella	1	ALP
Odachan	1	KNR
Olantha	2	MPM
Omanakkuttan	3	MPM
Oorpandy	1	KKD
Orkyma	2	KNR
Orumundakan (black)	1	ALP
Orumundakan (white)	1	ALP
Otta	1	KKD
Ottacheera	1	KNR
Padannavella	2	MPM
Palakkadan	1	ALP
Palakkadan (black)	1	ALP
Palakkadanmatta	1	MPM
Palakkadanvedandam	1	KKD
Palkaima	1	KNR
Pallipurampokkali	3	EKM
pandichambavu	1	TVM
Pandivella	1	ALP
Paramchitteni	2	MPM
Parambucheera	1	MPM
Parapilarppan	1	TVM
Pathonpatahra	1	TSR
Peruvayan	1	TSR
Peruvazha	1	TSR
Ponkuruka	3	EKM
Poochempan	1	TSR
Pookilakodiyan	1	TSR
Pookkulathari	1	PKD
Poothakali	1	TSR
Puncha	1	KSD
Punchanellu	1	KNR
Punchaparuthu	1	KKD
Punjakayama	1	WYD
Raajani	2	WYD
Redponmani	1	PKD
Rocket	2	MPM
Rosematta	1	KNR
Sundari	1	WYD
Swarnavalan	1	IDK
Thaithigham	1	WYD
Thalanellu	1	TSR
Thalavirichanvella	1	PKD
Thalavirichankaruthothu	1	PKD
Thondi	6	WYD
Thonnuranthondi	1	WYD
Thottathavalakkannan	1	KSD

Table 4. Contd.

Thottathilvithu	1	KSD
Thouvvan	1	KSD
Thuvinellu	1	PKD
Tiruvonam	1	ALP
Undachitteni	2	TSR
Undakaima	4	KSD
Undakkunji	1	KNR
Undamashuri	2	PKD
Urunikaima	2	WYD
Vadanakkudipokkali	1	EKM
Valeecha	1	WYD
Valiavattan	2	MPM
Valiyaittikkannan	1	EKM
Varukkunnanellu	2	KSD
Vasaramundan	1	ALP
Vayalithoova	1	ALP
Veliyan	7	WYD
Vella	1	MPM
Vellachettadi	1	MPM
Vellachitteni	1	TSR
Vellakaravala	1	ALP
Vellakattan	1	MPM
Vellakkodiyan	1	MPM
Vellakkoli	1	TSR
Vellakoyala	1	KNR
Vellakuttadan	1	KKD
Vellamthangi	1	MPM
Vellama	1	TSR
Vellamodan	1	PKD
Vellamthangi	1	TSR
Vellamundakan	1	KTM
Vellapandi	2	TSR
Vellaperuvalan	1	KNR
Vellapokkali	1	EKM
Vellapperuvasan	1	IDK
Vellaryan	2	MPM
Velumbazha	1	WYD
Velutha	1	MPM
Veluthakalluruli	1	MPM
Veluthamundodan	1	KSD
Veluthavattan	1	MPM
Veluthadichal	1	KSD
Veluthakoyyala	1	KNR
Veluthon	1	KNR
Venda	1	MPM
Vettakkalchettivirippu	2	EKM
Vetteri	1	PKD
Virippu (Koothumundakan)	1	ALP
Virippu -Kottuvalli	1	EKM
Vrischikapandi	2	MPM
Vyttilakuruka	2	EKM

Table 5. District wise collections of rare rice landraces of Kerala, India.

Cultivar name	No. of accessions collected	District(s) of collection
Allikkannan	2	KNR, KSD
Anakkodan	4	EKM,PKD,TSR
Aruvakkali	3	MPM, PKD
Chemban	2	PKD, TSR
Chenkayama	4	KSD, MPM, PKD
Chennellu	9	KNR, KSD, WYD
Cherumani	2	ALP, MPM
Chettadi	5	KKD, MPM, PKD
Chettivirippu	14	ALP. EKM
chuvannachitteni	4	EKM, MPM,PKD
Karuthachitteni	2	MPM, PKD
Kattamodan	5	MPM, TSR
Kochuvithu	2	ALP,KLM
Kothambarikaima	2	KNR. KSD
Kunjinellu	8	KNR, KSD, WYD
Kunjukunju	3	EKM, IDK, MPM
Kuruka	5	ALP, EKM
Kuttadan	4	MPM, TSR
Marathondi	3	IDK, WYD
Mashuri	2	KNR, PKD
Moonnamvila	3	KNR, WYD
Munda	3	KKD, WYD
Mundakakkutty	2	KKD, PKD
Mundakan (black)	2	KLM, TSR
Mundon	4	KNR, WYD
Nambiarvattam	2	ALP,KNR
Onavattan	3	KSD, WYD
Palthondy	2	IDK, WYD
Pandi	3	MPM, TSR
Parambuvattan	3	MPM, TSR
Pokkali	31	ALP. EKM
Ponnaryan	4	EKM, MPM,PKD
Rajakaima	2	KNR, KSD
Thavalakkannan	6	MPM, TSR, WYD
Thekkan	3	KSD, MPM
Thekkancheera	3	KNR, MPM, TSR
Thonnuran	6	KNR, KSD, MPM
Vaikatharyan	2	ALP, PTA
Valichoori	4	EKM, WYD
Varinellu	2	KNR, TSR
Vattan	3	MPM, TSR
Vellakaima	2	KNR, TSR
Vellakkokkan	2	MPM, PKD
Vellapponni	3	IDK, PKD
Vellari	4	KKD, MPM
Virippu	4	ALP, EKM

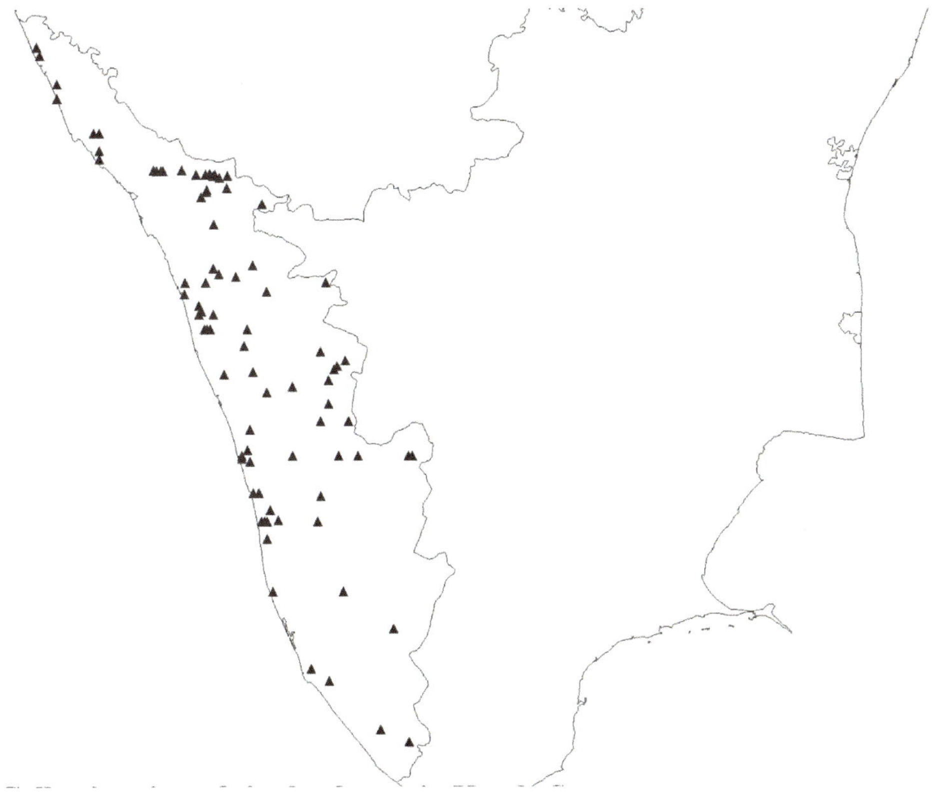

Figure 1. Collection sites of rice landraces in Kerala State.

Figure 2 Panicle variability in upland rice landraces

Figure 3 Grain variability in upland rice landraces

Figure 4 Variability in husk colour in rice landraces

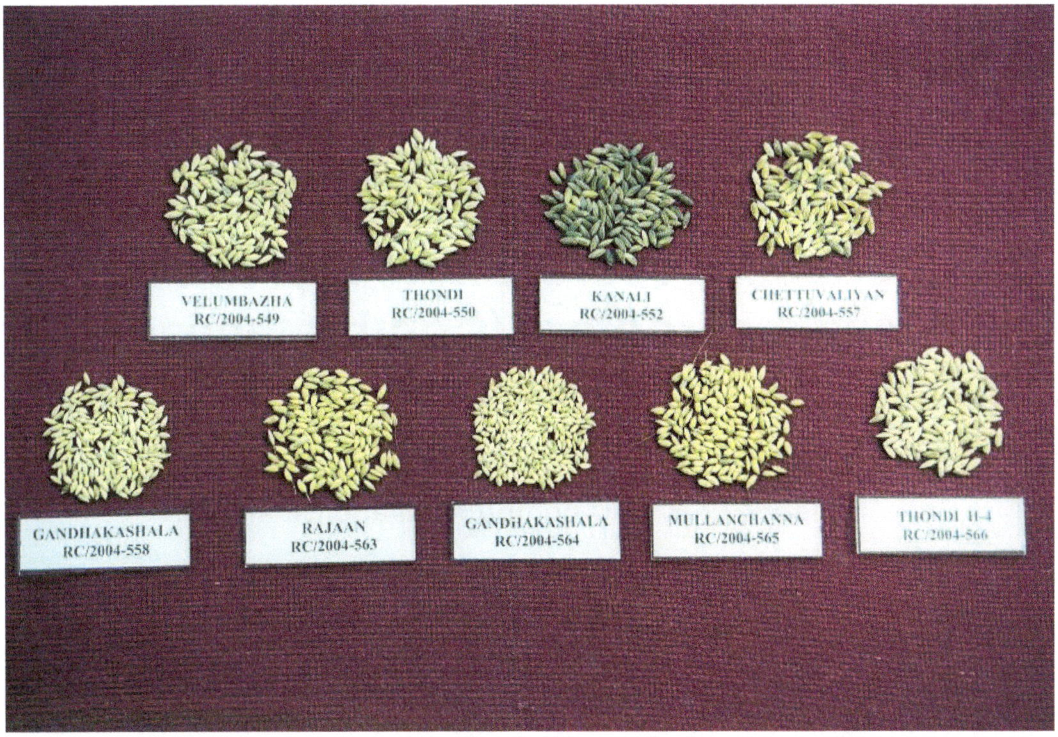

Figure 5 Variability in husk colour in rice landraces collected from Wyanad district of Kerala

If South Asia region is the primary centre of diversity for rice with seven species of *Oryza* (Chang, 1975, 1976; Kwangtung Agricultural and Forestry College, 1975) and the Jeypore tract (Orissa) is the secondary centre of diversity for rice and its close wild relatives (Sampath and Govindaswamy, 1958), then certainly, the humid tropical coastal and mid-lands of Kerala in the southern Western Ghats can be considered as one of the secondary centres of diversity for rice and its close wild relatives in addition to the existence of 623 named landraces.

REFERENCES

Chang TT (1975) Exploration and survey in rice. In: Frankel, OHand Hawkes JG, (eds.) Crop Genetic Resources for Today and Tomorrow: Cambridge University Press, Cambridge, UK. Pp 159-165.

Chang, TT (1976). The origin, evolution, cultivation, dissemination and diversification of Asian and African rices. Euphytica 25:425-441.

Hijmans RJ, Guarino L, Rojan E (2002). DIVA-GIS, Version 7.2.A geographic information system for the analysis of biodiversity data.Manual. International Potato Center, Lima, Peru

Kwangtung Agricultural and Forestry College (1975). The species of wild rice and their geographical distribution in China. ActaGeneticaSinica. 2:31-36.

Joseph JK, Abdul NM (1998) Collection of rice germplasm from Malabar, Kerala. Indian J. Plant Genetic Res. 11(2):173-181

Manilal KS (1990). Ethnobotany of the rices of Malabar In: Contribution to Ethnobotany of India. Botanical Survey of India, Calcutta Pp 243-253.

Sampath S, Govindaswamy S (1958). Wild rices of Orissa, their relationship to cultivated rices. Rice News Teller 6:17-20.

4

Relationship between stand regime and population structure of *Pseudocedrela kotschyi* (Meliaceae) and *Terminalia macroptera* (Combretaceae) in the Biosphere Reserve of Pendjari (Benin, West Africa)

Assédé E. P. S*, Adomou A. C. and Sinsin B.

Laboratory of Applied Ecology, Faculty of Agronomic Sciences, University of Abomey-Calavi, 03 BP 1974, Cotonou, Benin.

One of the major reasons for reforestation failure using tropical species is misunderstanding of the neighbourhood relationship within tree populations. This study used the natural stand structures of *Terminalia macroptera* and *Pseudocedrela kotschyi*, two socio-economically important species, to design enrichment planting. A comparative analysis of the population structure of *P. kotschyi* and *T. macroptera* in two regimes (pure and mixed) was performed. However, results indicated that young *T. macroptera* individuals were predominant in both stand regimes. In the same way, no significant difference was found between diameter size classes with respect to stand regime. Trees showed weak density and a random pattern with the nearest neighbour distances varying between 5.67 (pure) and 7.01 m (mixed). *P. kotschyi* young individuals were also found to be predominant irrespective of stand regime; yet, the diameter size class distributions revealed significant variations with respect to stand regime. Trees had a higher density and stronger clumped pattern in pure stand as compared to mixed stand. The nearest neighbour distances ranged from 2.97 (pure) to 4.87 m (mixed). Our findings highlighted the relevance of taking into account relationships between stand regime and tree population structure while designing artificial plantings.

Key words: Savanna woodland, density, spatial distribution, diameter class structure.

INTRODUCTION

Plant communities represent a complex system where a plant interacts with its neighbourhood. While there is still considerable scientific debate over ecosystems functioning (Graae et al., 2004; Klimas et al., 2007; Adjonou et al., 2009), understanding these systems is made tricky by the complexity of interaction between their different components.

Terminalia macroptera Guill. & Perr. (Combretaceae) and *Pseudocedrela kotschyi* (Schweinf.) Harms (Meliaceae) are two common savanna woodland tree species. They are economically and ecologically valuable. The understanding of their population structure and how it is altered by stand regime (important to design effective conservation plans) is still limited. Despite these paucities of information on their population structures, these species are being used in forest enrichment, thus leading to a low success rate. *T. macroptera* and *P. kotschyi* generally grow in relatively pure stand, but some times, they are together within the same habitat (Louppe et al., 2008). Despite the medicinal and therapeutic importance of *P. kotschyi* (Georgewill and Georgewill, 2009; Ayo et al., 2010) and *T. macroptera* (Conrad et al., 2001; Silva et al., 2002), few studies were focused

*Corresponding author. E-mail: emimiss@yahoo.fr.

Abbreviation: **BRP**, Biosphere Reserve of Pendjari.

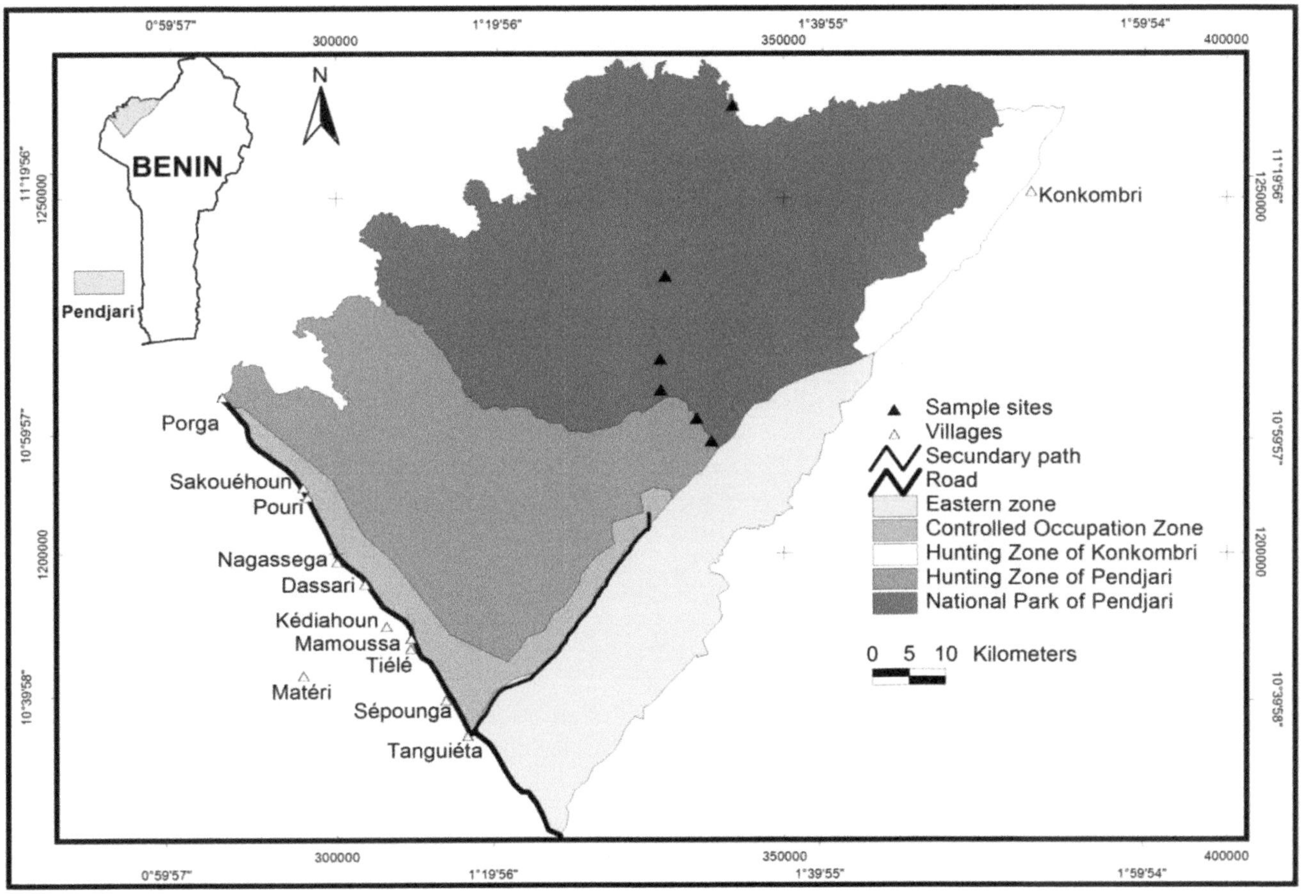

Figure 1. Biosphere Reserve of Pendjari with localisation of sample sites.

on population structure.

Conservation of any species requires knowledge on its abundance and population structure in different habitats and/or stand regimes (Peters, 1996). Since vital demographic rates are normally stage-dependent (Harper, 1977), determining the structure of a population is a first step that when combined with demographic data (Condit et al., 1998), can be the basis for management decisions (Bruna and Kress, 2002). Stand spatial structure is of prime importance in individual-tree growth models, where a local environment determines competition, growth, death and regeneration (Goreau et al., 1999). Individual plants with different neighbourhoods will have different growth, spatial patterns and population density (Pacala and Deutschman, 1995). So, plant spatial patterns heavily influence plant population dynamics (Bolker and Pacala, 1997), whereas plant demography generates and modifies plant spatial patterns (Barot et al., 1999). It is one of the key ecological parameters for guiding sustainable management of any given species (Peters, 1996). Therefore, detailed knowledge on species population structure behaviour is essential (Sunderland and Drasnfield, 2002) for sustainable management of the ecosystems of the two species.

In this study, we compared the population structure and spatial arrangement of *P. kotschyi* and *T. macroptera* in two stand regimes: pure and mixed stands in Biosphere Reserve of Pendjari. The objectives were: (a) to assess the density, diameter size class structure and spatial distribution of *P. kotschyi* and *T. macroptera* in pure and mixed stands (b) to compare the average nearest neighbour distance in both regimes and (c) to use outputs to help plan management strategies for both species.

MATERIALS AND METHODS

Study area

Field surveys were carried out within the Pendjari National Park, an integrally protected core zone (2,660 km²) within the Biosphere Reserve of Pendjari (BRP), Northwest Benin (10°30′ to 11°30′ N; 0°50′ to 2°00′ E) (Figure 1). The BRP is located in the Mekrou-Pendjari phytodistrict (Adomou et al., 2006) within the Sudanian Region (White, 1983). Apart from Atakora chain (400 to 513 m above sea level) bordering the southern limits of the reserve, the study region's topography ranges between 150 and 200 m above sea level (Delvingt et al., 1989). In this region, the rainfall regime is unimodal with mean annual rainfall being approximately 1000 mm (between May and October) and a dry season lasting for seven

Photo 1. Pure stand of *T. macroptera* a: individuals; b: bark; c: leave.

months (from October to April). Mean annual temperature ranges between 25 and 33°C, and relative humidity varies from 27 to 85% (Verschuren, 1988). Soils are mainly ferruginous and the main vegetation types are savannas, woodlands and dry forests. (Delvingt et al., 1989).

Studied species

T. macroptera Guill. & Perr. (Combretaceae) (Photo 1) occurs from Senegal to North and South Nigeria and across Africa to Sudan and in Uganda. The tree is about 13 m high and usually occurs in moist or occasionally flooded sites. *T. macroptera* is a tropical tree with strong multiple-use characteristics. It is valued for both the high-quality oil extracted from its seeds and as a timber resource. The wood is said to produce a very good charcoal (Coulibaly, 2006). A root decoction is given in treating many illnesses (fevers, jaundice, syphilis, urethral discharge and urinary trouble). The leaves are used as a wash and fumigation in fevers and skin diseases. Flavones, steroids and other substances (antimicrobial) are reported as being present in the plant (Conrad et al., 2001; Silva et al., 2002).

P. kotschyi (Schweinf.) Harms (Meliaceae.) (Photo 2), commonly known as the dry zone cedar, is widespread in savanna woodlands (Shahina, 1989) from Eastern Senegal to Western Ethiopia and Uganda (Louppe et al., 2008). It is a medium-sized tree up to 20 m in height. The dry zone cedar commonly presents is only 6 to 9 m high due to fire damage, with girth less proportionately reduced. It bears an oblong or pyramidal crown with branches often ascending coppicing (Louppe et al., 2008). A wide variety of medicinal treatments are made from root (Okunade et al., 2007), bark (Kassim et al., 2009; Ayo et al., 2010) and leaves (Musa et al., 2008; Asase et al., 2008) of this plant. The stem and root barks of *P. kotschyi* also contain essential oils, exclusively comprising sesquiterpenoids with very low antiradical and antioxidant activities (Boyom et al., 2004).

Despite their high therapeutic importance, the cultivation of both species is not yet mastered.

Sampling design

Potential sites of the censused species were identified using a vegetation map. Each site was georeferenced using a Global Positioning System (GPS GARMIN 60, Resolution 5 m). Six stands were established to determine distribution and density patterns of *T. macroptera* and *P. kotschyi* at the landscape level. Stands were established in homogenous and comparable habitat types, comprising two stands in the pure population of each species and two others in the mixed population of both species. Stands were established randomly. Stands of pure populations were named T (for *T. macroptera*) and P (for *P. kotschyi*). In mixed populations (Photo 3), stands were named T' (for *T. macroptera*), and P' (for *P. kotschyi*).

Data collection

Data collection was conducted at the end of the rainy season (year 2010), coinciding with the top stage of the leafing phenology of both species. Within each stand, internal transects were installed every 50 or 25 m, depending on the size of the population. A tree-mapping process was performed following Klimas et al. (2007). The x-coordinate was estimated based on distance to the proximate transect. The accuracy of the x-coordinate was initially tested by measuring the distance with a metric tape until an error of 2 m was attained, subsequently using the estimated x-coordinate in place of a metre-tape measurement. The mapping process consisted of linking each tree with one of its neighbours by distance azimuth and slope. The tree slope was determined with a clinometer of SUUNTO (Klimas et al., 2007). In each stand, a systematic count of all *T. macroptera* and *P. kotschyi* trees with a diameter ≥ 1 cm at 1.3 m

Photo 2. Pure stand of *P. kotschyi* a: individuals; b: bark; c: leave.

Photo 3. Mixed stand of *T. macroptera* and *P. kotschyi.*

above the ground (D_{130}) was made. Individuals of $D_{130} \geq 10$ cm were inventoried, mapped, measured with ribbon "pi" (1 cm precision), and labelled to avoid replication and omissions in the six stands.

Data analysis

Tree fitting $D_{130} \geq 10$ cm included an adult category, while individuals of $D_{130} < 10$ cm included juveniles (seedlings (1cm $\leq D_{130} < 5$ cm)) and saplings (5 cm $\leq D_{130} < 10$ cm) (Rondeux, 1999). The diameter size-class distribution (SCDs) of *T. macroptera* and *P. kotschyi* was established for adult trees. Individuals of both species were grouped into diameter SCD of 10 cm in order to obtain enough diameter classes (Glèlè and Sinsin, 2009). Tree densities were established for SCD. The observed SCD was fitted to the Weibull theoretical distribution. This distribution was preferred because of its flexibility (Johnson and Kotz, 1970). Its density function, *f(x)* is expressed for a tree-diameter x, as follows:

$$f(x) = \frac{c}{b}\frac{(x-a)^{c-1}}{b}e^{-[x-a/b]^c}$$

Where x = tree diameter, a = 10 cm for the diameter structure, b = scale parameter linked to the central value of diameters and c = shape parameter of the structure.

Tree diameters were used to estimate the parameters b and c for each species based on the maximum likelihood method (Johnson and Kotz, 1970) in Minitab software. The log-linear analysis (Caswell, 2001) was performed in SAS (SAS Inc, 2003) for each case to test the adequacy of the observed structure to the Weibull distribution. The considered model described by Caswell (2001) is:

Log Frequency = F + F_classe + F_Adjustment + ε

F = mean frequency of the classes, F_classe = non-random gap linked to the differences in frequency between classes, F_Adjustment = non-random gap linked to differences between observed and theoretical frequencies and ε is the error of the model. The hypothesis of adequacy between both distributions is accepted if the P value (probability of the test) is higher than 0.05 (Bonou et al., 2009).

The number of tree in each subplot was used to compute tree density per species, per trees category (adult, sapling, and seedling) and per regime (pure and mixed stands). Two-way analysis of variance and a mixed model were used to test the differences between adult, sapling and seedling density within and between stand regimes (Zar, 1999). P values ≤ 0.05 were considered to be statistically significant in all cases.

In the case of difference, the variation rate (R) of diameter classes between regime stands was examined:

$$R = \frac{100(Di-Dj)}{Di} \quad \text{with } D_i > D_j$$

Where *Di* is average density of diameter class "i"; D_j is average density of diameter class "j"

Spatial pattern analysis is considered as a stochastic model ruling the locations of points in some topological set (Diggle, 1983). The toroidal shift method was used, since there is interest in testing the independence between groups generated by different point processes. This is the correct hypothesis for analysing the interaction between two species of plants (Diggle, 1983). The pair-correlation function *g(r)* (Stoyan and Stoyan, 1994) was used to investigate the type of distribution pattern shown by adults of *T. macroptera* and *P. Kotschyi.* The univariate point pattern was used to investigate the type of distribution pattern shown by each species (a) in a pure stand and (b) in the type of distribution of one species in the presence of the other species in the mixed stand. The bivariate point pattern was used to investigate (c) the type of association between adults of *T. macroptera* and *P. kotschyi* individuals in the mixed stand. These methods test (1) whether observed samples can be considered as having a regular, random, or aggregated pattern (= pattern 1), (2) the effect of one species on the spatial distribution of the other in mixed stand (pattern 1), and (3) whether the relationship between *T. macroptera* and *P. kotschyi* suggests spatial association, repulsion or independence (= pattern 2). The pair-correlation function *g(r)* is a non-accumulative version of Ripley's K-function (Stoyan and Stoyan, 1994) with $g(r) = (1/2\ \pi r)$ dK(r)/dr (Wiegand and Moloney, 2004). The bivariate $g_{12}(r)$ is an intensity-normalised neighbourhood density function indicating the density of neighbouring trees as a function of distance *r* from an average tree divided by the intensity *k* of the pattern. The *K* function *K(r)* was defined from the expected number of extra points within distance r of an arbitrary point of the process. Test significance was estimated through the Monte Carlo procedure. 500 simulations were used to compute tests at the 1% confidence level (Mariott, 1979). To determine statistical signi-ficance of the observed *g(r)* and $g_{12}(r)$, 1% simulation envelopes of a random labelling hypothesis null model (pattern 1) and independence hypothesis null model (pattern 2) were generated by 500 replicated Monte Carlo simulations of the null model (Goreaud and Pelissier, 2003). In the case of *g(r)* for a given scale r outside the simulation envelopes, the null hypothesis was rejected, that is, for univariate point patterns with intensity λ; $g(r) > \lambda$ indicates clumping, while values of $g(r) < \lambda$ indicate regularity. For a bivariate point pattern with intensity λ_2 of pattern 2, $g_{12}(r) > \lambda_2$ indicates a positive association (attraction), whereas $g_{12}(r) < \lambda_2$ indicates a negative association (repulsion).

Ads package of R-2.9.1 software was used to perform spatial analyses. A univariate point pattern was performed under complete spatial randomness (CSR) between points of a given kind, and a bivariate point pattern under the independence hypothesis between *T. macroptera* and *P. kotschyi* individuals. The distribution of nearest neighbour distances (nndist) was determined for each species within the two regime types using spatstat, an R package for spatial point pattern analysis (Baddeley and Turner, 2005). The T value of students was calculated and associated with the p value to determine whether the average nndist was significantly different between habitats after testing the normality of each distribution using Minitab 14. 118 trees were sampled in the pure population and 128 trees in mixed stand in case of *P. kotschyi*, whereas 143 in the pure population and 289 in mixed stand were considered for *T. macroptera*. P values ≤ 0.05 were considered as being statistically significant.

RESULTS

Effect of stand regimes on *P. kotschyi* and *T. macroptera* size-class distribution

Log-linear analysis computed for *P. kotschyi* and *T. macroptera* in pure and mixed stands indicated a good fitness of the observed distribution to the Weibull distribution (P>0.05). The SCD (size-class distribution) of *P. kotschyi* in pure stand revealed an inverse "*J*" shape with a decrease in the number of individuals in the larger diameter classes (Figure 2a). The shape parameter *c* was closed to 1. Trees with 10 to 35-cm dbh were the best represented with most of the trees in the first class (10 to 15). Individuals of 50 < D_{130} > 55 cm were scarce, while those with D_{130} higher than 55 cm were nearly absent. In contrast, the SCD was bell-shaped with 1 < c < 3.6 for *T. macroptera* in pure stand

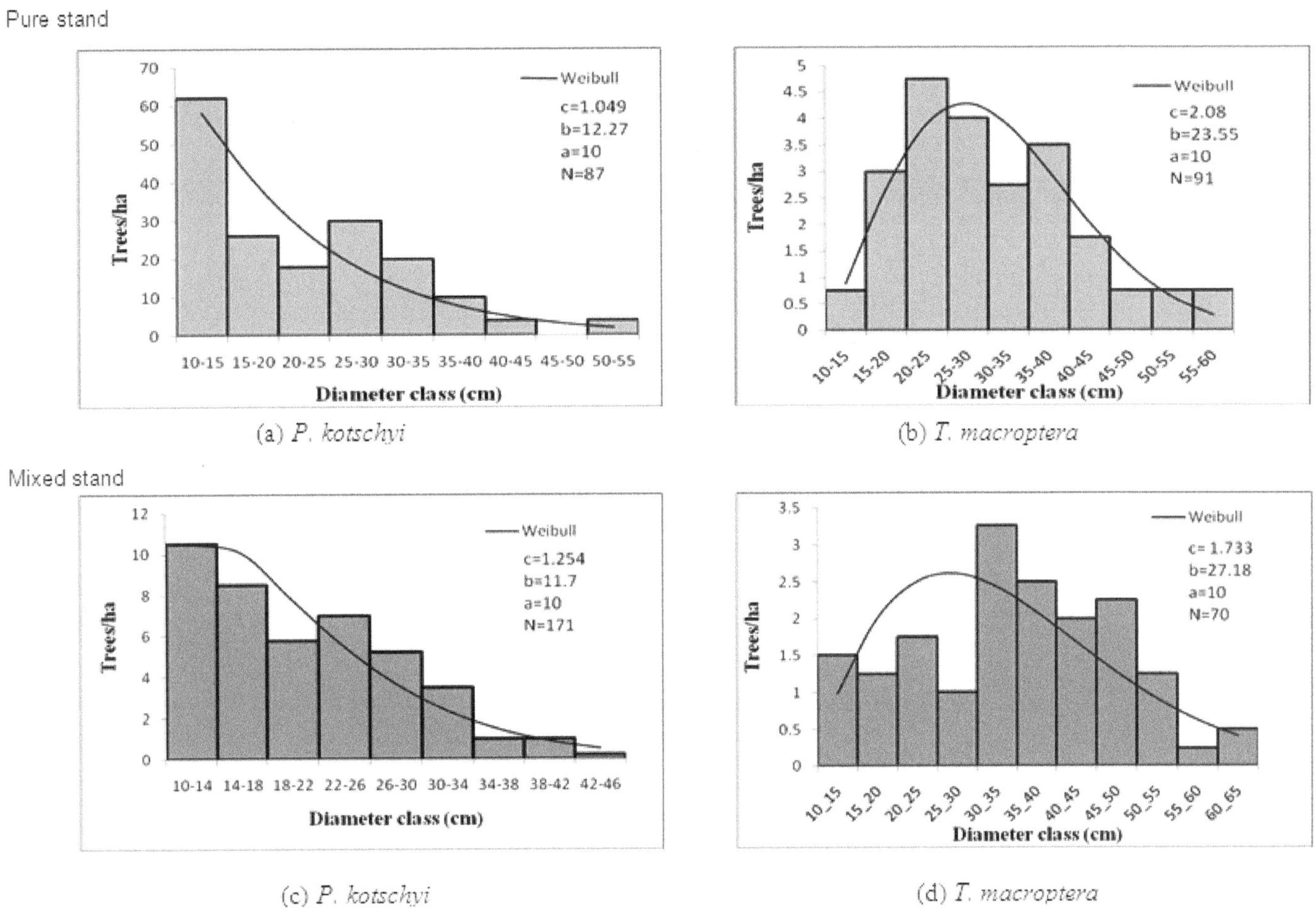

Figure 2. Diameter structure of *T. macroptera* and *P. kotschyi* population within the regime stand.

(Figure 2b). The 20 to 30 classes were the best represented. In the mixed stand, the SCDs of *P. kotschyi* and *T. macroptera* were skewed to the right ($1 < c < 3.6$) (Figure 2c and d). Individuals of *P. kotschyi* with 10 cm < D_{130} > 30 cm were the best represented with absence of D_{130} > 46 cm in this stand. *T. macroptera* were best represented in 30 to 35 classes and weakly represented with D_{130} higher than 55 cm.

Effect of stand regimes on *P. kotschyi* and *T. macroptera* density

Although the distribution of the density of *T. macroptera* stems (adult, sapling and seedling) revealed significant difference ($P < 0.005$), there is no difference in the regime as well as the interaction diameter regime ($P > 0.05$). The average density of diameter classes was significantly equal in pure and mixed stands (Table 1). The average density of adult varied from 21.62 ± 1.6 trees/ha (pure stand) to 16.12 ± 1.94 trees/ha (mixed stand). In contrast, the density of *P. kotschyi* stems varied greatly within as well as between the regimes stand ($P < 0.05$) (Table 1). Contrary to the seedlings, adult and sapling tree densities differed according to stand regimes ($P < 0.05$), with pure stand having the highest values (Table 2 and 3). Moreover, the average density of *P. kotschyi* individuals was reduced at least by 40% in each diameter class from pure to mixed stands. The most affected diameter class was the saplings. The average density of saplings in mixed stand was barely 13% of pure stand (Table 3).

Effect of stand regimes on *P. kotschyi* and *T. macroptera* spatial distribution

The spatial distribution of *P. kotschyi* and *T. macroptera* was different according to stand regime (Figures 3 and 4). The individuals of *P. kotschyi* with $D_{130} \geq 10$ cm in pure stand were aggregated towards the radius and more strongly at small distances (0 to 3 m) (Figure 4a).

In mixed stand, *P. kotschyi* was weakly aggregated between 0 and 5 m. This distribution gradually became regular from 15 m (Figure 4e). Thus, *P. kotschyi*

Table 1. Two-way ANOVA on SCDs and regime stands of *P. kotschyi* and *T. macroptera*.

	df	Sum sq.	Mean sq.	F	P
		T. macroptera			
Diameter classes	2	611.17	05.583	16.677	0.003 **
Stand regimes	1	0.75	0.750	0.041	0.846
Diameter classes - stand regime	2	2.63	1.313	0.071	0.931
Residuals	6	109.94	18.323		
		P. kotschyi			
Diameter classes	2	19140.3	9570.2	16.122	0.004 **
Stand regimes	1	19811.2	19811.2	33.374	0.001 **
Diameter classes - stand regimes	2	9486.0	4743.0	7.99	0.020 *
Residuals	6	3561.7	593.6		

*: Significant code.

Table 2. Comparison of stand regimes based on diameter class density of *P. kotschyi*: results of paired t-test.

Dbh classes	df	SD	T	P
dbh ≥ 10	2	2.7	87.6	0.0001* **
5 < dbh < 10	2	44.1	4.0	0.05*
1 < dbh < 5	2	6. 1	1.5	0.27

*, Significant code.

Table 3. Variation of diameter class densities of *P. kotschyi* individuals from pure to mixed stand.

Parameter	dbh ≥ 10	5 < dbh < 10	1 < dbh < 5
H1 (trees/ha)	173.7	117	10
H2 (trees/ha)	36.4	14.9	5.6
R (%)	79	87	44

H1: Average density in pure regime stand; H2: average density in mixed regime stand.

individuals showed a clumped pattern within a short radius, which gradually becomes random for medium radiuses. Results from univariate analysis of *T. macroptera* individuals in pure stand (Figure 4b) suggested a remarkable aggregation pattern within a 3 m radius. For medium radiuses above 3 m (5 < r < 15 m), a regular pattern appeared and became more evident between 5 and 15 m.

In mixed stand, the univariate pattern analyses of *T. macroptera* individuals revealed a random pattern especially for a 5m radius and above (Figure 4f). Before this distance, the pair density function (g(r)) did not overlap the simulation envelopes, suggesting a clear repartition of *T. macroptera* individuals. However, the bivariate analysis (g_{12}(r)) function among *P. kotschyi* and *T. macroptera* individuals exhibited absence of spatial association between the species (Figure 4c and d). *P. kotschyi* individuals in mixed stand were randomly distributed around *T. macroptera* individuals.

Effect of stand regimes on nearest neighbour distance

Contrary to *T. macroptera* (P > 0.05), the average nearest neighbour distance of *P. kotschyi* individuals differed according to stand regimes (P < 0.001). The average distance between two individuals of *P. kotschyi* was higher in mixed stand than in pure stand (Table 4).

DISCUSSION

Regime effect on diameter classes distribution

Assessing the stand regimes' effect on a species based on SCDs and spatial structure may be a rapid and useful approach. However, given the longevity of large trees and the scarcity of appropriately long-term demographic data, standing size distributions are a potential tool for predicting species' responses to changes in population structure (Feeley et al., 2007). The predominance of *P. Kotschyi* juveniles in pure as well as in mixed stand may indicate absence of a stand regime effect.

However, in pure stand, diameters reach 55 cm, while the largest diameter does not exceed 42 cm in mixed stand. Large discrepancies observed between *P. Kotschyi* populations SCDs following a stand regime firstly appeared to be linked to a stand regime effect. Observed difference was much more significant on saplings and adults. If we assumed that seedling was the first growth step of the species, this implies that the presence of *T. Macroptera* had no effect on the transition from germination to seedling of *P. Kotschyi* individuals, since germinated seed survival is more survival is more

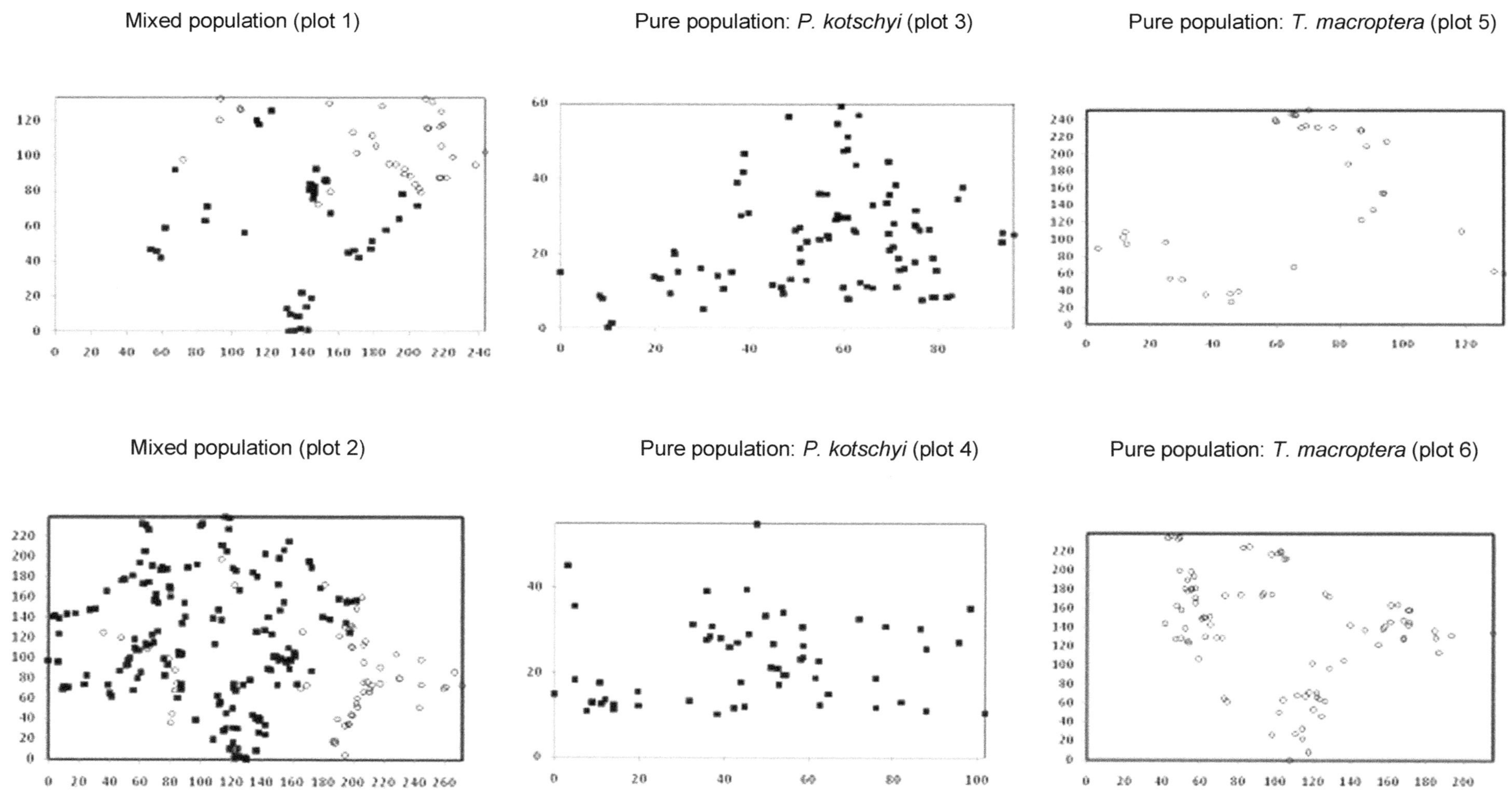

Figure 3. Spatial distribution of *P. kotschyi* (closed rectangles) and *T. macroptera* (open circles) individuals ≥ 10 cm dbh in each of the six study plots.

related to reserve storages (Dalling and Hubbell, 2002; Schiøtz et al., 2006). Moreover, *P. Kotschyi* seedling would be much more related to the fire and water influence (Louppe et al., 2008). For cons, the seedling recruitment would be marked by important nutritional needs and thus competition for lightand food (Molofsky and Augspurger, 1992; Burslem, 1995); still, both species were heliophilous plants (Louppe et al.,

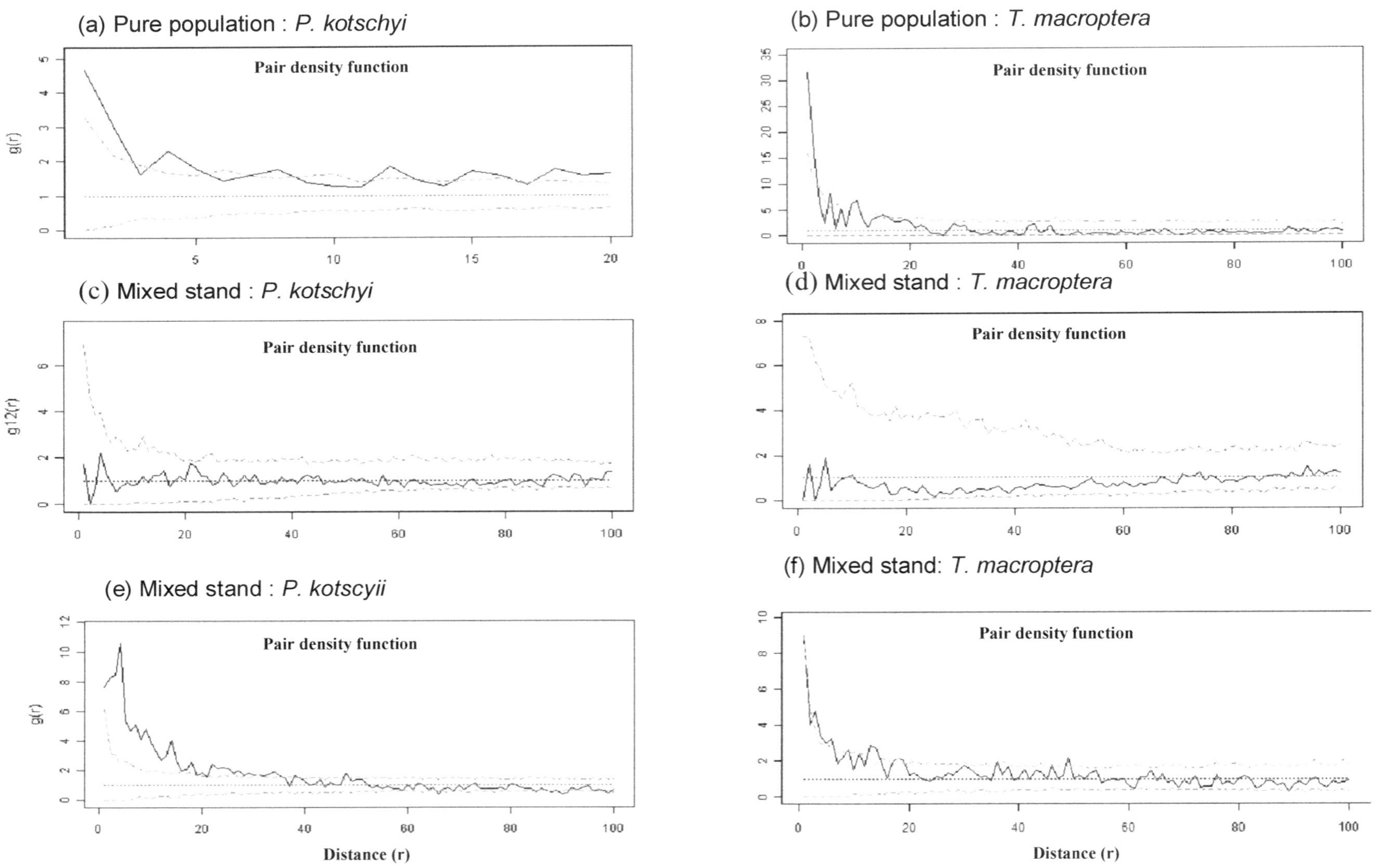

Figure 4. Point pattern analyses of *P. kotschyi* and *T. macroptera* in pure and mixed stands. (a) and (e) present the results of the univariate analyses of *P. kotschyi* in pure and mixed stands, respectively. (b) and (f) present the results of the univariate analyses of *T. macroptera* in pure and mixed stands respectively. (c) and (d) present the results of the bivariate analyses of *P. kotschyi* and *T. macroptera*, respectively in mixed stand. Solid black lines ( ): statistics calculated from the data (that is, g(r) values), green dashed lines ( ): simulation envelopes being the 995 lowest and highest values of the 999 simulations of the null model, horizontal dashed lines in red ( ): average statistics taken from the 500 simulations of the null model (the theoretical expectation).

Table 4. Difference between average nearest neighbour distance in pure and mixed stand.

Plant species	Regime	Mean	SD	T	P
T. macroptera	Pure	5.67	5.47	-1.82	0.07
	Mixed	7.01	6.08		
P. kotschyi	Pure	2.97	2.56	-5.75	0.0001*
	Mixed	4.81	4.04		

T: t-value of student; P: probability; *: significant level; Pure: pure stand; Mixed: mixed stand.

2008). Conversely, *T. macroptera* is known to develop a fairly impressive crown (Arbonnier, 2008). There was competitive advantage for species that responded well to high light levels and which are capable of rapid growth (Brown, 1998). The shade created by the *T. macroptera* crown impeded normal growth of *P. kotschyi* individuals within 5 m of radius and may have led to removal of some individuals because more seedlings died in apparent low-light areas (Tesfaye et al., 2002; Schiøtz et al., 2006).

Regarding *T. macroptera* as a *P. kotschyi* competitor, mixed stand would have a negative effect on *P. kotschyi.* The reduction of over 80% of sapling and 79% of adult density from pure to mixed stand supposed *P. kotschyi* in mixed stand to be facing vulnerability in some stage of its development. In effect, tree seedling establishment is an important part of population dynamics, as the recruitment of seedlings determines the composition of future populations (Schiøtz et al., 2006). The variation in density of trees could be linked to tree mortality as a result of the combined effects of fire and predators (Jachmann and Croes, 1991). However, this hypothesis was partially rejected because the trunk of both species (*T. macroptera* and *P. kotschyi*) was known as fire-resistant due to their thick, corky bark (Louppe et al., 2008).

The analysis suggests that in the long term, the mixed regime of *T. Macroptera* and *P. Kotschyi* would disappear in favour of the pure population of *T. Macroptera.* Then, the question is how these two species could be together across the generations. Under constant conditions, two species utilising and limited by a common resource cannot coexist in a limited system (Pulliam, 2000). But if there is assurance that the system has existed for decades, and has persisted through time without fail, it would be in perfect equilibrium. Indeed, most of the theory of community ecology has been built around the notion that the presence of a species in a given area indicates that the species is somehow adapted to local conditions and then can coexist with other species (Pulliam, 2000). However, the factors analysed in this study were not sufficient enough to highlight all reasons of the variation in densities. Regarding the c value, SCD suggests *T. macroptera* and *P. kotschyi* are facing vulnerability of trees in some stage of their development, whatever the stand regime. Conversely, large discrepancies were found between *T. macroptera* and *P. kotschyi* nearest neighbour distance (nndist) following stand regimes. Despite the usefulness of SCD to make reasonable decisions in forestry management (Husch et al., 2003), the SCD is not the best predictor of future trends in population change (Feeley et al., 2007; Fandohan et al., 2010). Like density, the size-class distribution and seedling establishment of a plant species may be affected by many ecological factors including seed dispersal, soil parameters, light tolerance and herbivores effect (Böhning-Gaese et al., 1999; Tesfaye et al., 2002; Wadt et al., 2005; Kainer et al., 2007; Gaoue and Ticktin, 2007). Additional studies are crucial to appreciate the role of *T. macroptera* in *P. kotschyi* population SCD in mixed stand.

Regime effect on spatial distribution

In pure stand, both species were found to be aggregated at a weak radius. Contrary to *P. Kotschyi*, *T. macroptera* became regular above a 5m radius. Dispersal limitation was known as a major factor affecting the degree of species clumping (Hardy and Sonké, 2004). *P. Kotschyi* seedling had a long taproot making their establishment difficult. To survive, the plant develops root suckers by which it can spread fairly well, resulting in clumps of trees (Louppe et al., 2008). The mainstream dispersal factor of *T. macropptera* (the wind) (Louppe et al., 2008) explains in part it random distribution. It was aggregated only at the first 3 m. Also, at adult stand, each individual develops an important crown and root (Arbonnier, 2008) that prevent inhibition of the growth of any kind of individual or species. So, its nearest neighbour distance (nndist) was higher than the *P. Kotschyi* nndist. This distribution type of both species explained the absence of association between them in mixed stand. Specifically, *T. macroptera* conserves the same characteristics in both stands (pure and mixed). It would mean that only *P. kotschyi* distribution was influenced by the presence of *T. macroptera* individuals. In effect, *P. kotschyi* clumped in pure stand rather than in mixed stand. However, the assemblage of two species may be viewed as resulting from niche-assembly and dispersal-assembly rules (Hubbell, 2001). *T. macroptera* and *P. kotschyi* were known to colonise the same habitat: Savanna woodland in lowland. However, even within an apparently homogenous habitat, species spatial distribution is often non-random. This may result from our inability to detect finer details of habitat heterogeneity (Hardy and Sonké, 2004). As said above, studies on spatial structure of a population may be coupled with soil and environmental analyses to detect probable habitat heterogeneity.

Implications to management strategies

Regarding the spatial distribution and SCD pattern of *T.*

macroptera and *P. kotschyi* under pure and mixed stands, we are not able to make a conclusion on the ideal formula of management strategies. However, findings suggest that *P. kotschyi* probably would present better evolution in pure stand. Even if the stand regime appears to be without an effect on *T. macroptera*, it is obvious that pure stand would contribute to the best development of *T. macroptera*. The great reduction of sapling density from pure to mixed stand suggests that in order to have a good recruitment and individuals with a large diameter, *P. kotschyi* artificial planting must be in pure stand. Subject to complementary studies on the disseminators of both species and soil conditions, the strategies of reforestation and artificial plantings will have to respect a low density for *T. macroptera* (21.62 ± 1.59 trees/ha trees/ha) and a strong density for *P. kotschyi* (147 ± 38.18 trees/ha) in pure stand. It resulted a nearest neighbour distance estimated at 5.67 m for *T. macroptera* and 2.97 m for *P. Kotschyi*.

Thus, our results demonstrated some of the ways in which spatial patterns and size-class distribution for the given species may differ across habitat type. They clearly illustrate the importance of pure stand in management strategy for *T. macroptera* and *P. kotschyi*.

ACKNOWLEDGEMENTS

We thank project BIOTA-West Africa III for financial assistance. We are indebted to the staff of the BRP who granted us authorization for field works. We are also grateful to local and field guides who spontaneously provided us with assistance during the data collection phase from around the BRP.

REFERENCES

Adjonou K, Bellefontaine R, Kokou K (2009). Les forêts claires du Parc national Oti-Kéran au Nord-Togo: structure, dynamique et impacts des modifications climatiques récentes. Sécheresse 20:1-10.

Adomou AC, Sinsin B, Van der Maesen JLG (2006). Phytosociological and chorological approaches to phytogeography: A meso-scale study in Benin. Syst. Geogr. *Pl*. 76:155-178.

Arbonnier M (2008). Arbres, arbustes et lianes des zones sèches d'Afrique de l'Ouest. Troisième édition, Cirad, Paris, p. 576.

Asase A, Kokubun T, Grayer RJ, Kite G, Simmonds MSJ, Oteng-Yeboah AA. Odamtten GT (2008). Chemical constituents and antimicrobial activity of medicinal plants from Ghana: *Cassia sieberiana*, *Haematostaphis barteri*, *Mitragyna inermis* and *Pseudocedrela kotschyi*. Phytother. Res. 22:1013-1016.

Ayo RG, Audu OT, Ndukwe GI, Ogunshola AM (2010). Antimicrobial activity of extracts of leaves of *Pseudocedrela kotschyi* (Schweinf.) Harms. Afr. J. Biotechnol. 9(45):7733-7737.

Baddeley A, Turner R (2005). Spatstat: An R package for analyzing spatial point patterns. J. Stat. Software 12(6):1548-7660.

Barot S, Gignoux J, and Menaut JC (1999). Demography of a Savanna palm tree: Predictions from Comprehensive Spatial Pattern Analyses. Ecology 80:1987-2005.

Böhning-Gaese K, Gaese BH, Rabemanantsoa SB (1999). Importance of primary and secondary seed dispersal in the Malagasy tree Commiphora guillaumini. Ecology 80:821–832.

Bolker B, Pacala SW (1997). Using moment equations to understand stochastically driven spatial pattern formation in ecological systems. Theor. Popul. Biol. 52:179-197.

Bonou W, Glèlè-Kakaï R, Assogbadjo AE, Fonton HN, Sinsin B (2009). Characterisation of *Afzelia africana* Sm. habitat in the Lama forest reserve of Benin. Forest. Ecol. Manag. 258 (7):1084-1092.

Boyom FF, Fotio D, Zollo PHA, Agnaniet H, Menut C, Bessiere JM (2004). Aromatic plants of tropical central Africa. Part XLIV. Volatile components from *Pseudocedrela kotschyi* (Schweinf.) Harms growing in Cameroon. Flav. Fragr. J. 19:9-11.

Brown ND (1998). Degeneration versus regeneration-logging in tropical rain forests. In: Goldsmith FB (eds.) Tropical Rain Forest, a Wider Perspective, Chapman and Hall, London, pp. 43-47.

Bruna EM, Kress WJ (2002). Habitat fragmentation and the demographic structure of an Amazonian understory herb (*Heliconia acuminata*). Conserv. Biol. 16:1256-1266.

Burslem DFRP (1995). Differential responses to nutrients, shade and drought among tree seedlings of lowland tropical forest in Singapore. In: Swain MD (eds.) The Ecology of Tropical Forest Seedlings. Parthenon, Carnforth, pp. 211-244.

Caswell H (2001). Matrix Population Models: Construction Analysis and Interpretation, 2nd ed. Sinauer Associates, Inc., Sunderland, Massachusetts, p. 722.

Condit R, Sukumar R, Hubbell SP, Foster R (1998). Predicting population trends from size distributions: a direct test in a tropical tree community. Am. Nat. pp. 495-509.

Conrad J, Vogler B, Reeb S, Klaiber I, Papajewski S, Roos G, Vasquez E, Setzer CM, Kraus W (2001). Isoterchebulin and 4, 6-O-Isoterchebuloyl-d-glucose, Novel Hydrolyzable Tannins from *Terminalia macroptera*. J. Nat. Prod. 64(3):294-299.

Coulibaly E (2006). Savoirs et savoir -faire des anciens métallurgistes d'Afrique occidentale: procédés et techniques de la sidérurgie directe dans le Bwamu (Burkina Faso et Mali). Karthala, Paris, p. 266.

Dalling JW, Hubbell SP (2002). Seed size, growth rate and gap microsite conditions as determinants of recruitment success for pioneer species. J. Ecol. 90:557-568.

Delvingt W, Heymans JC, Sinsin B (1989). Guide du Parc National de la Pendjari. – CECA-CEE-CEA, Bruxelles, p. 125.

Diggle PJ (1983). Statistical analysis of spatial point patterns. Academic Press, London, UK.

Fandohan AB, Assogbadjo AE, Glèlè Kakaï R, Sinsin B, Van Damme P (2010). Impact of habitat type on the conservation status of tamarind (Tamarindus indica L.) populations in the W National Park of Benin. Fruits 65(1):11-19.

Feeley JK, Davies SJ, Nur Supardi Noor MD, Kassim AR, Tan S (2007). Do current stem size distributions predict future population changes? An empirical test of intraspecific patterns in tropical trees at two spatial scales. J. Trop. Ecol. 23:191-198.

Gaoue OG, Ticktin T (2007). Patterns of harvesting foliage and bark from the multipurpose tree *Khaya senegalensis* in Benin: Variation across ecological regions and its impacts on population structure. Biol. Conserv. 137:424-436.

Georgewill UO, Georgewill OA (2009). Effect of extract of *Pseudocedrela kotschyi* on blood glucose concentration of alloxan induced diabetic albino rats. Eastern J. Med. 14:17-19.

Glèlè-Kakaï R, Sinsin B (2009). Structural description of two Isoberlinia dominated vegetation types in the Wari–Maro Forest Reserve (Benin). S. Afr. J. of Bot. 75:43–51.

Goreaud F, Pélissier R (2003). Avoiding misinterpretation of biotic interactions with the intertype K12-function: population independence vs. random labelling hypotheses. J. Veg. Sci. 14:681-692.

Goreaud R, Courbaud B, and Collinet F (1999). Spatial structure analysis applied to modeling of forest dynamics: a few examples. In: Amaro, A. & Tome M. (eds). Proceedings of the IUFRO Workshop, Empirical and Process Based Models for Forest Tree and Stand Growth Simulation, Novas Tecnologias, Oerias, PT, pp. 155-172.

Graae BJ, Hansen T, Sunde PB (2004). The importance of recruitment limitation in forest plant species colonization: a seed sowing experiment. Flora 199:263-270.

Hardy OJ, Sonke B (2004). Spatial pattern analysis of tree species distribution in a tropical rain forest of Cameroon: assessing the role of limited dispersal and niche differentiation. Forest. Ecol. Manag. 197:191-202.

Harper JL (1977). Population Biology of Plants. Academic Press, New York.
Hubbell SP (2001). A Unified Neutral Theory of Biodiversity and Biogeography, Princeton University Press.
Husch B, Beers T and Kershaw J (2003). Forest Mensuration. Eds Ronald Press Company, London, UK.
Jachmann H, Croes T (1991). Effects of browsing by elephants on *Combretum/Terminalia* woodland at the Nazinga Game Ranch, Burkina Faso, West Africa. Biol. Conserv. 57:13–24.
Johnson NL, Kotz S (1970). Distributions in Statistics: Continuous Univariate Distributions. John Wiley & Sons, New York.
Kainer KA, Wadt LHO, Staudhammer CL (2007). Explaining variation in Brazil nut fruit production. Forest. Ecol. Manage. 250:244-255.
Kassim OO, Loyevsky M, Amonoo H, Lashley L, Ako-Nai KA, Gordeuk VR (2009). Inhibition of *in-vitro* growth of Plasmodium falciparum by Pseudocedrela kotschyi extract alone and in combination with *Fagara zanthoxyloides* extract. Trans. R. Soc. Trop. Med. Hyg. 103:698-702.
Klimas CA, Kainer AK, and Wadt LHO (2007). Population structure of *Carapa guianensis* in two forest types in the southwestern Brazilian Amazon. Forest. Ecol. Manage. 250:256-265.
Louppe D, Oteng-Amoako AA, Brink M (2008). Plant resources of tropical Africa. Ed. CTA. Wagningen, Germany, p. 705.
Manlaya RJ, Cadeta P, Thioulouseb J, Chotte J (2000). Relationships between abiotic and biotic soil properties during fallow periods in the Sudanian zone of Senegal. Appl. Soil Ecol. 14:89-101.
Mariott FHC (1979). Barnard's Monte Carlo Tests: how many simulations? Appl. Stat. 28:75-77.
Molofsky J, Augspurger CK (1992). The effect of leaf litter on early seedling establishment in a tropical forest. Ecology 73:68-77.
Musa YM, Haruna AK, Ilyas M, Yaro AH, Ahmadu AA, Usman H (2008). Phytochemical, analgesic and anti-inflammatory effects of the ethyl acetate extract of the leaves of *Pseudocedrela kotschyii*. Afr. J. Trad. Compliment. Alt. Med. 5:92-96.
Okunade MB, Adejumobi JA, Ogundiya MO, Kolapo AL (2007). Chemical, phytochemical compositions and antimicrobial activities of some local chewing sticks used in South Western Nigeria. J. Phytopharmacother. Nat. Prod. 1(1):49-52.
Pacala SW, Deutschman DH (1995). Details that matter: the spatial distribution of individual trees maintains forest ecosystem function. Oikos 74:357-365.
Peters CM (1996). Observations on the sustainable exploitation of non timber tropical forests products. In: Ruiz PM, Arnold JEM (eds). Current Issues in non-timber forest products research, Cifor - ODA, Bagor, Indonesia, pp. 19-39.
Pulliam RH (2000). On the relationship between niche and distribution. Ecol. Lett. 3:349-361.
Rondeux J (1999). La mesure des arbres et des peuplements forestiers. Les Presses Agronomiques de Gembloux, 2e édition, Gembloux, Belgique, p. 521.
Schiøtz M, Boesen MV, Nabe-Nielsen J, Sørensen M, Kollmann J (2006). Regeneration in *Terminalia oblonga* (Combretaceae) - A common timber tree from a humid tropical forest (La Chonta, Bolivia). Forest. Ecol. and Manage. 225:306–312.
Shahina G (1989). Savanna Plants. An Illustrated Guide. Macmillan Publishers Ltd., London, p. 105.
Silva O, Ferreira E, Vaz Pato M, Canica M, Gomes ET (2002). *In vitro* anti-Neisseria gonorrhoeae activity of *Terminalia macroptera* leaves FEMS. Microbiol. Lett. 217:271-274.
Stoyan D, Stoyan H (1994). Fractals, Random Shapes and Point Fields. Methods of Geometrical Statistics. John Wiley & Sons, Chichester, New York, p. 389.
Sunderland TCH, Drasnfield J (2002). Species profiles rattans (Palmae: Calamoideae). In: FAO-INBAR-SIDA (Eds.), Rattan: Current Research Issues and Prospects for Conservation and Sustainable Development, Non-Wood Forest Product Bull. petin No. 14.
Tesfaye G, Teketay D, Fetene M (2002). Regeneration of fourteen tree species in Harenna forest, southeastern Ethiopia. Flora 197:461-474.
Verschuren J (1988). Notes d'écologie, principalement des mammifères du Parc National de la Pendjari, Bénin. Biologie. 58:185-206.
Wadt LHO, Kainer KA, Gomes-Silva DAP (2005). Population structure and nut yield of *Bertholletia excelsa* stand in southwestern Amazonia. Forest. Ecol. Manage. 211:371-384.
White F (1983). The vegetation of Africa: a descriptive memoir to accompany the UNESCO/AETFAT/UNSO. UNESCO. Nat. Resour. Res. 20:1-356.
Wiegand T, Moloney K (2004). Rings, circles and null1 - models for point pattern analysis in ecology. Oikos 104:209-229.
Zar JH (1999). Biostatistical Analysis. 4th ed. Prentice-Hall, New Jersey.

5

Estimation of blue sheep population parameters in the Dhorpatan Hunting Reserve, Nepal

Achyut Aryal[1,2]*, Sven Gastaur[2], Steffen Menzel[2], Til Bahadur Chhetri[2] and Jack B. Hopkins[3]

[1]Ecology and Conservation Group, Institute of Nature Sciences, Massey University, New Zealand.
[2]The Biodiversity Research and Training Forum (BRTF), Nepal.
[3]Department of Ecology, Montana State University, Bozeman, MT 59717-3460, USA.

Blue sheep *(Pseudois nayaur)* are the main prey species of snow leopards *(Uncia uncia)* and are a highly coveted species in Nepal. Currently, the conservation of blue sheep in Nepal is a national concern because it is not clear if populations were affected by the Maoist insurgency beginning in 1996. For this study, we estimated population parameters for blue sheep in the Dhorpatan Hunting Reserve and compared these estimates to studies conducted pre-insurgency. We found that the number (206 sheep) and density (1.8 sheep/km) of sheep in the Barse and Phagune Blocks (study area) have not changed since 1993 and 1976, respectively. A mean of 7 animals/herd from 29 herds were classified by sex. The mean sex ratio was higher for 2006 and 2008 counts (4.82 males/female, SD = 2.52) than pre-insurgency counts (0.82 males/female, SD = 0.06) however there was no significant difference between sex ratios. The proportion of ewes in the population was lower than a 2006 count, but similar to pre-insurgency counts. Population age structure was less variable in 2008 than previous counts. More young rams and fewer ewes were counted since 2006, which may have been due to misclassification of females and/or less poaching pressure on young males. Recent hunting pressure by insurgents may have reduced ewe population numbers, and as a result, lambs numbers.

Key words: Blue sheep, Dhorpatan Hunting Reserve, herd size, population density, population size, *Pseudois nayaur,* sex ratio.

INTRODUCTION

Blue sheep *(Pseudois nayaur)* are the main prey species of snow leopards (*Panthera unica*) in the high altitude biomes of Eurasia and their presence is often an indicator of snow leopards occupancy (Oli et al., 1993). Since blue sheep are the main prey species of snow leopards and other predators in the Himalayas, blue sheep conservation is currently a priority in Nepal.

Estimating blue sheep population parameters and determining causes of mortality are imperative for the proper management of blue sheep, snow leopard, and other carnivore populations in Nepal. Although 10,000 blue sheep were estimated to live in Nepal (Shackleton, 1997; Wegge and Oli, 1997), little is currently known about their populations.

*Corresponding author. E-mail: savefauna@yahoo.com

Historically, blue sheep have been harvested in the Dhortpantan hunting reserve (DHR), an international hunting hotspot for those seeking to kill a trophy animal in the Himalaya Range (Wegge 1976; FAO 1980). Each year, the government of Nepal authorizes a blue sheep harvest however it is currently unknown if the harvests are sustainable.

Due to political instability over the recent years (1996-2007 Maoist insurgency period) the DHR went unprotected. The field office moved to the city creating more opportunity for poaching blue sheep as well as other species in the DHR. In order to evaluate the impact this political strife had on the blue sheep population in the DHR, a comparative study was conducted. Specifically, we estimated the abundance, density, heard size, sex ratio, and age structure of blue sheep in the DHR and compared our results to past studies: Wegge (1976), Austegard and Haugland (1993), and Karki and Thapa (2006).

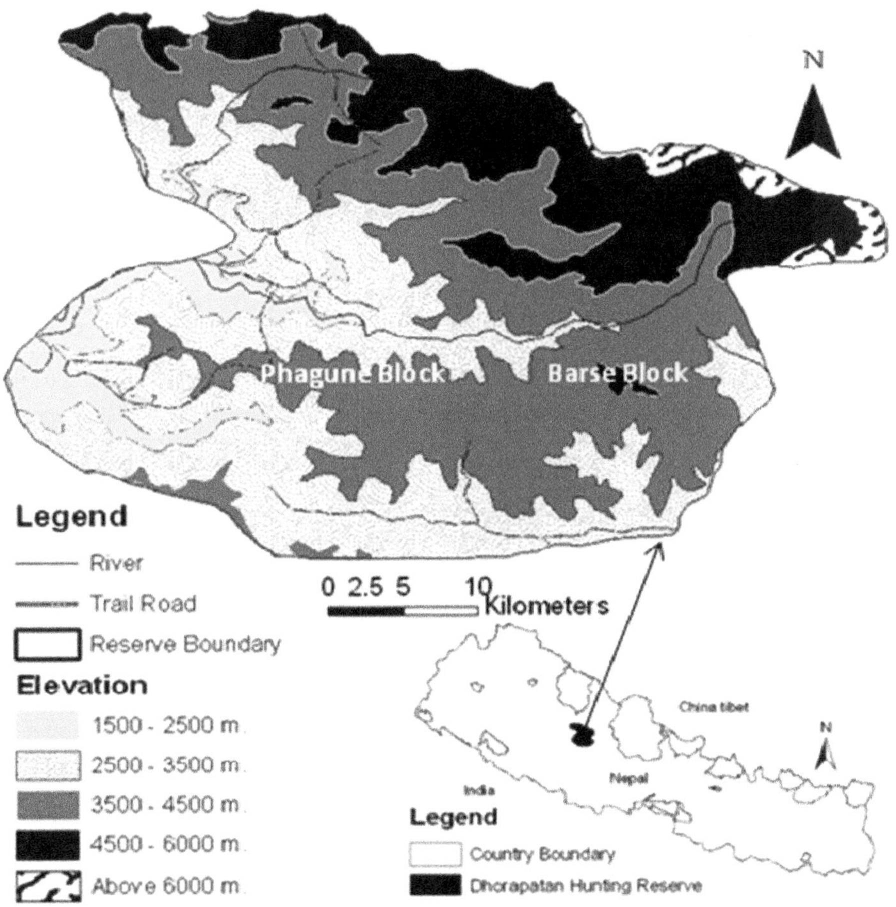

Figure 1. Study area (Barse and Phagune Blocks) in the Dhortpantan hunting reserve.

MATERIALS AND METHODS

Study area

The Dhorpatan hunting reserve is located in the Baglung District in the Dhaulagiri Himalaya of western Nepal (23°30'N-28°50'N, 82°50'E-83°15'E; Figure 1). The Reserve is divided into 7 blocks, covering 1,325 km^2, and ranges from 2,850 to 5,500 m in altitude (Wegge, 1979). The Phagune and Barse Blocks (study area) cover 265 km^2 of the Reserve (Figure 1) and are assumed to have high poaching pressure because they are easily accessible. Currently, blue sheep occupy only 115 km^2 of the study area because they do not use forested (50%) or rugged alpine habitats (25%) (Wegge, 1979).The DHR is located in front of a high saddle connecting the high Dhaulagiri and Hiuchuli and is shielded by several pasturelands south of Utter Ganga. As a result, the study area receives less precipitation than others areas of the Nepal Midlands (Stainton, 1972).

The DHR is characterized by a diversity of plant species. Vegetation found in the drier climatic belt to the north as well as species from the humid zone to the south are represented in the study area. Falling in a transition zone, the dry northern elements are more pronounced at higher altitudes and on southeast aspects. Shaded riparian habitats at low elevation are composed of well-developed mixed hardwood stands made of fir (*Abies spectabilis*) and hemlock (*Tsuga dumosa*), while birch (*Betula utilis*) and rhododendron (*Rhododendron campanulatum*) cover the upper northern slopes to tree line (3,050-3,660 m). The southern slopes (3,500-2,440 m) are comprised of a wide belt of sparse oak forest (*Quercus semecarpifolia*), interspersed with isolated blue pine (*Pinus excelsa*) and rhododendron (*Rhododendron arboreum*) (Wegge, 1976).

In addition to blue sheep and snow leopards, a variety of other ungulates and carnivores inhabit the Reserve: ungulates include, goral (*Nemorhaedus goral*), Himalayan tahr (*Hemitragus jemlahicus*), wild boar (*Sus scrofa*), Himalayan musk deer (*Moschus chrysogaster*), serow (Capricornis sumatraensis) and Indian muntjac (*Muntiacus muntjak*); and carnivores include, leopards (*Panthera pardus*), lynx (*Felis lynx*), wild dog (*Cuon alpinus*), red fox (*Vulpes vulpes*), and wolf (*Canis lupus*). In addition, omnivorous Asian black bears (*Ursus thibetanus*) inhabit forested areas in the DHR and red pandas (*Ailurus fulgens*) are reported in the upper forests of the Lower Seng and Upper Bakre Valleys (Wegge, 1976; Fox, 1985) where they primarily feed on bamboo (*Bambuseae*).

Survey methods

Blue sheep were surveyed using the 'Fixed-Point Count Method' (Jackson et al., 1996; Karki and Thapa, 2006; Schaller, 1973; Wegge, 1976). Population counts were conducted from April – May 2008 using binoculars (8–42 X) and a spotting scope (15–45X). Surveys were conducted from high vantage points along ridgelines in the morning more when animals were likely to be feeding and before viewing was restricted by fog, and in the late afternoon. Blue sheep were classified by sex and separated into the following age

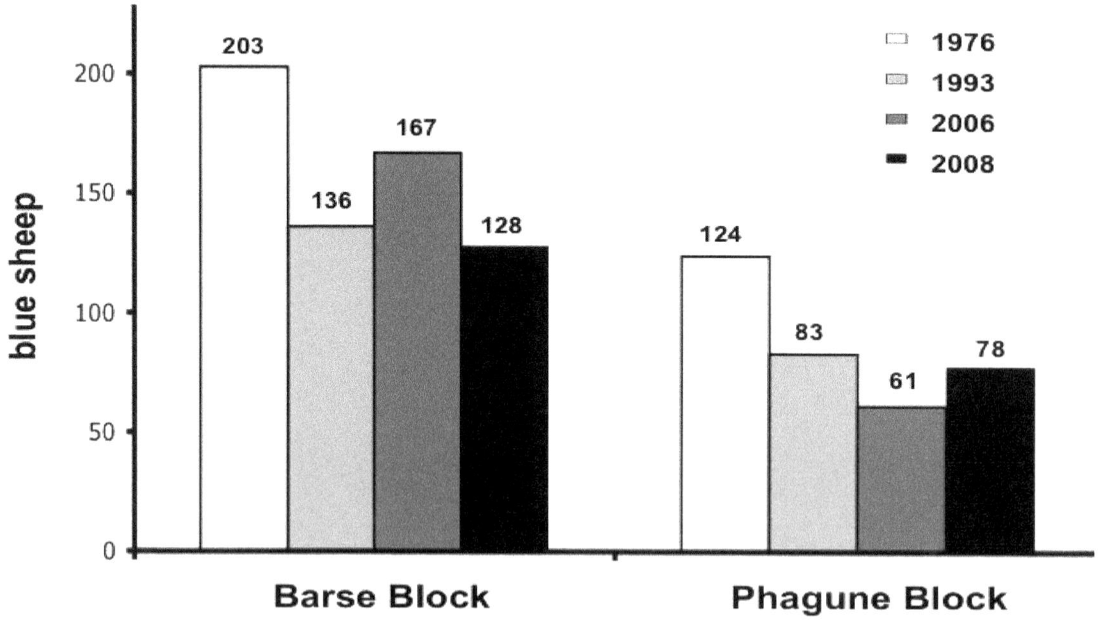

Figure 2. Blue sheep counts in the Barse and Phagune blocks of the DHR, 2008.

Table 1. Blue sheep counts by sex in each block (Sex Ratio = males/female).

Area	Phagune 1976	Barse 1976	Total	Phagune 1993	Barse 1993	Total	Phagune 2006	Barse 2006	Total	Phagune 2008	Barse 2008	Total
Rams	55	88	143	34	67	101	46	152	198	54	101	155
Ewes	69	115	184	49	69	118	15	15	30	24	27	51
Total	124	203	327	83	136	219	61	167	228	78	128	206
Sex ratio	0.80	0.77	0.78	0.69	0.97	0.86	3.06	10.13	6.60	2.25	3.74	3.04

classes according to Wegge (1976): lambs, yearlings, ewes, and rams (Class I: 15-35 cm horns, 2-3 yrs; Class II: 30-45 cm horns that curve backward, 4-7 yrs; Class III: 45+ cm horns curved, 7+ yrs). We used Student's t-test to determine if parameter estimates were significantly different between blocks and a Pearson's chi-square test to compare studies. Significance for all tests was determined using $\alpha = 0.05$.

RESULTS

Population size and density

After correcting for double counts, we counted 128 and 78 blue sheep in the Barse and Phagune Blocks, respectively (N = 206). Although population counts were significantly different ($\chi^2 = 37.59$, $p < 0.05$, df = 3), the total population ($\chi^2 = 1.12$, p = 0.57, df = 2) and number of sheep counted between blocks (Barse: $\chi^2 = 5.91$, p > 0.5, df = 2; Phagune: $\chi^2 = 3.59$, p = 0.17, df = 2) have not changed since 1993. Historically, there were more sheep in the Barse Block (t = 3.32, p = 0.02) (Figure 2).

We estimated population density at 1.8 blue sheep/km^2 (206 blue sheep/115 km^2), while Wegge (1976), Austegard and Haugland (1993), and Karki and Thapa (2006) reported 2, 2.3, and 1.4 blue sheep/km^2, respectively. There was no significant change to population density during the last 4 decades ($\chi^2 = 0.23$, p = 0.97, df = 3).

Heard size

A mean of 7 animals/herd from 29 herds (SD = 5.5, range = 1-37) were classified by sex. There was no significant difference in average herd size in each block (Barse: χ^2 = 2.06, p > 0.05, df = 11; Phagune: χ^2 = 4.31, p > 0.05, df = 8).

Sex ratio

The sex ratio in 2008 was 2.25 males/female and 3.74 males/female in Phagune and Barse Blocks, respectively (Table 1). These ratios were lower than 2006, but higher than the pre-insurgency counts. Sex ratios are

Figure 3. Estimated proportions of blue sheep by sex and age in each block of the study area, 2008.

significantly different throughout years (χ^2 = 15.77, p = 0.001 df = 3) however the average sex ratio since 2006 (mean = 4.82, SD = 2.52) was not higher than pre-insurgency counts (mean = 0.82, SD = 0.06; t = 2.25, p = 0.27).

Population age structure

Evidence suggests the population age structure in 2008 was less variable then previous records. Although ewes were counted in the highest proportion (30%) in 2008 (Figure 3), and this has historically been the case in the study area (Figure 4), ewes were more abundant during pre-insurgency counts. The proportions of trophy rams (Class III) and ewes have declined since 2006, but were similar to pre-insurgency counts (rams: χ^2 = 5.25, p = 0.07, df = 2; ewes: χ^2 = 0.06, p = 0.97, df = 2) (Figure 4).

The proportion of lambs was significantly different among counts (χ^2 = 22.02, p < 0, df = 3) (Figure 4), and besides Karki and Thapa (n = 7) (2006), we counted fewer lambs (n = 16) than previous studies. The proportions of class I and II rams were higher than previous studies, but not significantly different (Class I: χ^2 = 5.27, p = 0.15, df = 3; Class II: χ^2 = 2.8, p = 0.42, df = 3) (Figure 4).

DISCUSSION

Population numbers declined after 1976 and have remained relatively constant since 1993 (Table 1). The recent increase in the proportion of sub-adult rams (Figure 4) may be related to a reduction in poaching since the insurgency period however this increase is not significant and poaching data is incomplete. Another potential explanation for this increase is that livestock depredations have caused herders to poison blue sheep predators (e.g. wolves and leopards). We note that wild boar (*Sus scrofa*), were observed above tree line, suggesting recent predator loss in the DHR.

The variation in heard size may have been due to human presence, predation, or our survey coinciding with the breeding season. In addition, herd sizes may have been influenced by the distribution and quality of forage because patchy, poor quality vegetation cannot support large herds (Wilson, 1981).

Legal harvest of trophy rams in the early 1970s may have influenced our sex ratio estimates (Table 1). The low ratio of males to females in Phagune and Barse historically may have also resulted from poaching or selective hunting (Wegge, 1979) however the government's hunting quota has been historically low.

Although ewe proportions have remained relatively constant throughout the years, female counts have declined this may have been due to females being misclassified in the field as young males.

Local people preferentially hunt large males because they yield larger quantities of meat and their behavioural characteristics make them easier to hunt (Wegge, 1979). Local people complained that since the DHR headquarters moved in 2001, more illegal hunting has occurred in the DHR. They also mentioned that during the insurgency period, the DHR, especially the Barse and Phagune blocks, were the main shelter for Maoist rebels. Hunting pressure by insurgents may have had an additive effect on ewes, thus decreasing their numbers as well as lambs counted during recent surveys. Reports indicate livestock herders may have also responsible for illegal

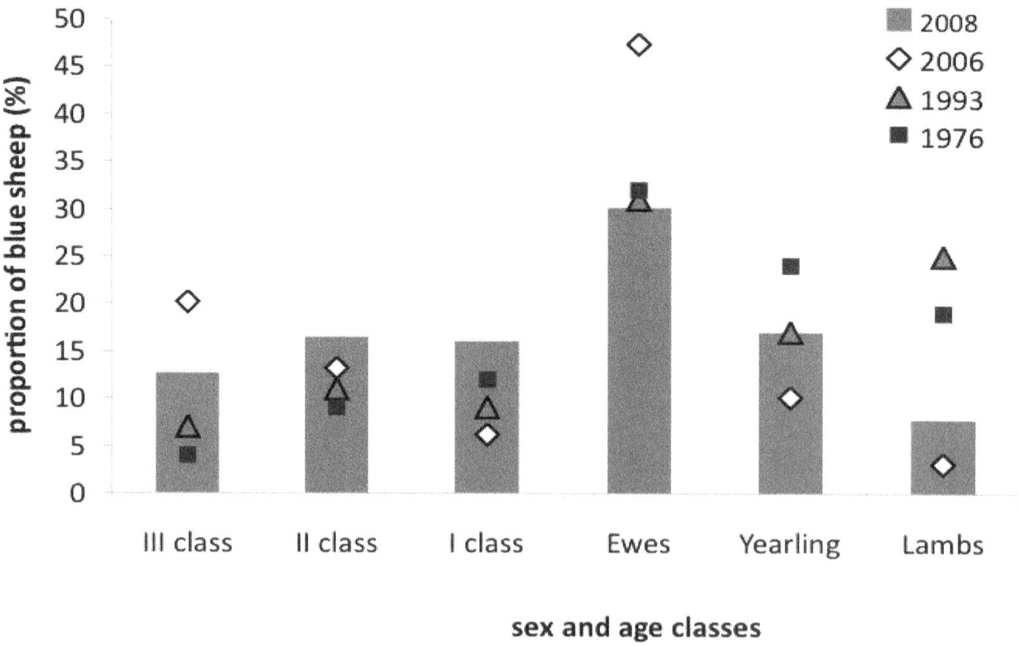

Figure 4. Estimated proportions of blue sheep by sex and age in the study area over the past 30+ years.

hunting during summer months as well as during winter months near villages.

ACKNOWLEDGMENT

We would like to thank the Department of National Parks and Wildlife Conservation for granting us permission to conduct the study, the Snow Leopard Conservation Grants Program for their financial support, and IDEA WILD for equipment.

REFERENCES

Austegard, Glenn O, Svein H (1993). Trophy Hunting in Nepal: a case study on Blue sheep in Dhorpatan Shikar Reserve. A thesis submitted in partial fulfillment of the requirements for the M.Sc. degree, Agriculture University of Norway.

FAO (1980). National parks and wildlife conservation Nepal. Project findings and recommendations. FAO, Rome p. 63.

Fox JL (1985). An observation of lynx in Nepal. J. Bombay Nat. Hist. Soc. 82: 394.

Geist V (1971). Mountain Sheep. A Study in Behavior and Evolution. University of Chicago Press, Chicago.

Jackson R, Hunter DO (1996). (Second Edition). Snow Leopard Survey and Conservation Handbook. International Snow Leopard Trust, Seattle, Washington and U.S. Geological Survey, Fort Collins Science Center, Colorado. 154p. + appendices.

Karki J, Thapa B (2006). Status of Blue sheep and Himalayan tahr in DHR. DNPWC, Nepal.

Murphy EC, Whitten KR (1976). Dall sheep demography in McKinley Park and a reevaluation of Murie's data. J. Wildl. Manage. 40(4): 597-609.

Oli M, Rogers ME (1997). Seasonal pattern in group size and population composition of blue sheep in Manang, Nepal. J. Wildl. Manage. 60(4): 797-801.

Oli MK, Taylor IR, Rogers ME (1993). Diet of the snow leopard (Panthera uncia) in the Annapurna Conservation Area, Nepal. J. Zool. Lond. 231: 365-370.

Schaller GB (1998). Wildlife of the Tibetan steppe. University of Chicago Press, Chicago, IL, USA.

Shackleton DM (1997). Wild Sheep and Goats and their relatives: Status Survey and Conservation Action Plan. IUCN/SSC Caprinae Specialist Group. IUCN, Gland, Switzerland and Cambridge, UK.

Shrestha R (2007). Coexistence of wild and domestic ungulates in the Nepalese Trans-Himalaya: Resource competition or habitat partitioning?. PhD thesis. Norwegian University of Life Sciences, Norway.

Wegge P (1976). Himalayan shikar reserves; surveys and management proposals. Field Document No. 5. FAO/NEP/72/002 Project, Kathmandu pp. 96.

Wegge P (1979). Aspects of the population ecology of blue sheep in Nepal. J. Asian Ecol. 1: 10-20.

Wegge P, Oli M (1997). Nepal. in Shackleton, D.M (Editor). Wild sheep and Goats and their relatives, status survey and action plan for caprinae. IUNC, Gland, Switzerland and Cambridge, UK pp. 213-239.

Wilson PN (1981). Ecology and habitat utilization of blue sheep (Pseudois nayaur) in Nepal. Biol. Conserv. 21: 55-74.

Woodgerd W (1964). Population dynamics of bighorn sheep on wildhorse Island. J. Wildl. Manage. 28: 381-391.

Variation of turbulent mixing parameters at the Egyptian Mediterranean coast

S. H. Sharaf El Din[1]*, F. M. Eid[1], N. N. Saad[2], K. A. Alam El Din[1], M. El Sharkawy[2]

[1]Oceanography Department, Faculty of Science, Alexandria University, Egypt.
[2]National Institute of Oceanography and Fisheries, Alexandria, Egypt.

The analysis of the turbulent mixing structure is accomplished by the dynamical computations of the mixing parameters represented by: buoyancy frequency (N), viscous dissipation (ε), turbulent kinetic energy (E), Richardson number (Ri), Reynolds stress, diffusive salt flux (DSF), diffusive heat flux (DHF). In general the computations show that, the water column becomes more stable and less stratified at the deeper layers. Also, the dynamical processes of vertical turbulent mixing are stronger in the surface layers.

Key words: Mediterranean sea, turbulence mixing.

INTRODUCTION

The Egyptian Mediterranean Coast lies at the south east sector of the Levantine sub-basin, (Figure 1) from 25° 30' E to 34° E and extends northward to 33°N. Its water volume is about 224, 801, 55 km^3, and has a surface area of about 154,840 km^2.

Relatively, few efforts have been carried out to study the dynamical process of vertical turbulent mixing in front off the Egyptian Coast. Sharaf El-Din (1964) investigated the mixing processes in the North Sea and computed the Eddy coefficient of viscosity. Morcos and Hassan (1976) and Mohamed (1999) studied the dynamical process of mixing qualitatively by studying the existing different water masses. The mixing process and circulation pattern off Egyptian Mediterranean Coast was studied by El-Sharkawy (2007). The dynamics and mixing of the Eastern Mediterranean outflow in the Tyrrhenian basin was studied by Sparnocchia (1999). Nabil (2009) studied the turbulent vertical mixing parameters such as Richardson Number (Ri), Reynolds Stress (J_m) and Diffusive Salt Flux (J_s) at the central area of the Red Sea. Also, Johnson (1996); Rossa and Lueckb (2005); Munka and Wunschb (1998); Kobayashia et al. (2006); Hornea et al. (1996); MacPhee-Shaw (2006) and Yoshida and Oakey (1996) studies the vertical turbulent mixing using the numerical models at different parts of Oceans and Seas.

The aim of the present study is to investigate the variation of dynamical processes of vertical turbulent mixing water of the Egyptian Mediterranean Coast.

MATERIALS AND METHODS

The temperature and salinity data from 160 thousands stations through the Mediterranean Sea representing the time interval from the early 20th century till 2002; however, data only collected from 1970 - 2002 was used in this research. The measurements have been carried out by various instruments as CTD, MBT, XBT, and sampling bottles at the standard depths. This large data archive has been subject to sophisticated statistical modeling to render various types of error to the minimum. The average monthly climatologically data of 1' x 1' grid has been used in the proceeding calculations carried out in this work.

Four stations off the Egyptian Mediterranean Coast, Figure 1 named C12, C11, C10, and C1 were carried out to study the current field by the Eulerian method. Most of the hourly measurements of the current were provided by Arrays of Acoustic Doppler Current Profilers (ADCP); however, in some cases Recording Current Meters (RCM) was used.

Methods of analysis of turbulent mixing structure

Viscous dissipation (ε):

$$\varepsilon = K_m \left(\frac{\partial \left\| \vec{V} \right\|}{\partial z} \right)^2 \qquad 1$$

Where; ε is viscous dissipation of turbulent kinetic energy, K_m is the

*Corresponding author. E-mail: nabilelbatikhy@yahoo.com.

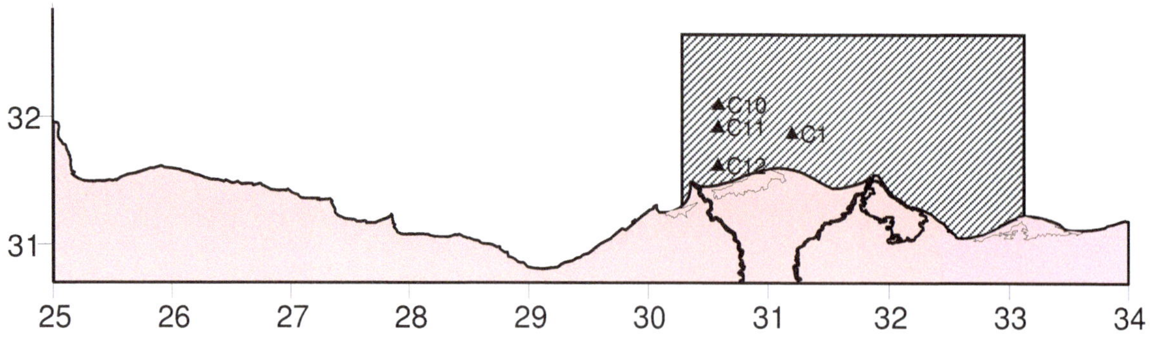

Figure 1. Area of study and location of current meter stations

Eddy Coefficient of viscosity; *V* is Northward velocity of current. Viscous Dissipation defines the rate of conversion of kinetic energy by viscosity (turbulent friction) to heat per unit mass (Proudman, 1953). This dynamical parameter is of negligible magnitude; however, it is a basic characteristic of mixing and the most important mixing parameter (Peters, 1999). The Eddy viscosity (K_m) is defined in terms of the molecular viscosity (μ), and density (ρ) by:

$$K_m = \mu / \rho \qquad 2$$

Turbulent kinetic energy (E) per unit mass (Pond and Pickard, 1983), which is a measure of turbulence

$$E = (U')^2 + (V')^2 \qquad 3$$

Where U' and V' are the deviation of eastward and northward velocity about its mean values respectively.

Brunt-vaisala frequency (N):

$$N = \sqrt{\frac{-g}{\rho}\frac{\partial \rho}{\partial z}} \qquad 4$$

Also, known as Buoyancy frequency. This is the natural frequency for the propagation of internal gravity waves (g) (Pond and Pickard, 1983), and is also a measure of dynamic stability.

Richardson number (Ri):

$$Ri = -\frac{N^2}{\left(\frac{\partial \|\vec{V}\|}{\partial z}\right)^2} \qquad 5$$

It is a non-dimensional indicator of the ensemble average flow state, but owing to the averaging, it is only loosely related to the dynamical instability and vertical turbulent mixing. However, values of *Ri* below the critical value (*Ri* = 0.25) Kelly (2005) and Robert (2005) indicate the occurrence of dynamical instability in the water column associated with strong processes of vertical turbulent mixing (Pond and Pickard, 1983).

Reynolds stress (J_m):

$$J_m = -\rho K_m \frac{\partial \|\vec{V}\|}{\partial z} \qquad 6$$

Which defines the vertical turbulent flux of momentum (Peters, 1999)?

Eddy diffusivity (K_s):

$$K_s = 0.2 \times \frac{\varepsilon}{N^2} \qquad 7$$

This defines the turbulent exchange coefficient of Salt (Peters, 1999).

Diffusive salt flux (J_s):

$$J_s = -10^{-3} \rho K_s \left(\frac{\partial S}{\partial z}\right) \qquad 8$$

This defines the vertical turbulent transport of salt (Peters, 1999).

Diffusive heat flux (J_h):

$$J_h = -\rho C_p K_h \left(\frac{\partial T}{\partial z}\right) \qquad 9$$

This defines the vertical turbulent transport of heat (Molinari et al., 1986), where the specific heat C_p=3.94 k j / kg c°, and the coefficient of heat diffusivity $K_h = 1 \times 10^{-4}\ m^{-2}s^{-1}$

RESULTS AND DISCUSSION

The dynamical computations for turbulence mixing parameters are carried out for the four stations C12, C11, C10 and C1 at three levels: near surface, intermediate depth, and near bottom. The depth of the three levels selected

Table 1. The three levels depths selected through the four casts of current stations.

	C12 m	C11 m	C10 m	C1 m
near surface	3	14	42	8
intermediate depth	10	46	170	34
near bottom	18	94	298	58

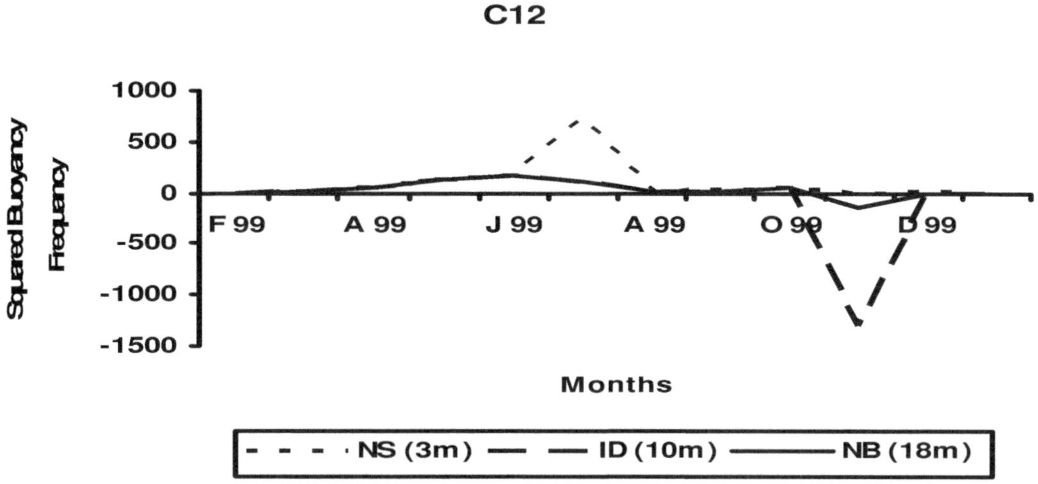

Figure 2a. Squared buoyancy frequency (10^{-6} s^{-2}) at stations C12.

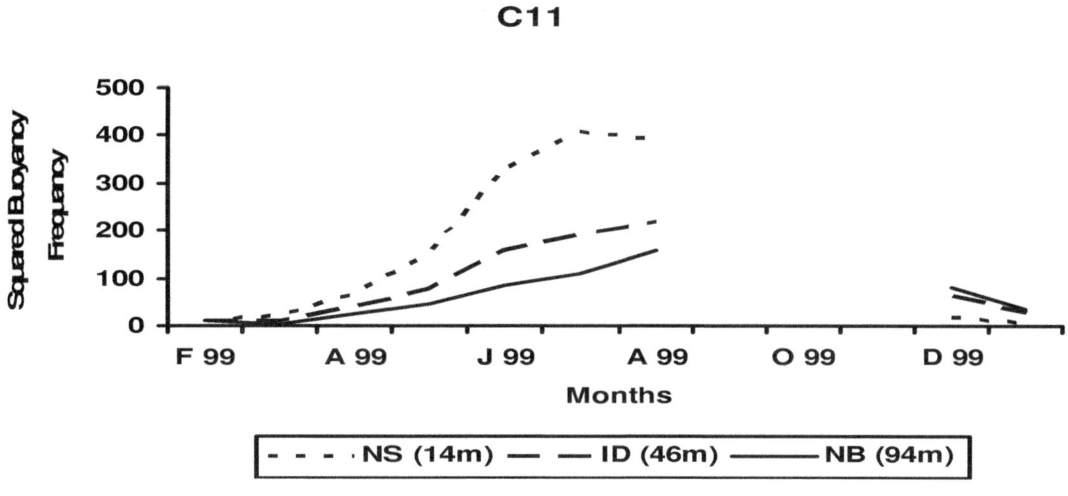

Figure 2b. Squared buoyancy frequency (10^{-6} s^{-2}) at stations C11.

through the four stations was tabulated in Table 1.

Brunt-Vaisala frequency or the buoyancy frequency (*N*)

As the static stability is positive for stable conditions and zero or negative for unstable conditions, the result values of N^2 at all stations attained positive values during the whole year except at station C12, it had negative values during the period from November to February. Theses negative values, appeared during the cold period, related to occurrence of dynamic instability in the water column associated with strong processes of vertical turbulent mixing, Figure (2a). At C12, values of N^2 vary from -1.28 × 10^{-3} s^{-2} at 10 m (intermediate depth) during November to 7.19 × 10^{-4} s^{-2} at 3 m (near surface) during July with an annual vertical average value -6.64 × 10^{-5} s^{-2}. The water column was stable at station C11 Figure (2b). Values of N^2 vary from 4.98 × 10^{-6} s^{-2} during March at 94 m (near bottom) to 410 × 10^{-6} s^{-2} during July at 14 m (near surface) with an annual vertical average value 1.12 × 10^{-4} s^{-2}.

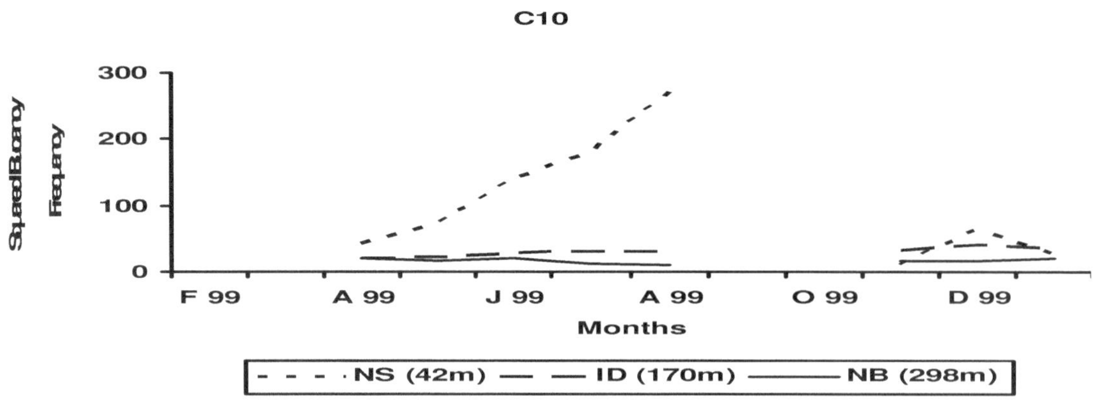

Figure 2c. Squared buoyancy frequency (10^{-6} s^{-2}) at station C10.

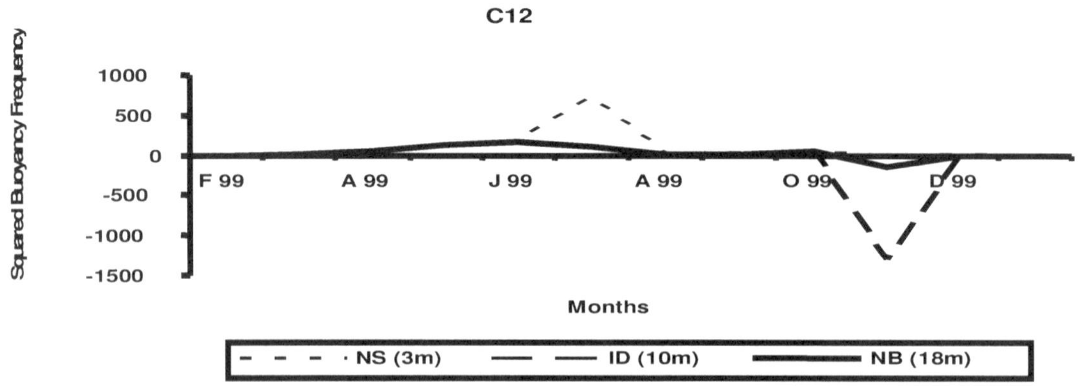

Figure 2d. Squared buoyancy frequency (10^{-6} s^{-2}) at station C1.

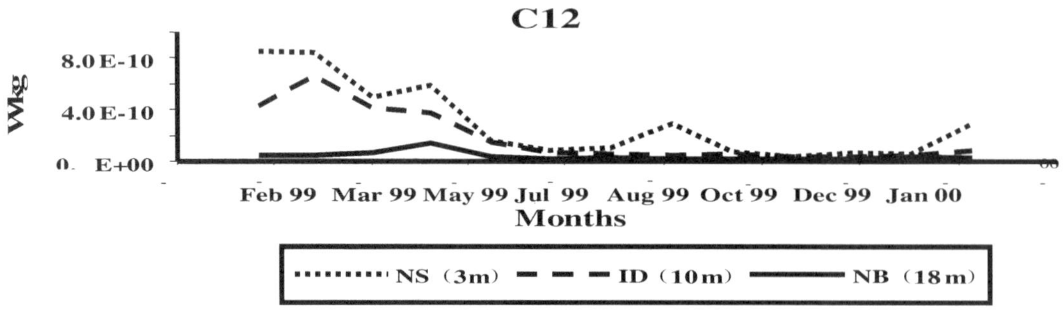

Figure 3a. Dissipated energy (w/kg) at station C12.

At station C10, Figure (2c), the water column also showed dynamic stability as values of N^2 vary from 9.59 × 10^{-6} s^{-2} during November at 42 m (near surface) to 2.66 × 10^{-4} s^{-2} during August at 42 m also with an annual vertical average value 4.58 × 10^{-5} s^{-2}. The water column was unstable (N^2 negative) at station C1, Figure (2d) during November and December, and values N^2 varies from -5.47 × 10^{-6} s^{-2} during December at 8 m (near surface) to 3.72 × 10^{-4} s^{-2} during July at 34 m (intermediate depth) with an annual vertical average value 9.53 × 10^{-5} s^{-2}.

The viscous dissipation

At station C12, Figure (3a), ε attained its greatest values during February which reached 8.52 × 10^{-10} W/kg at 3 m (near surface). The minimum value of ε was 1.68 × 10^{-11} W/kg attained during November at 18 m (near bottom), and the annual vertical average value was 1.77 × 10^{-10}

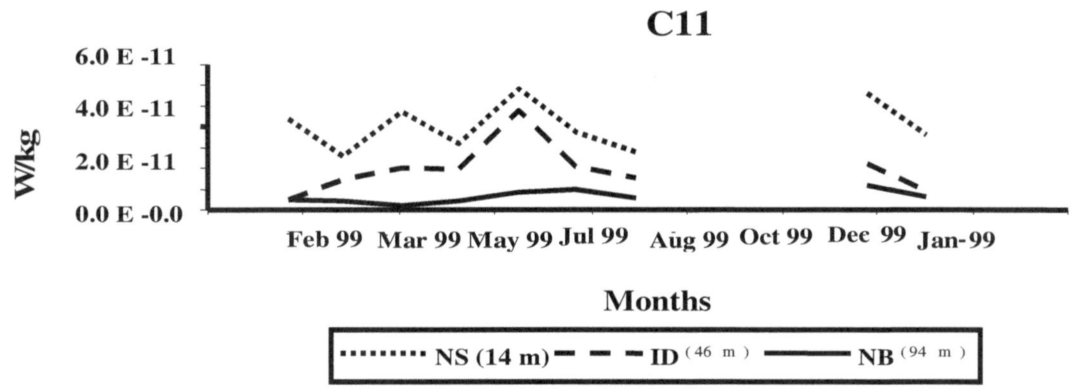

Figure 3b. Dissipated energy (w/kg) at station C11.

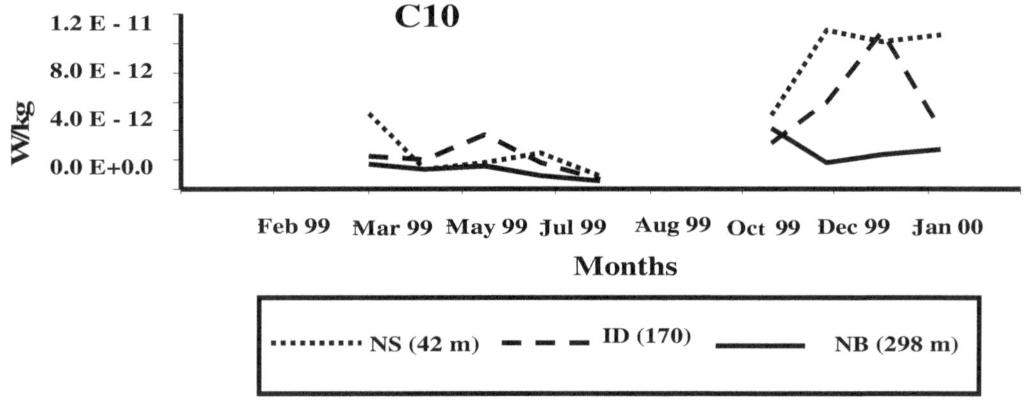

Figure 3c. Dissipated energy (w/kg) at station C10.

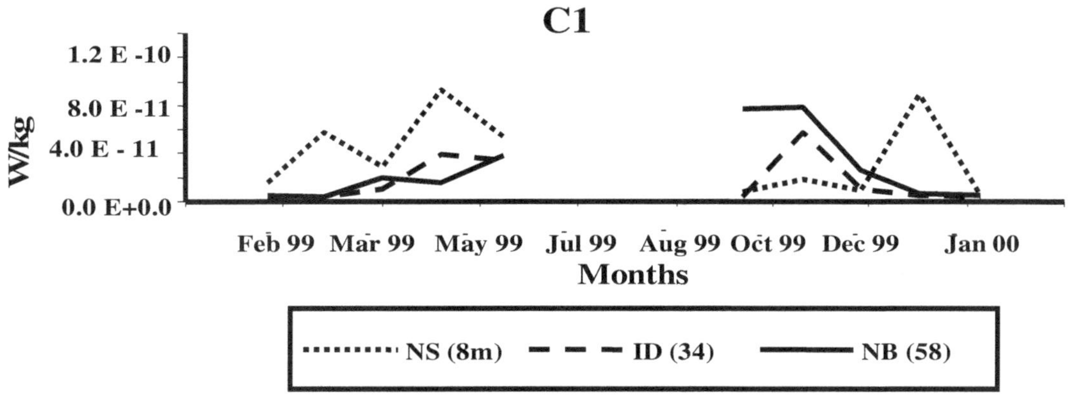

Figure 3d. Dissipated energy (w/kg) at station C1.

W/kg.

For the deeper stations C11, C10, and C1, (Figures 3b, c and d), it can be noticed that ε had relatively smaller magnitudes. For instance, at station C11, values ε range from 2.04×10^{-12} W/kg during April at 94 m (near bottom) to 5.81×10^{-11} W/kg during June at 14 m (near surface) with an annual vertical average value 2.18×10^{-11} W/kg. The viscous dissipated energy, at the deep station C10, was the weakest among the four stations as ε ranged from 5.76×10^{-13} W/kg during August at 298 m (near bottom) to 1.09×10^{-11} W/kg during December at 42 m (near surface), with an annual vertical average value

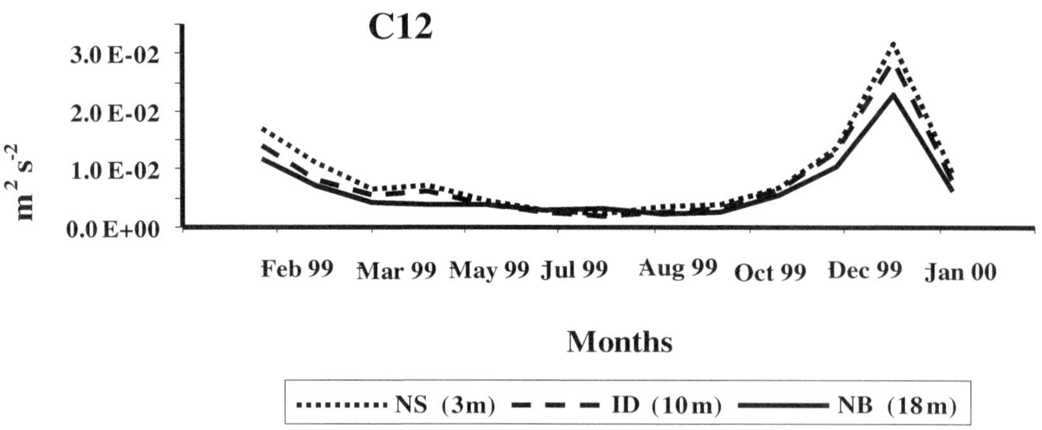

Figure 4a. Turbulent kinetic energy (m^2s^{-2}) at station C12.

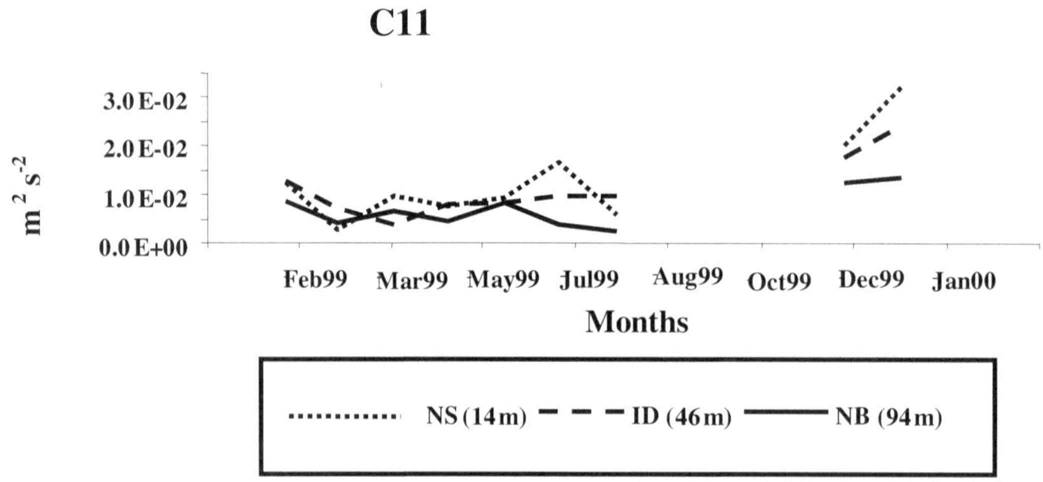

Figure 4b. Turbulent kinetic energy (m^2s^{-2}) at station C11.

3.69×10^{-12} W/kg.

The dissipation of turbulent kinetic energy by viscosity was relatively higher at station C1. Values of ε range from 3.4×10^{-12} W/kg during February at 34 m (intermediate depth) to 1.23×10^{-10} W/kg during August at 8 m (near surface) with an annual vertical average value 3.21×10^{-11} W/kg. It can be noticed that the maximum viscous dissipation always occur near the surface, that is, ε decreases with depth. This showed that the processes of vertical turbulent mixing were stronger in the upper layers.

The turbulent kinetic energy

Figures 4a revealed that at station C12, *E* ranged from 2.05×10^{-3} m^2s^{-2} at 10 m (intermediate depth) during August to 3.18×10^{-2} m^2s^{-2}at 3 m (near surface) during January, with an annual vertical average value 8.03×10^{-3} m^2s^{-2}. The behavior of *E* at station C11 was quite analogous; (Figure 4b), as *E* ranges from 2.36×10^{-3} m^2s^{-2} at 94 m (near bottom) during August to 3.22×10^{-2} m^2s^{-2}at 14 m (near surface) during January, with an annual vertical average value 1.04×10^{-2} m^2s^{-2}.

The computations revealed that *E* decreased significantly with depth. This can be seen from the distribution of *E* at the deep station C10, (Figure 4c), where *E* ranges from 8.19×10^{-4} m^2s^{-2} at 298 m (near bottom) during August to 8.26×10^{-2} m^2s^{-2}at 42 m (near surface) during December, with an annual vertical average value 1.17×10^{-2} m^2s^{-2}. At the station C1 (Figure 4d), *E* ranged from 2.14×10^{-3} m^2s^{-2} at 58 m (near bottom) during August to 3.47×10^{-2} m^2s^{-2}at 8 m (near surface) during December, with an annual vertical average value 8.62×10^{-3} m^2s^{-2}.

This temporal distribution of *E* shows that turbulence was stronger near the surface and decrease with depth, that is, the motion of water was more laminar at deeper layers. The minimum values of *E* were attained during August while the maximum ones were attained during December and January, that is, turbulence is weaker in

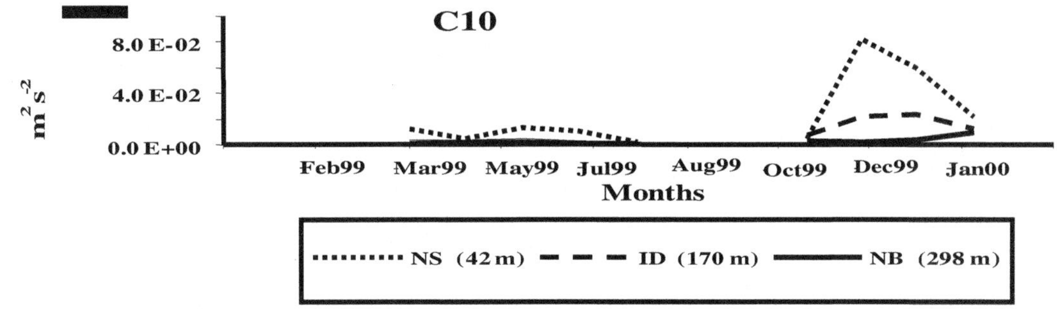

Figure 4c. Turbulent kinetic energy (m^2s^{-2}) at station C10.

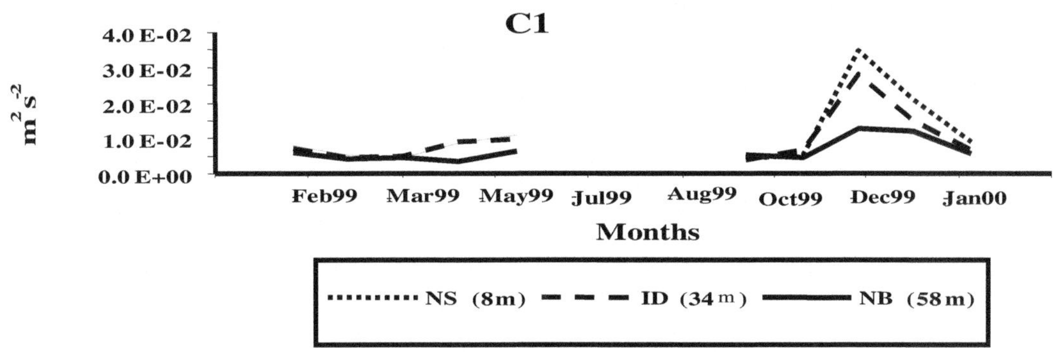

Figure 4d. Turbulent kinetic energy (m^2s^{-2}) at station C1.

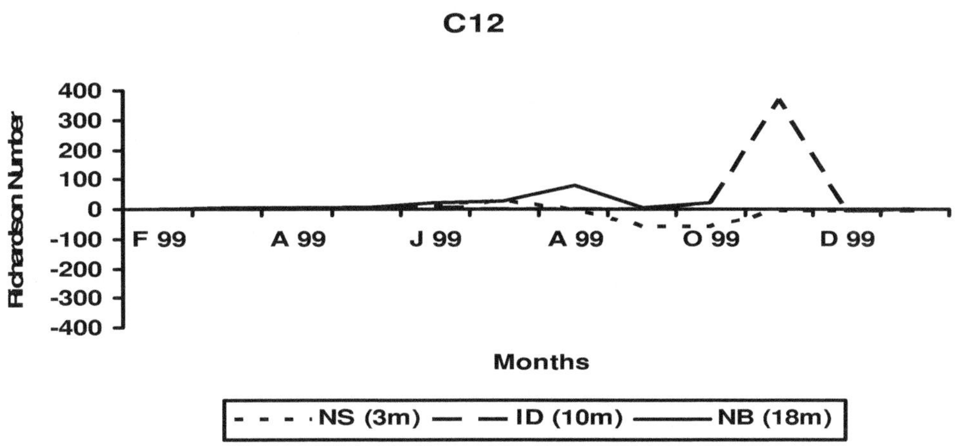

Figure 5a. Richardson number at station C12.

summer and stronger in winter.

The Richardson number *Ri*

From Figure (5a), the distribution of *Ri* at the shallow station C12 is characterized by possessing a maximum value of 367, attained during November at 10 m (near surface), with an annual vertical average value of 2.29. Negative values of *Ri* remarked at 3 m between august and December at C12 resulted from negative values of N^2 which indicated unstable density gradients with active convective overturning. Under such circumstances, *N* does not have an accepted physical meaning and the magnitude of negative *Ri* is not generally of interest.

Stations C11 and C10 (Figures 5b and c) are charac-

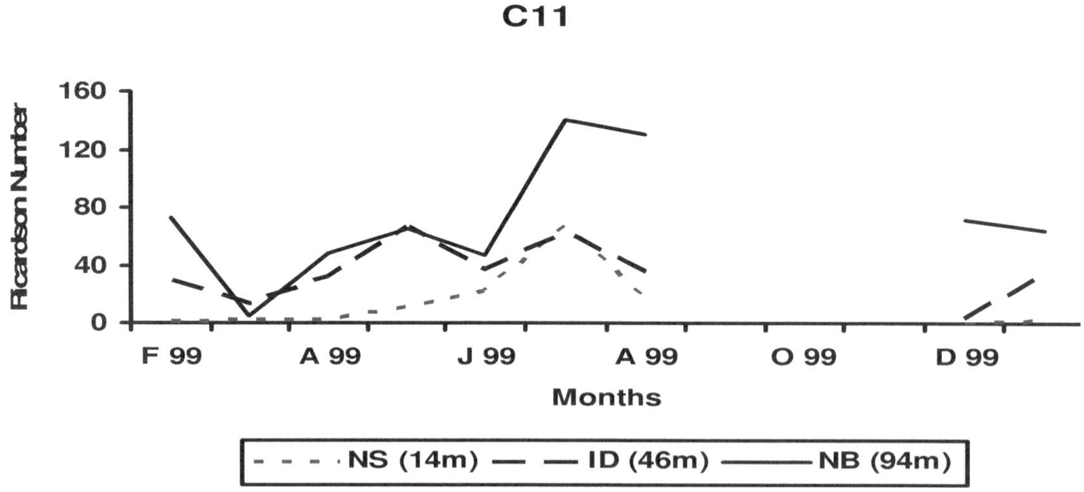

Figure 5b. Richardson number at station C11.

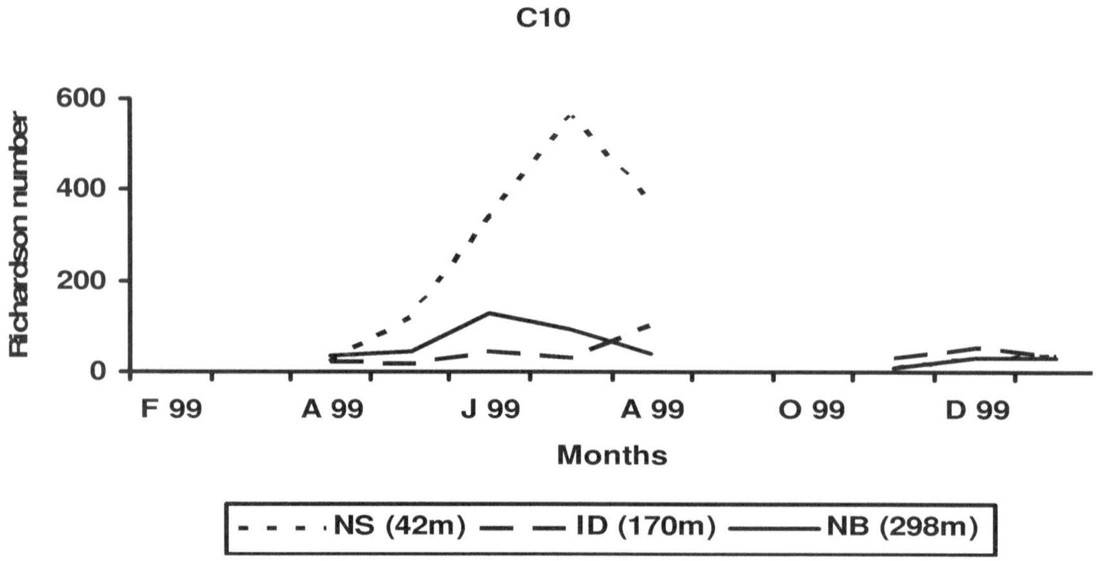

Figure 5c. Richardson number at station C10.

terized by possessing only positive values of *Ri*. Values of this dynamical parameter vary at station C11 from 0.54, attained during December at 14 m (near surface) to 140, attained during July at 94 m (near bottom) with an annual vertical average value 41.4. At the deep station C10, *Ri* varies from 9.30 attained during November at 42 m (near surface) to 558 attained during July at 42 m also with an annual vertical average value 8.51. A very small negative value of *Ri*, occurred again at station C1, (Figure 5d) during the cold months. *Ri* varied from 0.07 during May at 34 m (intermediate) to 228, attained during October at 34 m (intermediate) with an annual vertical average value 29.4. The behavior of *Ri*, on contrary to that of N^2, ε, *E*, shows no time or spatial dependence.

The Reynolds stress J_m

At the shallow station C12, J_m varies from -1.6×10^{-6} Pa, attained during August at 18 m (near bottom) to 8.07×10^{-6} Pa, attained during March at 3 m (near surface), with an annual vertical average value 2.29×10^{-6} Pa. Hereinafter, The negative sign indicates that the vertical transport was directed upwards and vice versa, (Figures 6a). The maximum upward Reynolds stress, at the station C11, (Figure 6b) occurred during January at 14 m (near surface) with magnitude -1.1×10^{-6} Pa, the maximum downward momentum flux occurs during March at 14 m also with a magnitude 3.8×10^{-6} Pa, while the annual vertical average value of J_m was.72×10^{-7} Pa.

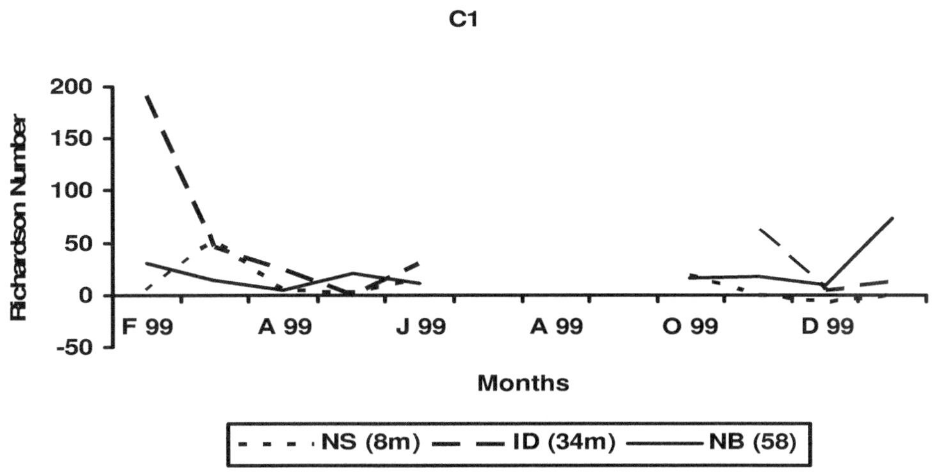

Figure 5d. Richardson number at station C1.

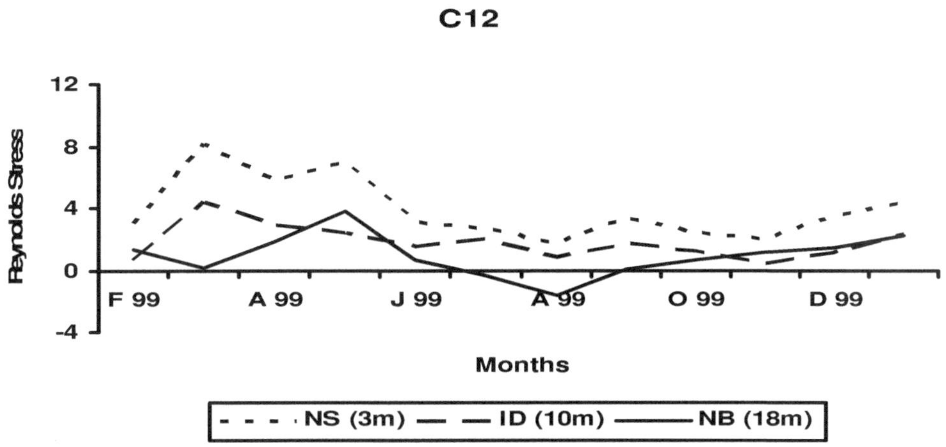

Figure 6a. Reynold's Stress (10^{-6} Pa) at station C12.

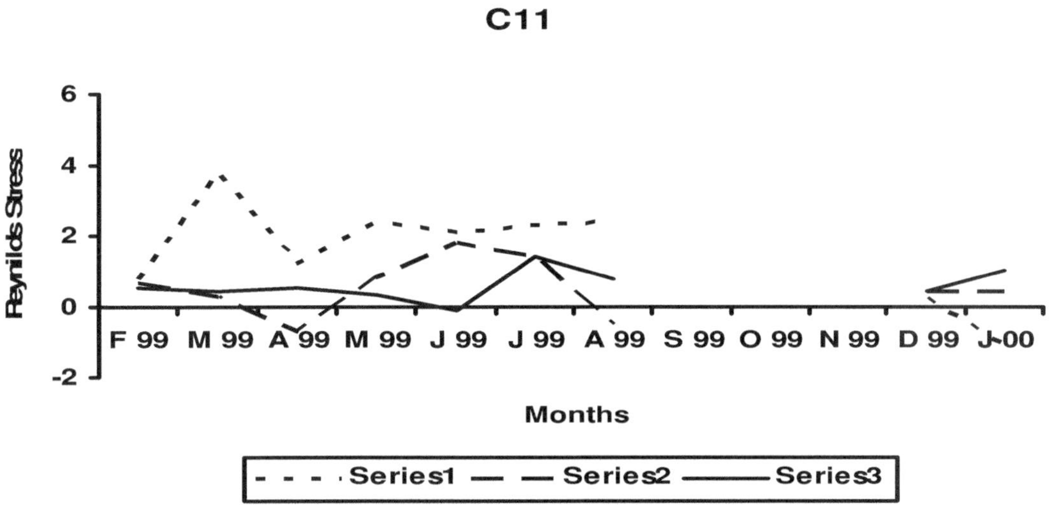

Figure 6b. Reynold's Stress (10^{-6} Pa) at station C11.

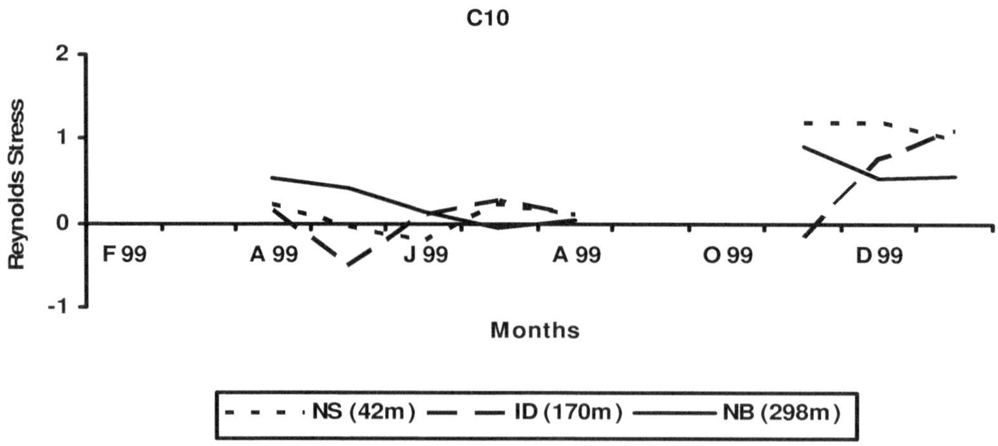

Figure 6c. Reynold's Stress (10^{-6} Pa) at station C10.

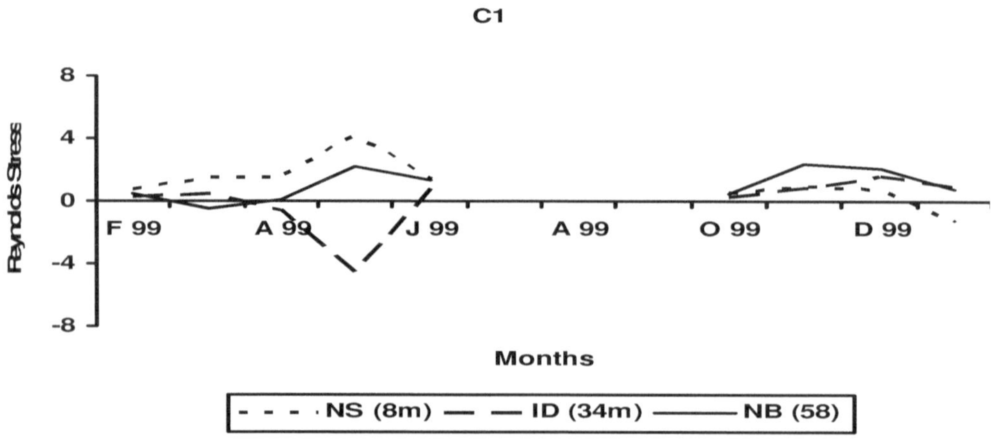

Figure 6d. Reynold's Stress (10^{-6} Pa) at station C1

Relatively weaker momentum fluxes occurred at the deep station C10, (Figure 6c) as J_m varied from -4.99 × 10^{-7} Pa, attained during May at 170 m (intermediate depth) to 1.2 × 10^{-6} Pa attained during November at 42 m (near surface), with an annual vertical average value 4.09 × 10^{-7} Pa. For station C1, (Figure 6d), a maximum upward momentum flux occurred during January at 8 m (near surface) with magnitude -1.25 × 10^{-6} Pa, the maximum downward momentum stress occurred during August at 58 m (near bottom) with a magnitude of4.39 × 10^{-6} Pa, while the annual vertical average value was 1.13 × 10^{-6} Pa.

The Eddy diffusivity

Figure 7a, showed the dynamical computations of Eddy diffusivity. At the four stations, the magnitude of K_s decreased relatively with depth. The temporal distribution of K_s at the shallow station C12 showed that K_s had maximum magnitude of 2.0 × 10^{-6} m^2s^{-1}, attained during April at 3 m (near surface), and a minimum magnitude of -1.0 × 10^{-6} m^2s^{-1} attained during December at 3m Also, with an annual vertical average value -0.3 × 10^{-6} m^2s^{-1}. Negative values of K_s result from negative values of N^2 which were associated with dynamical instability.

The minimum magnitude of K_s at the station C11, (Figure 7b), was 0.01 × 10^{-6} $m^2s^{-1,}$ attained during August at 94 m (near bottom), while the maximum magnitude was 1.22 × 10^{-6} m^2s^{-1} attained during January at 14 m (near surface) with an annual vertical average value 1.46× 10^{-7} m^2s^{-1}. Magnitudes of K_s were relatively weaker at the deep station C10, Figure (7c) as the minimum magnitude is 7.12 × 10^{-10} m^2s^{-1} during August at 42 m (near surface), the maximum magnitude is 0.07 × 10^{-6} m^2s^{-1} attained during January at 42 m (near surface) with an annual vertical average value of 7.14 × 10^{-8} m^2s^{-1}. At station C1, Figure (7d) the minimum magnitude of K_s is-2.7 × $10^{-6}m^2s^{-1}$ attained during May at 34 m (near surface), the maximum magnitude was 2.8 × 10^{-6} m^2s^{-1} attained

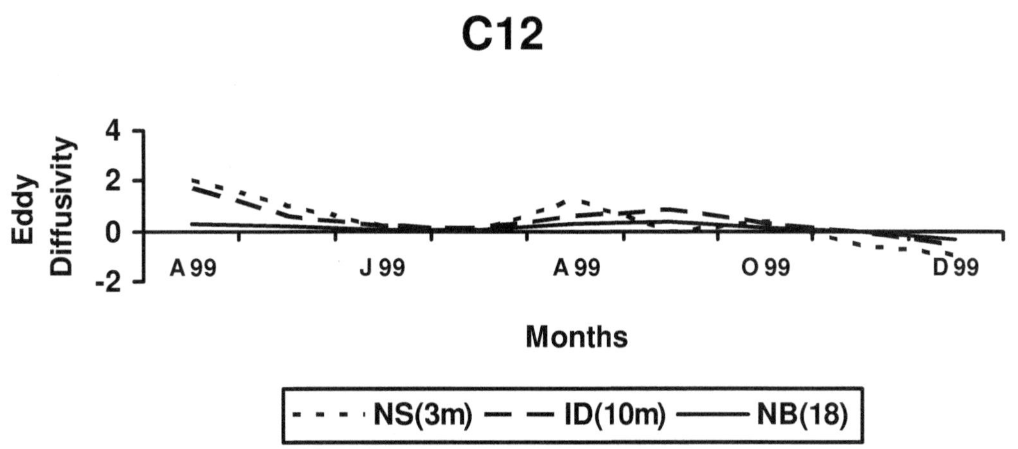

Figure 7a. Eddy diffusivity (10^{-6} m^2 s^{-1}) at station C12.

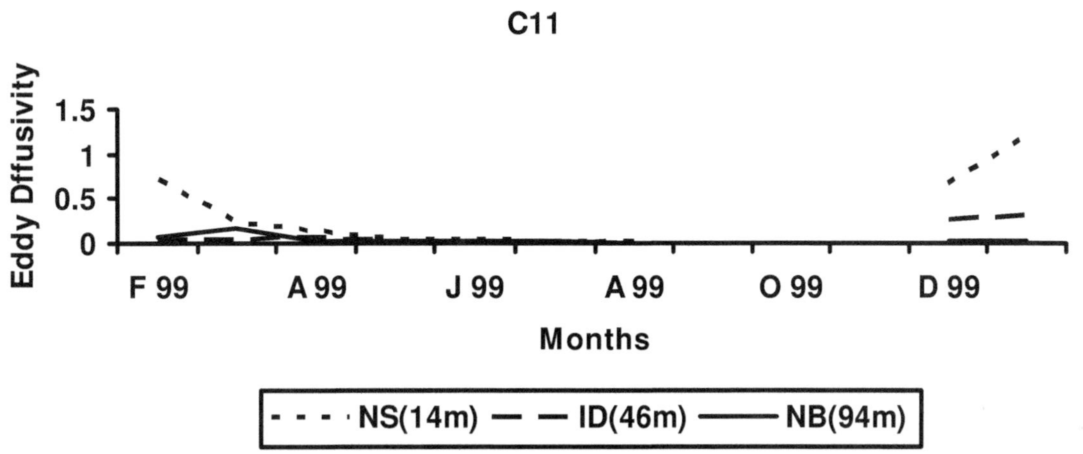

Figure 7b. Eddy diffusivity (10^{-6} m^2 s^{-1}) at station C11.

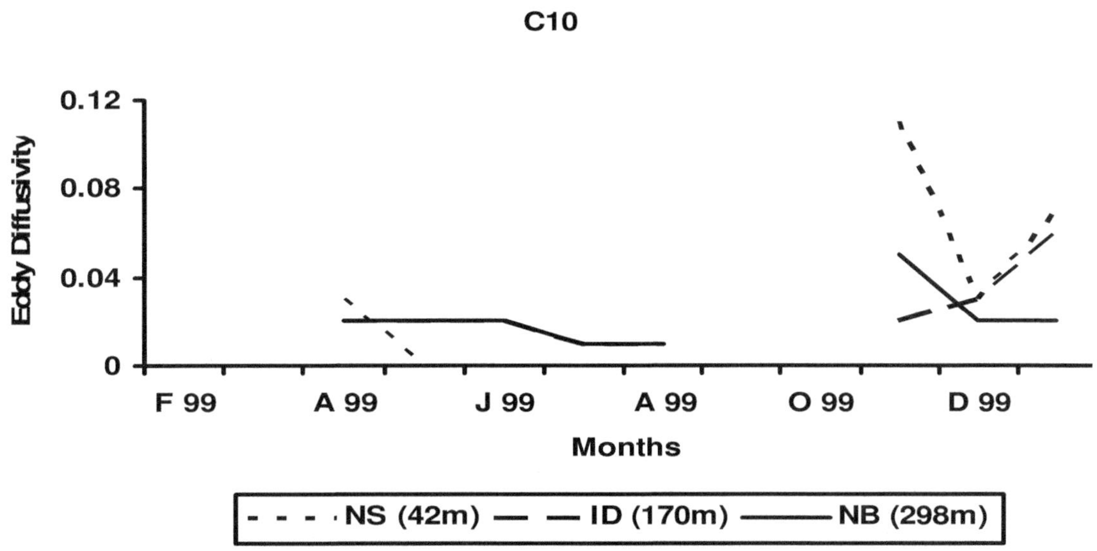

Figure 7c. Eddy diffusivity (10^{-6} m^2 s^{-1}) at station C10.

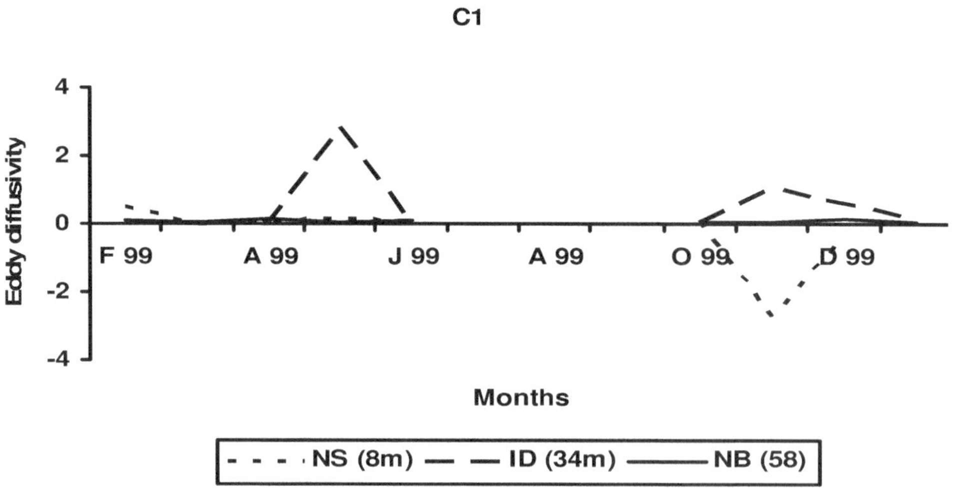

Figure 7d. Eddy diffusivity (10^{-6} m^2 s^{-1}) at station C1.

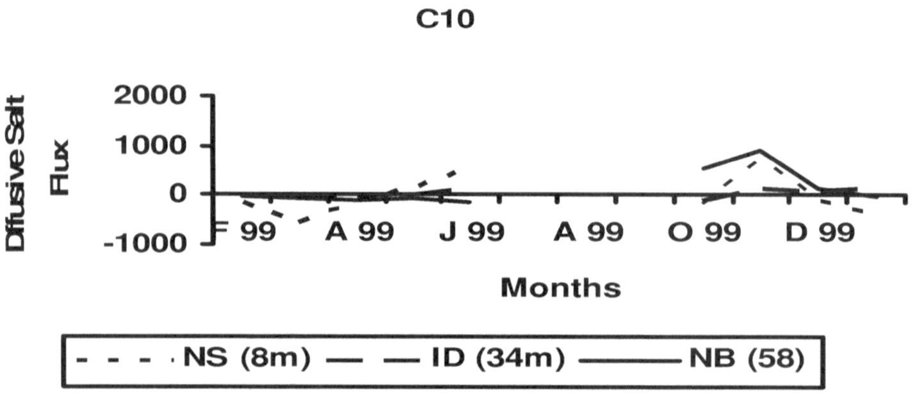

Figure 8a. Diffusive salt flux (10^{-12} kg m^2 s^{-1}) at station C10.

during January at 8 m (intermediate), while the annual vertical average value -7.04×10^{-7} m^{2}s^{-1}.

The diffusive salt flux

The numerical results showed that the distribution of J_s was analogous to that of K_s as both distributions reveal that the magnitudes of these two mixing parameters increased significantly during winter, and also decreased relatively with depth.

From Figure 8a the shallow station C12, J_s varied from a minimum of -2.49×10^{-8} kg m^{-2}s^{-1} attained during March at 3 m flowing upwards to a maximum of 2.78×10^{-9} kg m^{-2}s^{-1} attained during February, at 3 m also flowing downwards, with an annual vertical average value of 1.49×10^{-8} kg m^{-2}s^{-1}. Magnitudes of DSF at this shallow station C12 (21 m) are significantly greater than other stations due to the existence of strong vertical gradient of salt in the water column. Strong vertical gradients of salt could result from evaporation which had greater effect on the upper layer or could result from near boundary currents. Yoshida et al (1999) found that diffusivity and mixing processes were stronger near the bottom of the shallow Georges Bank (Canada) and that mixing was maximum when the flow is strongest.

For the deeper stations C11, C10, and C1, Figure (8b) it can be noticed that J_s has smaller magnitudes. For instance, at station C11, J_s ranged from a minimum of -1.98×10^{-10} kg m^{-2}s^{-1} attained during February at 14 m directed upwards to a maximum of 1.26×10^{-9} kg m^{-2}s^{-1} attained during January at 14 m also directed downwards, with an annual vertical average value 8.87×10^{-11} kg m^{-2}s^{-1}. The vertical diffusive salt flux at station C10, (Figure 8c) ranged from a minimum of -66×10^{-12} kg m^{-2}s^{-1} attained during December at 170 m to a maximum of 1.8×10^{-10} kg m^{-2}s^{-1} attained during November at 42 m, with an annual vertical average value of 4.73×10^{-11} kg m^{-2}s^{-1}. At station C1, (Figure 8d) J_s ranges from a minimum of -5.34×10^{-10} kg m^{-2}s^{-1} attained during March

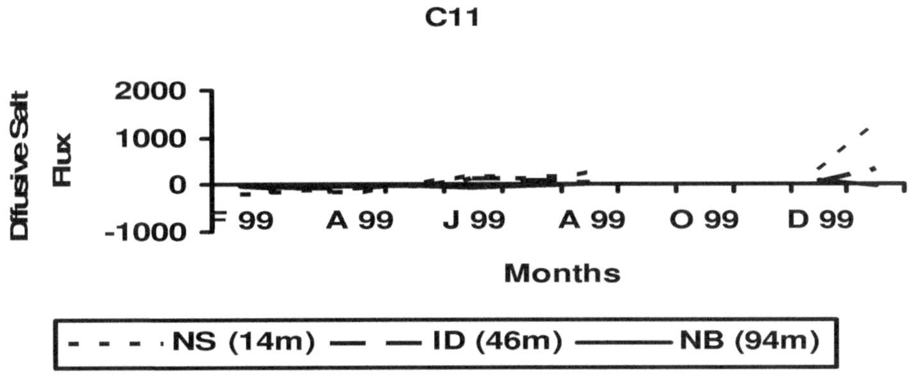

Figure 8b. Diffusive salt flux (10^{-12} kg m^2 s^{-1}) at station C11.

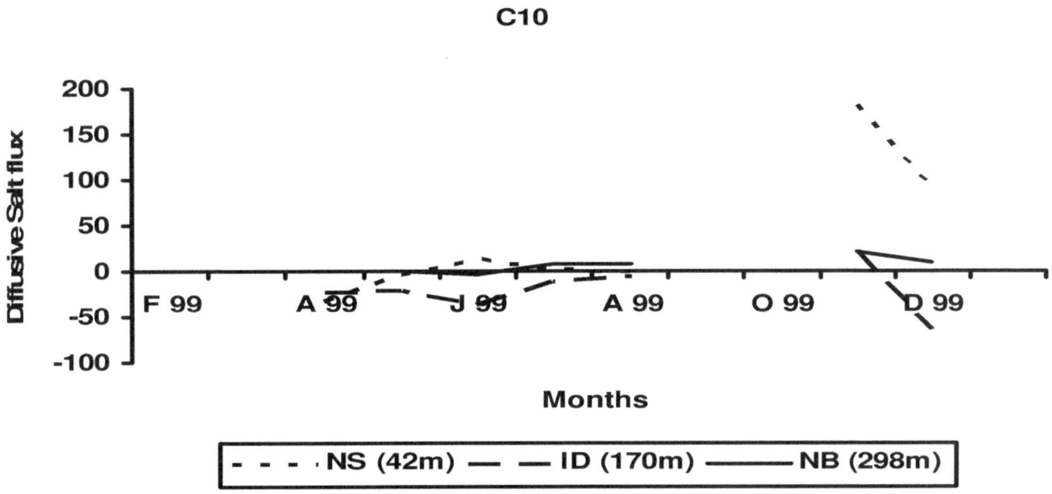

Figure 8c. Diffusive salt flux (10^{-12} kg m^2 s^{-1}) at station C10.

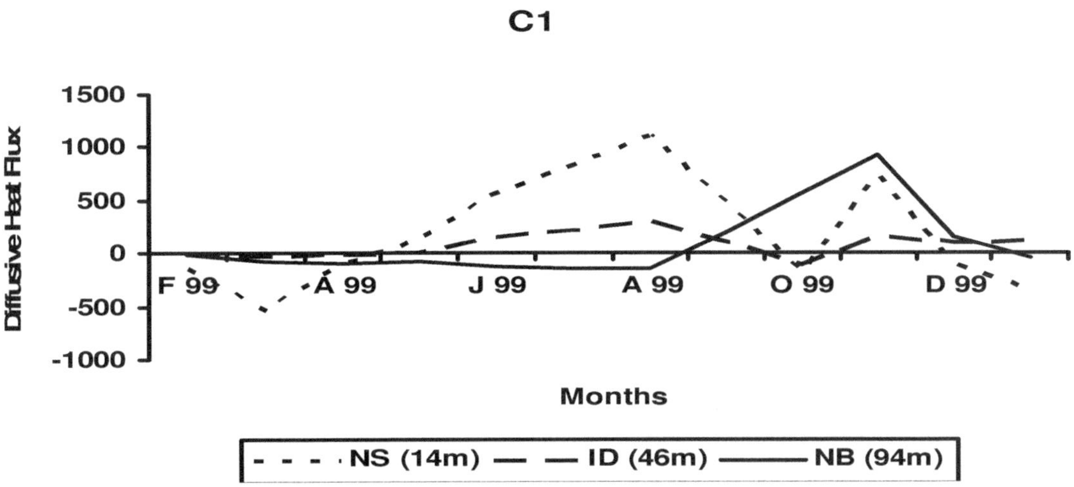

Figure 8d. Diffusive salt flux (10^{-12} kg m^2 s^{-1}) at station C1.

at 8 m to a maximum of 1.07 × 10^{-9} kg m^{-2}s^{-1} attained during August at 8 m also with an annual vertical average value 8.18 × 10^{-11} kg m^{-2}s^{-1}.

The diffusive heat flux

The temporal distribution of DHF at station C12, (Figure 9a)

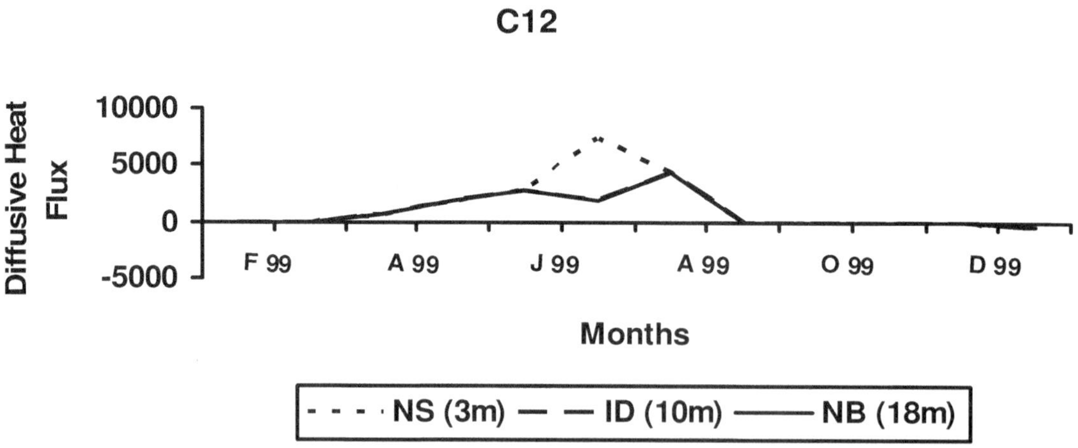

Figure 9a. Diffusive heat flux (10^{-6} j m^2 s^{-1}) at station C12.

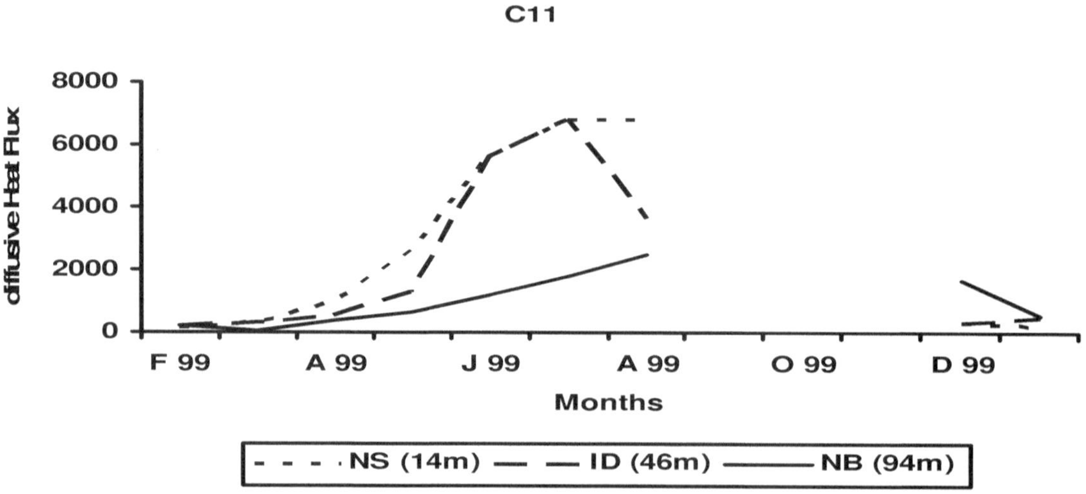

Figure 9b. Diffusive heat flux (10^{-6} j m^2 s^{-1}) at station C11.

had a minimum of -3.38×10^{-4} j $m^{-2}s^{-1}$, attained during January at 3 m which mean losses of temperature from the sea, and a maximum value of 7.27×10^{-3} j $m^{-2}s^{-1}$ attained during July at 3 m also, (that is, sea water gained temperature from surrounding environments) with an annual vertical average value of 9.32×10^{-4} j $m^{-2}s^{-1}$..

Only downward fluxes of heat occur at station C11, Figure (9b) as J_h varied from a minimum of 4.87×10^{-5} j $m^{-2}s^{-1}$ attained during March at 94 m to a maximum of 6.83×10^{-3} j $m^{-2}s^{-1}$ attained during August at 14 m, with an annual vertical average value of 2.05×10^{-3} j $m^{-2}s^{-1}$. The vertical turbulent transport of heat at station C10, Figure (9c) ranges from a minimum of 240×10^{-6} j $m^{-2}s^{-1}$ attained during April at 170 m (intermediate), to a maximum of 4.24×10^{-3} j $m^{-2}s^{-1}$ attained during August at 42 m, with an annual vertical average value of 8.31×10^{-4} j $m^{-2}s^{-1}$.

The temporal behavior of DHF at station C1, (Figure 9d) was analogous to other three stations as values of J_h ranged from a minimum of -7.1×10^{-3} j $m^{-2}s^{-1}$ attained during November at 34 m (intermediate depth) to maximum of 6.59×10^{-3} j $m^{-2}s^{-1}$ at 34 m attained during August, with an annual vertical average of 1.16×10^{-3} j $m^{-2}s^{-1}$. Thus, the temporal distribution of DHF, similarly to the buoyancy frequency, posse's maximum values during summer (July and August) which indicated that the two mixing parameters were related strongly to the heating by solar radiation. Also, the DHF generally decreased with depth, and was generally positive, that is, the heat flux was directed downwards.

Summary and Conclusion

The dynamical computations for turbulence mixing parameters are carried out for the four stations C12, C11, C10

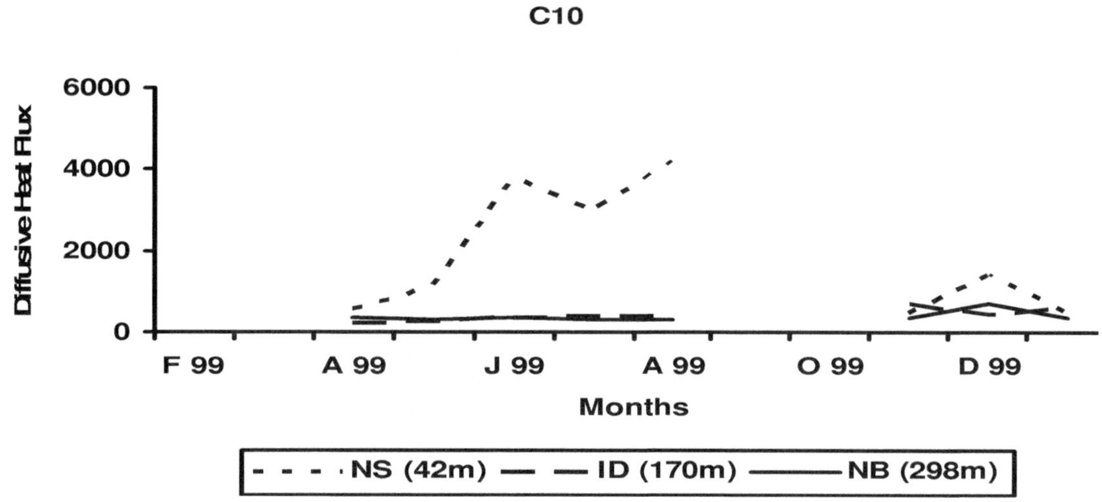

Figure 9c. Diffusive heat flux (10^{-6} j m^2 s^{-1}) at station C10.

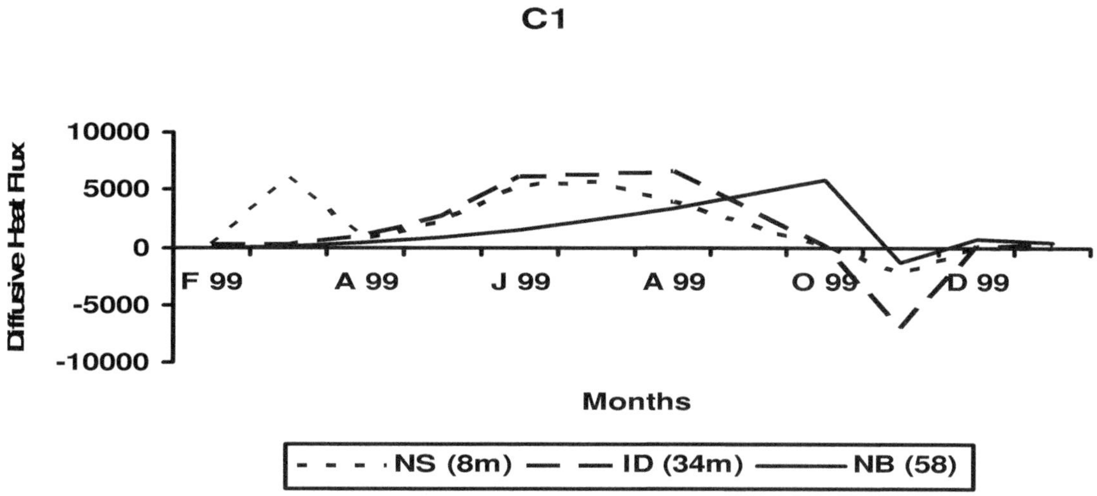

Figure 9d. Diffusive heat flux (10^{-6} j m^2 s^{-1}) at station C1.

and C1 at three levels: near surface, intermediate depth and near bottom In general, it is noticed that the greatest values of N^2 are attained during summer, in coincidence with the solar radiation, and the lowest values occur during autumn and winter. Also the absolute value of N^2 decreases with depth indicating that lower layers are less stratified and more homogenous. It can be noticed that the maximum viscous dissipation always occur near the surface, that is, ε decreases with depth. This shows that the processes of vertical turbulent mixing are stronger in the upper layers. This temporal distribution of *E* shows that turbulence is stronger near the surface and decrease with depth, that is, the motion of water is more laminar at deeper layers. The minimum values of *E* are attained during August while the maximum ones are attained during December and January, that is, turbulence is weaker in summer and stronger in winter. The behavior of *Ri*, on contrary to that of N^2, ε, *E*, shows no time or spatial dependence. The maximum upward Reynolds stress J_m is observed near surface during winter season while downward momentum flux occurs at near bottom during summer season. The dynamical computations of the magnitude K_s at the four stations decrease relatively with depth. The temporal distribution of DHF, similarly to the buoyancy frequency, posse's maximum values during summer (July and August) which indicate that the two mixing parameters are related strongly to the heating by solar radiation. Also, the DHF generally decreases with depth, and is generally positive, that is, the heat flux is directed downwards.

REFERENCES

El-Sharkawy MS (2007). Mixing processes and circulation pattern off

Egyptian Mediterranean Coast. Ph.D. Thesis, Oceanographic Department, Faculty of Science, Alexandria University.
Hornea EPW, Lodera JW, Naimeb CE, Oakeyb NS (1996). Turbulence dissipation rates and nitrate supply in the upper water column on Georges Bank. Deep Sea Research part II. Top. Stud. Oceanogr. 4(37-8): 1683-1712.
Johnson GC (1996). Stress on the Mediterranean outflow plume: Part II. Turbulent dissipation and shear measurements. J. Phys. Oceanogr. 24(10): 2084-2092.
Kelly AK (2005). Comparison of Modeled and observed sea surface temperature in the Eastern Mediterranean sea. Program of Analysis and Modeling in the Aegean and Eastern Mediterranean Sea. Program period May 2004 – April 2006.
Kobayashia S, Simpsona JH, Fujiwarab T, Porburghc KJ (2006). Tidal stirring and its impact on water column stability and property distribution in a semi-enclosed shelf sea (Seta Island, Japan). Continental Shelf Res., 26(11): 1295-1306.
MEDAR Group, MEDATLAS/2002 database (2002): Mediterranean and Black Sea database of temperature, salinity and biochemical parameters (4 CDs, IFREMER.
MacPhee-Shaw E (2006). Boundary-Interior Exchange: Reviewing the idea that internal wave mixing enhances lateral dispersion near continental margins. Deep Sea Res., Part II, Top. Stud. Oceanogr. 53(1-2): 42-59.
Mohamed EEE, El- Sharkawy MS, Saad NN , Anwar H (1999). A study of circulation, water masses, and Mixing processes in the southern Mediterranean off the Egyptian Coast during Autumn. J. KAU: Mar. Sci., 10: 3–15.
Molinari RL, Swallow J, Festa JF (1986). Evolution of the near surface thermal structure in the Western Indian Ocean during FGGE, 1979. J. Mar. Res., 44: 739-762.
Morcos SA, Hassan HM (1976). The water masses and circulation in the South-Eastern Mediterranean. Acta Adriatica, 18:12.
Munka W, Wunschb C (1998). Abyssal Recipes II: Energetics of tidal and wind mixing. Deep Sea Res. Part I, Oceanogr. Res. Papers 45(12): 1977-2010.
Peters H (1999). Spatial and temporal variability of mixing in anestuary. J. Mar. Res., 57: 805-845.
Proudman J (1953). Dynamical oceanography. John Wiley & Sons, inc.
Pond S, Pickard G (1983). Introductory oceanography. Pergamon Press
Robert HS (2005). Introduction of Physical Oceanography, Chapter 8, Copyright 1997 – 2005.
Rossa T, Lueckb R (2005). Estimating turbulent dissipation rates from acoustic backscatter. Deep Sea Res. Part I: Oceanogr. Res. Papers 52(12): 2353-2365.
Sharaf El-Din SH (1964). The circulation and mixing processes in the river Mersey and the Irish Sea. Ph. D. Thesis, Liverpool University.
Sparnocchia S, Gasparini GP, Astraldi M, Borghini M, Pistek P (1999). Dynamics and mixing of the Eastern Mediterranean outflow in the Tyrrhenian basin. J. Marine Syst., 20(1-4): 301-317.
Nabil NS, El-Sharkawy MS (2009). Evaluation of Some Dynamical Parameters at the Central Red Sea during Early Summer. J. KAU: Mar. Sci., Vol. 21 (under press).
Yoshida J, Neil SO (1996). Characterization of vertical mixing at a tidal-front on Georges Bank. Deep Sea Research Part II: Top. Stud. Oceanogr., 43(7-8): 1713-1744.

7

Patterns of diversity, abundance and habitat associations of butterfly communities in heterogeneous landscapes of the department of atomic energy (DAE) campus at Kalpakkam, South India

T. Ramesh[1], K. Jahir Hussain[2], M. Selvanayagam[1], K. K. Satpathy[2]* and M. V. R. Prasad[2]

[1]Loyola Institute of Frontier Energy (LIFE), Loyola College, Chennai- 600 034, India.
[2]Environmental and Industrial Safety Section, Safety Group, Indira Gandhi, Centre for Atomic Research, Kalpakkam-603102, India.

The diversity of butterflies inhabiting the department of atomic energy campus at Kalpakkam was recorded through a modified line transect methodology by setting a permanent line transect of 300m and recoding all species of butterflies observed within a five meter distance around the observer. Five habitats within the campus viz., Garden, Scrub jungle, Riparian woods, Sandy area and *Casuarina* plantation (Monoculture) were evaluated for analysis of the association of the butterfly species with the habitat. A total of 1908 individuals representing 55 species were observed across the five habitat types. Out of these, members belonging to the family Nymphalidae was the most common with 20 species being recorded accounting for 36.3% of total species and 53.6% of total number of individuals collected. The maximum diversity and abundance was observed in the scrub jungle and garden area; these two habitats sharing 29 species among themselves. The species accumulation curve and rarefaction curves computed indicated the likelihood of encountering more number of species in the campus had inventory been more rigorous and extended. The butterfly species viz., *Danaus chrysippus*, *Castalius rosimon*, *Tirumala septentrionis*, *Ariadne merione*, *Appears libythea* and *Cepora nerissa* preferred scrub jungle and garden habitats than the other habitats. The species profile of butterfly communities associated with different habitat and the importance of avian predation in the campus were also discussed in detail.

Key words: Butterfly, species composition, habitat association, DAE campus.

INTRODUCTION

Butterflies are taxonomically well studies group, which have received a reasonable amount of attention throughout the world (Ghazoul, 2002). Many of butterfly species are strictly seasonal and prefer only a particular set of habitats (Kunte, 1997) and they are good indicators in terms of anthropogenic disturbance and habitat quality (Kocher and Williams, 2000). Butterfly community assembly and the factors which influence it, have long been a topic of interest to ecologists and conservationists. Human dominated landscape form a substantial and ever-increasing amount of the earth's land surface. These modified habitats often negatively influence butterfly species and their dynamics (Gascon et al., 1999; Rickets et al., 2001).

Among insects, butterflies are the most studied group. In southern India, butterfly species have been documented since the turn of the 19th century (Bingham 1905, 1907, Williams, 1927). Later, Larsen (1987) made a detailed survey of butterflies of Nilgiri Mountains and recorded nearly 300 species including endemics. Many researchers have been significantly contributed to our understanding of butterfly diversity and abundance (Kunte et al., 1999; Arun and Azeez, 2003; Eswaran and Pramod, 2005; Xavier; 2006; Pramod kumar et al., 2007; Krishnakumar et al., 2008) on aspects such as habitat association, effect of disturbance and area clearance

*Corresponding author. E-mail: satpathy@igcar.gov.in.

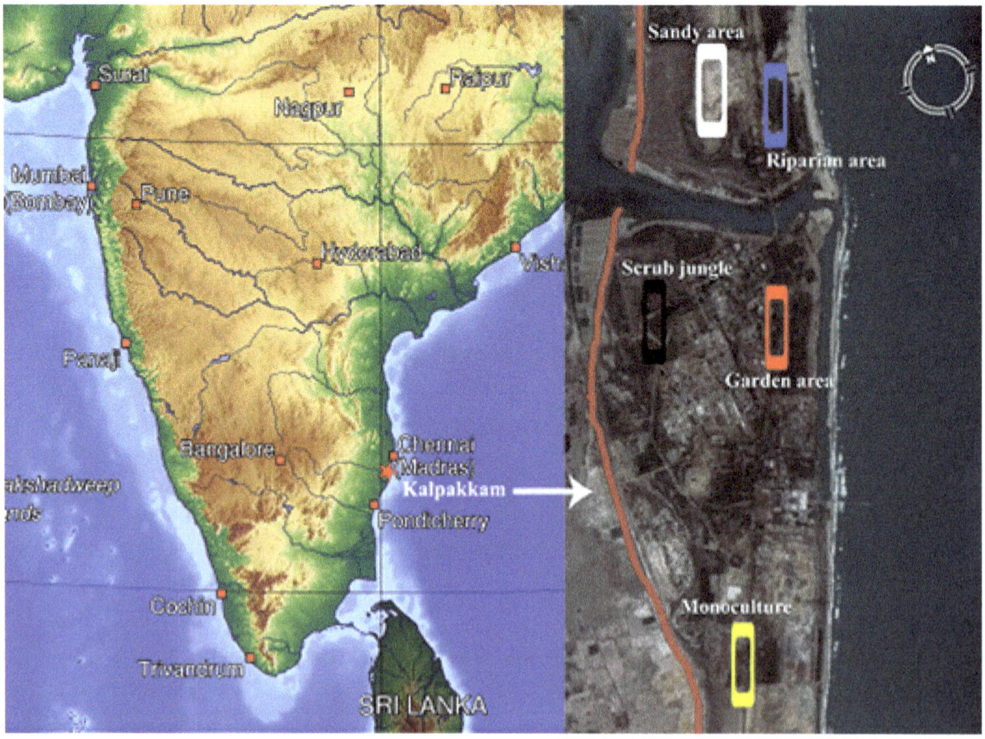

Figure 1. Study area.

(Soubadra Devy and Priya, 2001; Sreekumar and Balakrishnan, 2001; Dolia et al., 2008); seasonal abundance and migration patterns (Kunte, 1997; Arun, 2002; Kunte, 2005; Padhye et al., 2006) and on conservation (Mathew and Binoy, 2002; Mathew and Anto, 2007; Kunte, 2008) from Western Ghats. However, very little attention has been given to eastern plain areas of southern India. The purpose of present investigation is to understand the effects of habitat characteristic on butterfly diversity, community composition and density.

MATERIALS AND METHODS

Study area

The DAE campus at Kalpakkam (12° 33' N and 80° 11' E) encompasses seashore and a vast plain area (2500 acres) of the Bay of Bengal. The coastal system forms the complex natural site where intense interactions occur among land, sea and atmosphere. The unique interaction throws biological consortia peculiar to this ecosystem. It spreads through the biologically diverse and productive habitat of native flora and fauna and aesthetically blended with introduced vegetation. All the study areas are located in DAE campus, Kalpakkam, Tamil Nadu (Figure 1).

Inventory

Prior to butterfly census, two observers spent one year in the field (between April, 2007 to April, 2008) constructing a pictorial monograph (Jahir et al., 2008) for field identification and familiarizing themselves with the local butterfly fauna.

Monitoring (Line transect)

Butterfly species density was assessed quantitatively across different habitats. The entire campus was divided into five different habitats, which were divided on the basis of vegetation and soil type. Sampling was carried out at different habitats during (June, 2008 to February, 2009). Modifications of the line transect count as per. Kunte; (1997) was used to determine butterfly richness and abundance. In this method permanent 300m line transect was setup in each habitat. The transect in each habitat was slowly traversed at a uniform pace for 30min at each habitats from 8.30 to 11.30 h during good weather period (no heavy rain or strong winds). Butterfly species were recorded around a radius of five meter from the observer covering his either sides, above and front. This is a suitable method for surveying butterflies in a wide range of habitats including tropical forest (Walpole and Sheldon, 1999; Caldas and Robbins, 2003; Koh and Sodhi, 2004). All individuals were identified in the field using standard guides (Gunathilagaraj et al., 1998; Kunte, 2000; Jahir et al., 2008).

Habitat characterization

The main type of vegetation in DAE campus is dry evergreen and scrub comprising of members predominantly belonging to the families: *Poaceae, Fabaceae, Cyperaceae, Asteraceae, Euphorbiaceae, Verbenaceae, Solanaceae, Rubiaceae, Convolvulaceae and Amaranthaceae* (Gajendiran and Ragupathy, 2002).

Scrub jungle

The dry areas of the DAE premises shelter scattered patches of *Prosophis juliflora* (Sw.) DC. plants. Some of the dominant tree species are: *Acacia polyacantha* ((Willd.), Samanea saman (Jake.) Merr. and *Albanian lebbeck* (L.) Benth. Predominant herbs and shrubs are: *Lantana camera* (L.), *Canthium dicoccum* (Gaertn.), *Euphorbia antiquorum* (L.), *Martynia annua* (L.), *Calotropis gigantean* (L.), *Capparis Sp., Toddalia asiatica* (L.), *Tragia involucrate* (L.), *Crotalaria verucosa* (L.), *Heliotropium indicum* (L.), *Urena lobata* (L.), *Solanum Sp., Tephrosia purpurea* (L.), *Anisomelous Sp., Zizyphus oenoplia* (L.), *Ocimum basilicum* (L.), *Achyranthes aspera* (L.), *Croton bonplandianum* (Baill), *Solanum trilobatum* (L.), *Leucas aspera* (Willd.) Link., *Mimosa pudica* (L.), *Gomphrena serrata* (L.), *Vernonia cinerea* (L.), *Clitoria ternatea* (L.), *Tinospora cordifolia* (L.) and Common twiners.

Riparian area

The riparian area is relatively dense with *Pandanus fascicularis* (Lam.), Typha *angustata (Bory & Chaubard), Acacia auriculiformis Cunn. ex Benth., Cassia siamea Lamk., Ficus benghalensis* angustata (Bogy & Chaubard), *Prosophis juliflora* (Sw.) DC., *Casuarina equisetifolia (L.), Terminalia arjuna* (L.) along with some nectar fetching plants like *Catharanthus roseus* (L.) G.Don, *Anacardium occidentale* (L.) and *Tephrosia purpurea* (L.). Grasses are also abundant in this location

Garden area

Even though this area witness comparatively high anthropogenic pressure, it is a unique habitat with natural vegetation blended with introduced ornamental flora. Major part of this area comprises *Tridax procumbens* (L.), *Gomphrena serrata* (L.), *Cosmos sulphureus* Cav., *Ricinus communis* (L.), along with common garden species like *Ixora coccinea* (L.), *Ixora pavetta* Andr. and some moist deciduous tree species.

Sandy area

This habitat is a typical costal sand stretch with meager vegetation which consist of *Calotropis gigantean* (L.), *Catharanthus roseus* (L.) G. Don, *Tribulus terrestris* (L.) along with *Acacia auriculiformis* Cunn. ex Benth., and *Casuarina equisetifolia* (L.).

Monoculture

The southern area of campus is dense with *Casuarina equisetifolia* (L.) monoculture plantation. Canopy cover in plantation surveyed was high (60-80%), even though some weedy plants *Catharanthus roseus* (L.) was observed. This area had high anthropogenic disturbance.

Data analysis

Two components, namely α-diversity and β-diversity were calculated. The alpha statistic of Fisher et al; (1943) is often considered the best diversity index for many communities of species, including *Lepidoptera* (Robinson and Tuck, 1993; Wolda et al., 1994, Chey et al., 1997). The Shannon diversity index is also widely used for comparing diversity between various habitats (Clarke and Warwick, 2001). Apart from above indices Simpson indices was also calculated as measure of α-diversity. Beta diversity (between the habitats) captures a fundamental aspect of species diversity and spatial replacement in species identity between any two or more areas. We calculated some classical indices viz., Morista-Hor n similarity index, Sorenson classic and Jaccard classic indices to measure beta diversity based on habitat raw data. The Michaelis-Menten estimate was based on the mean species accumulation curve after 50 randomization of samples order and several diversity estimators were also calculated using the software Estimates (Colwell; 1997). Sample based on rarefaction was calculated using Biodiversity Pro software version 2 (Neil Mcaleece et al., 1997).

RESULTS AND DISCUSSION

A total of 55 species belonging to 5 families of order *Lepidoptera* were recorded (Appendix 1). Out of these, *Nymphalidae* were the most common with 20 species, followed by the *Pieridae* (15 species*), Lycaenidae* (11 species), and the least number of species was observed in *Hesperiidae* by only 4 species and *Papilionidae* by only 5 species. *Nymphalidae* was the dominant family accounting for 36.3% of species and 53.6% of individuals recorded (Table 1). A similar pattern of predominance of *Nymphalidae* was also reported by different researchers (Kunte; 1997; Kunte, et al., 1999; Eswaran and Pramod, 2005; Dolia, et al., 2008; Krishna kumar, et al., 2008; Soubadra devy and Priya, 2001; Pramod kumar et al., 2007; Padhye, et al., 2008) from Western Ghats. However, with respect to Eastern plains of Southern India the information on butterfly diversity and bioecology is sketchy and sporadic. India has more than 1,400 species of butterflies, 330 of them in the Western Ghats alone, and of which 37 are endemic (Kunte, 2008). In the present investigation *Pachliopta hector,* a scheduled species which is protected by Wildlife protection act was also documented.

In total, 1908 individuals from 45 transects were observed and identified. A large number was observed at the scrub jungle habitat (649 individuals, 34% of total) while equal numbers were recorded at garden area (632, 33%) and very less numbers were recorded in the sandy area (173, 9%) and monoculture habitat (157, 8.2%). Generally, simple comparison of absolute species number between samples is used most of the time as diversity measure. We also calculated Fisher's alpha diversity and Shannon diversity indices as a measure of diversity within a habitat since these indices incorporate both species richness and abundance into a single value. The Fisher alpha diversity indicated the following habitats in a decreasing order of diversity; scrub jungle (9.12) riparian woods (8.72), garden area (8.29), sandy area (6.26) and monoculture (4.46). The Shannon's diversity index showed the same pattern with minor variations. The Simpson and Shannon J (evenness) indices revealed that in scrub jungle the individuals among species were not evenly distributed during the survey period indicating that some species were more abundant than the others. This reflects on the difference in the efficiency of different butterfly species to efficiently use the habitat. The abundance of individuals of a species at any given point on a temporal scale is again dependent on various biotic and abiotic environmental factors. Hills richness

Table 1. Family wise composition of number, percentage of species and individuals observed.

Family	Number of species (%)	Number of individual (%)
Hesperiidae	4 (7.2)	12 (0.6)
Papilionidae	5 (9)	143 (7.4)
Lycaeniidae	11 (20)	199 (10.4)
Pieridae	15 (27.2)	530 (27.7)
Nymphalidae	20 (36.3)	1024 (53.6)
Total	55	1908

Table 2. Diversity index and abundance scores for butterfly communities along habitats.

Habitat	Fisher alpha	Shannon H'	Shannon J'	Simpson's (D)	Hill's H0	Individuals
Garden area	8.29	1.30	0.84	0.06	36	632
Scrub jungle	9.12	1.20	0.75	0.12	39	649
Riparian woods	8.72	1.23	0.82	0.08	31	297
Sandy area	6.26	1.05	0.79	0.13	21	173
Monoculture	4.46	0.90	0.74	0.19	16	157

indices indicated that scrub jungle was the richest (39 species) followed by garden area (36 species), riparian woods (31 species), sandy area (21 species), and monoculture (16 species) as shown in Table 2.

The structural complexity of habitat and diversity of vegetation forms have been shown to be correlated with animal and insect species diversity (Gardner et al., 1995). Southwood (1975) suggest that the herbivores are more influenced by the food quality. Host plants are utilized only when sufficient adult resources (nectar) are also available (Grossmueller and Lederhouse, 1987). Successful butterfly habitat must therefore include sufficient larval and adult food resources. In the present study, the maximum number of species and individuals were observed in scrub jungle and garden area, where availability of diverse plants and access to host plants viz., *Tragia involucrate, Crotalaria verucosa, Heliotropium indicum, Tridax procumbens, Ricinus communis, Leucas aspera, Mimosa pudica, Gomphrena serrata, Vernonia cinerea, Chromolaena odorata, Anisomelous sp., Lantana camara* and ornamental flowering plants promoted the butterfly richness and density. Most of these plants provide rich nectar sources to adult butterflies. Comparatively the other habitats especially, monoculture and sandy area have lesser density of vegetation. These habitats being highly disturbed due to anthropogenic activities could also account for lower butterfly colonization.

The butterfly distribution are expected to cover with the distribution of their host plants even at small scales and type of vegetation may reflect difference in the composition of butterfly communities among habitats at the generic and family level (Beccaloni; 1997). Our results also conformed the above finding; at the scrub jungle *Danaus chrysippus*, *Danaus genutia*, *Tirumala septentrionis* and *Euploea core* of subfamily Danainae contributed 38% of total density of that particular habitat. This is due to the fact that the two abundant host plants of that habitat namely *Heliotropium indicum* and *Crotalaria verucosa* are major source of Pyrollizine alkaloids, which is a precursor of Danaids pheromone cum defense chemical called Danaidone (Boppre et al.,1978). Interestingly, Danaidone is not available from the plants that the larvae feed on and must be obtained by the adults after emergence. And this task is accomplished by feeding on plants that do contain these substances. Studies elsewhere have shown that males that do not accumulate Danaidone are consistently refused by the females. So it seems that the females not only seek chemicals for their own survival and their progeny, but also indirectly select genes that are correlated well with good foraging ability is an excellent strategy for survival of the species. It is principally from these plant species that the male Danaids obtain most of the chemicals required to synthesize their pheromones. All *Danaids* are irresistibly attracted to these plants. Hence, the adult Danaid density was more at scrub jungle where these plants are abundant. This dominance of particular taxa was clearly reflected in Shannon's J evenness index.

Beta diversity is essentially a measure of how different (or similar) a range of habitats are in terms of the variety and abundance of species found in them. An approach to the measurement of beta diversity is to investigate the degree of association or similarity of habitats or samples using standard ecological technique of ordination and classification (Southwood, 1978). A vast range of similarity indices are available, however, some of the oldest similarity coefficients are also the most useful (Magurran, 1988). Particularly widely used are the Jaccard and Sorenson index. The shared species statistics between pairs of the five habitats are provide in the Table 3. The

Table 3. Shared species statistics and similarity coefficients between pairs of the five habitats.

First sample	Second sample	Shared species	Jaccard classic	Sorensen classic	Morisita-Horn
Garden area	Scrub jungle	29	0.63	0.773	0.607
Garden area	Riparian woods	21	0.457	0.627	0.698
Garden area	Sandy area	16	0.39	0.561	0.433
Garden area	Monoculture	15	0.405	0.577	0.554
Scrub jungle	Riparian woods	24	0.522	0.686	0.707
Scrub jungle	Sandy area	18	0.429	0.6	0.444
Scrub jungle	Monoculture	15	0.375	0.545	0.749
Riparian woods	Sandy area	17	0.486	0.654	0.812
Riparian woods	Monoculture	13	0.382	0.553	0.778
Sandy area	Monoculture	9	0.321	0.486	0.571

number of species observed in each habitat and the number of species encountered in both the habitats were compared. The garden area and scrub jungle showed highest number of shared species (29 species), because these areas had comparatively similar plant composition and provide perennial nectars sources for adult butterflies. Even though the sandy area and monoculture habitats shared nine general species, they comprise of a relatively contrast set of species (multifamily) due to their difference in vegetation type. Seventeen species of butterflies were seen in both the riparian habitat and sandy area and this is reflected in the Jaccard value. From the results of the Sorenson index, it recognized as one of the best indices of shared species measure, it is clear that garden/scrub jungle and scrub jungle/ riparian habitat were represented by relatively same species and abundance profile. Similarly the high value for riparian habitat and sandy area was due to their similarity in species assemblage. The Morisita Horn index is highly sensitive to the species assemblage and abundance of the most abundant species between each pair of habitats. This index reveled that riparian woods and sandy area were similar in their abundant species profile. In these habitats, the most abundant species was mottled emigrant which was not common to other habitats.

Species accumulation curve is a approach by plotting the cumulative number of species collected against the sampling effort (months/sample unit), which describes species richness in a local homogenous assemblages as a function of sampling effort. This technique is very much useful for standardizing the sample size. In the present study the species accumulation curve could not attain asymptote even after nine sampling efforts (especially scrub jungle habitat), which indicated that, with increase in sampling efforts there is more likelihood of adding new records of species (Figure 2), from the study area.

As discussed by Colwell and Coddington (1994), the problem of estimating the true number of species shared by two (or more) habitats or biotops based on sample data presents a difficult but important challenge. Most widely used approach for estimating species richness is based on extrapolation from known species accumulation curve. Michaelis-Menten type model describes well about the accumulation of species records as the number of sampling attempt increases. In rarefaction curve Michaelis-Menten Mean (MMMean) and Coleman curve were used as estimators of species richness. MMMean and Coleman curve were fitted to sampling data after randomizing them 50 times using the procedure of Colwell (1997). In the present study, the curve indicated that more number of species observed in scrub jungle from the early effort. However, after the second effort very less numbers were observed (Figure 3).

To compare sample-based rarefaction curves, expected expected species of the sample (habitat) are plotted against number of individuals. This plot provides a measure of species diversity, which is robust to sample size effects, permitting comparison between communities. Steeper curves indicate more diverse communities. The steeper curve was observed for scrub jungle and garden area which were equally richer and more abundant when compared to all other habitats. At community level the sandy area and monoculture were less diverse (Figure 4). Richness estimators were highly influenced by rare species. In our study, out of 55 species, eight species were singleton species and three species were represented by two individuals (doubletons). The larger the number of singletons within a sample, for a given number of doubles, the greater would be the difference between observed and the true species richness for the assemblages sampled (Senthilkumar, 2003).

To find out the abundance pattern, the rank abundance curve was plotted using grand mean of the all species. The common species are displayed on the left, and the rare species are on the right. While ranking overall abundance (all habitats pooled together) and ten species were more common followed by few abundant species and large proportion of rare species was observed (Figure 5). Species abundance of DAE butterflies has showed that, only a small proportion (18%) of the assemblage occurred in relatively high number. This is a reflection of a situation where one or a few factors dominate the ecology of a community (Magurran, 1988). Dominance of a species in an ecosystem reveals its

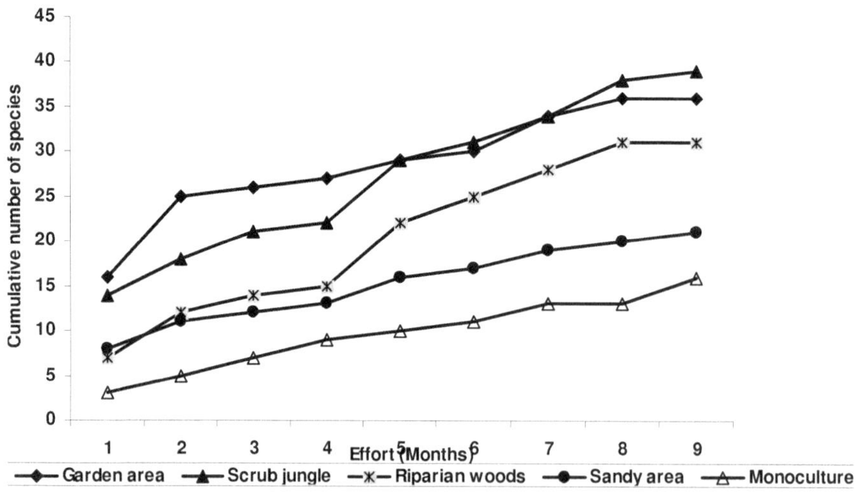

Figure 2. Species accumulation curves for five habitats.

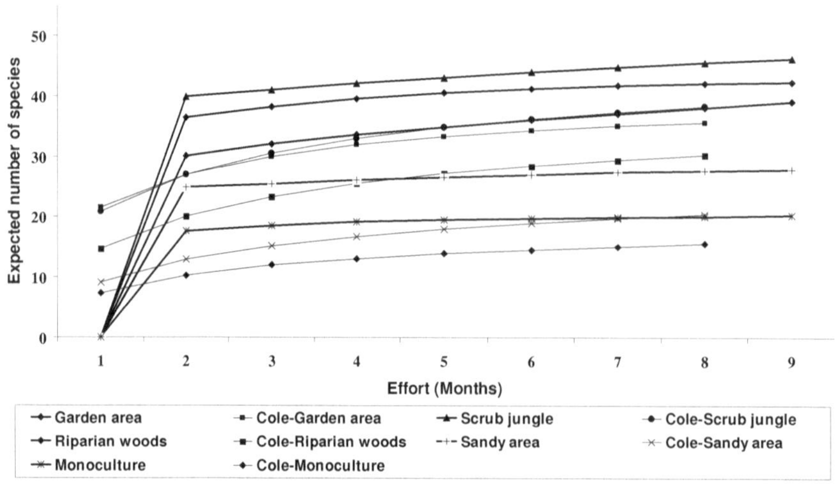

Figure 3. Rarefaction curve for five different habitats.

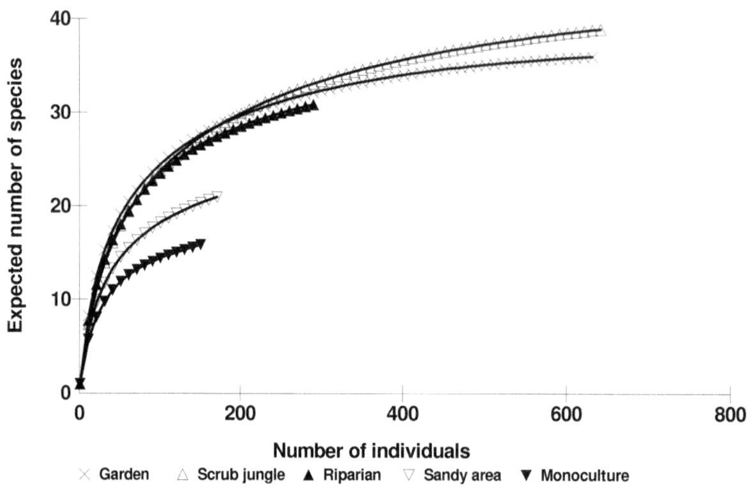

Figure 4. Sample based rarefaction curve for the five different habitats.

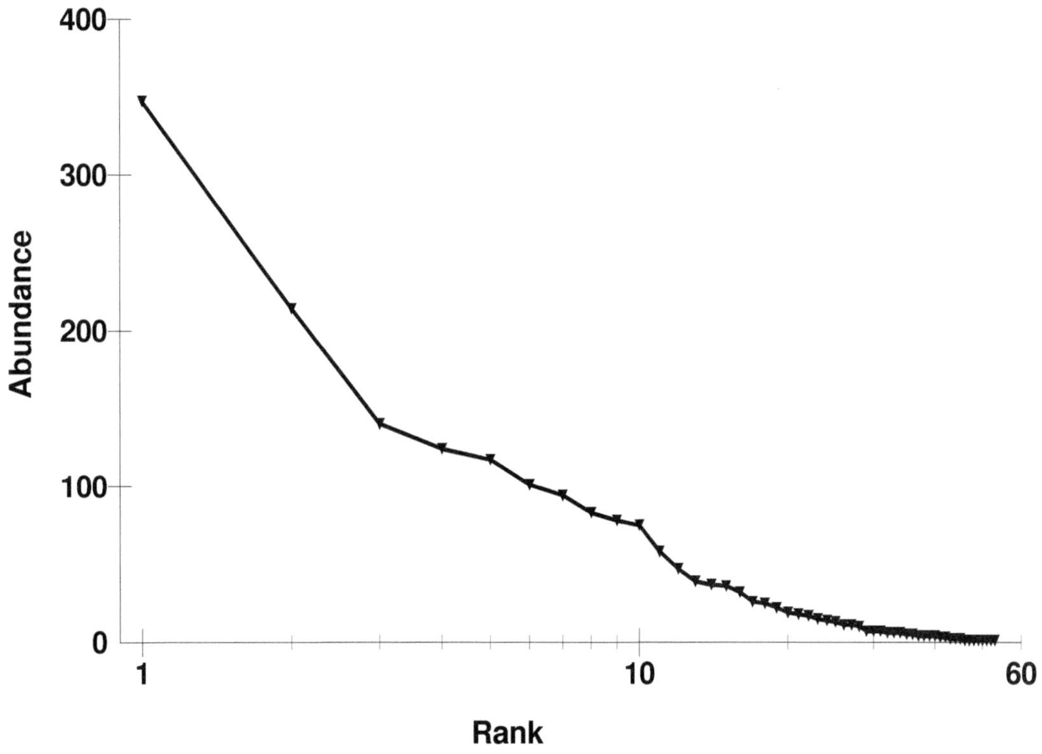

Figure 5. Over all species rank abundance.

survival superiority over other species. In the present study the overall dominance by *D. chrysippus* was due to its chemical defence against predations. The larvae of this species often feed on plants containing poisons and unpalatable substances. Instead of eliminating or detoxifying chemicals, they assimilate them in their body fluids and retain them till their adulthoods. The male adult also assimilates a toxin called danaidone from plants. Females prefer to mate with males who exude significant amounts of danaidone from their hair pencils. In addition, the spermatophore (sperm package) that the male passes over to the female during mating also contains danaidone as well as sperm and other nutrients. The danaidone is incorporated into the eggs as well as into the female's tissues. This is not only to increase the chances of survival of the eggs that are laid, but alsomakes the female less palatable to predators. In the study area the most important predators are birds, small mammals, reptiles, predating insects and spiders (Jahir et al., 2008). Birds can be considered the most important predators on adult, mobile butterflies (Bowers et al., 1985). During our intense field survey we observed birds species viz., *Merops orientalis, Merops philippinus, Dicrurus macrocercus, Dicrurus leucophaeus, Dicrurus aeneus, Artamus fuscus, Lanius cristatus, Terpsiphone paradise, Coracias benghalensis* and *Hirundo daurica* are actively preying on palatable butterfly species. Hence the predatory pressure imposed by avian fauna on butterfly community is one of the important governing factors of butterfly population. Being an unpalatable prey species to birds, *D. chrysippus* has greater survival superiority. Similarly *Acraea violae* also possesses plant acquired chemical defense which provided the survival ability than other species in the heterogeneous ecosystem. In the present study, it is evident that apart from vegetation type, predation by birds is also a major factor influencing butterfly community.

Habitat association of butterflies can be directly related to the availability of food plants (Thomas; 1995).The criterion of representative diversity which is based on the recognition of assemblages of species that are typical for specific habitats seems more interesting (Dufrene and Legendre, 1997). A potentially useful tool is the single species approach based on focal or surrogate species, which can indicate ecological change, patterns of richness or habitat type (Caro and Doherty, 1999; Fleishman, et al., 2000). Each habitat has a specific set of micro environment suitable for a species. In the present investigation species such as *Cepora nerissa* were observed only in the garden area and *Appias libythea* were observed only in the scrub jungle habitat, indicating its preference towards particular habitat. Percentage frequencies of selected species are given in (Table 4). Out of 13 species, seven species were predominant at the garden area and four species were very common at the undisturbed scrub jungle. These two habitats were strongly

Table 4. Percentage composition of selected butterfly species associated with five habitats.

Species name	Garden area	Scrub jungle	Riparian woods	Sandy area	Monoculture
Appias libythea	0	100	0	0	0
Cepora nerissa	100	0	0	0	0
Castalius rosimon	49	30	6	10	5
Pachliopta hector	60	16	9	0	15
Euploea core	62	19	17	2	0
Junonia atlites	78	11	0	11	0
Tirumala septentrionis	62	15	14	3	8
Leptosia nina	83	10	7	0	0
Colotis amata	91	0	9	0	0
Ariadne merione	26	72	2	0	0
Phalanta phalantha	11	81	0	0	8
Danaus chrysippus	15	53	13	5	14
Danaus genutia	25	63	6	3	3

strongly supporting the butterfly community. From our observation we conclude that, even in the small study area butterfly communities varied significantly among different habitats. In a heterogeneous ecosystem like DAE campus, vegetation type played a major role in diversity and density patterns of butterfly communities. Moreover the avian predation also has a profound effect on abundance pattern of butterfly communities.

ACKNOWLEDGMENTS

Authors are thankful to Dr. Krushnamegh Kunte, Post-doctoral Research Fellow, FAS Center for Systems Biology, Harvard University, Cambridge, USA, for help in identification and reviewing manuscript. Authors are grateful to Dr. Baldev Raj, Director, IGCAR and Dr. P. Chellapandi, Director, Safety Group, IGCAR for their continuous encouragement and support.

REFERENCES

Arun PR (2002). Butterflies of Siruvani forest of Western Ghats, with notes on their seasonality. Zoo's print. J. 18(2): 1003-1006.

Arun PB, Azeez PA (2003). On the butterflies of Puyankutty forest, Kerala, India. Zoo's print. J. 18(12): 1276-1279.

Beccaloni GW (1997). Vertical stratification of the ithomiine butterfly (Nymphalidae: Ithomiinae) mimicry complexes: the relationship between adult flight height and larval host-plant height. Biol. J. Lin. Soc. 62: 313–341.

Bingham CT (1905). Fauna of British India, Butterflies Vol. 1. Taylor and Francis, London.

Bingham CT (1907). Fauna of British India, Butterflies Vol. 2. Taylor and Francis, London.

Boppre M, Robert L, Petty, Schneider D, Meinwald J (1978). Behaviorally Mediated Contacts Between Scent Organs: Another Prerequisite for Pheromone Production in Danaus chrysippus Males (Lepidoptera). J. Comp. Physiol (A)., 126: 97-103.

Bowers MD, Brown IL and Wheye D (1985). Bird predation as a selective agent in a butterfly population. Evolution 39: 93–103.

Caldas A , Robbins RK (2003). Modified Pollard transects for assessing tropical butterfly abundance and diversity. Biol. Conserv. 110: 211-219.

Caro TM, Doherty GO (1999). On the use of surrogate species in conservation biology. Conserv. Biol. 13: 805- 814.

Chey VK, Holloway JD , Speight MR (1997). Diversity of moths in forest plantations and natural forests in Sabah. Bull. Entomol. Res. 87: 371-385.

Clarke KR, Warwick RM (2001). Changes in marine communities: an approach to statistical analysis and interpretation, 2nd edition, PRIMER-E: Plymouth.

Colwell RK (1997). EstimateS version 7.5. Statistical estimation of species richness and shared species from samples. Available from http://viceroy.eeb.uconn.edu/estimates.

Colwell RK, Coddington JA (1994). Estimating terrestrial biodiversity through extrapolation. Philos. Trans. R. Soc. Lond. B. 345: 101–118.

Dolia J, Devy MS, Aravind NA, Kumar A (2008). Adult butterfly communities in coffee plantations around a protected area in the Western Ghats, India. Anim. Conserv. 11: 26–34.

Dufrene M, Legendre P (1997). Species assemblages and indicator species: the need for a flexible asymmetrical approach. Ecol. Monogr. 67: 345-366.

Eswaran R, Pramod P (2005). Structure of butterfly community of Anaikatty hills, Western Ghats, Zoo's print. J. 20(8): 1939-1942.

Fisher RA, Corbet AS, Williams CB (1943). The relation between the number of species and the number of individuals in a random sample of an animal population. J. Anim. Eco., 12: 42-58.

Fleishman E, Murphy DD, Brussard PF (2000). A new method for selection of umbrella species for conservation planning. Ecol. Appl., 10: 569-579.

Gajendiran N , Ragupathy S (2002). The macroflora of east coast at Kalpakkam. Report submitted to Director IGCAR.

Gardner SM, Cabido MR, Valladares GR and Diaz S (1995). The influence of habitat structure on arthropod diversity in Argentine semi-arid Chaco forest. J. Veg. Sci. 6: 349–356.

Gascon C, Lovejoy TE, Bierregaard RO, Malcolm JR, Stouffer PC, Vasconcelos HL, Laurance WF, Zimmerman B, Tocher M , Borges S (1999). Matrix habitat and species richness in tropical forest remnants. Biol. Conserv. 91: 223–229.

Ghazoul J (2002). Impact of logging on the richness and diversity of forest butterflies in a tropical dry forest in Thailand. Biodivers. Conserv. 11: 521–541.

Grossmueller DW, Lederhouse RC (1987). The role of nectar source distribution in habitat use and oviposition by the tiger swallowtail butterfly. J. Lepid.' Soc. 41(3):159–165.

Gunathilagaraj K, Perumal TNA, Jayaram K, Ganesh KM (1998). Some South Indian butterflies: field guide. Published under project lifecape, Indian Acedamy of Science, Bangalore. pp: 274.

Jahir HK, Satpathy KK, Prasad MVR, Sridharan VT, Ramesh T, Selvanayagam M (2008). Faunal Diversity Assessment at Department of Atomic Energy (DAE) Campus, Kalpakkam. p 268.

Kocher SD, Williams EH (2000). The diversity and abundance of North American butterflies, vary with habitat disturbance and geography. J. Biogeogr. 27: 785-794.

Koh LP, Sodhi NS (2004). Importance of reserves, fragments and parks for butterfly conservation in a tropical urban landscape. Ecol. Appl. 14: 1695–1708.

Krishnakumar N, Kumaraguru A, Thiyagesan K, Asokan S (2008). Diversity of papilonid butterflies in the Indira Gandhi wildlife sanctuary, Western Ghats, southern India. Tiger Paper 35: 1-8.

Kunte K (1997). Seasonal patterns in butterfly abundance and species diversity in four tropical habitats in the northern Western Ghats. J. Biosci. 22: 593-603.

Kunte K, Joglekar A, Utkarsh G, Pramod P (1999). Patterns of butterfly, bird and tree diversity in the Western Ghats. Curr. Sci. India, 29: 1-14.

Kunte K (2000). Butterflies of Peninsular India, Universities Press Limited. Hyderabad. India. 254p.

Kunte K (2005). Species composition, sex-ratios and movement patterns in Danaine butterfly migrations in southern India. J. Bombay Nat. Hist. Soc., 102(3): 280-286.

Kunte K (2008). The Wildlife (Protection) Act and conservation prioritization of butterflies of the Western Ghats, southwestern India. Curr. Sci. India 94: 729-735.

Larsen TB (1987). The butterflies of the Nilgiri Mountains of Southern India (Lepidoptera: Rhopalocera). BNHS, 84: 291-316.

Magurran AE (1988). Ecological Diversity and its Measurement. Chapman and Hall, London.

Mathew G, Anto M (2007). In situ conservation of butterflies through establishment of butterfly gardens: A case study at Peechi, Kerala, India. Curr. Sci. India, 93(3): 337-347

Mathew G, Binoy CF (2002). Migration of butterflies (Lepidoptera: Rhopalocera) in the new Amarambalam reserve forest of the Nilgiri biosphere reserve. Zoo's Print J. 17(8): 844-847.

Neil McAleece PJD, Lambshead, Paterson PLJ (1997). Biodiversity Pro (version 2). The Natural History Museum, London.

Padhye AD, Dahanukar N, Paingankar M, Deshpande M , Deshpande D (2006). Season and Landscape wise distribution of butterflies in Tamhini, Northern, Western Ghats, India. Zoos Print. J. 21(3): 2175-2181.

Pramod KMPM, Hosetti BB, Poomesha HC, Raghavendra GHT (2007). Butterflies of the tiger lion safari, Thyavarekoppa, Shioga, Karnataka. Zoo's print. J. 22(8): 2805.

Rickets TH, Daily GC, Ehrlich PR, Fay JP (2001). Countryside biogeography of moths in a fragmented landscape: biodiversity in native and agricultural habitats. Conserv. Biol., 15: 378–388.

Robinson GS, Tuck KR (1993). Diversity and faunistics of small moths (microlepidoptera) in Bornean rainforest. Ecol. Entomol. 18: 385-393.

Senthilkumar N (2003). Biodiversity studies of Tettigoniids of Tamil Nadu and bio ecological characteristics of cynocephalus maculates, PhD, thesis submitted to university of madras, Chennai, Tamil Nadu, India.

Soubadra DM, Priya D (2001). Response of wet forest butterflies to selective logging in Kalakad–Mundanthurai Tiger Reserve: Implications for conservation. Curr. Sci. India 80(3): 400-405.

Southwood TRE (1975). The dynamics of insect populations: In Insects Science and Society. Academic Press, New York. pp: 151-199

Southwood TRE (1978). Ecological Methods with Particular Reference to the Study of Insect Populations. Chapman and Hall, London. pp: 524 .

Sreekumar PG, Balakrishnan M (2001). Habitat and altitude preferences of butterflies in Aralam Wildlife Sanctuary, Kerala, Trop. Ecol. 42(2): 277-281.

Thomas JA (1995). The ecology and conservation of Maculinea arion and other European species of large blue butterfly. In: A.S. Pullin (ed.) Ecology and Conservation of Butterflies. Chapman and Hall, London. pp: 180-210

Walpole MJ, Sheldon IR (1999). Sampling butterflies in tropical rain-forest: an evaluation of a transect walk method. Biol. Conserv. 87: 85–91.

Williams CB (1927). A study of butterfly migration in south India and Ceylon, Green, J.C.F. Fryer and W. Ormiston. Trans. Ent. Soc. Lond. 75: 1-33.

Wolda H, Marek J, Spitzer K, Novak I (1994). Diversity and variability of Lepidoptera populations in urban Brno, Czech Republic. Eur. J. Entomol. 91: 213-226.

Xavier A (2006). Butterfly fauna of government arts and science college campus, Kozhikode, Kerala. Zoo's Print. J. 21(5): 2263-2264.

Appendix 1. Butterflies sighted from the area during the study period.

S. no	Family/Subfamily	Scientific name	Common name
	Hesperiidae		
1	Hesperiinae	Parnara bada (Moore, 1878)	Common straight swift
2	Hesperiinae	*Suastus gremius* (Fabricius,1798)	Indian palm pop
3	Pyrginae	*Gomalia elma* (Trimen, 1862)	African mallow skipper
4	Pyrginae	*Spialia galba* (Fabricius, 1793)	Indian grizzled skipper
	Lycaeniidae		
5	Curetinae	*Curetis thetis* (Drury, 1773)	Indian sunbeam
6	Polyommatinae	*Azanus ubaldus* (Stoll, 1782)	Bright babul blue
7	Polyommatinae	*Castalius rosimon* (Fabricius, 1775)	Common pierrot
8	Polyommatinae	*Catochrysops strabo* (Fabricius,1793)	Forget me not
9	Polyommatinae	*Chilades lajus* (Stoll, 1780)	Lime blue
10	Polyommatinae	*Everes lacturnus* (Godart, 1824)	Indian cupid
11	Polyommatinae	*Jamides celeno* (Cramer, 1775)	Common cerulean
12	Polyommatinae	*Leptotes plinius* (Fabricius, 1793)	Zebra blue
13	Polyommatinae	*Megisba malaya* (Horsfield,1828)	Malayan
14	Theclinae	*Spindasis vulcanus* (Fabricius,1775)	Common silverline
15	Unassigned	*Pseudozizeeria maha* (Kollar,1844)	Pale grass blue
	Nymphalidae		
16	Biblidinae	*Ariadne merione* (Cramer, 1777)	Common castor
17	Danainae	*Danaus chrysippus* (Linnaeus, 1758)	Plain tiger
18	Danainae	*Danaus genutia* (Cramer, 1779)	Striped tiger
19	Danainae	*Euploea core* (Stoll, 1780)	Common crow
20	Danainae	*Tirumala limniace* (Cramer, 1775)	Blue tiger
21	Danainae	*Tirumala septentrionis* (Butler, 1874)	Dark blue tiger
22	Heliconiinae	*Acraea violae* (Fabricius, 1793)	Tawny coster
23	Heliconiinae	*Phalanta phalantha* (Drury,1773)	Common leopard
24	Limenitidinae	*Neptis hylas* (Linnaeus,1758)	Common sailer
25	Nymphalinae	*Hypolimnas bolina* (Linnaeus, 1758)	Great eggfly
26	Nymphalinae	*Hypolimnas misippus* (Linnaeus, 1764)	Danaid eggfly
27	Nymphalinae	*Junonia orithya* (Linnaeus, 1758)	Blue pansy
28	Nymphalinae	*Junonia iphita* (Cramer,1779)	Chocolate pansy
29	Nymphalinae	*Junonia atlites* (Linnaeus, 1763)	Grey pansy
30	Nymphalinae	*Junonia lemonias* (Linnaeus, 1758)	Lemon pansy
31	Nymphalinae	*Junonia almana* (Linnaeus, 1758)	Peacock pansy
32	Nymphalinae	*Junonia hierta* (Fabricius 1798)	Yellow pansy
33	Nymphalinae	*Vanessa cardui* (Linnaeus, 1758)	Painted lady
34	Satyrinae	*Melanitis leda* (Linnaeus, 1758)	Common evening brown
35	Satyrinae	*Mycalesis perseus* (Fabricius, 1775)	Common bush brown
	Papilionidae		
36	Papilioninae	*Atrophaneura aristolochiae* (Fabricius, 1775)	Common rose
37	Papilioninae	*Graphium Agamemnon* (Linnaeus, 1758)	Tailed joy
38	Papilioninae	*Papilio polytes* (Linnaeus, 1758)	Common mormon
39	Papilioninae	*Pachliopta hector* (Linnaeus, 1758)	Crimson rose
40	Papilioninae	*Papilio demoleus* (Linnaeus, 1758)	Lime butterfly

Apendix 1. Cont'd.

	Pieridae		
41	Coliadinae	*Catopsilia crocale* (Cramer, 1775)	Common emigrant
42	Coliadinae	*Catopsilia pyranthe* (Linnaeus, 1758)	Mottled emigrant
43	Coliadinae	*Eurema hecabe* (Linnaeus, 1758)	Common grass yellow
44	Pierinae	*Anaphaeis aurota* (Fabricius,1793)	Pioneer
45	Pierinae	*Appias libythea* (Fabricius. 1775)	Striped albatross
46	Pierinae	*Colotis amata* (Fabricius, 1775)	Small salmon arab
47	Pierinae	*Colotis danae* (Fabricius, 1775)	Crimson tip
48	Pierinae	*Colotis eucharis* (Fabricius, 1775)	Plain orange tip
49	Pierinae	*Colotis etrida* (Boisduval, 1836)	Little orange tip
50	Pierinae	*Cepora nerissa* (Fabricius, 1795)	Common gull
51	Pierinae	*Delias eucharis* (Drury , 1773)	Common jezebel
52	Pierinae	*Hebomoia glaucippe* (Linnaeus,1758)	Great orange tip
53	Pierinae	*Leptosia nina* (Fabricius,1793)	Psyche
54	Pierinae	*Pareronia valeria* (Cramer,1776)	Common wanderer
55	Pierinae	*Ixias pyrene* (Linnaeus, 1764)	Yellow orange tip

8

Population census of critically endangered *Dactylorhiza hatagirea* (D. Don) Soo in Suru valley, (cold desert region, Jammu and Kashmir, India)

Tsewang Rinchen, Shreekar Pant* and Mohmad Anwar

Conservation Ecology Lab, Centre for Biodiversity Studies, School of Biosciences and Biotechnology, BGSBU University, Rajouri-185 131, J and K, India.

***Dactylorhiza hatagirea* (D. Don), Soo (Panja, Salam-panja or Hath-panja) is a critically endangered native and high value medicinal plant of the Himalayan Region. The several medicinal properties of the roots of this species have increased its risk of extinction due to pressures for utilization. It is also subjected to browse by cattle's. The species also does not regenerate well in nature which is another risk factor. Therefore, an attempt has been made to investigate the population ecology of the species as a foundation for its conservation. Forty eight sites in the Suru valley, a part of the cold desert region were surveyed but only in sixteen sites was the target species present. The abundance of the species, impacts of over exploitation and its current status indicated that it may soon be extirpated from the Suru valley. So a plan for conserving the remaining number of the species is presented. It could provide a template for conservation in other locations where the species is at risk.**

Key words: Angulakpa, cold desert, population ecology, hath-panja, wanglak.

INTRODUCTION

Dactylorhiza hatagirea (D.Don), Soo (Panja, Salam-panja or Hath-panja), locally known as 'wanglak' or 'angulagpa' in various parts of Ladakh, is a critically endangered species (CAMP status), and rare (IUCN status) (Kala, 2000; Samant et al., 2001). It is a native and near endemic to Indian Himalayan region (Samant et al., 1998; Badola and Aitken, 2003; Ved et al., 2003). It extends to Pakistan, Afghanistan, Nepal, Tibet and Bhutan. In India, it is reported from Jammu and Kashmir (Dhar and Kachroo, 1983), Uttarakhand (Hajra and Balodi, 1995) and Himachal Pradesh (Aswal and Mehrotra, 1994). Generally, it is widely and narrowly distributed between 2996 and 4120 m amsl in open, grassy slopes and alpine meadows. However, the population of the species is small due to high anthropogenic pressures (photoplate A and B).

Since the time immemorial, the species is used in various Indian systems of Medicine that is, Ayurveda, Siddha, Unani, and also used in some traditional medicinal systems that is, Amchi medicinal system. It is widely used to cure dysentery, diarrhoea, chronic fever, cough, stomachache, wounds, cuts, burns, fractures, and general weakness particularly in debilitated women after delivery and to increase regenerative fluids. Tubers of *D. hatagirea* are rich in starch, mucilage, sugar, phosphate, chloride and glucoside-loroglossin (CSIR, 1996). In Ladakh, it is locally called 'Wanglak' or 'angulakpa' due to its palmately lobed tubers that stimulate a hand with fingers open, and yields 'salap', which is used as farinaceous food, expectorant, aphrodisiac and used as nervine tonic. Local inhabitants collect the roots, dry it and then take with milk to increase vigourness. Due to its high medicinal and edible value, the species has great demand in national and international market (Badola and Pal, 2002; Olsen and Helles, 1997). Further, extraction of

*Corresponding author. E-mail: shreekarpant.2@rediffmail.com.

Plate A. Site showing anthropogenic pressure **B**. Uprooted individuals of *D. hatagirea* in Suru valley.

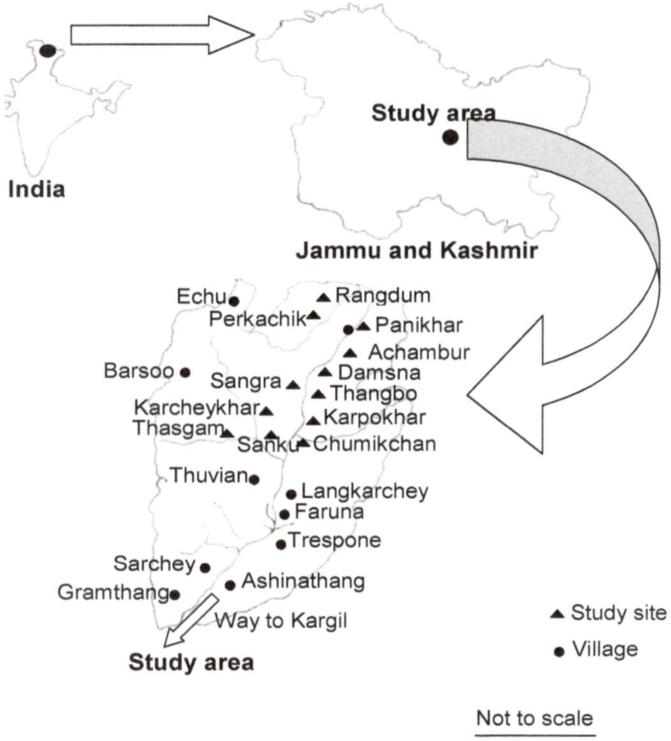

Figure 1. Suru valley, a part of cold desert region of India .

raw material from its wild population is the only source for meeting the market demand. According to a report, the annul demand of the species is approximately 5000 tons (Kala, 2004). This leads over-exploitation of the species from wild. The main objective of this research was to assess the population status of *D. hatagirea* in Suru valley and to identify factors responsible for its depletion.

METHODS

This study was conducted during April to October, 2010 in the Suru valley cold desert region located in Kargil district of Ladakh province of Jammu and Kashmir (Figure 1).

The altitude ranges between 2100 to 4200 m amsl and the area supports diverse habitats, species and communities of the sub-alpine and alpine zones. Local inhabitants are totally dependent on the locally available bio-resources to meet out their domestic needs.

Identification and selection of sites and habitats

The sites representing *D. hatagirea* were selected along transacts between 2911 to 3990 m amsl. A total of 11 transects namely Sanku (2911 to 3525 m); Chumikchan (3309 to 3525 m); Karcheykhar and Karpokhar (3270 to 3887 m); Thangbo (2742 to 3598 m); Sangra (3057 to 3228 m); Damsna (3165 to 3990 m); Achambur (3193 to 3447 m); Panikhar (3390 to 3963 m); Parkachik

Table 1. Site characteristic and densities of *D. hatagirea* and its major associates in Suru valley, J and K.

S/N	Altitude (m)	Aspect	Habitat (s)	Coordinates	*D. hatagirea* (Ind m²)	Densities of associate species (Ind m²)
1	3951	NE	Meadows	N 34°16.486' E 075°57.927'	1.90	*Polygonum* sp. (3.3), *S.chrysanthemoides* (2.85), *M. longifolia* (2.5)
2	3990	NE	Riverine	N 34°15.991' E 075°58.053'	2.20	*S. chrysanthemoides* (4.4), *Equisetum* sp. (17.2), *R. hirtellus* (2.45)
3	3088	NE	Meadows	N 34°16.161' E 075°58.478'	1.60	*Equisetum* sp. (16.2), *P. haxandrum* (1.95), *Polygonum* sp. (2.2)
4	3109	NE	Meadows	N 34°16.681' E 075°58.840'	1.40	*Geranium* sp. (2.45), *Polygonum* sp. (2.1), *Equisetum* sp. (6.85), *S. chrysanthemoides* (3.1)
5	2998	NE	Meadows	N 34°15.692' E 075°57.187'	1.10	*Polygonum* sp. (3.1), *Equisetum* sp. (10.85), *S.chrysanthemoides* (5.75)
6	3416	NE	Meadows	N 34°16.467' E 075°56.192'	8.00	*S. chrysanthemoides* (4.95), *Gentiana* sp. (3.25), *M. lupilina* (3.7)
7	3443	NE	Meadows	N 34°16.834' E 075°56.504'	0.44	*M. lupilina* (5.6), *Trifolium* sp. (3.5), *Gentiana* sp. (2.4)
8	3525	NE	Meadows	N 34°16.623' E 075°56.754'	0.85	*M. lupilina* (2.5), *Trifolium* sp. (1.05), *S. chrysanthemoides* (1.55)
9	3034	NE	Meadows	N 34°17.011' E 075°57.227'	3.00	*Equisetum* sp. (4.55), *Oxytropis lapponica* (1.75), *M. lupalina* (4.2), *Polygonum* sp. (2.85)
10	2996	NE	Meadows	N 34°17.168' E 075°57.541'	2.45	*S. chrysanthemoides* (2.6), *Geranium* sp. (2.45), *Equisetum* sp. (5.5), *Polygonum* sp. (4.35)
11	2974	NE	Meadows	N 34°17.353' E 075°57.729'	1.75	*Polygonum* sp. (3.9), *Senecio chrysanthemoides* (2.7), *Geranium* sp. (1.3)
12	2961	NE	Meadows	N 34°17.381' E 075°57.984'	1.30	*Equisetum* sp. (3.45), *Oxytropis lapponica* (1.15), *Polygonum* spp. (3.1)
13	2948	NE	Meadows	N 34°17.378' E 075°57.998'	6.10	*S.chrysanthemoides* (3.7), *Polygonum* spp. (2.25)
14	3018	NE	Meadows	N 34°16.958' E 075°57.821'	3.65	*Polygonum* sp. (2.95), *M. lupalina* (2.75), *Equisetum* sp. (2.95)
15	3044	NE	Meadows	N 34°16.727' E 075°57.818'	3.40	*Geranium* sp. (1.5), *Polygonum* sp. (2.75), *M. lupalina* (4.5)
16	2911	NE	Meadows	N 34°17.239' E 075°58.181'	3.50	*Equisetum* sp. (3.95), *Polygonum* sp. (2.6), *S. chrysanthemoides* (2.75)

(3579 to 3845 m); Thasgam (3058 to 3990 m); and Rangdum (3134 to 3990 m) selected to collect the information. Observations on altitude, coordinates, aspect, slope and habitats were taken. Sites having high percentage of moisture were considered as moist habitat.

Survey, sampling, identification and analysis of data

The field surveys and samplings were conducted during April to October, 2010 within the selected sites in the Suru valley. In each site, a plot of 50 × 50 m was laid and randomly, 20 quadrat of 1 × 1 m in each plot were laid for herbs.

For the size, number of quadrats and collection of data standard ecological methods (Curtis and Intosh, 1950; Smith, 1957; Misra, 1968; Kersaw, 1973; Dombois and Ellenberge, 1974; Dhar et al., 1997; Joshi and Samant, 2004; Samant and Joshi, 2005) were followed.

To identify the factors responsible for depletion of population, field observations were made and old knowledgeable persons were interviewed with the help of semi-structured open ended questionnaire to know the current status, utilization pattern including traditional/indigenous practices.

RESULTS

Forty eight sites were sampled. Out of which *D. hatagirea* was found only in sixteen sites. The physical and vegetation characteristics of the sixteen sites were represented in Table 1. Almost in all reported sites, *Dactylorhiza* showed least density except in two sites (8.00 and 6.1 ind/m^2; whereas among other associates, the maximum density was of *Equisetum* sp. (3.45 to 17.2 ind/m^2) followed by *Polygonum* sp. (2.6 to 4.35 ind/m^2), *Senecio chrysanthemoides* (2.05 to 5.75 ind/m^2), *Medicago lupalina* (1.75 to 5.6 ind/m^2), *Ranunculus hirtellus* (1.5 to 4.55 ind/m^2) and *Podophyllum hexandrum* (1.95 to 3.65 ind/m^2). While other species showed comparatively less densities. Further, anthropogenic and grazing pressures were also recorded from all the sites.

In the questionnaire survey of the study area, we collected the information from only 75 and 25% male and female, respectively. Out of total informants, only 41.6 and 31% males and females, respectively were well versed with the use of the species and they also know potential sites of the target species where they used to visit in the past. But according to the respondents, it was clears that the target species facing grave threat due to the over exploitation and illegal collection made by the local inhabitants and also the outsiders.

DISCUSSION

Out of 48 sites, the target species that is, *D.*

hatagirea was present only in 16 sites having a very low density except in two sites that is 8 and 6.1 ind/m^2 (Table 1). The density values are more or less similar to the study conducted in Tungnath area that is, 1.8 and 0.7 ind/m^2 of Garhwal Himalaya (Giri et al., 2008). A very few studies have been conducted in the other parts of Himalaya that is, Uttarakhand (Nautiyal and Nautiyal, 2004; Bhatt et al., 2005) but this part of Suru valley, Kargil District, a proposed part of CDBR, is untouched or unexplored and the information presented in the paper is the first hand information of this area.

According to the local inhabitants, Amchies (local herbalists) used to visit potential sites almost thrice every year for collection of medicinal plants. But from the past few years, they visit either alone or along with the researchers for the collection of high value medicinal plants for use and research purposes. Further, some local inhabitants collect this high value medicinal plant for illegal trading. The local inhabitants could collect Rs.100 to 200 per kg of dried roots of *D. hatagirea.* For 1 kg of dried roots, 90 to 100 mature plants were exploited. As a result, so many areas were seen there, though *D. hatagirea* was, one time, present in abundance, but now a few individuals of the species were seen. This indicates that if the casual factors continue to operate, this species may become extinct within a few years.

During the survey, it was also observed that local inhabitants carry their livestock in the higher regions of the valley for grazing. This is another level of disturbance because due to the grazing and trampling the underground part of the *D. hatagirea* get exposed or removed. These levels of disturbances like grazing pressure, over exploitation and unawareness of proper procedure of collection and propagation, etc. are the other major factors for declining of species from its natural habitats.

Therefore, it needs further research to overcome with these problems and also needs to promote cultivation, propagation, awareness and conservation of this species though people participation and of course through various conservation methods like, *in-situ* and *ex-situ*.

ACKNOWLEDGEMENTS

Authors are thankful to Vice Chancellor and Registrar, BGSB University for providing facilities and also thankful to the Forest Department, Kargil for providing map and other necessary facilities, during field surveys and stay. Ministry of Environment and Forests (File no.No.8/22/08-CS/BR/48) is greatly acknowledged for financial assistance. Authors greatly acknowledge the anonymous reviewers for their valuable comments.

REFERENCES

Aswal BS, Mehrotra BN (1994). Flora of Lahaul-Spiti. Bishan Singh Mahendra Pal Singh, Dehra Dun, India.

Badola HK Pal M (2002). Endangered Medicinal plant in Himachal Pradesh. Curr. Sci., 83(7): 797-798.

Badola HK, Aitken S (2003). The Himalayas of India: A treasury of Medicinal plant under siege. Biodiversity, 4: 3-13.

Bhatt A, Joshi SK, Gairola S (2005). *Dactylorhiza hatagirea* (D.Don) Soo. A West Himalayan Orchid in Peril. Curr. Sci., 89(4): 610-612.

CSIR (1996). Council of Scientific and Research Institute. The wealth of India (Raw material) - Council of Scientific and Industrial Research, New Delhi, India. Vol., 7.

Curtis JT, Intosh Mc (1950). The interrelation of certain analytic and Phytosociological characters. Ecology, 31: 434-455.

Dhar U, Kachroo P (1983). Alpine flora of Kashmir Himalaya. Scientific Publishers, Jodhpur, India.

Dhar U, Rawal RS, Samant SS (1997). Structural diversity and representativeness of forest vegetation in a protected area of Kumaun Himalayan, India: implication for conservation. Biol. Conserv., 6: 1045-1062.

Dombois M, Ellenberge H (1974). Aims and Method of Vegetation Ecology. John Wiley, USA.

Giri D, Arya D, Tamta S, Tewari LM (2008). Dewaling of an endangered orchid *Dactylorhiza hatagirea* (D.Don) Soo; A case study from Tungnath Alpine medows of Garhwal Himalaya, India. Natl. Sci., 6(3): 6-9.

Hajra PK, Balodi B (1995). Plant Wealth of Nanda Devi Biosphere Reserve. - Botanical Survey of India, Calcutta, India.

Joshi HC, Samant SS (2004). Assessment of forest vegetation and prioritization of communities for conservation in a part of Nanda Devi Biosphere Reserve, West Himalaya, India. Int. J. Sust. Dev. Wor. Ecol., 11: 326-336.

Kala CP (2000). Status and conservation of rare and endangered medicinal plants in the Indian Trans-Himalayas. Biol. Conserv., 93(3): 371-379.

Kala CP (2004). Assessment of species rarity. Curr. Sci., 86: 1058-1059.

Kersaw KA (1973). Quantitative and Dynamic Plant Ecology. Second edition. Edward Arnold limited, London.

Misra R (1968). Ecological Work Book. Oxford and IBH publishing Company, Calcutta.

Nautiyal MC, Nautiyal BP (2004). Agro-techniques for high altitude Medicinal and Aromatic plants (Silver Jubilee publication of HAPPRC). Bishen Singh Mahendra Pal Singh, Dehra Dun, India.

Olsen CS, Helles F (1997). Medicinal plants, Markets and margins in the Nepal Himalaya: Trouble in Paradise. Mt. Res. Dev., 17(1): 363-74.

Samant SS, Dhar U, Palni LMS (1998). Medicinal plants of Indian Himalayan: Diversity, Distribution Potential Values. HIMAVIKAS Publication. No.13, Gyanodaya Prakashan., Nainital, India.

Samant SS, Dhar U, Palni, LMS (2001). Himalayan medicinal plants-Potential and prospects. Gyanodaya Prakashan, Nainital.

Samant SS, Joshi HC (2005). Plant Diversity and Conservation status of Nanda Devi National Park and comparisons with Highland National Park of Indian Himalayan region. Int. J. Biol. Manag., 1: 65-73.

Smith PG (1957). Quantative Plant Ecology. Academic press, New York.

Ved DK, Kinhal GA, Prabhakaran KVR, Ghate U, Sankar RV, Indresha JH (2003). CAMP Report: Conservation assessment and Management prioritization for Medicinal plant of Jammu and Kashmir, Himachal Pradesh and Uttaranchal. *In*: Workshop at Shimla, Himachal Pradesh.

Taxonomic revision of genus *Hordeum* L. (Gramineae) in Egypt

Wafaa. M. Amer[1]*, Ahmad. K. Hegazy[1] and Safwat. A. Azer[2]

[1]Department of Botany, Faculty of Science, University of Cairo, Giza 12613- Egypt.
[2]Flora and Phytotaxonomy Researches Department, Agriculture Research Center, Dokki 12618, Giza- Egypt.

This study was conducted to revise the taxonomic identity and clarify inter- and intra-specific relationships among the studied wild and cultivated *Hordeum* taxa in Egypt. The studied taxa included *Hordeum marinum* subsp. *gussoneanum*, *H. marinum* subsp. *marinum*, *H. Hordeum leporinum*, *Hordeum glaucum*, *Hordeum spontaneum* and *Hordeum vulgare*. The results revealed that unweighted pair group method with arithmetic mean (UPGMA) dendrogram showed a considerable degree of dissimilarity among the studied *Hordeum* taxa at 1.50 dissimilarity distance. It divided the studied taxa into four groups at 1.33. Group 1 included *H. marinum* subsp. *gussoneanum* and *H. marinum* subsp. *marinum*. Group 2 included *H. leporinum*. Group 3 included *H. glaucum*. Group 4 included *H. spontaneum* and *H. vulgare* which is related to the other groups (1, 2, and 3). Moreover, by using UPGMA dendrogram, *H. leporinum* and *H. glaucum* are located in transitional position between *H. marinum* and both of *H. vulgare* and *H. spontaneum* forming a transitional step between the taxa of Group 1 and 4. This work provided a taxonomic key and separated the two subspecies of *H. murinum* into different species namely *H. glaucum* and *H. leporinum*. Moreover, the morphological similarities among *Hordeum* taxa have been proved.

Key words: Taxonomy, *Hordeum marinum, Hordeum leporinum, Hordeum glaucum, Hordeum spontaneum, Hordeum vulgare,* unweighted pair group method with arithmetic mean (UPGMA) dendrogram, Egypt.

INTRODUCTION

The variability in this genus has been a source of controversy among botanists for over a century. There is still no complete agreement as to the status of the various forms. This is clearly expressed by the more than forty taxa that exist (Bothmer and Jacobsen, 1985; Bothmer, 1992; Kochieva et al., 2001).

The taxonomic delimitation of the genus is still not satisfactorily resolved and there is still much disagreement among botanists and cytogeneticists concerning generic concepts. Based on morphology, genus *Hordeum* L. is divided into four sections: *Hordeum, Anisolepis, Critesion* and *Stenostachys*. Section *Hordeum* consisted of perennial and annual species and has a Eurasian distribution ranging from the Mediterranean area to West-Central Asia and included three species namely: *Hordeum vulgare* L., *Hordeum bulbosum* L. and *Hordeum murinum* L. (Bothmer and Jacobsen, 1985; Zohary and Hopf, 1988; Bothmer, 1992).

Section *Anisolepis* Nevski included perennial and annual species native to South and North America and represented eight species (Bothmer and Jacobsen, 1985). Section *Critesion* (Rafinesque) Nevski characterized by setaceous, long and spreading glumes and included six species (Bothmer and Jacobsen, 1985).

Section *Stenostachys* Nevski is the largest group of taxa in genus *Hordeum* L. with wide distribution. This section included eleven species, among of them: *H. marinum* Huds. with two subspecies: subsp. *gussoneanum* (Parl.) Thell. and subsp. *marinum* (Bothmer and Jacobsen, 1985). Bothmer (1992) treated genus *Hordeum* as thirty two species among of them: *H. marinum* (subsp. *Marinum* and

*Corresponding author. E-mail: wafaa_amer@hotmail.com

Table 1. Number of surveyed localities (L), fresh examined specimens (F.S) and checked herbarium specimens (H.S) of studied *Hordeum* L. taxa in Egypt.

***Hordeum* taxa**	**L**	**F.S**	**H.S**
H. vulgare	3	30	20
H. spontaneum	0	0.0	12
H. marinum subsp. *marinum*	3	30	18
H. marinum subsp. *gussoneanum*	3	30	13
H. murinum glaucum	3	30	30
H. murinum leporinum	3	30	22
Total number	15	150	115

subsp. *gussoneanum*), *H. murinum* (subsp. *leporinum*, subsp. *glaucum* and subsp. *murinum*), and *H. vulgare* (subsp. *spontaneum* and subsp. *vulgare*).

Feinbrun-Dothan (1986) treated genus *Hordeum* as seven species among of them: *Hordeum spontaneum, Hordeum glaucum, H. marinum,* and *H. vulgare.* Humphries (1996) treated genus *Hordeum* as eleven species among of them: *Hordeum spontaneum*, *H. vulgare*, *H. murinum* (subsp. *murinum*, subsp. *glaucum* and subsp. *leporinum*), and *H. marinum*. Stace (1997) treated genus *Hordeum* as 10 species among of them: *H. vulgare*, *H. murinum* (subsp. *murinum*, subsp. *glaucum* and subsp. *leporinum*), *H. marinum* and *Hordeum geniculatum*.

Many taxonomists treated *Hordeum murinum* complex into different species: Baum and Bailey (1984a, b, 1989) treated *H. murinum* group as three species namely: *H. glaucum, H. murinum* and *Hordeum leporinum* (var. *leporinum* and var. *simulans*). Rajhathy and Morrison (1962) treated *H. murinum* group as two species (*H. glaucum* and *H. murinum*).

Jasska (1992) treated it as three species namely: *H. glaucum*, *H. murinum* and *Hordeum hrasdanicum* Gandilyan. Later, Mizianty (2006), El-Rabey and Al-Malki (2011) and Sardou et al. (2011) treated it as two different species. On the other hand, Booth and Richards (1976), Melderis (1985), Bothmer et al. (1991), Boulos (1995, 2005, 2009), Jacobsen and Bothmer (1995), Humphries (1996), and Jahan and Vahidy (2007, 2011) treated *H. murinum* complex as three subspecies of one species namely: *H. murinum* (subsp. *leporinum*, subsp. *glaucum* and subsp. *murinum*).

In Egypt, *Hordeum* taxa were treated by many authors among of them: Täckholm (1974) treated *Hordeum* L. taxa as six species (*H. spontaneum* C. Koch, *H. murinum* L., *H. leporinum* Link, *H. glaucum* Steud., *H. marinum* Huds., and *H. geniculatum* All.). Last treatments of *Hordeum* L. taxa in Egypt (Cope and Hosni, 1991; El-Hadidi and Fayed, 1994/1995; Boulos, 1995, 2005, 2009) treated the genus as four species; some of these species included infra-specific taxa. These species are: *H. vulgare* L. represented as four cultivars (Giza 123, Giza 124, Giza 125, and Giza 126), *H. spontaneum* C. Koch., *H. marinum* (subsp. *gussoneanum* and subsp. *marinum*), and *H. murinum* (subsp. *glaucum* and subsp. *leporinum*).

The present study aimed to derive data from macro- and micro-morphological attributes and used to revise the taxonomic identity of *Hordeum* taxa in Egypt. The retrieved data will be subjected to statistical analysis to clarify inter- and intra-specific relationships among the studied *Hordeum* taxa and confirmed the separation of *H. murinum* complex into *H. glaucum* and *H. leporinum.* Moreover, taxonomic key, for *Hordeum* taxa in Egypt, constructed to distinguish among studied taxa

MATERIALS AND METHODS

Field trips were carried out and 150 examined specimens were collected from 15 localities and 115 herbarium specimens were checked at Egyptian herbaria (Table 1). Phytogeographical distribution (longitude and latitude) of the studied *Hordeum* taxa were identified (Table 2). The fresh and herbarium specimens were subjected to taxonomic revision based on macro- and micro-morphological characters.

Distribution map is provided based on the specimen's distribution for each species/subspecies according to the ecological territories. Within each territory, specimens are distributed according to the sequence of the location from West to East, and from North to South (Figure 1). *Hordeum* specimens are distributed mainly along the Mediterranean Sea Strip (Map 1 and Table 2). However *H. marinum* subsp. *gussoneanum* (28° 59′ 18″ N° to 31° 11′ 54″ N°) and *H. murinum* L. subsp. *glaucum* and *H. murinum* L. subsp. *Leporinum* (29° 49′ 11″ N° to 31° 33′ 49″ N°) distributed southern of this limited area.

Data collection

Data collections for investigation and identification criteria of the studied taxa were based on the authentic flora and taxonomic (Cope and Hosni, 1991; Boulos, 1995, 2005, 2009). The identification of the studied *Hordeum* L. taxa was based on macro- and micro-morphological comparison among the examined specimens.

Voucher specimens kept at the studied herbaria among of them: Flora and Phytotaxonomy Researches Department Herbarium Agriculture Research Center a (CAIM), and Cairo University Herbarium (CAI). Description of *H. spontaneum* C. Koch, in this work, was based on the data retrieved from the herbarium, due to the lack of fresh specimens during the study years.

Statistical analysis

Statistical analysis of the differential characters was carried out. The output retrieved in dendrogram form was used to construct

Table 2. Phytogeographical distribution (latitude and longitude) of studied *Hordeum* L. taxa in Egypt.

***Hordeum* Taxa**	**Latitude (N°)**	**Longitude (E°)**
H. vulgare	30° 10′ 48″ to 31° 38′ 12″	25° 53′ 53″ to 29° 02′ 48″
H. spontaneum	30° 39′ 16″ to 30° 49′ 06″	27° 15′ 55″ to 33° 48′ 30″
H. marinum subsp. *marinum*	30° 59′ 30″ to 31° 27′ 02″	30° 03′ 03″ to 31° 34′ 33″
H. marinum subsp. *gussoneanum*	28° 59′ 18″ to 31° 27′ 02″	30° 03′ 03″ to 31° 34′ 33″
H. glaucum	29° 49′ 11″ to 31° 27′ 02″	25° 46′ 25″ to 34° 14′ 27″
H. leporinum	29° 49′ 11″ to 31° 27′ 02″	27° 15′ 55″ to 34° 14′ 27″

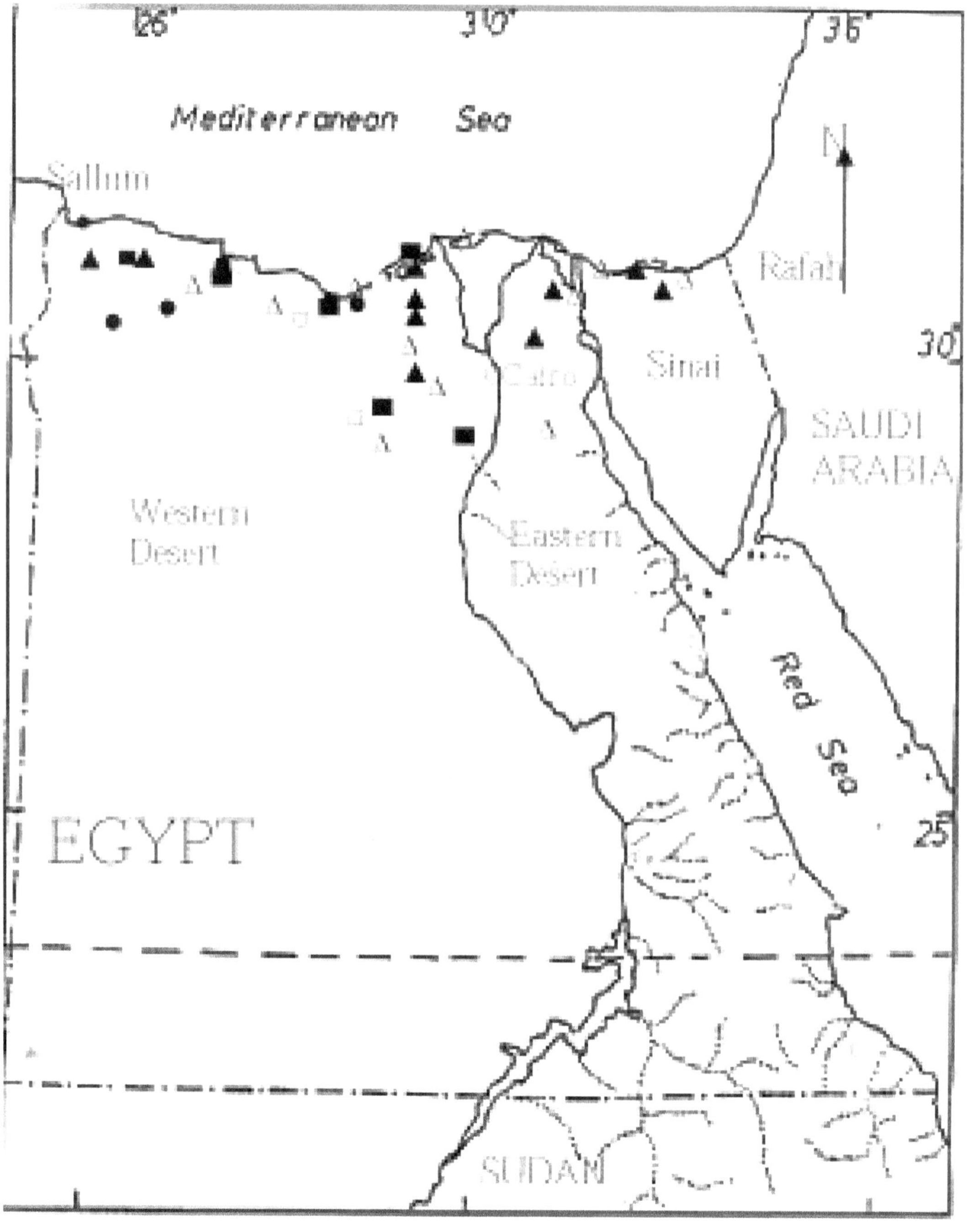

Figure 1. Distribution of *Hordeum* taxa collected during this study; *H. marinum* subsp. *gussoneanum* (■); *H. marinum* subsp. *marinum* (□); *H. glaucum* (▲); *H. leporinum* (△) and *H. vulgare* (●).

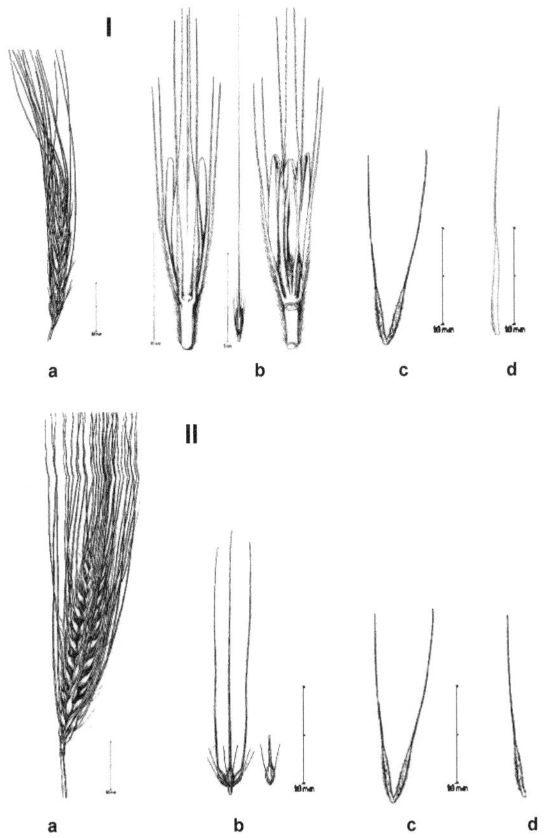

Figure 2. Detailed line drawing of *H. spontaneum* (I) and *H. vulgare* (II): a = spike, b = triplet of spikelets, c = two glumes of spikelets, = glume of the lateral spikelets (after Bor, 1968, 1970).

inter- and intra-specific relationships among the studied *Hordeum* taxa. The constructed dendrogram was based on macro- and micro-morphological data (Table 4). The macro-morphological data included habit, leaf color and spike shape and color.

In addition to, micro-morphological data included anther length and color; glume shape and surface; rachilla color and surface. Each taxon is provided with a detailed line drawing for the following characters: spike morphology, triplet of spikelets (figure 2,3 and 4) and lateral glumes to distinguish between wild and cultivated *Hordeum* taxa. The use of substantial number (1 = presence, 0 = absence) of attributes twenty five macro- and micro-morphological characters for the studied wild and cultivated *Hordeum* taxa (Table 4). The data treated as a binary character in a data matrix (Table 4) using NTSYS-PC version 1.50 program (numerical taxonomy and multivariate analysis system; Rohlf, 1988). Clustering was performed using the unweighted pair-group arithmetic mean method (UPGMA) (Sneath and Sokal, 1973; Dunn and Everitt, 1982). The output was plotted in the form of dendrogram (Figure 5).

RESULTS

Diagnostic features of the studied *Hordeum* taxa

Hordeum L., Sp. Pl. ed. 1: 84 (1753), Gen. Pl. ed. 5: 37 (1754)

Annual or perennial herbs with flat leaf-blades and auricled base. Inflorescence an erect or nodding distichously compressed spike, spikelets are in triplets, each with one floret. The central spikelet is bisexual, the laterals are bisexual, male or sterile. Cultivated barley, each triplet consisting of a central sessile spikelet flanked by two sessile ones, while, in wild barley the central sessile spikelet flanked by two pedicelled lateral ones.

Glumes narrow, 1 to 3-veined, long awned, sometimes expanded at the base. Lemma of the bisexual florets 5-veined awned or awnless. Palea 2-nerved, narrowly ovate and keeled. The grain with adhered lemma and palea, the later may be free. Stamens 3, Lodicules 2, wedge-shaped or spathulate and densely ciliate in cultivated species, while, in the wild species are lanceolate, or with an additional lateral tooth slightly ciliate above. Fruit elliptic caryopsis, deeply furrowed ad-axially, subtended by apical hairy appendage. Key to the studied *Hordeum* taxa is shown in Table 3.

***H. vulgare* L., Sp. Pl., ed. 1, 84 (1753). Syns. *H. hexastichon* L., Sp. Pl., ed. 1, 85 (1753); *H. sativum* Pers., Syn. Pl. 1: 108 (1805):** Annual grass, culms solitary or loosely fasciculate, up to one meter tall, erect smooth and glabrous. Leaf-blades flat, linear-acute up to 45 cm long, glabrous, scaberulous margins on both surfaces. Spike, excluding the awn, 6 to 10 x 1.5 to 2 cm, square or hexagonal in section, all spikelets fertile. Central spikelet sessile, glumes flattened, usually 10 to30 mm long. Lemma glabrous or scabrid, mostly near the apex, awn of lemma 30 to 180 mm long, usually scabrid. Anthers 6 to 10 mm long, yellowish. Rachilla up to half the length of the palea, with long or short hairs. Lateral spikelets usually sessile. Glumes linear-lanceolate produced into a fine, scabrid awn. Lemmas broadly ovate-lanceolate, strongly 5-nerved, armed with a stiff awn up to 15 cm long.

***H. spontaneum* C. Koch, Linnaea, 21: 430 (1848). Syns. *H. ithaburense* Boiss., Diagn. Pl. Or. Nov. Ser. 1, 13: 70 (1873); *H. spontaneum* var. *ithaburense* (Boiss.) Nab. in Publ. Fac. Sci. Univ. Masaryk 111: 32 (1929):** A robust, annual grass. Culms usually solitary, erect, or loosely fasciculate, up to 80 (to 100) cm tall, glabrous, with brown nodes. Leaf-blades linear-acuminate, flat, somewhat flaccid, up to 15.0 x 0.4 to 0.8 cm. Leaf-blades scaberulous on the upper surface and30 mm long. Lemma glabrous or scabrid, mostly near the apex, awn of lemma 30 to 180 mm long, usually scabrid. Anthers 6 to 10 mm long, yellowish. Rachilla up to half the length of the palea, with long or short hairs. Lateral spikelets usually sessile. Glumes linear-lanceolate produced into a fine, scabrid awn. Lemmas broadly ovate-lanceolate, strongly 5-nerved, armed with a stiff awn up to 15 cm long.

***H. spontaneum* C. Koch, Linnaea, 21: 430 (1848). Syns. *H. ithaburense* Boiss., Diagn. Pl. Or. Nov. Ser. 1, 13: 70 (1873); *H. spontaneum* var. *ithaburense* (Boiss.) Nab. in Publ. Fac. Sci. Univ. Masaryk 111: 32 (1929):** A robust, annual grass. Culms usually solitary,

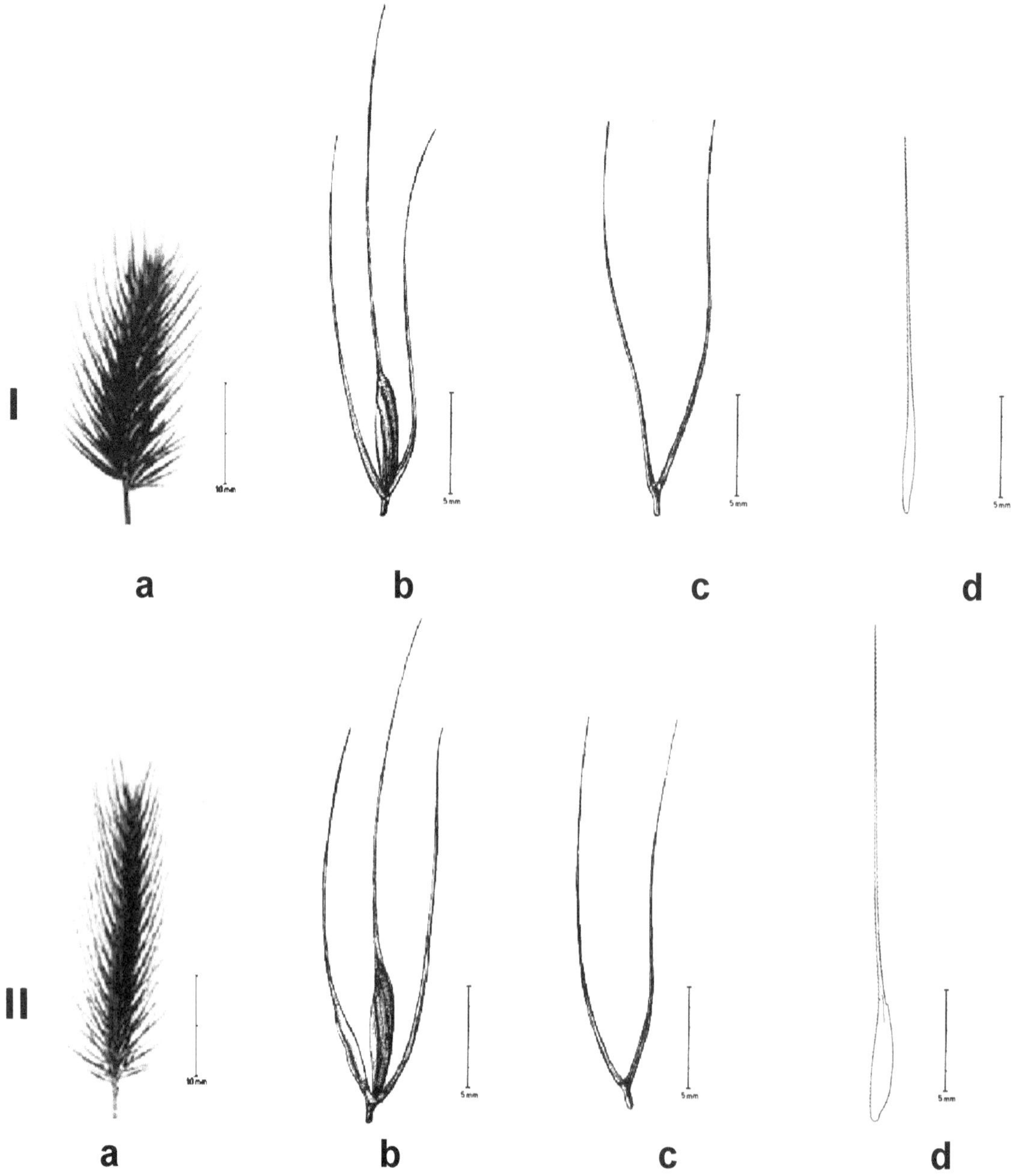

Figure 3. Detailed line drawing of *Hordeum marinum* subsp. *gussoneanum* (I) and *H. marinum* subsp. *marinum* (II): a =: spike, b = lateral of spikelet, c = lateral glumes with rachis, and d = inner glume of the lateral spikelets.

erect, or loosely fasciculate, up to 80 (to 100) cm tall, glabrous, with brown nodes. Leaf-blades linear-acuminate, flat, somewhat flaccid, up to 15.0 x 0.4 to 0.8 cm. Leaf-blades scaberulous on the upper surface and glabrous lower one, margins rough scabrid. Leaf sheath auricled at the mouth. Ligules about 1.5 mm long. Spike (excluding the awns) 4 to 10.0 x 0.7 to 1.0 cm, dense, erect, and distichously. Spikelets one flowered.

Rachis dens silky-villose along margins, fragile, and readily disarticulating. Rachilla prolongation more than 1/2 length of lemma. Glumes linear-lanceolate, silky-villose, each ending in a fine awn longer than glume proper. Central spikelet hermaphrodite and sessile.

Lemma lanceolate, ending in a long tough awn; awn stout, flattened, scabrous, straight, 8 to 14 (to 20) x 0.15 cm. Lateral spikelets staminate, neuter or vestigial, subtended by short pedicel. Lemma awnless, with rounded-obtuse apex.

Hordeum spontaneum C. Koch, not traced in Egypt during the course of study. However, the last collections from this nearly extinct species were recorded during 1988 in Wadi Habis 60 km southern Mersa Martuh city

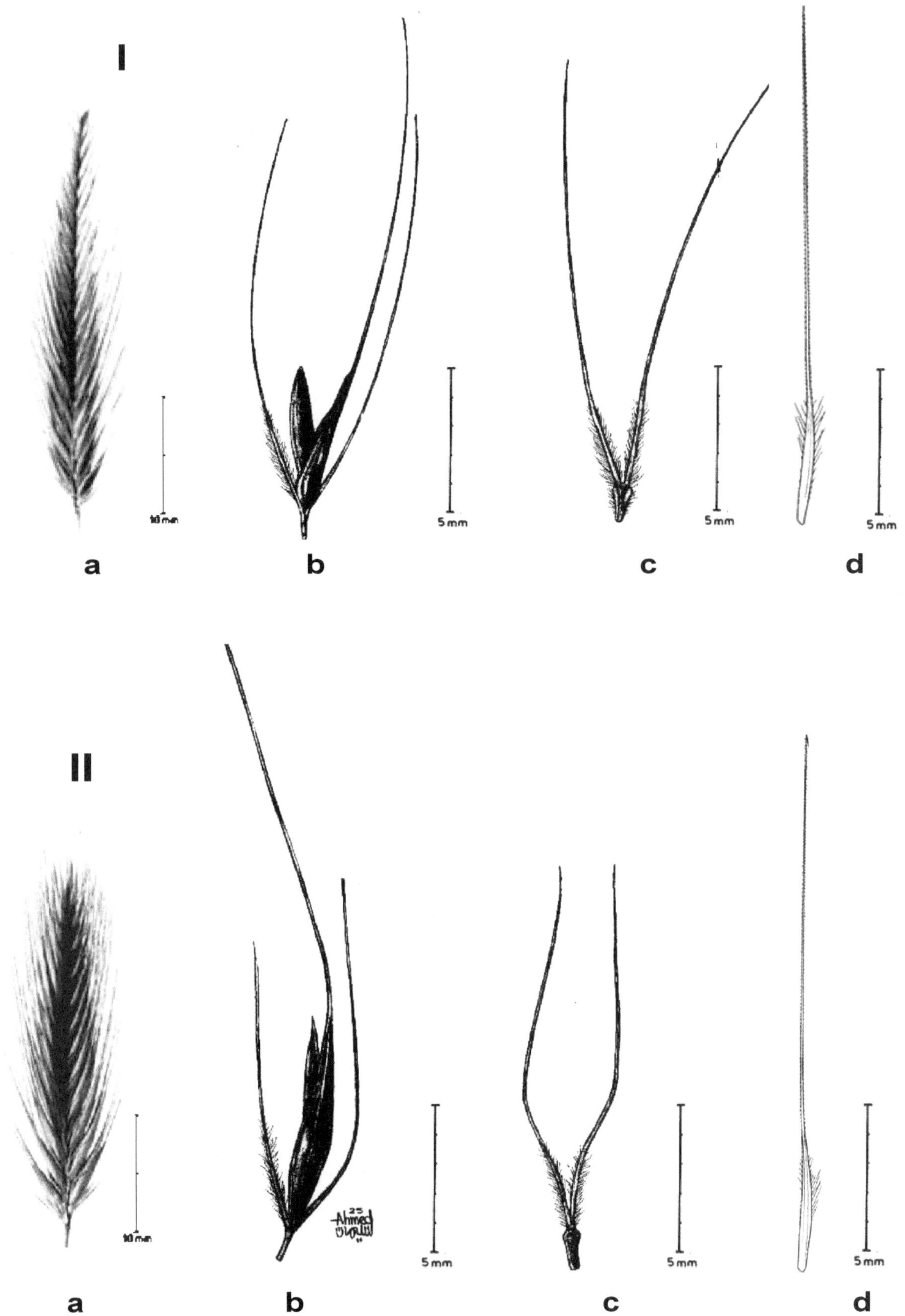

Figure 4. Detailed line drawing of *H. glaucum* (I) and *H. leporinum* (II): a = spike, b = lateral of spikelet, c = lateral glumes with rachis, d = inner glume of the lateral spikelets.

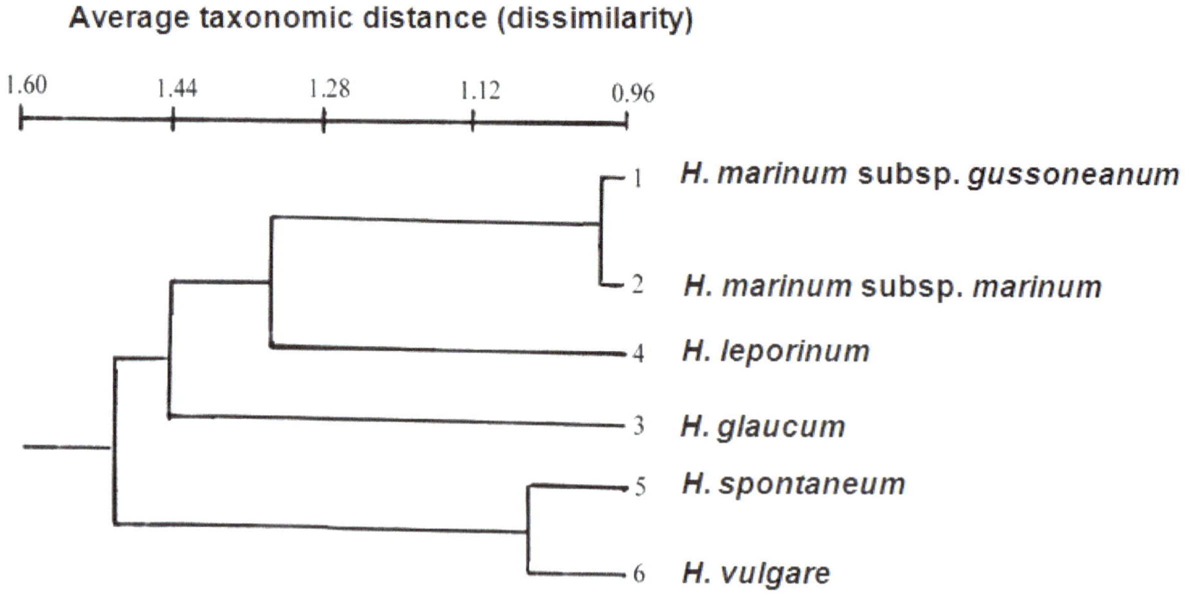

Figure 5. UPGMA dendrogram based on the studied 25 macro-and micro-morphological characters illustrating average taxonomic distance (dissimilarity) among *Hordeum* taxa in Egypt.

Table 3. Key to the studied *Hordeum* taxa.

1	Spike in four or six-rowed, rachis tough not breaking and articulating at maturity.	*H. vulgare*
+	Spike in two rows, rachis fragile articulating at maturity.	2
2	Awn of central lemma stout, 5-12 cm long; lemmas of the two lateral spikelets awnless	*H. spontaneum*
+	Awn of central lemma slender, up to 3 cm long; lemmas of the two lateral spikelets awned.	3
3	Glumes of central spikelets glabrous; spike up to 3 cm, central spikelet sessile.	4. *H. marinum*
+	Glumes of central spikelets hairy; spike up to 8 cm, central spikelet pedicel.	5. *H. murinum*
4	Inner glume of the lateral spikelets winged on one side	*H. marinum* subsp. *marinum*
+	Both glumes of the lateral spikelets wingless	*H. marinum.* subsp. *gussoneanum*
5	Rachilla yellowish–brown and ciliated; anther of central spikelet blackish, heteromorphic and not exserted at anthesis	*H. glaucum*
+	Rachilla green and non-ciliated; anther of central spikelet yellowish or brownish, isomorphic and exserted at anthesis	*H. leporinum*

along the western desert. Taxonomic revision and macro- and micro-morphological description of *H. spontaneum* C. Koch, were based on authentic herbarium samples deposited in the Egyptian herbaria.

H. marinum **Huds., Fl. Angl., ed. 2, 1. 57 (1778). Syn. *H. maritimum* With., Arr. Br. Pl. ed. 2, 1: 127 (1787); Fl. Orient. 5: 687 (1884):** Annual grass with loosely tufted of solitary culms, erect or ascending. Stem 10 to 40 (to 60) cm tall, smooth and glabrous. Leaf-blades linear acuminate, 6 to 8 x 0.15 to 0.4 cm, scaberulous to smooth on the margins.

Sheaths auricled at the mouth, glabrous on the upper part of the stems. Spike 2 to 5 cm long (excluding awns), very bristly oblong, tapered toward the apex, greenish or purplish. Spikelets one flowered glabrous or hairy. Central spikelet sessile and much longer than the lateral spikelets. Glumes of the central spikelet equal, scabrid, setaceous, 10 to 25 mm long. Lemma, elliptic-oblong, acute, sessile, tipped when an awn up to 25 mm long, 5-nerved. Palea narrows, as long as the lemma, 2-nerved. Lateral spikelets glumes scabrous, bristle-like, unequal, the lower with an awn up to 25 mm long, the inner with a wing (0.6 to 1.4 mm) broad at the base and with the awn

Table 4. Summary table of 25 macro- and micro-morphological characters studied among *Hordeum* taxa: *H. marinum* subsp. *gussoneanum* (a), *H. marinum* subsp. *marinum* (b), *H. glaucum* (c), *H. leporinum* (d), *H. vulgare* cv. Giza 123 (e) and *H. spontaneum* (f) with their code for statistical analysis (1=presence, 0=absence).

Taxa	(a)	(b)	(c)	(d)	(e)	(f)
			Spike			
1. Length (including awns)	Short (0)	Short (0)	Short (0)	Short (0)	Long (1)	Long (1)
2. Shape	Dense (1)	Dense (1)	Not loose (1)	Loose (0)	Dense (1)	Dense (1)
3. Rachis	Brittle (1)	Brittle (1)	Brittle (1)	Brittle (1)	Tough (0)	Brittle (1)
4. Rows	Two-rows (1)	Two-rows (1)	Two-rows (1)	Two-rows (1)	Four-six rows (0)	Two-rows (1)
5.anthesis	Green (1)	Green (1)	Glaucous (0)	Green (1)	Green (1)	Green (1)
			Anther			
6. Length	Isomorphic (1)	Isomorphic (1)	Heteromorphic (0)	Isomorphic (1)	Isomorphic (1)	Isomorphic (1)
7. Color	Blackish (1)	Yellowish (0)	Blackish (1)	Yellowish (0)	Yellowish (0)	Yellowish (0)
8. Anthesis	Not exerted (1)	Not exserted (1)	Not exserted (1)	Exserted (0)	Exserted (0)	Exserted (0)
			Lateral spikelets			
9.pedicel	Pedicelled (1)	Pedicelled (1)	Pedicelled (1)	Pedicelled (1)	Sessile (0)	Pedicelled (1)
10. Awned-lemma	Awned (1)	Awned (1)	Awned (1)	Awned (1)	Awned (1)	Awnless (0)
11. Length (excluding awns)	Same as central spikelet (1)	Same as central spikelet (1)	Same as central spikelet (1)	Longer than central spikelet (0)	Same as central spikelet (1)	Same as central spikelet (1)
12. Glumes size	Isomorphic (1)	Heteromorphic (0)	Isomorphic (1)	Heteromorphic (0)	Isomorphic (1)	Isomorphic (1)
13. Outer glume	With broad base (1)	Without broad base (0)	Without broad base (0)	Without broad base (0)	Without broad base (0)	Without broad base (0)
14. Inner glume	Wingless (0)	Winged (1)	Wingless(0)	Wingless (0)	Wingless (0)	Wingless (0)
15.fertility	Sterile (0)	Sterile (0)	Sterile (0)	Sterile (0)	Fertile (1)	Sterile (0)
16. Palea	Glabrous (1)	Glabrous (1)	Pilose (0)	Pilose (0)	Glabrous (1)	Glabrous (1)
17. Palea and lemma	Glume like (1)	Not so (0)	Not so (0)	Not so (0)	Not so (0)	Not so (0)
			Rachilla			
18. Prolongation- color	Green (1)	Green (1)	Yellowish-brown (0)	Green (1)	Green (1)	Green (1)
19. Surface	Non-ciliated (0)	Non-ciliated (0)	Ciliated (1)	Non-ciliated (0)	Ciliated (1)	Ciliated (1)
Central spikelet						
20. Glumes surface	Glabrous (1)	Glabrous (1)	Ciliated (1)	Ciliated (1)	Ciliated (1)	Ciliated (1)
21. Awn length	Short (0)	Short (0)	Short (0)	Short (0)	Long (1)	Long (1)
22. Pedicel	Sessile (0)	Sessile (0)	Pedicelled (1)	Pedicelled (1)	Sessile (0)	Sessile (0)
			Leaf			
23. Glabrous blade	Not (0)	Not (0)	Not (0)	Not (0)	Glabrous (1)	Glabrous (1)
24. Auricle	Shot (0)	Shot (0)	Long (1)	Long (1)	Long (1)	Long (1)
25. Color	Green (1)	Green (1)	Bluish-green (0)	Green (1)	Green (1)	Green (1)

about 20 mm long. Lemma c.5 mm long, shortly awned.

***H. marinum* subsp. *marinum*. Syns. *H. marinum* Huds., Fl. Angl., ed. 2, 57 (1778), *H. maritimum* With., Arr. Br. Pl. ed. 2, 1: 127 (1787):** Summer or winter annual species. Basal leaf-sheaths with dense short hairs 0.1 to 0.25 mm. Glumes subulate and heteromorphous, inner glume of lateral spikelets expanded into a wing 0.6 to 1.4 mm wide at base.

***H. marinum* subsp. *gussoneanum* (Parl.) Thell., Naturf. Ges. Zürich 52: 441 (1908). Syns. *Hordeum hystrix* Roth, Cat. Bot. 1: 23 (1797); *Hordeum gussoneanum* Parl., Fl. Palerm. 1: 246 (1845); *H. geniculatum* All., Fl. Pedem. 2: 259 (1785):** Similar to subsp. *marinum* but leaf-sheath with hairs 0.5 to 1.0 (to 1.2) mm. Inner glume of lateral spikelets 0.2 to 0.6 (to 0.7) mm wide, wingless, subulate (slightly swollen).

***H. glaucum* Steud., Syn. Pl. Glum. 1: 352 (1854). Syn. *H. murinum* subsp. *glaucum* (Steud.) Tzvelev, Act. Nov. Sist. Vyssh. Rast. 8: 67 (1971):** Leaves glaucous. Spikes 3.0 to 7.0 cm long and not loose. Palea of central spikelet 0.7 to 0.8 times as long as palea of lateral spikelets. Central spikelet glumes subulate, lemma oblong-elliptic, 5 to 6 mm long, tipped with an awn 10 to 25 mm long. Lateral spikelets pedicelled. Glumes subulate; lemma 5 to 8 mm long.

Upper glumes similar to those of the central spikelet. Lower glumes narrower than the upper ones. Lemma longer than that of the central spikelet. Prolongation of rachilla of lateral spikelets stout, yellowish-brown. Rachilla yellowish–brown and ciliated with hairs of 0.25 to 0.75 mm long. Anther of central spikelet blackish, heteromorphic (0.2 to 0.5 mm), not exserted at anthesis.

***H. leporinum* Link, Linnaea 9: 133 (1835). Syn. *H. murinum* subsp. *leporinum* (Link) Arcang., Fl. Ital. 805 (1882):** Leaves green and glabrous on both surface. Spike 4 to 8 x 0.8 to 1.0 cm and loose. Palea of central spikelet 0.7 to 0.9 times as long as palea of lateral spikelets. Central spikelet glumes 25 mm and ciliated on margins, awned 25 mm, lemma and palea are 10 mm long. Lateral spikelets: outer glume 25 mm scarbid; inner glume 20 mm ciliated on both margins; pedicel 2.0 mm, awn 30 mm; lemma and palea are 13 mm. Grain 6.0 x 2.0 mm. Stamens yellowish. Anthers of central spikelet 0.7 to 1.4 mm; Rachilla prolongation slender, green and non-ciliated. Anther of central spikelet yellowish or brownish, isomorphic (0.7 to 1.4 mm), exserted at anthesis.

Major differences between wild cultivated *Hordeum* taxa

The data retrieved in this revision revealed that, the major differences between cultivated and wild *Hordeum* taxa can be summarized as follows: Inflorescence of cultivated barley (*H. vulgare*) is six-rowed, while wild species (*H. marinum, H. murinum* and *H. spontaneum*) are two-rowed barley. *H. spontaneum* is distinguished from the other wild species by awns lateral spikelets are reduced while the central lemma posses long awn. *H. marinum* is characterized by glabrous central glumes and sessile central spikelet. These characters were not observed in *H. murinum* (hairy central glumes and pedicel central spikelets).

Themacro-andmicro-morphologicaldifferences between *H. marinum* subsp. *gussoneanum* and *H. marinum* subsp. *marinum* are shown in Table 4. Whereas, the main differences between the two subspecies: inner glumes of the lateral spikelets winged in subspecies *marinum* but not in subsp. *gussoneanum*. Moreover, palea and lemma reduced to glume like in subsp. *gussoneanum*, while anther color blackish on subsp. *gussoneanum* and yellowish in subsp. *marinum* (Table 4).

The major differences between the *H. glaucum* and *H. leporinum* are: rachilla surface ciliated in subspecies *glaucum*, glabrous in *H. leporinum* and glumes of lateral spikelets isomorphic in *H. glaucum*, but heteromorphic in *H. leporinum*. In addition to, anther length and color are heteromorphic and blackish in *H. glaucum* but isomorphic and yellowish or brownish in *H. leporinum* (Table 4).

Morphological similarities among the studied *Hordeum* taxa

The dendrogram (Figure 5) produced by the statistical analysis, based on twenty five macro- and micro-morphological characters (Table 4) showed that, wild and cultivated *Hordeum* taxa had a highest average taxonomic distance value of 1.50.

At point 1.33, (Table 5 and Figure 5) the studied *Hordeum* taxa was separated into four groups (G1, G2, G3 and G4). The first group (G1) included *H. marinum* subsp. *gussoneanum* and *H. marinum* subsp. *marinum*. The second group (G2) included *H. leporinum*. The third group (G3) included *H. glaucum*. The forth group (G4) included *H. spontaneum* and *H. vulgare*.

DISCUSSION

Genus *Hordeum* L. was one of the tribe Triticeae genera subjected to several studies, but still there is no complete agreement about its taxonomic delimitation as well as the status of the various forms. The variability in this genus has been a source of controversy among botanists for over a century. Some authors treat the genus to include more than forty taxa among of them: Bothmer and Jacobsen (1985), Bothmer (1992), and Kochieva et al. (2001). The taxonomic delimitation of the genus is still not satisfactorily resolved and there is still much disagreement among botanists and cytogeneticists concerning generic concepts.

Several different proposals for the classification of the

Table 5. Grouping of the studied *Hordeum* taxa based on the UPGMA dendrogram retrieved from the statistical analysis of 25 macro-and micro- morphological characters (d.d = dissimilarity distance, G = group).

***Hordeum* taxa**	**Groups at (1.33 d.d)**
H. marinum subsp. *gussoneanum and H. marinum* subsp. *marinum*	G1
H. leporinum	G2
H. glaucum	G3
H. spontaneum and *H. vulgare*	G4

H. murinum complex members have been put forward. On the basis of morphological characters, this study placed *H. leporinum* and *H. glaucum* in different groups (G2 and G3) (Figure 5 and Table 5). This was confirmed by the dissimilarities in some macro- and micro-morphological characters (Table 4). Among of these dissimilar characters are; anther length of central and lateral spikelets isomorphic in *H. leporinum* and heteromorphic in *H. glaucum*; glumes size of lateral spikelets heteromorphic in *H. leporinum* and isomorphic in *H. glaucum* and rachilla surface non ciliated in *H. leporinum* and ciliated in *H. glaucum.*

Our results are in agreements with cytological studies, where, the two species have different chromosomes numbers (2n=28 or 42 in *H. leporinum* and in *H. glaucum* 2n=14). This notable dissimilarity between two species was in agreements with many authors (Bowden, 1962; Bor, 1970; Cocks et al., 1976; Booth and Richards, 1976; Baum and Bailey, 1984a, b, 1989, 1990; Löve, 1984; Bothmer and Jacobsen, 1985; Melderis 1985; Giles and Lefkoviteh, 1986; Jasska, 1992; Bianchi et al., 2002; Mizianty, 2006; El-Rabey and Al-Malki, 2011; Sardou et al., 2011), where they treated these two subspecies as separate species (*H. glaucum* Steud. and *H. leporinum* Link.).

On the other hand, many authors (Bothmer, 1992; Jacobsen and Bothmer, 1992, 1995; Bothmer et al., 1995; Stace, 1997; Boulos, 1995, 2005; Provan et al., 1999; El-Rabey et al., 2002; Amirouche and Misset, 2003; Jahan and Vahidy, 2007, 2011; Eilam et al., 2010; Tanno et al., 2010; El-Rabey and Al-Malki, 2011; Mavi et al., 2011) treated the *H. murinum* as different subspecies (*glaucum* and *leporinum*).

Treatment of *H. marinum* subsp. *gussoneanum* and *H. marinum* subsp. *marinum* in this study in a same group G1 (Table 5 and Figure 5) confirmed by the studied macro- and micro-morphological similarities between them (Table 4). Among of these similar characters are: the isomorphic anther length; lateral spikelets pedicelled and non-ciliated rachilla surface in the two subspecies. This finding similarity was also confirmed by earlier taxonomist (Bothmer and Jacobsen, 1985; Mandaville, 1990; Cope and Hosni, 1991; Bothmer, 1992; Provan et al., 1999; Sahebi et al., 2004; Boulos, 2005; Jahan and Vahidy, 2010; Ourari et al., 2011).

The treatment of *H. spontaneum* and *H. vulgare*, in the same group (G4) (Table 5 and Figure 5) was attributed to the macro- and micro-morphological similarities between these two similar species, where this study treated these two species as two related species. This was attributed to the macro- and micro-morphological similarities between these two similar species (Table 4). Among of the similar characters are: anther isomorphic and yellowish; glumes sizes of lateral spikelets isomorphic and rachilla surface ciliated. This similarity between the two species of *H. vulgare* was confirmed by earlier treatments. This finding was in agreements with many investigators (Bor, 1970; Davis et al., 1985; Feinbrun-Dothan, 1986; Cope and Hosni, 1991; Humphries, 1996; Badr et al., 2000; Boulos, 2005).

Conclusion

The taxonomic revision based on morphological characters in this study revealed that, the position of *H. marinum* subsp. *gussoneanum* and *H. marinum* subsp. *marinum* in one group "G1" confirmed its taxonomic treatment as two subspecies of the same species. While, *H. leporinum* is located at "G3" and the *H. glaucum* is located in "G4", indicate the dissimilarity between these two species. Similarly, the position of *H. vulgare* in "G4" with *H. spontaneum* confirmed their genetic affinity accordingly; *H. spontaneum* is a promising wild crop-relative for future breeding programs for *H. vulgare. H. spontaneum* not traced in Egypt since the latest record in 1988; now it considered as nearly extinct species. The species was collected at that time from Wadi Habis 60 km southern Mersa Martuh city along the Western Desert. Its natural habitats removed during 1990s and substituted by the olive cultivations.

This study recommends that regulations should be developed to protect and conserve the potential species. Wild crop relatives among them must be subjected to national conservation programs. Accordingly, additional molecular researches are needed to fill gaps in our knowledge and to resolve the taxonomic limitations of *Hordeum* species in Egypt.

REFERENCES

Amirouche R, Misset MT (2003). Hordein polymorphism in diploid and tetraploid Mediterranean populations of the *Hordeum murinum* L. complex, Plant. Syst. Evol., 242: 83–99.

Badr A, Mu¨ller K, Scha¨fer-Pregl R, El- Rabey H, Effgen S, Ibrahim HH, Pozzi C, Rohde W, Salamini F (2000). On the origin and domestication history of barley (*Hordeum vulgare*), Mol. Biol. Evol., 17(4): 499–510.

Baum BR, Bailey GL (1984a). Taxonomic studies in wall barley (*Hordeum murinum*) and sea barley (*Hordeum marinum*). 1. Character investigation: assessment of new and traditional characters, Can. J. Bot., 62: 753-762.
Baum BR, Bailey GL (1984b). Taxonomic studies in wall barley (*Hordeum murinum*) and sea barley (*Hordeum marinum*). 2. Multivariate morphometrics, Can. J. Bot., 62: 2754-2764.
Baum BR, Baily LG (1989). Species relationships in the *Hordeum murinum* aggregate viewed from chloroplast DNA restriction fragment patterns, Theor. Appl. Genet., 78: 311-317.
Baum BR, Bailey GL (1990) Key and synopsis of North American *Hordeum* species, Can. J. Bot., 68: 2433–2442.
Bianchi DS, Senften JK, Felber F (2002). Isozyme variation of *Hordeum murinum* in Switzerland and test of hybridization with cultivated barley, Weed Res., 42: 325–333
Booth TA, Richards AJ (1976). Studies in the *Hordeum murinum* L. aggregate. I. Morphology, J. Linn. Soc. Bot.. 72: 149-159.
Booth TA, Richards AJ (1976). Studies in the *Hordeum murinum* L. aggregate: Disc electrophoresis of seed protein, J. Linn. Soc. Bot., 76: 115-125.
Bor NL (1970). *Flora Iranica*. In: Rechinger, KH. (ed.) Akademishe Druck- u. Verlagsantalt Graz-Austria, 70: 232-242.
Bothmer R (1992). The wild species of *Hordeum*: relationships and potential use for improvement of cultivated barley. In. Shewry, P.R. (ed). Barley: genetics, Biochemistry, Molecular Biology and Biotechnology, CAB international, pp.3-17.
Bothmer R, Jacobsen N (1985). Origin, taxonomy, and related species. In. Rasmusson, DC (ed). Barley. American Society of Agronomy, Crop Science Society of America, Soil Science of America, Publishers, Madison, Wisconsin. ASA Monograph, 26: 19-56.
Bothmer R, Jacobsen N, Seberg O (1991). A new combination in *Hordeum* (Triticeae, Poaceae), Nord. J. Bot., 11: 53.
Bothmer R, Jacobsen N, Baden C, Jorgensen RB, Linda-Laursen I (1995). An ecogeographical study of the genus *Hordeum*. International Plant Genetic Resources Institute book, 2nd edition. pp.1-129.
Bothmer R, Jacobsen N, Jørgensen RB, Nicora E (1982). Revision of the *Hordeum pusillum* group, Nord. J. Bot., 2: 307-321.
Boulos L (1995). Flora of Egypt checklist. Al Hadara Publishing Cairo Egypt, 1-283.
Boulos L (2005). Flora of Egypt Vol. IV. Al-Hadara Publishing, Cairo. Egypt, 1-617.
Boulos L (2009). Flora of Egypt checklist. Revised annotated edition. Al Hadara Publishing Cairo Egypt, p.292.
Cocks PS, Boyce KG, Kloot PM (1976). The *Hordeum murinum* Complex in Australia, Aust. J. Bot. 24: 651-662.
Cope TA, Hosni HA (1991). A key to Egyptian grasses. Royal Botanic Gardens, Kew, UK, pp. 1-74.
Davis PH, Mill RR, Ton K (1985). Flora of Turkey and East Aegean Islands. Edinburgh University Press, Great Britain, pp.262-268.
Dunn G, Everitt BS (1982). An introduction to mathematical taxonomy. Cambridge. New York.
Eilam T, Anikster Y, Millet E, Manisterski J, Feldman M (2010). Genome size in diploids, allopolyploids, and autopolyploids of Mediterranean Triticeae. J. Bot., ID 341380, pp.1-12.
El-Hadidi N, Fayed A (1994-95). Materials for excursion flora of Egypt, Cairo University Herbarium. Täeckh, 15: 1-233.
El-Rabey H, Badr A, Scha¨ fer-Pregl R, Martin W, Salamini F (2002). Speciation and species separation in *Hordeum* L. (Poaceae) resolved by discontinuous molecular markers, Plant Biol., 5: 567–575.
El-Rabey H, Al-Malki AL (2011). Application of randomly amplified polymorphic DNA (RAPD) markers and polyphenol oxidases (PPO) genes for distinguishing between the diploid (*glaucum*) and the tetraploid (*leporinum*) accessions in *Hordeum murinum* complex, Afr. J. Biotechnol., 10(61): 13064-13070.
Feinbrun-Dothan N (1986). Flora Palaestina. The Israel Academy of Science and Humanities, 4: 179-183.
Giles BE, Lefkoviteh LP (1986). A Taxonomic investigation of the *Hordeum murinum* Complex *(Poaceae),* Plant Syst. Evol., 153: 181-197.
Humphries CJ (1996). Flora Europaea 5. In Tutin et al. (eds.). Cambridge University Press, pp. 204-205.
Jasska V (1992). Isoenzyme variation in barley *(Hordeum* L.). 2. Aspartate aminotransferase and 6-phosphoglyconate dehydrogenase. Hereditas, 116: 29-35.
Jacobsen N, Bothmer R (1992). Supraspecific groups in the genus *Hordeum*. Hereditas, 116: 21-24
Jacobsen N, Bothmer R (1995). Taxonomy in the *Hordeum murinum* complex (Poaceae), Nord. J. Bot., 15 (5): 449-458.
Jahan B, Vahidy AA (2007). Giemsa N- banding pattern in some wild diploid species of *Hordeum*, Pak. J. Bot., 39(2): 421-429.
Jahan B, Vahidy AA (2010). Giemsa N-banding pattern in tetraploid taxa of *Hordeum marinum* Huds, Pak. J. Bot. 42(2): 833-838.
Jahan B, Vahidy AA (2011). Giemsa N- Banding Pattern in *Hordeum murinum*, Pak. J. Bot., 43(5): 2645-2651.
Kochieva EZ, Goryunova SV, Pomortsev AA (2001). RAPD analysis of the genome in species of the genus *Hordeum,* Russ. J. Genet., 37 (8): 905-910.
Löve A (1984). Conspectus of the *Triticeae*. Feddes Rep. 95: 425-521.
Mandaville JP (1990). Flora of eastern Saudi Arabia. Kegan Paul International London and New York jointly with the National Commission for the Wildlife Conservation and department Riyadh, pp.1-482.
Mavi DO, Doğan M, Cabi E (2011). Comparative leaf anatomy of the genus *Hordeum* L. (Poaceae), Turk. J. Bot., 35: 357-368
Melderis A (1985). *Hordeum* L. Flora of Turkey. In. Davis, P. H. (ed.). Edinburgh University Press. Great Britain, 9: 262-269.
Mizianty M (2006). Variability and structure of natural populations of *Hordeum murinum* L. based on morphology, Plant Syst. Evol., 261: 139-150.
Ourari M, Ainouche A, Coriton O, Huteau V, Brown S, Misset MT, Ainouche M, Amirouche R (2011). Diversity and evolution of the *Hordeum murinum* polyploid complex in Algeria. Genome, 54: 639–654.
Provan J, Russell JR, Booth A, Powell W (1999). Polymorphic chloroplast simple sequences repeat primers for systematic and population studies in the genus *Hordeum*, Mol. Ecol., 8: 505-511.
Rajhathy T, Morrison JW (1962). Cytogenetic studies in the genus *Hordeum*. 6. The *murinum* complex, Canad J. Genet. Cytol., 4: 240-247.
Rohlf JF (1988). NTSYS-pc-numerical taxonomy and multivariate analysis. New York.
Sahebi J, Rahiminejad MR, Ghahreman A (2004). A Taxonomic review of the *Hordeum marinum* Sensu Lato (Poaceae: Triticeae) In Iran, Iranian J. Sci. Technol., 28: (A1) 127-135.
Sardou MA, Baghizadeh A, Tavasoli A, Babaei S (2011). The use of microsatellite markers for genetic diversity assessment of genus *Hordeum* L. in Kerman province (Iran), Afr. J. Biotechnol., 10(9): 1516-1521.
Sneath PH, Sokal RR (1973). Numerical taxonomy. Freeman. San Francisco,pp. 573.
Stace C (1997). New flora of the British Isles 2nd edition. Cambridge University Press, pp 894-897.
Täckholm V (1974). Students' Flora of Egypt (2nd edition). Published by Cairo University. pp. 1-888.
Tanno K, Bothmer R, Yamane K, Takeda K, Komatsuda T (2010). Analysis of DNA sequence polymorphism at the cMWG699 locus reveals phylogenetic relationships and allopolyploidy within *Hordeum murinum* subspecies. Hereditas, 147: 34–42.
Zohary D, Hopf M (1988). Domestication of plants in the old world. Clarendon Press. Oxford, p.56.

10

Diversity and distribution of Mollusca in relation to the physico-chemical profile of Gho-Manhasan stream, Jammu (J & K)

K. K. Sharma, Komal Bangotra* and Minakshi Saini

Department of Zoology, University of Jammu, Jammu-180006, Jammu and Kashmir, India.

The present communication is aimed at assessing the molluscan diversity and distribution under the impact of some physico-chemical variables on them in Gho-Manhasan stream (a distributory of Chenab), Jammu (J & K). The study involved bottom sampling for 12 months (June, 2010 to July, 2011) from the banks of the stream and registered a total of 11 taxa of freshwater molluscs from class Gastropoda and Bivalvia representing 54.453 and 45.546% of the total population of molluscs. Numerically, *Pissidium mitchelli* (Bivalvia) was the most abundant taxa followed by *Melanoides tuberculata* (Gastropoda). *Lymnea luteola, Bellamya bengalensis, Physella acuta, Gyraulus ladacensis* showed less frequent appearance whereas *Lamellidens corrianus* and *Corbicula cashmeriensis* were recorded as rare taxa. Analysis of the data revealed that the physico-chemistry of the stream exercised profound effect on the diversity and distribution of the malacofauna. Significant changes in molluscan assemblages were primarily due to changes in the stream water quality. Coefficient of correlation (r) between molluscs and physico-chemical parameters revealed significant relationship with bivalves.

Key words: Physico-chemical profile, distributory, gastropoda, bivalvia, malacofauna and correlation.

INTRODUCTION

Molluscs are extremely important communities of many ecological communities. They prove immensely beneficial both economically and medicinally (Wosu, 2003). They have been important to humans throughout history as a source of food, jewellery, tools and even pets. Fresh water molluscs play significant role in public and veterinary health (Supian and Ikhwanuddin, 2002). Some fresh water snails are vectors of diseases of humans and livestock, serve as the intermediate hosts for a number of infections such as helminth diseases caused by trematodes (Abd El-Malek, 1958; Dazo et al., 1966; Barbosa and Barbosa, 1994; Brown, 1994; Karimi et al., 2004; Cañete et al., 2004; Kazibwe et al., 2006; Mostafa, 2009). The ecology of these organisms is considered to be affected by environmental factors like physico-chemical parameters (Garg et al., 2009), availability of food, competition, predator-prey interactions (Williams, 1970; Harman, 1972; McMahon et al., 1974; Lassen, 1975; Ofoezie, 1999), substrate architecture (Kershner and Lodge, 1990) and macrophytes (Bronmark, 1985; Costil and Clement, 1996; Ofoezie, 1999).

Although, molluscs are common components of the benthic communities, understanding their role in the aquatic ecosystems and their contribution to biomass production is deficient (Supian and Ikhwanuddin, 2002). Realising the ecological importance of this scantily explored phylum, the present work was undertaken to enlist the molluscan wealth of Gho-Manhanhasan stream in Jammu and to perceive the impact of some physico-chemical variables for a period of one year (July, 2010 to June, 2011). Further, the results obtained can make an effective contribution to holistic water quality assessment and the riverine management studies.

MATERIALS AND METHODS

Study area

Gho-Manhasan stream Figure 2, located at 32.56°N and 74.95°E in

*Corresponding author. E-mail: komal0911@gmail.com.

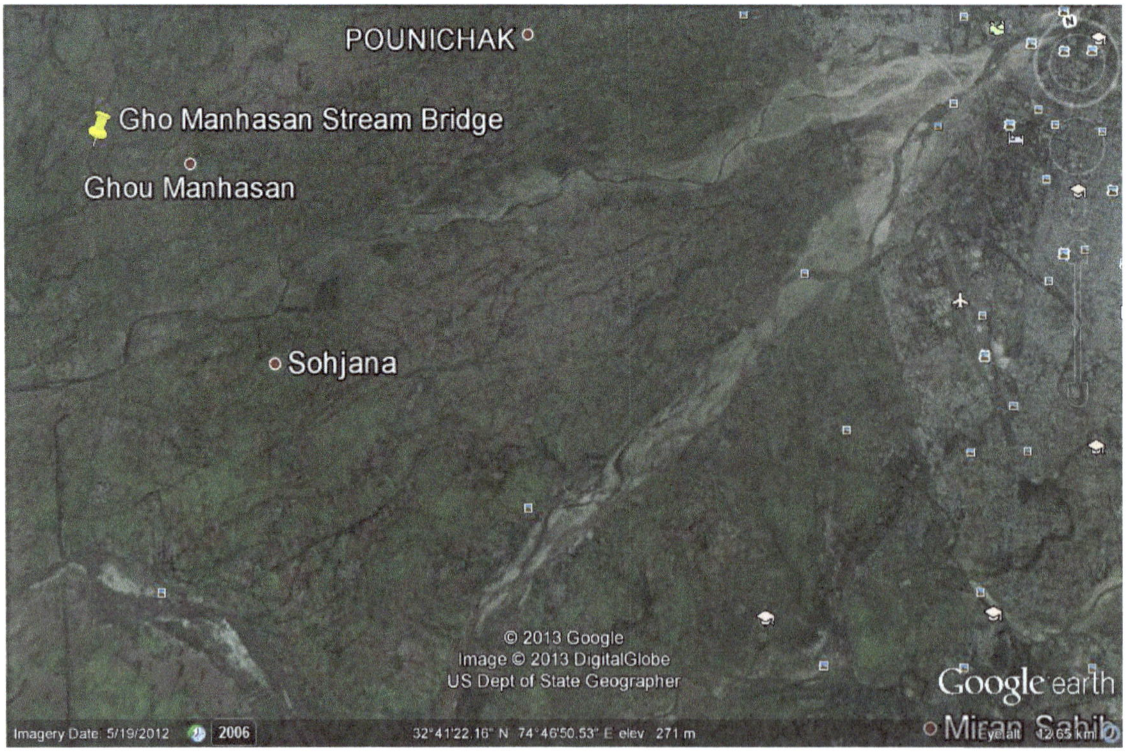

Figure 1. Map showing the study area.

Jammu (J & K), is a distributory of River Chenab, traverses through the Gho-Manhasan area supporting small villages along its sides. The stream has a great diversity of both flora and fauna (Figure 1). The molluscs constitute an important component of the macrobenthic invertebrate community of the stream (Bangotra, 2012). The stream is surrounded by agricultural fields along its entire length which affect the physico-chemistry of the stream (especially during monsoons) hence its inhabitants. Moreover, the stream receives sewage, domestic waste water and also faces anthropogenic pressure such as fishing, washing of clothes and utensils, cattle bathing, construction activities as stated by Bangotra (2012).

Collection of molluscs

Collection of molluscs was done from 4 littoral sites of the stream by collecting bottom samples by quadrant method using Ekman's dredge having an area of 232 cm^2. The samples were put into separate polythene bags and were brought to the laboratory for further studies. The samples were washed in the laboratory and the molluscs were isolated using sieve no. 40 with 256 meshes/cm^2 and preserved with 70% ethyl alcohol. Preserved samples of molluscs were identified with the help of Ward and Whipple (1959), Pennak (1978), Tonapi (1980) and Adoni (1985). The abundance of these organisms was calculated as number per square meter by applying the following formula:

N= O/A.S x 10,000 Welch (1948)

Where, N = no. of macrobenthic organisms/m^2; O = no. of organisms counted; A = area of metallic sampler in square meter; S = no. of samples taken at each stations.

For the analysis of physico-chemical parameters, the water samples were collected from 4 littoral sites of the stream and were brought to the laboratory for further experimental study. pH was measured using portable pH meter (Hanna model) at the study site. Dissolved oxygen, free carbon dioxide, carbonates and bicarbonates, chloride, calcium and magnesium were measured by following procedure recommended by APHA (1985). Physical parameters like water temperature, depth, transparency and velocity were also measured every month during the study period.

Statistical analysis

Statistical analysis of data was done using Margalef's Richness index (d'), Shannon-Weiner index (H'), Simpson's Dominance index (D) and Pielou's equitability index (J).

RESULTS

The data presented in the present manuscript comprises of the average of the actual number of molluscs/m^2 collected monthly from the 4 selected study sites.

A total of 1330.5 molluscs from two classes, viz., Gastropoda (7 species) and Bivalvia (4 species) were collected during the present study contributing 54.453 and 45.546% to the total annual molluscan count (Table 1 and Figure 3). Class Gastropoda included *Bellamya bengalensis* (Lamarck, 1822), *Melanoides tuberculata* (Muller, 1774), *Lymnea accuminata* (Lamarck, 1822), *Lymnea luteola* (Lamarck, 1822), *Physella acuta* (Draparnaud, 1805), *Gyraulus ladacensis* (Nevill, 1878) and *Indoplanorbis exustus* (Deshayes, 1834) whereas class Bivalvia comprised of *Pissidium mitchelli (Prashad)*,

Figure 2. Views of Gho-Manhasan stream showing natural habitat of molluscs.

Sphaerium indicum (Deshayes, 1854), *Corbicula cashmeriensis* (Desh.) and *Lamellidens corrianus* (Lea, 1834). *P. mitchelli* was found to be quantitatively richest taxa followed by *M. tuberculata* which is clearly depicted in Table 1 and Figure 4. The seasonal variations in the population density of molluscs is depicted in Figure 5 which shows that *M. tuberculata*, *L. accuminata*, *P. mitchelli* and *S. indicum* were collected frequently as compared to *B. bengalensis*, *L. luteola*, *P. acuta*, *G. ladacensis* and *I. exustus* and to the rarely recorded *C. cashmeriensis* and *L. corrianus*.

Peak in the population of gastropods was observed during March, 2011 whereas their lowest number was recorded during December, 2010. On the contrary, increase in population of bivalves was observed in winter with their maximum number during December, 2010 and their complete absence was recorded from April, 2011 to June, 2011. An overall decline in their population was also observed during rainy season (July, 2010 to September, 2010). The details of the physico-chemical parameters with range and standard deviation are shown in Table 2. Table 3 shows the correlation coefficient (r) values between physicochemical parameters and the population of gastropods and bivalves revealing significant correlation of bivalves with the stream water quality. This implies that the observed significant change in molluscan assemblages was due to the alterations in the abiotic factors.

DISCUSSION

In this piece of work, predominance of gastropods was registered over bivalves which is in agreement with Sharma et al. (2009) who also recorded similar predominance of gastropods while studying the distribution and ecology of some fresh water molluscs of the Jammu division of J&K state.

In consonance with the findings made in the present work, Subba (2003) reported ten freshwater molluscs from Ghodagodhi Tal Area (Kailali district) which included *B. bengalensis* f. *typica*, *Pila globosa*, *I. exustus*, *Lymnaea accuminata* f. *typica* and *L. accuminata refeseens*. Hussein et al. (2011) also recorded 13 snail species (9 families) from Qena Governorate, Upper Egypt represented by *Bellamya unicolor*, *M. tuberculata*, *Lymnaea natalensis*, *Physa acuta*, *Gyraulus ehrenbergi* and others.

The abundance of the malacofauna was observed to be favoured by the environmental conditions prevailing in the stream including abiotic and biotic components. Garg et al. (2009) also attributed the richness of molluscs observed in Ramsagar dam to the cumulative effect of alkaline nature of water, high calcium contents and the presence of macrophytic vegetation.

The highest gastropod count recorded during March, 2011 from the stream corresponds with Diab (1993) who reported higher snail abundance in Spring and low in Summer in Beheira Province. The presently recorded low count of gastropods agrees with El-Kady et al. (2000) who also recorded lowest number of snails during winter (January and February) in Sinai Peninsula. Supian and Ikhwanuddin (2002) reported that *M. tuberculata* is the commonest and most wide-ranging member of the family Thiaridae, found in almost any kind of freshwater. Peak in the population of *M. tuberculata* recorded during October, 2010 is in agreement with Hussein et al. (2011) who reported that this species showed a maximum cohort of small-sized individuals in October, 2009. According to Pointier et al. (1993), maximum reproduction of this species took place between June and November in France. Flores and Zaffaralla (2012) also cited Thiaridae as the most persistent and abundant macroinvertebrate family. Contreras-Arquieta (1998) reported that members of Thiaridae are quick colonizers, tolerant to habitat diversity and variability due to a very strong and thick shell; many forms are parthenogenetic females capable of multiplication in a short time, viviparous, operculate and have average longevity of five years. Highest count of *P. acuta* recorded in March, 2011 finds support from Hussein et al. (2011) who also reported a March-May peak in their population.

Strzelec and Królczyk (2004) indicated that many gastropod species are tolerant to most physico-chemical parameters and their occurrence is affected by the quality of bottom sediments and abundance of vegetation and reported that the most suitable substrate for snails in rivers is a sandy bottom covered with thin layer of organic silt. Lacoursiere et al. (1975) and Vincent et al. (1982) suggested that gastropod variability may be explained by

Figure 3. Molluscan fauna of Gho-Manhasan stream.

abiotic factors like depth, current and sediments. Jurkiewicz-Karnkowska (2011) opined that low food quality, especially in many semi-permanent habitats, could constrain the development of molluscan communities.

Temperature seems to be an important factor for the distribution of molluscs as gastropods showed a positive correlation (Table 3). Bivalves, however, exhibited a negatively significant correlation with water temperature. Further, temperature plays an important role in abundance of *Lymnaea* sp. (Cañete et al., 2004). Michael (1968), Dutta and Malhotra (1986) and Malhotra et al.

Table 1. Seasonal abundance of molluscan density in Gho-Manhasan stream from July, 2010 to June, 2011.

Name of organism	Jul, 2010	Aug, 2010	Sep, 2010	Oct, 2010	Nov, 2010	Dec, 2010	Jan, 2011	Feb, 2011	Mar, 2011	Apr, 2011	May, 2011	Jun, 2011	Total org/m^2	APC
Phylum – Mollusca														
Class - Gastropoda														
Family – Viviparidae														
Bellamya bengalensis	0	0	0	2.25	0	0	0	11.25	33.75	18	0	0	65.25	4.904
Family – Thiaridae														
Melanoides tuberculata	24.75	31.5	29.25	103.5	18	29.25	29.25	11.25	18	6.75	0	0	301.5	22.660
Family – Lymnidae														
Lymnea accuminata	13.5	2.25	0	9	24.75	0	38.25	27	0	4.5	20.25	2.25	141.75	10.653
L: luteola	0	0	2.25	4.5	0	0	0	0	4.5	0	0	0	11.25	0.845
Family – Physidae														
Physella acuta	2.25	0	0	0	0	0	0	4.5	90	0	0	0	96.75	7.271
Family – Planorbidae														
Gyraulus ladacensis	0	0	0	4.5	0	0	0	2.25	6.75	15.75	0	27	56.25	4.227
Indoplanorbis exustus	0	15.75	0	0	0	0	0	0	0	0	13.5	24.75	54	4.058
Total Gastropoda	40.5	49.5	31.5	123.75	40.5	29.25	67.5	56.25	153	45	33.75	54	724.5	54.453
Class – Bivalvia														
Family – Sphaeriidae														
Pissidium mitchelli	11.25	6.75	15.75	15.75	69.75	108.75	108	128.25	27	0	0	0	491.25	36.922
Sphaerium indicum	2.25	0	2.25	11.25	18	36	20.25	13.5	0	0	0	0	103.5	7.779
Family – Corbiculidae														
Corbicula cashmeriensis	0	0	0	4.5	0	0	0	0	0	0	0	0	4.5	0.338
Family – Unionidae														
Lamellidens corrianus	0	0	0	0	6.75	0	0	0	0	0	0	0	6.75	0.507
Total Bivalvia	13.5	6.75	18	31.5	94.5	144.75	128.25	141.75	27	0	0	0	606	45.546
Total Mollusca	54	56.25	49.5	155.25	135	174	195.75	198	180	45	33.75	54	1330.5	

APC = Annual percent contribution.

(1996) also recorded a positive correlation between molluscs and temperature, while a negative correlation between temperature and molluscs was noticed by Ricker (1952), Shrivastava (1956) and Vasisht and Bhandal (1979). This study revealed that molluscs were negatively correlated with depth of the stream. A similar negative correlation between water depth and some molluscs (*Cleopatra bulimoides*, *Theodoxus niloticus*, *Succinea cleopatra* and *Bulinus truncatus*) was reported by Hussein et al. (2011).

The results obtained in this survey revealed a positive correlation of malacofauna with transparency (Table 3). To our knowledge, the mussels are well known for their filtering ability which enhances water clarity of the aquatic media. Mcivor (2004) also discussed the potential

Table 2. Mean ± SD of the physico–chemical parameters recorded from the Gho-Manhasan stream.

Parameters	Minima	Maxima	Mean ± SD
Air temperature (°C)	11.0	39.0	14.75 – 0.21 ± 36.1 – 3.03
Water temperature (°C)	9.0	28.0	9.0 – 0 ± 28.37 – 3.20
Depth (cm)	8.8	48.0	10.75 – 1.51 ± 32.65 – 10.06
Transparency (cm)	2.4	48.0	5.5 – 0.19 ± 32.50 – 10.06
Speed (m/s)	0.11	0.41	0.16 –0.01 ± 0.33 – 0.05
Dissolved oxygen (mg/l)	4.4	22.0	5.5 - 0.48 ± 20.5 – 2.63
pH	6.8	8.1	6.87 – 0.07 ± 7.57 – 0.41
Free Carbon dioxide (mg/l)	0	14.0	2.36 – 0.92 ± 10.0 – 2.83
Carbonate (mg/l)	0	24.0	13.0 – 0 ± 24.0 – 3.0
Bicarbonate (mg/l)	189.1	884.5	337.02 – 12.41 ± 756.4 – 143.8
Calcium (mg/l)	8.82	64.96	11.02 – 0.56 ± 60.13 – 11.47
Magnesium (mg/l)	17.98	69.98	21.86 – 1.04 ± 66.21 – 6.43
Chloride (mg/l)	15.96	79.84	16.46 – 1.41 ± 73.84 – 8.64

Table 3. Correlation coefficient (r) between molluscs of Gho-Manhasan stream with some physico-chemical parameters.

Parameters	Gastropoda	Bivalvia
Water temperature	0.11794	-0.81949*
Depth	-0.02545	-0.32164
Transparency	0.263124	0.37901
Velocity	-0.20228	-0.55863*
Dissolved oxygen	-0.07476	-0.202
pH	0.18796	0.76517*
FCO_2	-0.17703	-0.30251
CO_3^{2-}	-0.16176	0.69611*
HCO_3^-	0.40366	0.35283
Ca^{2+}	-0.02713	0.31325
Mg^{2+}	0.11542	0.48197
Cl^-	0.1272	-0.13325

*Marked correlation was significant ($p < 0.05$).

role of fresh water mussels as "living biofilters" due to their key participation in removal of suspended particulate matter in the river ecosystems. As per his findings, the filtration rates of mussel populations in four British rivers indicated that mussel filtering removed between 7 and 30% of the particulate matter in a parcel of water travelling 10 km downstream.

The stream flow exerts greater influence on the distribution of these macrobenthic invertebrates which was concluded from this work as both gastropods and bivalves were negatively correlated with the stream velocity (Table 3). Appleton (1978) also described water flow in lotic environments as the most important factor for explaining the longitudinal distribution of the snails that are the intermediate hosts of *Schistosoma* spp. In a survey conducted by Giovani et al. (2005), the water flow and rainfall were designated as variables responsible for the distribution of population of *Physa marmorata* and *M. tuberculata*.

The pH recorded in the stream ranged from slightly acidic to highly alkaline and furnished a weakly positive relationship with gastropods, whereas a strongly significant correlation with bivalves as depicted in Table 3. However, a very weak and insignificant negative correlation between molluscs and pH recorded by Garg et al. (2009) suggests that molluscs were found to be independent of fluctuations with respect to pH value. The findings of Martins-Silva and Barros (2001) revealed that acidic pH is unfavorable to the occurance of molluscs.

Positive correlation of bicarbonate alkalinity with both the classes of molluscs and a significant correlation between carbonates and bivalves indicate that the alkaline

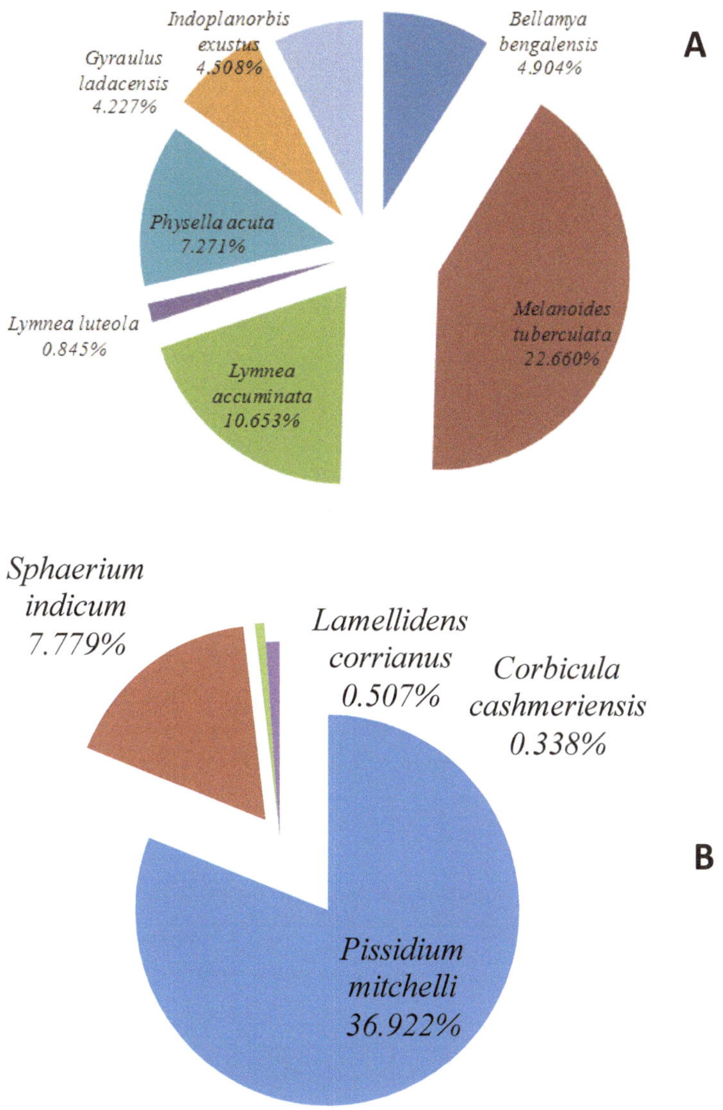

Figure 4. Pie chart showing annual percent contribution of different gastropods and bivalves to the total molluscan population.

conditions promoted abundant molluscan populations (Table 3). Sakhre and Joshi (2003) also supported this point of view by observing the greater molluscan population in alkaline lakes as compared to acidic-lakes.

The molluscs were found to exhibit a negative correlation with FCO_2 as shown in Table 3. Similar relationship between molluscs and FCO_2 was recorded by Garg et al. (2009) for the molluscs of Ramsagar Reservior (M.P).

Both gastropods and bivalves had a negative insignificant relationship with DO (Table 3) which finds support from Garg et al. (2009) report. Cheatum (1934) and Sharma (1986) documented that some molluscs can survive in very low oxygen conditions and have noted an inverse relationship between the two.

Gastropods were observed to show a very weak negative correlation with calcium as shown in Table 3 which gets support from the findings of Hussein et al. (2011) in which calcium concentration was negatively correlated with the abundance of *B. unicolor, Succinea cleopatra, P. acuta* and *Biomphalaria alexandrina* (decreased with increasing Calcium). Georgiev and Stoycheva (2010), however, recorded a total of 28 snail species on the Samothraki Island (North-Eastern Greece) and reported their abundance in humid environment, being particularly accumulated at places where artificial sources of calcium were available. Bivalves on the contrary, were found to show a moderately positive correlation with calcium. Several authors showed that calcium concentration is an important factor in controlling the abundance and distribution of molluscs in freshwater (Boycott, 1936; Macan, 1950; Mckillop and Harrison, 1972; Dussart, 1976, 1979; Supian and Ikhwanuddin, 2002). Aquatic molluscs are known to accomplish their

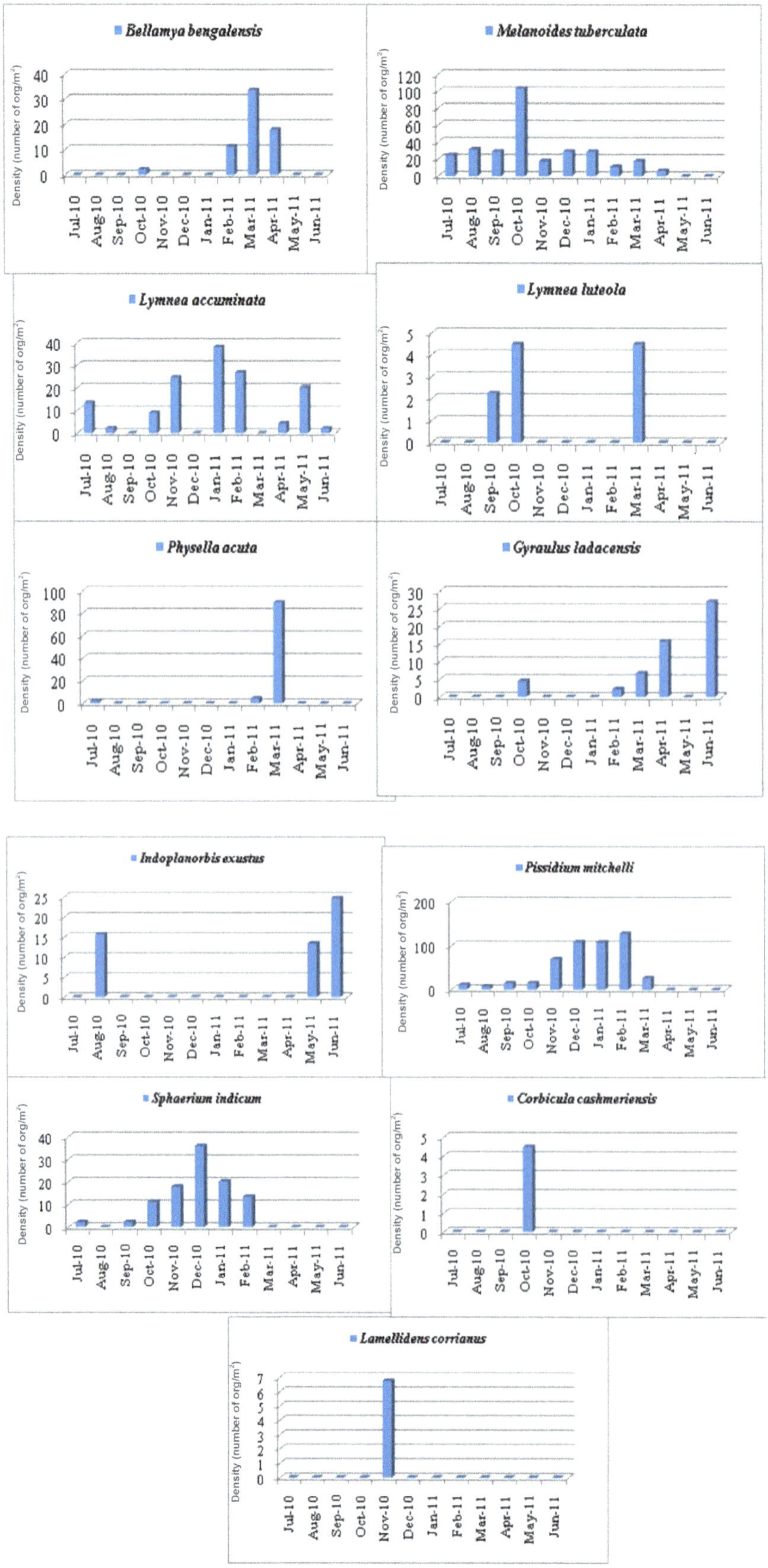

Figure 5. Seasonal variations in the density of each molluscan species (number of org./m^2) collected from Gho-Manhasan stream.

calcium requirements by absorption of calcium from the external medium (Jodrey, 1953; Kado, 1960; van der Borght, 1963) although calcium present in the food is also utilized (van der Borght and van Puymbroeck, 1966).

The correlation values presented in Table 3 reveal that chlorides and bivalves was negative. However, a positive correlation of chlorides with the gastropods suggests that the chloride content of the water body may favour their survival. The sources of chloride were the effluent and sewage discharge by residents of adjoining areas. Feema (1981) also reported high chloride content due to the discharge of contaminants like urea rich sewage. Other sources of chloride included surface run-off during monsoons and defaecation activities (human and cattle). Such conditions were considered ideal for the rapid multiplication of gastropod species. The present view is supported by Ndifon and Ukoli (1989) who also recorded frequent presence of thairid, *M. tuberculata* in environment receiving domestic sewage and animal as well as human excrements. Giovanelli et al. (2005) also associated abundance of *M. tuberculata* with faecal coliforms and high chloride concentration in the studied lotic systems.

Conclusion

The freshwater molluscs play a massive role in nature and help in assessment of ecological status of the water bodies. Being herbivores, they form the lower strata of aquatic trophic linkages and perform many other ecological activities. Hence, studies pertaining to their diversity, distribution and ecology become imperative. The results of the present study indicated that the diversity and distribution of the malacofauna of Gho-Manhasan stream, especially, a thiarid, *M. tuberculata* and bivalves intimately correlated with the physico-chemical regime of the stream. These species can be considered as bioindicators of pollution as they were found to respond prominently to nutrient inputs, discharge of sewage and excreta produced by animals and humans. A progressive increase in their number with increasing pollution load indicates that they possess great tolerance against the contaminants present in water and flourish well in their presence. Findings of the present work shall be utilized by future researchers and ecologists as supplementary information in public and veterinary health sciences, ecotoxicology, water quality assessment and river management studies.

REFERENCES

Abd El-Malek E (1958). Factors conditioning the habitat of Bilharziasis intermediate hosts of the family Planorbidae. Bull. Org. mond. Santé. Bull. World Health. Organ. 18:785-818.

Adoni AD (1985). Workbook on limnology. 216 pp.

APHA (1985). Standard methods for examination of water and wastewater. American Public Health Association, American Water Works Association and Water Pollution Control Federation, Washington, D. C.

Appleton CC (1978). Review of literature on abiotic factors influencing the distribution and life cycles of bilharziasis intermediate host snails. Malacol. Rev. 11:1-25.

Bangotra K (2012). Diversity of macrobenthic invertebrates associated with macrophytes in lotic and lentic water bodies of Jammu. M.Phil. Dissertation submitted to the University of Jammu, Jammu.

Barbosa FS, Barbosa CS (1994). The bioecology of snail vector for Schistosomiasis in Brazil. Cad. Saùde Pùbl., Rio de Janeiro, 10(2):200-209.

Bellamya bengalensis Lamarck (1882). In: The IUCN Red List of Threatened Species, Version 2012.2. <www.iucnredlist.org>. Downloaded on 29 March 2013.

Boycott AE (1936). The habitat of freshwater molluscs in Britain. J. Anim. Ecol. 5:116-186.

Bronmark C (1985). Freshwater snail diversity: effects of pond area, habitat heterogeneity and isolation. Oecologia 67:127-131.

Brown DS (1994). Freshwater Snails of Africa and their Medical Importance (2nd edn.). Taylor & Francis, London, pp. 609.

Cañete R, Yong M, Sánchez J, Wong L, Gutiérrez A (2004). Population dynamics of intermediate snail hosts of *Fasciola hepatica* and some environmental factors in San Juan Martinez Municipality, Cuba. Mem. Inst. Oswaldo Cruz, Rio de Janeiro, 99(3):257-262.

Cheatum EP (1934). Limnological investigation on respiration, annual migratory cycle and other related phenomena in freshwater pulmonate snails. Tans. Am. Microscope Soc. 53:348.

Contreras-Arquieta A (1998). New records of snail Melanoides tuberculata (Muller, 1774) (Gastropoda: Thiaridae) in the Cuatro Cienegas Basin, and its distribution in the state of Coahuila, Mexico. Southwest. Nat. 43(2):283-286.

Costil K, Clement B (1996). Relationship between freshwater gastropods and plant communities reflecting various trophic levels. Hydrobiologia 321:7-16.

Dazo BC, Hairston NG, Dawood IK (1966). The ecology of *Bulinus truncatus* and *Biomphalaria alexandrina* and its implications for the control of Bilharziasis in the Egypt- 49 Project Area. Bull. Org. mond. Santé Bull. World Health Organ. 35:339-356.

Diab MRM (1993). Biological studies on Trematode larvae and freshwater snails. M. Sc. Thesis, Vet. Med. Fac. Alexandria Univ. pp. 155.

Dussart GBJ (1976). The ecology of freshwater molluscs in North West England in relation to water chemistry. J. Molluscan Stud. 42:181-198.

Dussart GBJ (1979). Life cycles and distribution of the aquatic gastropod mollusc *Bithynia tentaculata* (L.) *Gyraulus albus* (Müller), *Planorbis planorbis* (L.) and *Lymnaea peregra* (Müller) in relation to water chemistry. Hydrobiology 67:233-239.

Dutta SPS, Malhotra YR (1986). Seasonal variations in the macrobenthic fauna of Gadigarh stream (Miran Sahib), Jammu. Indian J. Ecol. 13:138-145.

El-Kady GA, Shoukry A, Reda LA, El-Badri YS (2000). Survey and population dynamics of freshwater snails in newly settled areas of the Sinai Peninsula. Egyptian J. Biol. 2:42-48.

Feema (1981). Métodos de Análises Físicoquímicas da Água, Cadernos Feema, Série Didática 2(2):14-79.

Flores MJL, Zaffaralla MT (2012). Macroinvertebrate composition, diversity and richness in relation to the water quality status of Mananga River, Cebu, Philippines. Philippine Sci. Lett. 5(2):103-113.

Garg RK, Rao RJ, Saksena DN (2009). Correlation of molluscan diversity with physicochemical characteristics of water of Ramsagar reservoir. India. Int. J. Biodivers. Conserv. 6:202-207.

Georgiev D, Stoycheva S (2010). Notes on the ecology and species diversity of the inland molluscs of Samothrakhi Island (North-Eastern Greece). North-Western J. Zool. 1:71-78.

Giovanelli A, da Silva CLPAC, Leal GBC, Baptista DF (2005). Habitat preference of freshwater snails in relation to environmental factors and the presence of the competitor snail *Melanoides tuberculatus* (Müller, 1774). Mem Inst Oswaldo Cruz, Rio de Janeiro, 100(2):169-176.

Gyraulus ladacensis Nevill (1878). In: WMSDB: Worldwide Mollusc Species Data Base.

http://www.bagniliggia.it/WMSD/HtmFamily/PLANORBIDAE12.htm
Harman WN (1972). Benthic substrates: their effect on water Mollusca. Ecology 53:271.
Hussein MA, Obuid-Allah AH, Mahmoud AA, Fangary HM (2011). Population dynamics of freshwater snails (Mollusca: Gastropoda) at Qena Governorate, Upper Egypt. Egypt. Acad. J. Biol. Sci. 3(1):11-22.
Indoplanorbis exustus Deshayes (1834). Accessed through: World Register of Marine Species at http://www.marinespecies.org /aphia.php?p=taxdetails&id=716351. Downloaded on 29 March 2013.
Jodrey LH (1953). Studies on shell formation III. Measurement of calcium deposition in shell and calcium turnover in mantle tissue using the mantle-shell preparation and "Ca. Biol. Bull. mar. Biol. Lab. Woods Hole 104:398-407.
Jurkiewicz-Karnkowska E (2011). Effect of habitat conditions on the diversity and abundance of molluscs in floodplain water bodies of different permanence of flooding. Pol. J. Ecol. 59(1):165–178.
Kado Y (1960). Studies on shell formation in molluscs. J. Sci. Hiroshima Univ. Ser. Bi. 19:163-210.
Karimi GR, Derakhshanfar M, Paykari H (2004). Population density, trematodal infection and ecology of *Lymnaea* snails in Shadegan, Iran. Arch. Razi Inc. 58:125-129.
Kazibwe F, Makanga B, Rubaire-Akiiki C, Ouma J, Kariuki C, Kabatereine NB, Booth M, Vennervald BJ, Sturrock RF, Stothard JR (2006). Ecology of *Biomphalaria* (Gastropoda: Planorbidae) in Lake Albert, Western Uganda: snail distribution, infection with schistosomes and temporal associations with environmental dynamics. Hydrobiologia 568:433-444.
Kershner MW, Lodge DM (1990). Effect of substrate architecture on gastropod-substrate associations. J. N. M. Benthol. Soc. 9(4):319-326.
Lacoursiere E, Vaillancourt G, Couture R (1975). Relation entre les plantes aquatiques et les Gastéropodes (Mollusca: Gastropoda) dans la région de la centrale nucléaire de Gentilly I (Quèbec). Can. J. Zool. 53:1868-1874.
Lamellidens corrianus Lea (1834). In: IUCN Red List of Threatened Species. Version 2012.2. <www.iucnredlist.org>. Downloaded on 29 March 2013.
Lassen HH (1975). The diversity of freshwater snails in view of the equilibrium theory of island biogeography. Oecologia 19:1-8.
Lymnea acuminata Lamarck (1822). In: IUCN 2012. IUCN Red List of Threatened Species. Version 2012.2. <www.iucnredlist.org>. Downloaded on 29 March 2013.
Lymnea luteola (Lamarck, 1822) In: IUCN 2012. IUCN Red List of Threatened Species. Version 2012.2. <www.iucnredlist.org>. Downloaded on 29 March 2013.
Macan TT (1950). Ecology of freshwater in the English Lake District. J. Anim. Ecol. 19:124-146.
Malhotra YR, Sharma KK, Thakial MR (1996). Ecology of macroinvertebrates from a fish pond. Proc. Natl. Acad. Sci. India 66:53-59.
Martins-Siva MJ, Barros M (2001). Occurrence and Distribution of Fresh-Water Molluscs in the Riacho Fundo Creek Basin, Brasilia, Brazil. Rev. Biol. Trop. 49:3-4.
Mcivor AL (2004). Freshwater mussels as biofilters. Dissertation submitted to the University of Cambridge for the degree of Doctor of Philosophy, Pembroke College.
Mckillop WB, Harrison AD (1972). Distribution of aquatic gastropods across an interface between the Canadian Shield and limestone formation. Can. J. Zool. 50:1433-1445.
McMahon RF, Hunter RD, Russell-Hunter WD (1974). Variation in aufwuchs at six freshwater habitats in terms of carbon biomass and carbon: nitrogen ratio. Hydrobiologia 145:391-404.
Melanoides tuberculata Muller (1774). In: Red-rimmed melania - Wikipedia, the free encyclopedia http://en.wikipedia.org/wiki/Red-rimmed_melania.
Michael RG (1968). Studies on bottom fauna in a tropical freshwater pond. Hydrobiology 31:203-230.
Mostafa OMS (2009). Effect of salinity and drought on the survival of *Biomphalaria arabica*, the intermediate host of *Schistosoma mansoni* in Saudi Arabia. Egyptian Acad. J. Biol. Sci. 1(1):1-6.
Ndifon GT, Ukoli FMA (1989). Ecology of freshwater snails in south-western Nigeria. I: Distribution and habitat preferences. Hydrobiology 171:231-253.
Ofoezie IE (1999). Distribution of freshwater snails in the man-made Oyan Reservoir, Ogun State, Nigeria. Hydrobiology 416:181-191.
Pennak RW (1978). Fresh water invertebrates of United States. *Physella acuta* Draparnaud (1805). In: Wikipedia, the free encyclopedia http://en.wikipedia.org/wiki/Physella_acuta.
Pointier JP, Théron A, Borel G (1993). Ecology of the introduced snail *Melanoides tuberculata* (Gastropoda: Thiaridae) in relation to *Biomphalaria glabrata* in the marshy forest zone of Guadeloupe, French West Indies. J. Mol. Stud. 59:421-428.
Ricker WE (1952). The benthos of Cultus lake. J. Fish. Res. Bd. Canada, 9:204-212.
Sakhre VB, Joshi PK (2003). A Studied physicochemical limnology of Papnas: a minor wetland in Tuljapur town, Maharashtra. J. Aquat. Biol. 18: 93-95.
Sharma RC (1986). Effect of physico-chemical factors on benthic fauna of Bhagirathi river, Garhwal Himalayas. Ind. J. Ecol., 13: 133-137.
Sharma KK, Sharma SP, Sawhney N (2009). Distribution and ecology of some fresh water molluscs of the jammu division of j&k state. J. Environ. Bio-sci. 23 (2):179-181.
Shrivastava VK (1956). Studies on the freshwater bottom fauna of North India. Quantitative fluctuations and qualitative composition of benthic fauna in lake in Lucknow. Proc. Nat. Acad. Sci. 25: 406-416.
Sphaerium indicum Deshayes (1854). In: IUCN 2012. IUCN Red List of Threatened Species. Version 2012.2. <www.iucnredlist.org>. Downloaded on 29 March 2013
Strzelec M, Królczyk A (2004). Factors affecting snail (Gastropoda) community structure in the upper course of the Warta river (Poland). Biologia, Bratislava 59(2):159-163.
Subba BR (2003). Molluscan checklist of Ghodaghodi Tal area, Kailali District. Our Nature, 1:1-2.
Supian Z, Ikhwanuddin AM (2002). Population dynamics of freshwater molluscs (Gastropod: *Melanoides tuberculata*) in Crocker Range Park, Sabah. ASEAN Rev. Biodiv. Environ. Conserv. (ARBEC), pp. 1-9.
Tonapi GT (1980). Freshwater animals of India. An Ecological Approach. Oxford and IBH publishing co., New Delhi, Bombay, Calcutta, 341 pp.
Van der BO, van Puymbroeck S (1966). Active transport of alkaline earth ions as physiological base of the accumulation of some radio-nuclides in freshwater molluscs. Nature, Land. 204:533-535
Van der BO (1963). In- and out-fluxes of calcium ions in freshwater gastropods. Archs int. Physiol. Biochem. 71: 46-50.
Vasisht HS, Bhandal RS (1979). Seasonal variations of benthic fauna of some North Indian lakes and ponds. Ind. J. Ecol. 6:77-83.
Vincent B, Lafontaine N, Caron P (1982). Facteurs influencant la structure des groupements de macroinvertébrés benthiques et phytophiles dans la zone littorale du Saint-Laurent (Québec). Hydrobiology 97:63-73.
Ward HB, Whipple GC (1959). Freshwater Biology. John Wiley and Sons, Inc. New York, 2:1248.
Williams NV (1970). Studies on aquatic pulmonate snails in Central Africa. I: Field distribution in relation to water chemistry. Malacologia 10:153-164.
Wosu LO (2003). Commercial snail farming in West-Africa - A guide. Ap Express Publishers, NSukka-Nigeria.

Status of the Swayne's Hartebeest, (*Alcelaphus buselaphus swaynei*) meta-population under land cover changes in Ethiopian Protected Areas

Yosef Mamo[1], Girma Mengesha[2]*, Aramede Fetene[3], Kefyalew Shale[2] and Mezemir Girma[4]

[1]Hawassa University, P.O.box 5, Hawassa, Ethiopia.
[2]College of Forestry and Natural Resources, Hawassa University, Wondo Genet, P.O. Box 128, Ethiopia.
[3]Debere Markos University, P. O. box 269, Debre Markos, Ethiopia.
[4]Mezemir Girma, Southern Nation Nationalities People Region, P.O. Box 69, Hawassa, Ethiopia.

This study aims to understand status and population structures of Swaynes' Hartebeest (SHB) (*Alcelaphus buselaphus swaynei)* meta-population under land cover changes in Maze National Park (MaZNP), Nech Sar National Park (NNP) and Senkele Swayne's Hartebeest Sanctuary (SHBS) from 2008 to 2009. A total, with direct count method based on silent detection of vehicles along roads was used to count SHB in the Protected Areas (PAs) within 5 blocks of the entire SHBS. In each of the lager MaZNP and NNP, 6 blocks were randomly sampled following habitat types such as grasslands. One wildlife expert, six scouts and one researcher were assigned to each of the block for counting the Hartebeest during early morning and late in the afternoon. A SPSS, Excel software and Landsat satellite imagery of the PAs was used for the SHB populations and land cover data analysis. Of the 840 SHB individuals recorded in these PAs, 364 occurred in MaZNP, 464 in SHBS and 12 individuals in NNP. The adult male SHB was 47% (MaZNP), 39% (SHBS) and 42% in NNP. The relationship between the adult males and adult females is highly significant for the MaZNP (2t=0.969, P<0.01). The study revealed that SHB population size is increasing in MaZNP and SHBS. However, there is dramatic decrease in the NNP. Since 1970s, the species population size has fluctuated from 865 to 480 to 840 in the PAs. This might be associated to decreases in grasslands such as from 75 to 48% in the MaZNP and 37 to 34% in the NNP during 1986 to 2005. Conservation measures that increase the population size of this endangered species is urgently needed to conserve the endemic species in Ethiopia.

Key words: Population size, Protected Areas, Status, Swayne's Hartebeest.

INTRODUCTION

Though most antelope species still exist in large number in the Sub-Saharan Africa, three quarter of the hartebeests are declining (Eastes, 1999). The hartebeests are large antelopes grouped in the Bovidae family. Of the nine sub-species recognized, two are extinct and the remaining seven are confined to dramatically contracted habitats (Lewis and Wilson, 1979). Previously, they occupied in a wider areas of Morocco up to north eastern Tanzania to south of Congo and from Southern Angola to South Africa. However, habitat destruction has drastically reduced their range. Therefore, at present, hartebeest occurs only in parts of Botswana, Namibia, Ethiopia, Tanzania and Kenya (Refera, 2005).

The term meta-population was first defined by Levins (1969) as sub-populations that exist in discrete habitat

*Corresponding author. E-mail: gutgirma@gmail.com.

patches and the sub-populations may turnover with extinctions and re-colonization from other patches. Today, it has been used to define any population with spatial subdivision. In this paper, it is used to define spatial subdivision of Swaynei's Hartebeest (*Alcelaphus buselaphus swaynei)* in Ethiopia. Ethiopia's rugged topography and the varied climatic conditions have endowed it with spectacular wildlife species of scenery in Africa. Therefore, the country is responsible to protect these wildlife species that include endemic, endangered, vulnerable and rare species (Blower, 1969). Ethiopia harbors three sub-species of hartebeests (*Alcelaphus buselaphus)*: *A. b. lelwel*, *A. b. tora* and the endemic *A. b. swaynei* all of which are classified as endangered (Nobuko, 2004). The Swaynei's hartebeest (SHB), *A. b swaynei* is long-faced, having chocolate color of the three sub-species occurring in the country. SHB is normally seen in herds and characterized by a steeply sloping back. The hartebeest is a large ungulate ranging from 1.5 m to 2.45m in length and mass of 75 to 200 kg and both sex bears horns (Kingdon, 1989). Sexual maturity may occur as early as 12 month, but they attain maximum weight until 4 years (Kingdon, 1989). In the last few decades, at the beginning of the nineteenth century, SHB had been reported to be abundantly distributed from Somalia up to eastern part of the Rift Valley Lakes regions of Ethiopia. But currently, the number of SHB and its range has been drastically reduced in Ethiopia and believed to be exterminated outside its land. However, their population size has decreased and now endangered (Nobuko, 2004). Duckworth et al. (1992) described that the population size of SHB (*A. b. swaynei)* in Ethiopia was declining and it was a highly threatened endemic sub-species restricted to Ethiopia. The factors described for declining of the SHB population size includes human induced habitat destruction and fragmentations. Once, the total population of this hartebeest in Ethiopia was estimated to be 600 to 700 individuals (Bolton, 1973). Of these, large number of SHB populations existed in the Senkele Hartebeest Sanctuary (SHBS) during 1976 to 1988, ranging from 448 to 2379 individuals (Nobuko, 2004). With the objective to ensure the survival of the species, 90 individuals were transferred to Nech Sar National Park (NNP) and 120 to Awash National Park in 1974 from SHBS (Lealem, 1974). According to Duckworth et al. (1993), only 40 individuals of the species survived in NNP until 1992. Refera (2005) reported the occurrence of 242 to 277, less than 20 and 200 individuals of SHB in Maze National Park (MaZNP), NNP and SHBS, respectively. However, in 2008, the total number of individuals of SHB counted in SHBS was 283 for the wet season and 351 during dry season. In most part of the SHBS they used to occur, SHB showed a declining trend (Kumsa and Bekele, 2008). According to Vymyslická et al. (2010), 35 individuals of SHB existed in the NNP, where 23 individuals were in a reproductive herd (17 females and 6 males), three individuals in a temporary male bachelor group, and nine solitary males. Similarly, recent studies conducted on a five large antelope including SHB by Fetene et al. (2011) indicated existence of the 35 individuals in the NNP. Generally, studies have shown that the numbers of individuals of the SHB in the Ethiopian PAs were declining and continued to be rare. They were locally extinct in some of the PAs like in Awash National Park and at verge of extinction in the NNP. Consequently, the SHB is listed as an imminent danger of extinction and is completely protected by the law in Ethiopia (IUCN, 2002). At present, remnants of the endangered and endemic, SHB in Ethiopia are found in four separate localities, in SHBS, MaZNP, NNP and Alidige Wildlife Reserve (AWR). The major factors for its further decline, which hinders its conservation effort and increased threat, include high human population growth, intensive agriculture and livestock grazing that altered vegetation communities of the PAs. The SHB habitats have been surrounded by high human settlers and associated Livestock populations particularly with cattle. Competitions with the cattle for resource, mainly with grass, have increased in all the PAs that in turn deteriorated the grass and could increase in shrubby and other vegetation communities. In the MaZNP, the present population was estimated to 250 individuals but reliable data on trends, and status were not available. In this Park, the human population pressure was relatively less stressing though the number of livestock population was increasing. The human population pressure was more stressing in the SHBS as it was surrounded by human settler and high livestock population (Kumsa and Bekele, 2008). In the NNP, in addition to overgrazing with high livestock populations in one of the PAs, there was occasional poaching (Refera et al., 2003; Fetene et al., 2011). Thus, the population of SHB was not still doing well and categorized as globally-endangered species (IUCN, 2002). The factors such as intensive agriculture and livestock grazing within and around the PAs could alter land cover changes. We thus hypothesized that these factors may affect the land cover changes and could be attributed to the further decline in the population size, alteration in age and sex structures of the SHB in the PAs. The pressure on this species is still increasing due to a high demand to use the relatively intact vegetation remains within and around the PAs. Thus, understanding of the current status, population size, and age and sex structures of the species under changing land cover in the PAs is vital to prioritize and strengthen its conservation endeavors. Furthermore, it may be very difficult to prioritize conservation program for the species without well-documented scientific investigation of the current status, the dynamic population size, age and sex composition as related to land cover change. Therefore, the aim of this study is to investigate the current status and population structure of SHB in three Ethiopian PAs, namely MaZNP, SHBS and NNP in relation to land cover changes in and around the PAs.

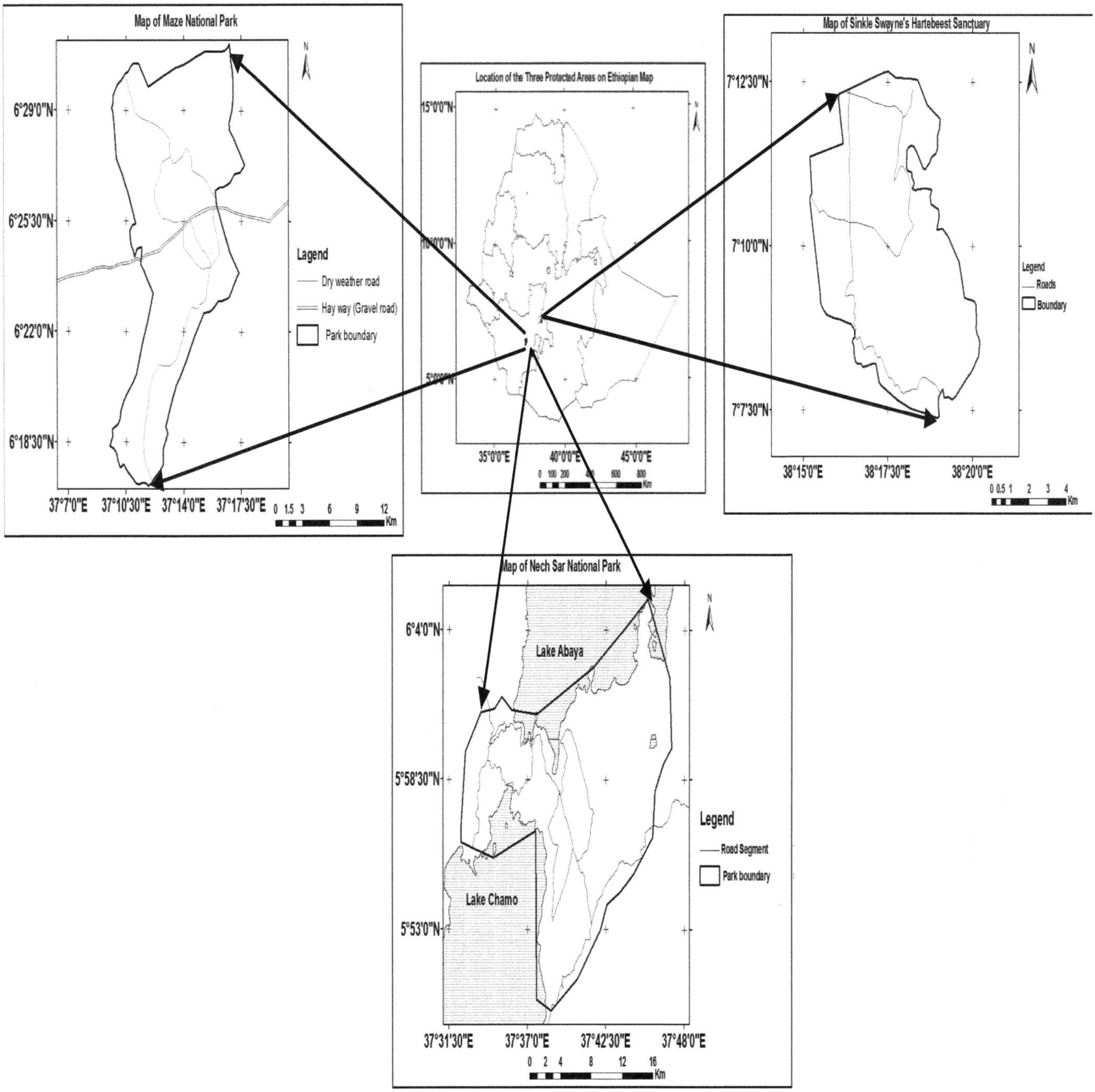

Figure 1. Map showing the study areas.

MATERIALS AND METHODS

Description of the study area

This study was conducted in MaZNP, SHBS and NNP (Figure 1). The MaZNP is found in Gamo Gofa Zone of Southern Nation Nationalities and People Regional States (SNNPRS) at 470 km from Addis-Ababa. MaZNP used to be a controlled hunting area and currently upgraded to a National Park. The total area of the Park is 202 km². Geographically, it is located at 6°17'31'' to6°31'30'' N latitude and 37°09'00" to 37°18'00" E longitudes with an altitudinal ranges from 1000 to 1200 m a.s.l (Figure 1). The socio-economic condition of the people is mainly agro-pastoralist. There is no meteorological station for the researchers to determine the climatic condition of the area. However, it is reported that, the rainfall pattern around Morka (the small town near the Park) is bi-modal type. There is a short rainy season from March to April while the main rainy season is from June to September. With regard to temperature condition, hot months are reported from November to February and cold months from June to September (Refera, 2005). The minimum temperature at Morka is 15.3°C and the maximum is 35.5°C (Refera et al., 2003). The majority of the National Park area is characterized by plain topography and is dominated by *Compretum-Terminalia* community where the Maze River is the largest water system passing through it. The SHBS occurs between Oromiya and (SNNPRS) and covers an area of 57 km² (Information from Park office, 2011). It is found at 320 km from Addis Ababa in the western side of the Great Ethiopian Rift Valley (Eshetu, 2003). The sanctuary is one of the principal wildlife Conservation areas of Ethiopia (Hillman, 1993). Geographically, it is found between 7° 7'15" to 7°12'45" N latitude and 38°5'15" to 38°20'15" E longitude with the altitude ranges varying from 2000 to 2100 m a.s.l (Figure 1) (Messana and Netserab, 1994; Tekele, 1996). Vegetation types in the sanctuary are divided into savanna woodland, valley complex and grassland. The average rainfall at Senkele, measured over four years (1991 to 1994) is 1,116 mm (Kumsa, 2006). It has a moderately bimodal pattern of rainfall typical of the 'Woinadega' agro-ecological zone of Ethiopia. The three-month dry season, from November to January is followed by the 'Belg' rains, from March to April. During the dry season, cloud cover is at a minimum and wind speeds are at their annual maximum contributing to the high levels of evapo-transpiration associated with tropical climate patterns. The SHBS was established in 1976 to save the most viable population of the SHB, which are endemic and endangered sub-species (EWCO, 1990; Hillman, 1993; Messana and Netserab, 1994; Tedela, 1995). The extent of the Sanctuary has been declining from 200 km² in 1972 to about 58 km² in 1973, then to 36.4 km² and 28 km² (Messana and Netsereab, 1994). This reduction in area could be associated with human induced land cover changes where the surrounding farmers involved in agriculture and livestock converting land cover of the SHBS. But at present, its area has increased from 36 km² to 57 km² (Park office report, 2011).

The NNP is found at the center of Ethiopian rift valley floor between two lakes of Abaya and Chamo, 500 km south of Addis-Ababa in SNNPRS (Figure 1). It covers an area of 514 km², of which 78 km² is fresh water habitat and 436 km² of terrestrial habitats, composed of mosaic forest, open woodland and savannah grasslands (Fetene et al., 2011). NNP was established in 1974 and geographically located between 5°51' to 6°05' N Latitude and 37°32'-37°48' E Longitude with an altitudinal range of 1100 to 1650 m a s l (Bolton, 1973; Duckworth et al., 1992; Andarge, 2001; Svialek, 2008; Shimeles, 2010; Fetene et al., 2011). The climate of the study area is characterized by a relatively hot climatic condition with low and unevenly distributed rainfall pattern (Aregu and Demeke, 2006; Datiko, et al., 2007).

Studies also indicated that the NNP is rich in biodiversity harboring more than 90 mammalian and 350 species of birds. Moreover, the Park is used as destination point of many Palaearctic and intra-Africa migrants (Duckworth et al., 1992; Duckworth et al., 1993; Whitaker, 2007). There are two main river systems that flow through the park and form riverine forests and its woodlands. The Arba Minch ground water forest is a unique forest formation in Africa and is found in the western part of NNP. It serves as source of numerous natural springs locally known as 'Arba Minch' meaning forty springs. In the far eastern part of the park hot springs bubble to the surface (Duckworth et al., 1992; Andarge, 2001; Negussie, 2008). The soil of Nech Sar plains and surrounding bush land is black cotton soil with high clay content. According to Bolton (1973), the rugged mountains parts of the park have brown calereous loam derived from volcanic rocks. The soil ranges from patchy dark clay alluvial soils along the river basin to dominating dry cotton soils with poor drainage (Negussie, 2008).

Materials

In this study, 7x35 magnifying binocular, GPS, digital camera, and field guide for identification of the mammals, topographic map of the study area, satellite imagery data and field data sheets were used.

Swaynes' Hartebeest (SHB) population

Before conducting preliminary survey and actual data collection, all available literatures related to SHB and major habitats of the species were searched. The literatures were used to determine the previous population size, status and trends of the SHB population in the three Ethiopian PAs.

Preliminary survey to the PAs was carried out for ten days before the actual field data collection during February 2008. Based on the survey and literatures searched, the boundaries of different habitats in the Parks were identified and number of sampling blocks was determined. The number of blocks was determined separately for the three PAs depending on the size, type of vegetation and other land features of the PAs. First, each of the three PAs was divided into blocks following Norton Griffiths (1978), Sutherland (1996), Gebre (2000), Gebre and Yirga (2004) and Kumsa and Bekele (2008). For the smaller area, Senkele Hartebeest Sanctuary (57 km²), 5 blocks covering the entire census zone was used to count the SHB (Gebere, 2000). However, for the larger, MaZNP (202 km²) and NNP (436 km² of the terrestrial habitat), each of the PAs was divided into 6 sampling blocks (Norton-Griffiths, 1978) based on land cover features such as grassland, woodland, Shrublands and Valley that was also used as boundary demarcation. The area of the sample blocks for MaZNP and NNP was designed to cover 50% the Park areas, MaZNP at least 17 km² covered in each of the block and 40% of the area for NNP with at least 29 km² of one sample block.

The size of NNP is large but it is made 40% because the SHB in the Park is very low and found in restricted areas as compared to the denser SHB in the MaZNP. Therefore, counting of the lower densely populated SHB in NNP was easy by keeping the area searched lower. For counting the hartebeest, total count with direct observation based on silent detection was used following the methods described in Norton Griffith (1978), Melton (1983), Caughly and Sinclair (1994) and Sutherland (1996) and Gebre (2000). For the total count, various census groups or team of observers consisting of thirty persons were formed to count the SHB and complete counting within short time period.

The team for all the blocks of one of the PAs is composed of 20 scouts, five wildlife experts and five of the researchers. These counted the animals simultaneously so as to reduce or avoid double counting. In each of the block, a team of five to six persons (consisting of one wildlife expert, five experienced wildlife scouts and one of the researchers) was assigned depending on the size of the block. The block is bounded by patrolling roads or paths constructed within the protected areas in various directions along the different land features and these were used as boundary of the sample block: in case where no road or path, rivers, and ridge tope were used as the boundary of the block. The counting team approached the block in the different directions along the patrolling road constructed. Then the team members estimated SHB populations found in herds or counted those encountered in isolation. Similar counting method was used for the remaining PAs. In addition, vehicle was used to count SHB along the roads, paths and tracks constructed for patrolling purpose within and around the sample blocks in all the PAs. While counting, efforts were made to count or estimate the number of individuals of a herd.

During the field assessment indirect evidence of the animals (footprint, drooping and feeding signs) were also collected. Counting was carried out early (from 6:00pm to 10:00am) in the morning and late in the afternoon from 4:00pm to 6:00pm when the animals were active for sixteen days; eight days during the dry and eight days of the wet seasons. Of these, ten days was used for the on foot count and 6 days for the vehicle count. Herd estimation or counting of individuals seen was conducted using unaided eyes or binoculars on foot form vehicle on the counting blocks. The number of hartebeest counted and estimated was recorded on the data sheet prepared for this purpose.

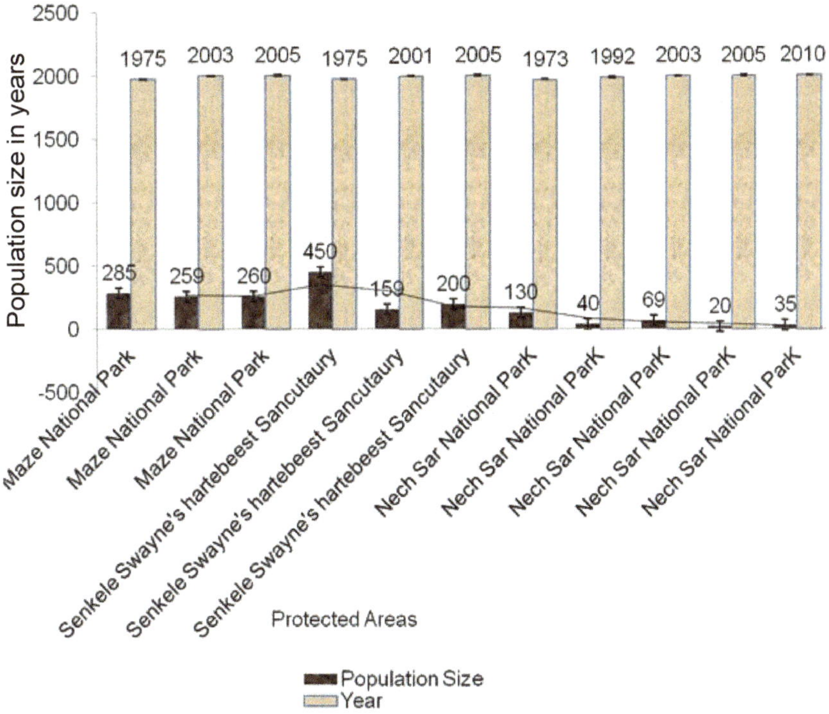

Figure 2. Swayne's Hartebeest Population size of the three protected areas during early 1970's and 2000's. (Sources: Bolton, 1973; Stephenson, 1975; Duckworth et al., 1992, 1993; Refera, 2005 Vymyslická et al., 2010, Fetene et al., 2011).

Population sex and age structures

During counting, the herd's sex and age composition was carefully observed. Sex identification was made for adults and sub-adults by observing primary sex characteristics. Indirect characteristics such as smaller body size in adult females as compared to adult males and reproductive behavioral characteristics were observed based on Lewis and Wilson (1979) and Messana (1993). Horns size and shape were also noted as horn in adult SHB females are more slender and less marked than in adult males (Gebre, 2000). The young male also has smaller body and horn sizes as that of the female but the presence of male reproductive organ such as scortum in young male and or udder around the genitalia in female aided in identification. While counting, four age groups were identified namely: Calf, Juvenile, Sub adult and adult. The horns can be compared based on Kok design (Kok, 1975). We classified the age group of the SHB according to Gebre (2000) in months: Calf (0 to 9), Juvenile (9 to 18), Sub-adult (18 to 30), and Adults (above 30 months).

Land cover change

For the land cover change identification, four imageries were acquired from Land Cover Facilities Archives for the periods 1986 and 2005 (NASA, 2005). Then a supervised image processing was undertaken using GRASS 6.4, which is an opens source GIS.

DATA ANALYSIS

A SPSS 15 software package was used to analyze the previous SHB population size, depict their trends from literatures, and to compute the correlation between the adult male and adult females of the PAs (Gebere, 2000; Mengesha and Bkele, 2008; Kumsa and Bekele, 2008). Moreover, both the SPSS and excel statistical software was used to estimate and analyze the population size from the current data and the different sex and age categories. The land cover analysis was carried out based on satellite imagery of Landsat (NASA, 2005).

RESULTS

This study revealed that the population size of SHB was fluctuating during the past 37 years (1973 to 2010) in the MaZNP, SHBS and NNP (Figure 2). They were decreasing from 865 individuals during 1975 with highest number in SHBS (450 individuals) to 480 individuals in 2005 with the highest number of individuals in the MaZNP (260 individuals). Between 1975 and 2010, the number of SHB was dramatically declining in the NNP from130 individuals in 1973 to 35 individuals in 2010 (Figure 2). Presently, a total of 840 individuals of SHB were recorded in the three Ethiopian protected areas. Of these, 364 occurred in the MaZNP, 464 in the SHBS and 12 individuals in the NNP (Table 1).

There were more adult male SHB recorded as compared to the female of the species consisting of 48% in MaZN, 39% in SHBS and 42% in NNP in the present study (Figure 3). The number of Juveniles were low in all

Table 1. Current population size, sex and age structures of Swayne's Hartebeest in the three Ethiopian Pas.

PAs	Sex and age categories	N	df	Sum	Mean	95% Confidence Interval of the difference	
						Lower	Upper
Maze National Park	Adult male	6	5	172.00	28.67	-10.45	67.80
	Adult female	6	5	49.00	8.17	1.54	14.85
	Sub-adult	6	5	51.00	8.50	2.00	14.99
	Juvenile	6	5	25.00	4.17	1.24	7.09
	Calf	6	5	34.00	5.67	2.11	9.22
	Unknown	6	5	33.00	5.50	-2.34	13.35
	Sub-total	-	-	364	60.6	-5.88	127.00
Senkele Hartebeest Sanctuary	Adult male	6	4	180.00	30.33	2.54	49.46
	Adult female	6	4	154.00	26.00	5.03	56.57
	Sub-adult	6	4	71.00	11.83	8.62	19.78
	Juvenile	6	4	49.00	8.17	7.96	11.64
	Calf	6	4	10.00	1.67	-2.30	6.30
	Unknown	6	-	.00	.0000	.000	.000
	Sub-total	-	-	464	78	21.85	143.75
Nechsar National Park	Adult male	6	5	5.00	.8333	-.1985	1.86
	Adult female	6	5	4.00	.6667	-.1902	1.52
	Sub-adult	6	5	1.00	.1667	-.2618	.595
	Juvenile	6	5	1.00	.1667	-.2618	.595
	Calf	6	5	1.00	.1667	-.2618	.595
	Unknown	6	0	.00	.0000	.0000	.0000
	Sub-total	-	-	12	2.00	-1.17	5.17
	Total	-	-	840	140.6	14.9	276

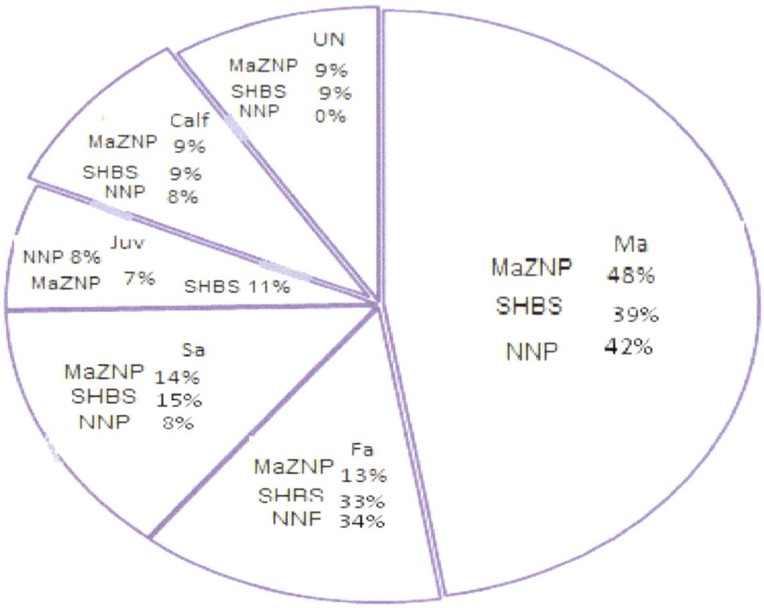

Figure 3. Percentage of sex and age composition of the SHB in the three protected areas. Note: MazNP=Maze National Park, SHBS=Swayne's Hartebeest Sanctuary, NNP= Nech Sar National Park.

Table 2. Relationships between the adult male and adult female of the three Ethiopian protected Areas.

Sex categories		Maze NP		Senkele SHBS		Nech Sar NP	
		AM	AF	AM	AF	AM	AF
AM	A Pearson correlation	1.00	.969**	1.00	.268	1.00	-.332
	Sig. (2-tailed)	-	.001	-	.663	-	.520
	N	6.00	6.00	5.00	5.00	6.00	6.00
AF	Pearson Correlation	.969**	1.00	.268	1.00	-.332	1.00
	Sig. (2-tailed)	.001	-	.663	-	.520	-
	N	6.00	6.00	5.00	5.00	6.00	6.00

Note: ** Correlation is significant at the 0.01 level (2-tailed) tailed). MazNP=Maze National Park, NechsarNP=National Park, AM= Adult male. AF=Adult female.

Table 3. The land cover changes between 1986 and 2005 in hactor and % in the three protected areas. Percentages are shown in the parenthesis.

PAs	Senkele SHBS		NNP		MaZNP	
Year	1986	2005	1986	2005	1986	2005
Grass	4,008(71)	3,823(67)	17,610 (37)	16,230 (34)	16,340 (75)	10,462(48)
Shrub	537 (9)	821 (14)	4,516 (9)	8,944 (19)	1,496 (7)	6,710 (31)
Woodland	1,128 (20)	1,029(18)	6,674 (14)	4,305 (9)	2,510 (12)	988 (5)
Forest	-	-	4,990 (10)	3,890 (8)	1,343 (6)	3,529 (16)
Water	-	-	14,265 (30)	14,687 (31)	-	-
Total area	5,673	5,673	48,056	48,056	21,689	21,689

the PAs studied (7% in MaZNP, 11% in SHBS and 8% in NNP) (Figure 3). The computed Pearson correlation for the relationship between the adult male and adult female SHB across the blocks showed high significant difference (2t= 0.969, P ≥ 0.01 level) for the MaZNP (Table 2). However, there was no significant difference between the adult male and adult female in SHBS and NNP (Table 2). Analysis of the available land cover map, namely, 1986 and 2005, of the MaZNP, SHBS and NNP have shown changing land cover of the PAs (Table 3, Figures 4, 5 and 6).

Tables 3, Figures 4, 5 and 6 show the major types of land or vegetation cover in the three PAs during 1986 and 2005. In MaZNP, from 1986 to 2005 the grassland decreased from 75 to 48%. On the other side, the shrub land increased from 7 to 31%. The total area covered by forest and woodland in 1986 is 18% of the area. In 2005, the forest and woodland cover was 21% of the area. In SHBS, both in 1986 and 2005, the major land cover was grassland. This class mainly includes grass covered areas and area used as cultivated land and open areas. The second major class of land cover is the woodland. The most important changes observed from 1986 to 2005 are an increase of the shrub land from 9 to 14%. In the NNP, the key land cover type during 1986 and 2005 is the grassland covering 37 and 34% of the area, respectively. The total area covered by the forest and woodland in 1986 and 2005 was 24 and 17% of the area, respectively. An increase of the area covered by shrub was observed from 1986 to 2005 from 9 to 19% of the area.

Based on personal observation during the survey and also literatures searched, it was recognized that the size of the PAs was changing by re-demarcation in efforts for the conservation of the species by concerned authorities. Thus, there was better conservation endeavors in the SHBS and MaZNP. For instance, there was an increase in the size of SHBS from 36 to 57 km² and human activates was kept low in the MaZNP. Moreover the team observed poachers with automatic guns running to hide themselves from the teams in the MaZNP.

DISCUSSION

This study indicated that the number of individuals of SHB was fluctuating, increased in some of the years and decreased in the others between 1970s and 2010 (Figure 2). This is in agreement with the work of Easts (1999) that showed the decline in the individual number of the species and its fluctuation form one period in time to the other. This fluctuation and decline in number could be associated with the impacts of human activates that altered the historical vegetation cover of the Pas,

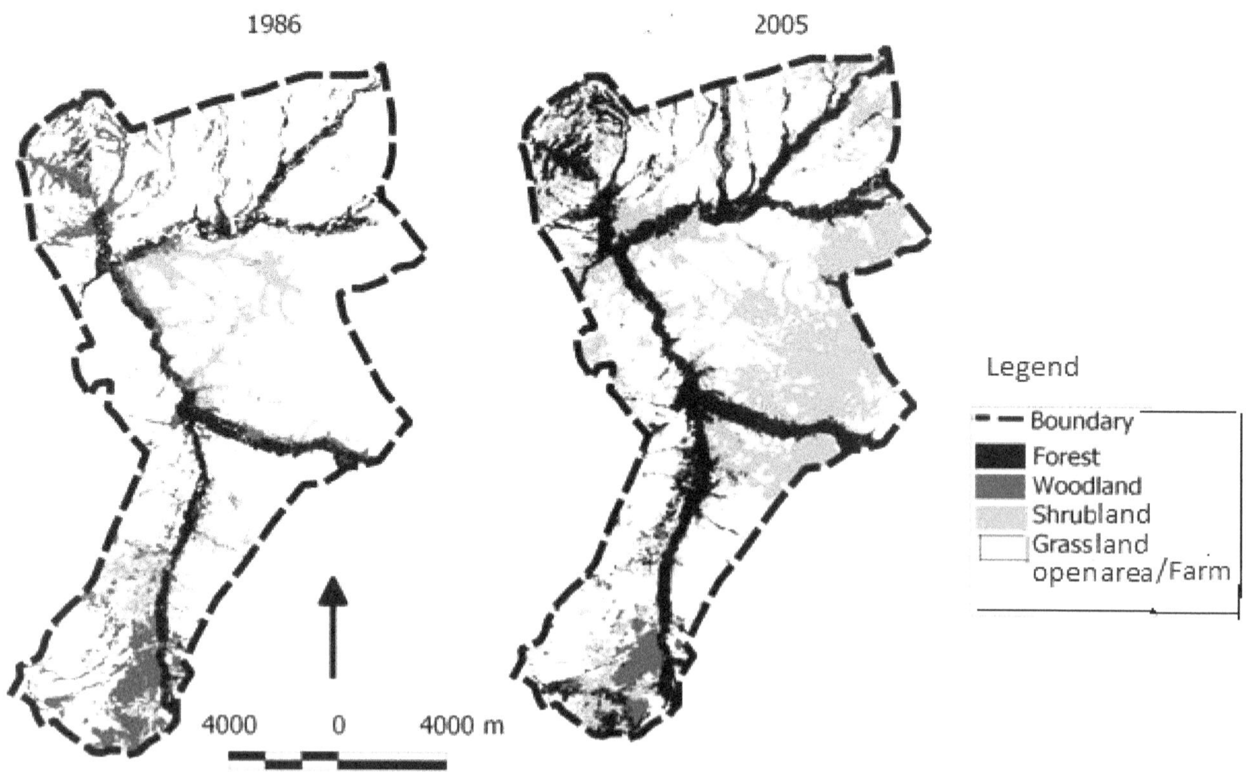

Figure 4. Maze National Park land cove types of the 1986 and 2005.

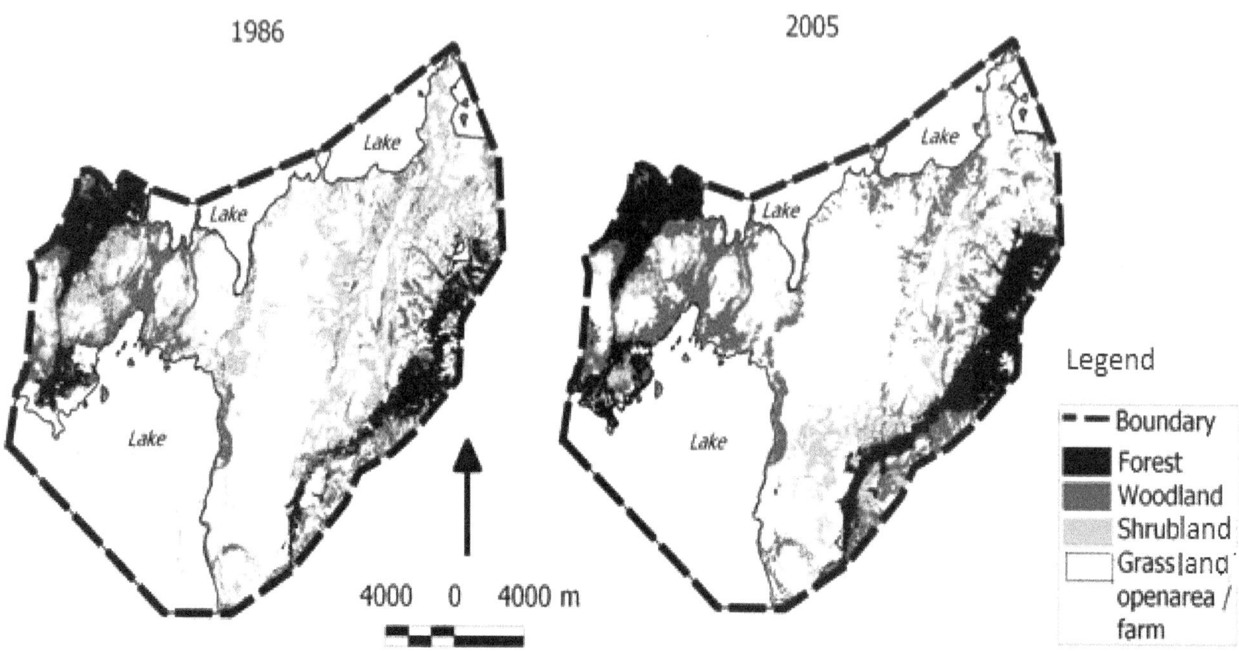

Figure 5. Nechsar land cover type of the 1986 and 2005.

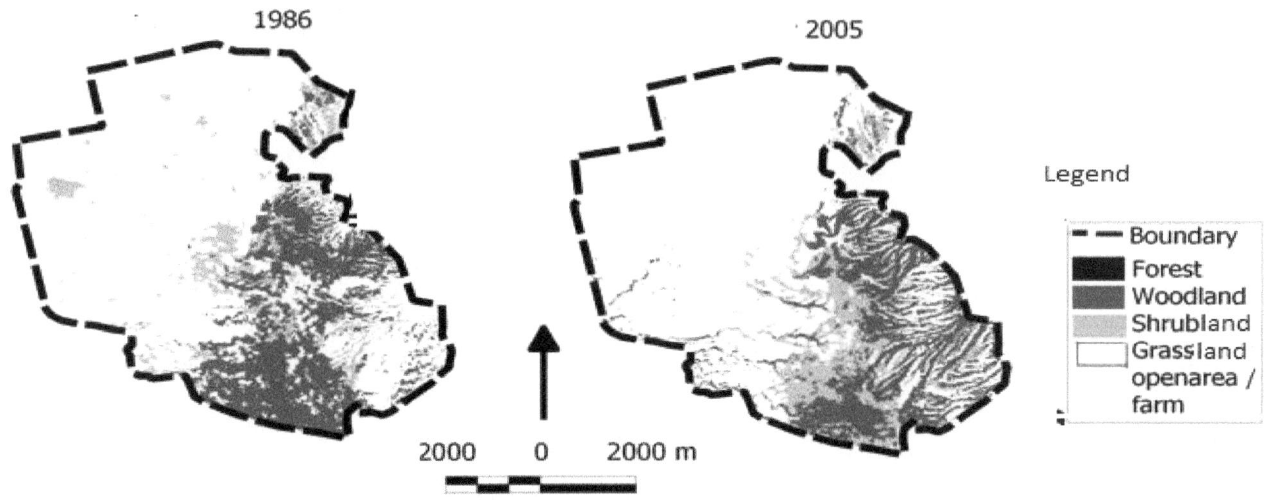

Figure 6. Senkele Swayne's Hartebeest sanctuary land cove type of the 1986 and 2005.

especially the grassland habitats into shrub and woodlands (Table 3, Figures 4, 5 and 6). This concurs with Bolton (1973), Stephenson (1975), Referea (2005), Kumssa and Bekele (2008) and Fetene et al. (2011) that indicated the impacts of human activities on the SHB in the PAs. Furthermore, the contraction and expansion of the size of the PAs might have also contributed to the fluctuation in the number of the hartebeest. Besides, currently there is an increase in the number of individuals of SHB in SHBS and MaZNP. This might be attributed to the conservation efforts in the PAs by the concerned authorities such as increase the size of SHBS and decreasing human activities that reduced grassland cover in the MaZNP. This is similar to the work of Andarge (2001), Refera (2005), Refera et al. (2003), Svialek (2008), Shimeles (2010), Fetene et al. (2011) and SHBS office report (2011) that showed an increase in the size of SHBS and reductions in the impact of human activities such as grazing in the PAs by their livestock resulted in an increase in population size of the SHB in the PAs. Moreover, the increase in the individual number of SHB in SHBS is in agreement with the work of Gebere and Yirga (2004) and Kumsa and Bekele (2008) that revealed increase of SHB in SHBS. Conversely, the SHB in the NNP dramatically decreased since early 1970s with increasing trends in some years to this study period (Figure 2 and Table 1). This is in line with Duckworth et al. (1992), Fetene et al. (2011) and Datiko and Bekele (2011) that showed the occurrence of 40 individuals in 1992, 35 individuals in 2011 and 12 individuals in 2011, respectively. One of the factors indicated for the dramatic decline in the NNP by the authors was overgrazing by the high livestock populations. This has reduced the grassland covers and increased the shrub lands leading to the dramatic decline in the meta-population of SHB in the NNP. Moreover, it has exposed the hartebeest to poaching and competition for grass with the livestock.

This agrees with Easts (1999), Jacobs and Schloeder (2001), Mengesha et al. (2009, 2008) that revealed the decrease in a number of individuals of a species in a protected area was associated to destruction of vegetation cover of the PAs.

The number of individuals of the male SHB was higher than the number of individuals of the females in the PAs (Figure 3). The highest proportion of adult male individuals of SHB in the three Ethiopian protected areas might be due to the adult male's ability to withstand and escape environmental threats such as decrease in grass cover and their better abilities to access such resources elsewhere compared to the females. This is similar to the work of Lewis and Wilson (1979), Messana (1993), Mamo et al. (2010), Fetene et al. (2011) and Datiko and Bekele (2011) that showed ability of larger adult male mammals to better tolerate environmental variables as compared to adult female of the species.

This study revealed that the number of individuals of adult males and females in the MaZNP showed a highly significant variation unlike the number of individuals of adult males and females in the SHBS and NNP. This variation might have occurred as a result of human activities that occasionally set fire that devastates the lowland vegetation cover of the MaZNP, the historical legal hunting and the current observed poaching activities by the local people called Bonna people. These could alter the behavior and distribution of the male and female animals within the MaZNP. This is in agreement with the work of Lewis and Wilson (1979), Messana (1993), Easts (1999) and Refera et al. (2003) that showed the effect of habitat destruction and poaching on the reproductive behavior and distribution of hartebeest in the protected area.

Conclusions

The study indicated that the numbers of individuals of SHB in the MaZNP, SHBS and NNP was fluctuating between the early 1970s and 2010. The trends in the population size of SHB in the PAs were unpredictable as their numbers increased in some of the years and decreased in others. The present investigation showed an increase in the numbers of individuals of SHB in the MaZNP and SHBS but dramatic decrease in the NNP. The factor identified was land cover change form more grass and woodland cover to shrub land type in the PAs. This meant that the original vegetation cover that support SHB was altered. The presence of more individuals of adult males as compared to the female and the fewer juveniles than adults, sub adults and calves in all the PAs showed declining population. Therefore, urgent conservation measures that reduce the threat to the endemic species and increases their population size is needed to conserve the species in Ethiopia.

ACKNOWLEDGEMENTS

Our sincere thanks go to Wondo Genet College of Forestry and Natural Resources for the financial support provided. We are also indebted to the Bureau of Hotels, Parks and Tourism of the Southern Nationalities Peoples' Regional State and the Maze, Nech Sar National Parks and Senkele Swayne's Hartebeest Sanctuary for their collaboration and assistance during field data collections. We also thank the wildlife experts and scouts at the protected areas for the unreserved support they provided during field data collection.

Finally we are very grateful to the anonymous referees for their critical and invaluable comments that greatly helped us for the improvement of the manuscript.

REFERENCES

Andargie T (2001). Floristic Composition and Ecology of the Savanna Grassland and Woodland. Vegetation in Nech Sar National Park and Its Conservation Status. M.Sc. Thesis. Addis Ababa University, Addis Ababa, pp. 1-68.

Aregu L, Demeke F (2006). Socio-economic Survey of Arba-Minch Riverine Forest and Woodland. J. Dry lands 1:194-205.

Blower J (1969). Wildlife conservation in Ethiopia. Walia 1:4 – 23.

Bolton M (1973). Hartebeest in Ethiopia. Oryx 12:99-108.

Caughley G, Sinclair AE (1994). Wildlife Ecology and Management. Blackwell Scientific Publications, Oxford, pp. 1-200.

Datiko D, Bekele A (2011). Population status and human impact on the endangered Swayne's hartebeest (*Alcelaphus buselaphus swaynei*) in Nechisar Plains, Nechisar National Park, Ethiopia. Afr. J. Ecol. 49:311–319.

Datiko D, Bekele A, Belay G (2007). Species composition, distribution and habitat association of rodents from Arba Minch forest and farmlands, Ethiopia. Afr. J. Ecol. 4:651–657.

Duckworth JW, Harrison DL, Timmins RJ (1993). Notes on a collection of small mammals from the Ethiopians Rift Valley. Mamma. 57:278–282.

Duckworth JW, Evans, MI, Safford, RJ, Telfer, MG Timmins, RJ, Zewdie C (1992). A Survey of Nechisar National Park, Ethiopia: Report of the Cambridge Ethiopia Ground- Water Forest Expedition 1990. International Council for Bird Preservation, U.K., pp. 5-50.

Easts R (1999). African Antelope Database 1998.IUCN/SSC Antelope specialist Group. IUCN, Gland, Switzerland, pp. 434.

Eshetu M (2003). Perception of local communities towards wildlife and protected areas of Ethiopia. Durell Institute of Conservation and Ecology, University of Kent United Kingdom, pp. 40-87.

EWCO-Ethiopian Wildlife Conservation Organization (1990). Wildlife Conservation Area Summary Sheet, EWCO, Addis Ababa, pp. 33.

Fetene A, Mengesha G, Bekele T (2011). Spatial Distribution and Habitat Preferences of Selected Large Mammalian Species in the Nech Sar National Park. Nat. Sci. 9:80-90.

Gebere B, Yirga, S (2004). Seasonal home range of Swayne's Hartebeest (*Acelaphus buseclaphus sywani*) in senkele swayne's Hartebeest Sanctuary. Ethiop. J. Sci. 27:121-126.

Gebre B (2000). Some patterns of population dynamics and distribution of Swayne's Hartebeest (*Alcelaphus buselaphus swaynei*) in Sinkelle Sanctuary. M.Sc. Thesis, Addis Ababa University, Addis Ababa, pp. 1-95.

Hillman JC (1993). Ethiopia. Compendium of Wildlife Conservation Information. The Wildlife Conservation Society, New York, pp. 1-100.

IUCN (2002). Red Data List of Threatened Animals. International Union for Conservation of Nature and Natural Resources; Gland, Switzerland, pp. 61.

Jacobs M, Schloeder C (2001). Impact of conflict and biodiversity in Ethiopia. Biodiversity support program, Washington DC, pp. 45.

Kingdon J (1989). East African mammals: An atlas of evolution in Africa. (Large Mammals). University of Chicago Press, Chicago, pp. 294-299.

Kok OB (1975). Behavior and ecology of the red hartebeest (*Alecephlaus buselaphus caama*) Orange Free State, South Africa, pp. 59.

Kumsa T (2006). Human-wildlife conflict and population status of Swayne's hartebeest (*Alcelaphus buselaphus swaynei*) in Senkele Swayne's hartebeest Sanctuary. MSc. thesis Addis Ababa University, Addis Ababa, pp. 114.

Kumsa T, Bekele A (2008). Population status and structure of the endangered Swayne's hartebeest, (*Alcelaphus buselaphus swaynei)*, in Senkele Swayne's Hartebeest Sanctuary, Ethiopia. Acta Zool. Sinica 54:569–575.

Lealem B (1974). Operation Swayne's hartebeest. *Orix*, 12:556-558.

Levins R (1969). Some demographic and genetic consequences of environmental heterogeneity for biological control. Bull. Entomol. Soc. Am. 15:237-240.

Lewis JW, Wilson DE (1979). The breeding strategy of Senkele Swayne's Hartebeest at Senkele Ethiopia. Waliyia 10:1-10.

Mamo Y, Pinard MA, Bekele A (2010). Demography and dynamics of Mountain Nyala (*Tragelaphus boxtoni*) in the Bale Mountain National Park. Curr. Zool. 56:660-669.

Melton DA (1983). Population Dynamics of waterbucks *Kobus ellipsierymnusin* Unflozi Game Reserve A. J. Ecol. 21:77-91.

Mengesha G, Bekele A (2008). Diversity, distribution and habitat association of large mammals of Alatish, North Gonder, Ethiopia. Act. Zool. Sin. 54:20-29.

Mengesha G, Sahale K, Bekle M, Deble M (2009). Alternative approaches to managing conflict over the natural resources. The case of Abijata-Shalla Lakes National Park. Pages 1- 48 in Tolera M, Lemenih M, Ellis F. J. Development–Oriented Interdisciplinary Thematic Action Research (DOIT-AR) Eds .2009. Hawassa University, Wondo Genet College, Wondo Genet., pp. 1-48.

Messana GHM (1993). The reproductive ecology of Swayne's Hartebeest (*Alcelaphus buselaphus swaynei*). PhD thesis, Wolfson College, University of Cambridge, Cambridge, pp.1-76.

Messana M, Netsereab B (1994). *The* Senkelle Swayne's Hartebeest Sanctuary Management Plan. EWCO, Addis Ababa, pp. 1- 25.

NASA (2005). Landsat programme, 2003. Landsat Sci. (L71169056_05620051124, p169r056_5dt19860128_z37, p168r55_5t860121, p168r054_5dt19860121_z37) SLC-OFF. USGS Sioux F alls

Negussie A (2008). Analysis of Land and Vegetation Cover Dynamics Using Remote Sensing & GIS Techniques. A Case Study of Nech

Sar National Park. MSc. Thesis, Addis Ababa University, Addis Ababa, pp. 1-98.
Nobuko N (2004). Resisting Imposed Wildlife Conservation: Arssi Oromo and The Senkele Swayne's Hartebeest Sanctuary, Ethiopia. Afr. Monog. 25:61-77.
Norton-Griffith M.N (1978). Counting Animals (2nd edition). African Wildlife Foundation, Nairobi, Kenya, pp.1-25.
Refera B (2005). Population status of Swayne's Hartebeest in Ethiopia.
Fifth Annual Sahleo-Saharan interest group Meeting report, Souss, Tunisia, pp. 10-15.
Refera B, Mathwos E, Mengistu W (2003). Population Status of Swayne's hartebeest, mammals and bird species richness, abundance and vegetation type study of Maze Wildlife Area. Ethiopian Environmental Protection Authority, Addis Ababa, pp. 2-46.
Shimeles S (2010). Vegetation Ecology, Forage Biomass and Nutritive Value of Selected Brows Species in Nech Sar National Park, Ethiopia. M.Sc. Thesis. Addis Ababa University, Addis Ababa, pp.1-68. Sutherland WJ (1996). Ecological Census Technique: a handbook. Cambridge University Press, Cambridge, pp. 1-268.
Svialek BJ (2008). Use of GIS Technologies in Biodiversity Conservation: Case Study of Vegetation and Soil Mapping in Nech Sar National Park, Ethiopia. M.Sc. Thesis. Czech University, Prague,. pp. 189.
Stephenson JG (1975). An investigation into with recommendation on the status of Swayne's Hartebeest in the Shashemene Area EWCO, Addis Ababa, pp. 7.
Tedela S (1995). Protected areas management crisis in Ethiopia. Walia, 16: 17-30.
Tekele F (1996). Community-Based Conservation of Senkele Swayne's Hartebeest Sanctuary. M.Sc. Thesis. International center for protected landscapes, University College of Wales, pp. 1-75
Vegetation in Nech Sar National Park and Its Conservation Status. M.Sc. Thesis. Addis Ababa University, Addis Ababa, pp. 1-68.
Vymyslická P, Hejcmanová P, Antonínová M, Stejskalová M, Svitálek J (2010). Daily activity pattern of the endangered Swayne's Hartebeest (*Alcelaphus buselaphus swaynei* Sclater, 1892) in the Nechisar National Park, Ethiopia. Afr. J. Ecol. 10.1111/j.1365-2028.2010.01243.x.
Whitaker R (2007). Sustainable Use of the Lake Chamo Nile crocodile Population. African Parks (Ethiopia) Nechsar National Park Project. Project Document, Arba Minch, pp. 1-45

Zooplankton abundance in the Massa Lagoon, Southern Morocco: Impact of environmental variables

H. Badsi[1]*, H. Oulad Ali[1], M. Loudiki[2], M. El Hafa[1], R. Chakli[1] and A. Aamiri[1]

[1]Laboratory of Oceanography and Limnology, Faculty of Sciences, Department of Biology, Ibn Zohr University, P.O. Box. 8106, Agadir 80060, Morocco.
[2]Laboratory of Biology and Biotechnology of Microorganisms Faculty of Sciences Semlalia Department of Biology, Cadi Ayyad University, Marrakech, Morocco.

The biological reserve of Massa, situated in the National Park of Souss Massa, certified Ramsar has a global interest in birdlife. The functioning of the lagoon is affected by the problems of eutrophication, declining water levels, lack of communication with the sea. This work represents a first contribution to the study of the plankton community living in the lagoon of this reserve in connection with the physical chemistry of water. The analysis of the physico-chemical and biological state shows a very advanced eutrophic water of the lagoon, characterized by high levels of orthophosphate and chlorophyll a. The structure of zooplankton populations is characterized by the dominance of species of eutrophic environments and tolerates changes in salinity.

Key words: Reserve of Massa, eutrophication, zooplankton, lagoon, physico-chemical.

INTRODUCTION

Aquatic ecosystems are closed or semi-closed as our site appears as a finite and vulnerable resource threatened by catches in the environment (Aspe et al., 1999) and are considered a siege of several biological, chemical and physical phenomena. Evaluation of factors controlling the distribution of zooplankton species is, by its central position and its dual control by resources and predation, a central point in the biological and ecological functioning of aquatic ecosystems must be taken into account in the context of the management of these environments through the manipulation of the food web (Pinnel-Alloul et al., 1996). It is difficult to identify the main causes that explain the variation in zooplankton abundance (Velho et al., 2001), while abiotic factors such as salinity Ayad (2002), precipitations, or turbidity, have been identified as critical factors in the development of zooplankton (Dejen et al., 2004). Other authors (Wetzel, 2001; Fernández-Rosado et al., 2001) have also mentioned the influence of environmental and biotic interactions on the composition, abundance and dynamics of zooplankton.

There has been no hydrobiological study of the Massa lagoon l. Its development and its conservation require the completion of such study.

The lagoon of the biological reserve of Massa is situated in the heart of the National Park of Souss Massa in the South of Morocco. It included in 2005 on the list of Ramsar sites and also as a natural environment that has international significance and one of the most important migration routes and stopover site of wintering birds. This area is also the last refuge for the last wild population of bald ibis (*Geronticus eremita*) in the world, which is a rare and endangered species.

The environment of the Massa lagoon was indeed more severely affected by pollution and in particular with the silting of the mouth, forming a barrier beach sand to prevent any communication with the sea, the installation of the dam Youssef Ben Tachfine upstream, recurrent droughts experienced by the region in recent decades, and agricultural pollution is ever increasing.

These hydrological and environmental issues have accentuated and accelerated the eutrophication of the downstream part of Massa lagoon (lack of oxygen, temperature increasingly, high algal development and mortality of fish). These threats to the area, declared the

*Corresponding author. E-mail: hind.badsi@hotmail.com.

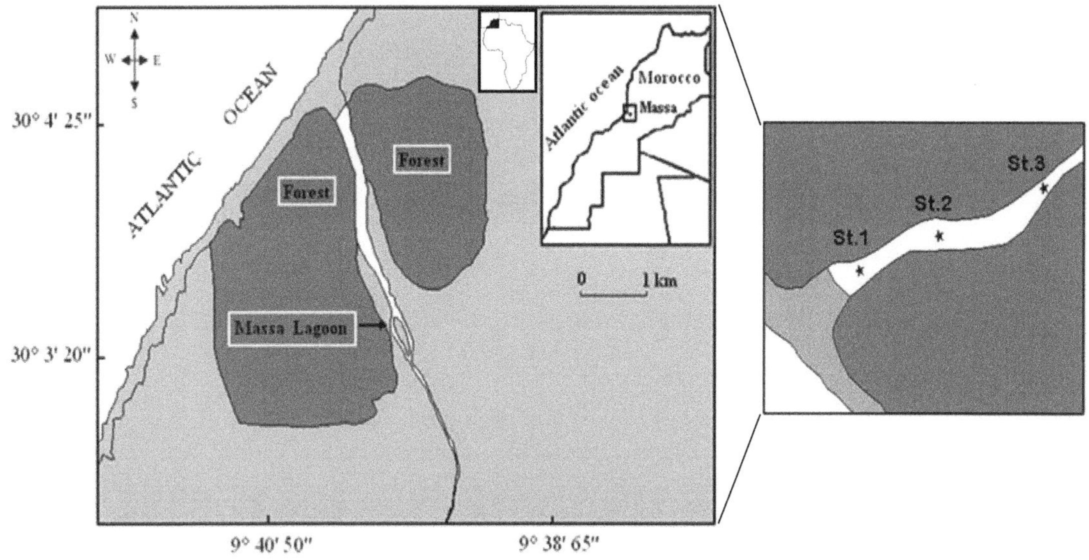

Figure 1. Location of the region of Souss Massa and the three study sites.

Table 1. Average values of physico-chemical of lagoon from February to July 2008.

Parameter		pH			Temperature (°C)			Salinity (mg/l)	
Sites	1	2	3	1	2	3	1	2	3
February	8.1	8.9	7.58	16.1	16.9	17	14	13.05	11.33
March	8.2	8.11	7.93	16.3	17.4	17.3	14.5	14	12.2
April	8.28	8.23	8.03	16.5	17.3	17	16	15	13.5
May	8.21	8.10	8.19	23.2	23.1	23.5	20	17	12.11
June	7.26	7.90	7.32	24.8	25.3	24.9	21.8	20.12	13
July	7.11	7.06	7.60	25.4	26.95	25.5	24.6	21.2	15.7

area protected by the Ramsar Convention in 2005, which include wetlands of Morocco, have consequences for the loss of biodiversity, thus leading to impairment of the biological equilibrium of the lagoon and even destruction of the ecosystem and serious disturbances in the intercontinental migration routes of birds.

Given this situation, we conducted an ecological analysis; through an initial study by the hydrobiological study of the physical chemistry of water and the study of zooplankton which is its key position in the trophic chain, gives a fundamental role in zooplankton and its study seem essential to better understand the functioning of aquatic systems. So its development for its conservation requires the realization of this kind of study.

MATERIAL AND METHODS

Study sites

The Massa lagoon is located along South-East coast of Morocco, about 60 km South of Agadir (Figure 1). The choice of sampling stations was conducted according to the gradient of salinity, agricultural areas, significant anthropogenic activities and places of wintering birds.

Station 1: (Lat. 30°04'20, 29" N: long.9°40'03, 23"O), downstream of the lagoon near the sea and where shorebirds are abundant and sedentary.

Station 2: located at 2 Km of the station 1, which is located (Lat. 30°03'53, 25"N: long. 9°39'51, 09" O). It is approximately in an intermediate zone, deeper, near the position of forest and is influenced by human activities.

Station 3: (Lat. 30°3'23, 04" N: long. 9°39'25, 91"O) and along the southern boundary. It has special features since and localized near farming areas of alfalfa and corn.

Sampling

At each site, qualitative samples have been realized over a period of 6 months (from February to July, 2008 Table 1a, b and c). The horizontal sampling, conducted in each site was created using are empty mesh 80 µm plankton net. At each site, we apply two vertical lines upward. The fauna recovered in the first line is filtered through a nylon sieve of 40 µm mesh vacuum and then placed in a pillbox and preserved in 5% formalin immediately after collection. The

Table 1(b). Average values of physico-chemical of lagoon from February to July 2008.

Parameter	Chlophyll a (mg/m^3)			Ammonium (mg/l)			Orthophosphates (mg/l)		
Sites	1	2	3	1	2	3	1	2	3
February	20	25	45	0.4	0.23	0.2	0. 09	0.12	0 .18
March	42	62	76.33	2.01	0.95	0.35	0.15	0.15	0.20
April	25	50	56.87	1.97	0.50	0.75	0.10	0.13	0.20
May	71.5	72	72	0.5	0.31	0.48	0.17	0.31	0.65
June	40	30	59	1.1	0.6	0.25	0.082	0.118	0.328
July	30	28	48	2.05	0.25	0.12	0.253	0.23	0.40

Table 1(c). Average values of physico-chemical of lagoon from February to July 2008.

Parameter	Dissolved oxygen (mg/l)			Chemical Oxygen Demand (mg/l)			Nitrate (mg/l)		
Sites	1	2	3	1	2	3	1	2	3
February	3.75	4.33	4.58	410	420	455	6.33	5.15	6.8
March	5.5	4.85	5.2	450	500	520	4.98	4.89	5.62
April	4.2	4.18	4.89	400	486	450	9.86	7.3	9.92
May	5.98	5.76	5.6	210	100	110	6.44	4.7	6.68
June	5.45	5.13	5.1	466	476	389	8.39	6.56	7.61
July	5.33	3	6.35	300	350	320	7.1	5.5	5.55

second feature is kept alive for the determination of some species difficult to identify after fixation (protozoans and rotifers) and also for observation after the return to the laboratory. The vertical sampling was done using a closing bottle with a capacity of two liters of type WILDCO valves and two samples filtered through a filter of 40 µm were fixed in 5% formalin and this type of sampling was also used for water samples for the physical chemistry study.

Abiotic variables

In situ measurements

Measurement *in situ* field measurements were for temperature, dissolved oxygen, salinity, pH and water transparency. Water temperature and dissolved oxygen using a thermo-oximeter, salinity using a salinometer, pH using a pH meter and transparency using a Secchi disk.

Laboratory analysis

Suspended solids (mg/l) (filtration filter fiber glass diameter equal to 45 µm). The chemical parameters, nutrients (mg/l) nitrates, nitrites, ammonium and orthophosphate conventional two methods (gravimetric, titrimetric, etc..) were used (Rodier, 1996).

Measures of chlorophyll a

Measuring the concentration of chlorophyll-a was achieved by filtration on a filter Whathman GF / C (0.45 mm), storage and extraction in 90% acetone for 24 h and kept at 4°C in protection from light. Chlorophyll was calculated using a spectrophotometer with a formula Lorenzen (1967).

Biotic variables

After homogenization of the sample, the counting of the number of specimens of each species was done in a Dollfus cuve under a stereoscopic microscope after staining with Rose Bengal (Ait salah, 1997). To facilitate the enumeration of species, we have added to the contents of tanks few milliliters of solution of alcohol 10% glycerol. After evaporation of the alcohol and water species are immobilized in glycerine which can count easily.Species identification was made with reference to the key determination (Pourriot et al., 1986) for Rotifers, (Dussart, 1967, 1969) for Copepods and Amoros (1984) for Cladocera.

Statistical analysis

To study the variations of zooplankton, a PCA was conducted. The principal component analysis (PCA) was one of the most used by Giraudel (2000). It is a factorial method of statistical analysis, linear and multidimensional part of descriptive statistics. It can synthesize, describe and classify data from one table to provide a summary.

RESULTS AND DISCUSSION

Physico-chemical environment

The physico-chemical results during the study period are presented in Table 1, while the inventories of zooplankton in the three sampling stations are summarized in Table 2. Temperature is a key factor regulating the growth of zooplankton populations Ban (1994). During the study period, water temperature ranged from 16, 1 to 26, 95°C with the minimum value in spring and the maximum in summer. It showed significant positive correlation with salinity (r = 0.65) and significant negative correlation with pH (r = -0. 65). The pH values indicate the nature of the water slightly alkaline. It showed significant positive correlation with suspended solids (r = 0.19) and negative correlation with water temperature (r = -0.65). The same values are recorded by (Chaouti 2005) at the Smir lagoon. The dissolved oxygen values ranged from 3 to 6.356 mg l^{-1}, showed significant positive correlation with chlorophyll a (r = 0.57) and negative correlation with chemical oxygen demand (r = -0.39).

The comparison of results shows a variability of these parameters. At site 2, the high value of the water temperature in summer (26, 95°C), due to arid climate, has caused a sharp decrease in dissolved oxygen (3 mg / l) and decreased pH. The interaction between these parameters at this station can be explained by microbial activity that was high in sludge sediment, resulting in a decrease in the concentration of dissolved oxygen and an increase of nitrate and phosphate, Kagalou et al. (2006) demonstrated that the decomposition of organic matter occurs at the sediment-water interface.

During the study period, the salinity of the lagoon ranged from 12, 01 to 24, 6 mg l^{-1}. A gradient of increasing salinity is observed from site 3 to 1. The high evaporation caused by the highest values was recorded in summer .The values the COD vary between 100 to 520 mg l^{-1}. It showed negative correlation with water temperature (r =-0. 48).The high values of COD were due probably to the addition of waste (Mishra and Saksena, 1993; Pandey et al., 1989). Nitrates are nitrogen form preferentially absorbed when the algae have the same time of NH_4^+ and NO_3^-Ait Salah (1997).The values ranged from 4. 08 to 8.64 mgl^{-1}.

The transparency (0.44 to 1.76 cm) showed significant positive correlation with water temperature (r = 0.41) and significant negative correlation with ammonium nitrogen(r = 0.62).The concentration of orthophosphate in the water of the lagoon varies from 0.09 to 0.65 mg l^{-1}. It increases upstream because of the prevailing agricultural effluents to the environment of the lagoon and the livestock are the main socio-economic activities. Indeed, in the irrigated valley Massa, 700 ha are devoted to vegetables and banana. The lagoon is eutrophic and shallow Ruggiero et al. (2004) note that in this type of ecosystem, the availability of phosphorus is mainly determined by the internal load of this element in the sediment and its mobility in the sediment depends on oxygen and redox potential. Suspended solids are crucial in the biogeochemical cycle of many pollutants in aquatic systems by limiting the chemical exchanges between the dissolved and particulate (Maldiney and Mouchel, 1995). The values fluctuated between 272.97 to 50 mg l^{-1}. Indeed, the ammonium ion content of the lagoon shows a decreasing gradient of the site 1 to 3 fluctuated between 0.1 to 2.05 mg l^{-1}. That is the excretions of shorebirds that are very abundant in the site1. Concentrations of chlorophyll a vary from 20 to 76.33 mg/m^3 which are due to significant development of algae. Comparison with other similar ecosystems indicates that the lagoon Massa levels of chlorophyll a were similar to that found in the Nador lagoon by Lefebvre et al. (1993). The evaluation of the trophic level of the Massa lagoon by model O. C D. E, class lagoon in a hyper-eutrophic state.

Zooplankton community

Zooplankton community: data represented in Table 2 reveals that greatest number of zooplankton were noted 54 species at Site 3, followed by Site 4 (45), Site 2 (37).

In the three stations, the rotifers form the bulk of the zooplankton population and they represent respectively 62, 59 and 50% of the total abundance (Figure 2). The Copepods forming the second group of zooplankton a qualitative point of view with a large diversity at the third station (26%), to the detriment of Protozoa and Ostracods are poorly represented. This high species richness downstream, indicating a favorable environment for the successive development of several species.

Figure 4 has shown comparatively high values of temperature; nitrate, salinity and low values of pH and DO are responsible for relatively higher abundance of species of zooplankton at Site I and II. In Site III zooplankton population is less as compare to Site I and II may be less nutrient are available there. The relative contributions of abiotic factors have been determined in the spatiotemporal distribution of zooplankton species collected at three sites.

Abundance of zooplankton

The abundance of zooplankton is largely dominated during this study by Rotifers and Copepods. The average relative importance of the major taxonomic groups in the three sampling sites is presented in Figure 3 In general, the zooplankton population was very low. Densities of zooplankton ranged between 551 and 4217 organism l^{-1} (Figure 3). Rotifers ranged between 503and 4198 organism l^{-1} and were dominant always during the study. The densities of Copepods are low and fluctued between 18 and 76 organism l^{-1}(Figure 3).

Distribution of rotifers

Rotifers are the most important animal group belonging to the ecological niche of small filters (Margalef et al., 1983). Rotifers are able to ingest small particles such as bacteria and organic detritus often abundant in eutrophic environments. According to Margalef (1983) and Pace (1984), a strong representation of rotifers in aquatic freshwater can be considered as an indicator of high biological trophic level. Brachionides is the most dominant presence of the genus *Brachionus* which is common in tropical waters. Gannon and Stemberger

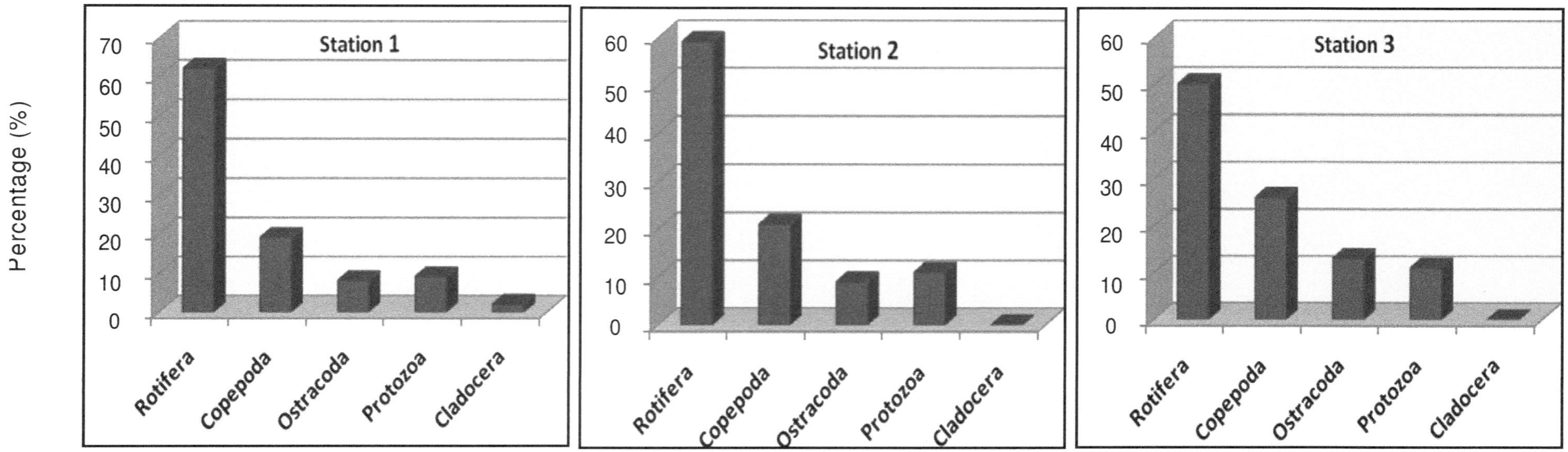

Figure 2. Relative importance of the abundance of the different zooplankton groups observed during the study in the Massa lagoon woks.

(1978), Sladecek (1983) and Maemets (1983), have established a relationship between a high number of species of the genus *Brachionus* and a high trophic level. In this study, *Brachionus plicatilis* (4198 org/l) was among the rotifers, the most common and quantitatively dominant species. These species are essentially herbivore which appears to corroborate the status assigned to hypereutrophic lagoon. Analysis of the stands at three stations to describe the composition and distribution of rotifers that are present in spring and are small numbers in summer and winter.

Overall, the rotifer appears to be continuing into the surface and, as pointed out, this is due to the wealth available algal surface, where chlorophyll a is important. Indeed one of the characteristics of the population of rotifers living in the lagoon is low fertility. There are one or two eggs, which may be a characteristic of eutrophic environments as already noted by several authors Lair et al. (1996); his feature would reflect the importance of using the detrital food despite a good use of algae food. The predominance of one Rotifer species was attributed to the fact that these organizations have a strategy (Amblard et al., 1998) they are opportunistic, small size, with short life cycles and high tolerance to a variety of environmental factors.

Distribution of copepods

In the Massa lagoon, the Copepods are poorly represented (Figure 3). *Halicyclops neglectus* which is a typical species of brackish coastal ponds and salt waters and supports a salinity of 34% according to Loffler (1961) in (Dussart, 1969) is always present during the study period with densities that exceed 38 org l^{-1}. The carnivore species copepod *Acanthocyclops robustus* (Geraldes and Boavida 2004) is dominant in spring with the high densities (57 org l^{-1}).

The Harpacticoida, with *Epactophanes richardi* species are present only at Site 3. It is, therefore, clear that there is a preference of these species to "low salinity" water. The literature calls them as species with wide ecological valence. The scarcity of Copepods is certainly not due to lack of food resources, but it could be attributed to the phenomenon of predation and competition with rotifers colonizing the environment throughout the study period with high densities (4200 org l^{-1}) and biomasses (178.36 µg. l^{-1}) which is calculated by the equations of length-weight regressions (Ruttner-Kolisko 1977).

The absence of oval females was noted. This is due to predation by larvae of the dipteran Chaoborus

Table 2. Occurrence of zooplankton sampling at three sites.+: Present - Absent

Nbr Species	1	2	3
Rotifera (35)	**33**	**26**	**20**
Ascomorpha sp	-	+	-
Asplanchna priodonta	+	+	+
Brachionus angularis	+	+	-
Brachionus calyciflorus	+	+	+
Brachionus caudatus	+	+	-
Brachionus leydigi	+	+	+
Brachionus plicatilis	+	+	+
Brachionus quadridentatatus	+	+	+
Brachionus rotundiformis	+	+	+
Brachionus rubens	+	+	+
Brachionus strita	+	+	-
Brachionus urceolaris	+	+	+
Cluniobicularis	-	+	+
Collotheca sp	+	+	+
Colurella sp	+	+	-
Eosophora	+	-	-
Epiphanes clavulata	+	-	-
Euchlanis sp	+	-	-
Gastropus sp	+	-	-
Hexarthra fennica	+	+	+
Hexarthra mira	+	+	+
Lecane inermis	+	-	+
Lecane luna	+	-	+
Lepadella patina	+	-	-
Notholca squamula	+	+	-
Philodina	+	+	+
Pompholyx complanata	+	+	+
Pompholyx solcata	+	-	-
Rotaria	+	+	+
Synchaeta lokowitziana	+	+	-
Synchaeta oblonga	+	+	+
Synchaeta pectinata.	+	+	+
Synchaeta stylata	+	-	+
Synchaeta tremula	+	+	-
Trichocerca rosea	+	+	-
Crustacea	**1**	**0**	**0**
Cladocera			
Daphnia longispina	+	-	-
Copepoda (13)	**10**	**9**	**7**
Acanthocyclops robustus	+	+	+
Calanus helgolandicus	+	+	-
Centropages typicus	+	+	-
Cryptocyclops bicolor	+	+	-
Epactophanes richardis	-	-	+
Eurytemora affinis	-	-	+
Halicyclops magniceps	+	+	+

Table 2. Contd.

Halicyclops neglectus	+	+	+
Macrocyclops albidus	+	+	+
Moraria sp	-	-	+
Oithona nana	+	+	-
Paracalanus typicus	+	-	-
Paracyclops fimbriatus	+	+	+
	1	**1**	**1**
Larvae (1)			
Nauplii	+	+	+
	4	**4**	**5**
Ostracoda (6)			
Cyprideis littoralis	+	+	+
Cypridopsis aculeata.	+	+	+
Cypridopsis vidua	+	+	+
Eucypris stagnalis	+	+	-
Hemicythere villosa	-	-	+
Herpetocypris chevreuxi	-	-	+
Protosoa (5)	**5**	**5**	**4**
Euplotes sp	+	+	+
Didinium sp	+	+	+
Zoothamnium sp	+	+	+
Vorticella sp	+	+	-
Codonella cratera	+	+	+

gender met at Station 3, which is a major predator of oval females, larger and more visible (Rabette and Lair, 1998).

Distribution of cladocera

The only species of Cladocera recorded at the site I was *Daphnia longispina*, which is very common in small collections of eutrophic water (Amoros, 1972) and very scarce in the media of high degree of eutrophication (Amengal, 1987).The scarcity of Cladocera is explained by the Planktivore predatory fish action inhabiting the lagoon as it has been demonstrated in several similar environments (Persson, 1988; Francisco, 1999) besides, increasing concentrations of suspended solids can block the filtering apparatus and impede their nutrition (Kirk et al., 1990).

Distribution of ostracoda

Ostracoda were recorded at all sites indicating that they are tolerant to the extreme environmental conditions prevailed in this region.

Statistical treatment of data

After logarithmic transformation of variables. The first and second axis of PCA accounted for 25, 35 and 19.7% of the total variance. The first factorial axis (PC1) stronglyassociated with chlorophyll a while the second factorial axis (PC2) closely related with nitrate nitrogen (Figure 4).

Total zooplankton density showed highly significant positive correlation with salinity (r = 0.68) so, the dynamics of zooplankton populations present in the Massa lagoon is controlled by this parameter. The selective effect of this parameter binding significantly reduces the richness and species diversity. Thus, the overall results of this study leads us to conclude that the salinity is, in conjunction with the resources, a factor in determining control and selectivity for the diversity and dynamics of plankton populations in Massa lagoon. A factorial axis 1 accounts for 34.02% of the total inertia and the factorial axis 2 accounts for 25.78% of the total inertia. The factorial axis 1 opposes spring in its positive side; in negative side opposes summer and winter.

The factorial axis 2 opposes spring and summer in its negative side (figure 5) and winter in positive side. The projection of both months and species in the factorial axes (Amengal, 1987; Amoros, 1972) shows that a

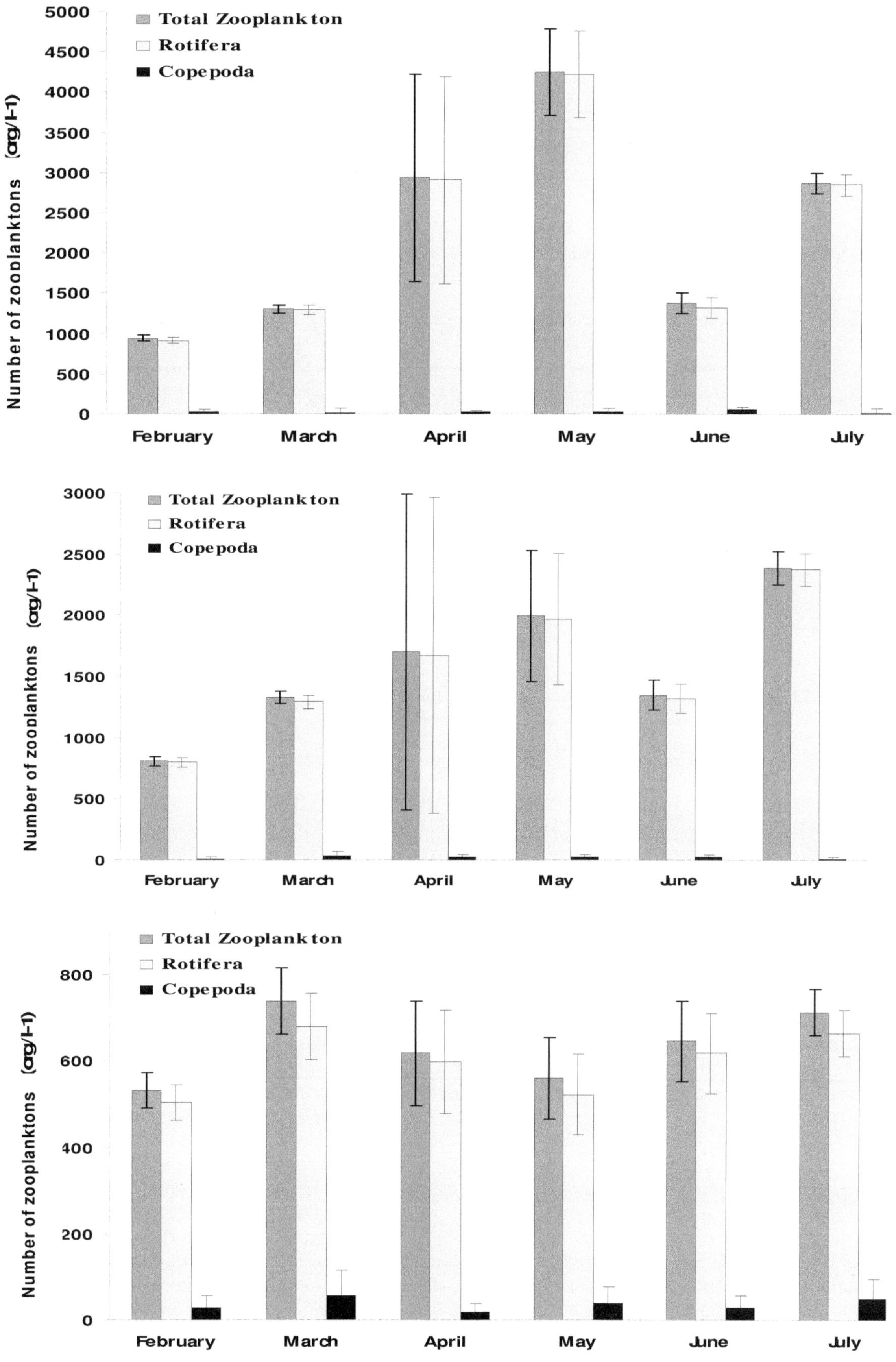

Figure 3. Average density of zooplankton groups (org/l^{-1}) determined in Massa lagoon during the study.

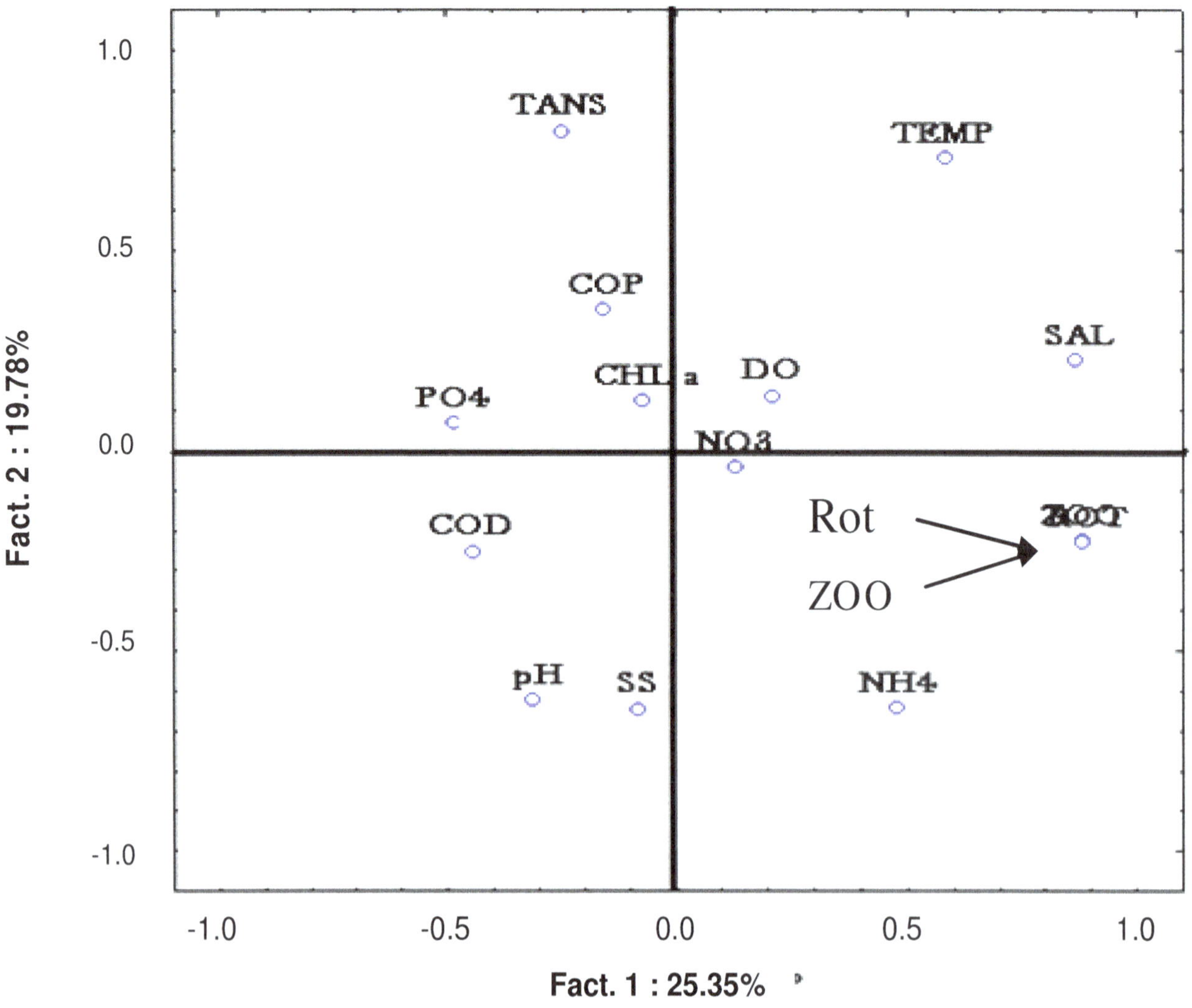

Figure 4. Principal Component Analysis (PCA) of physico-chemical variables and main total zooplankton. Abbreviations: CHL a; chlorophyll a, DO; dissolved oxygen, NH4; ammonium nitrogen, PO4; orthophosphates, COD; chemical oxygen demand, SS; Suspended solids,NO3; nitrate nitrogen, TRNS; nitrate nitrogen, TEMP; temperature, ZOO; total zooplankton. , Rot: Rotifer; COP: Copepod.

factorial axis opposes tree types of species: Spring represented by the following species Mv, Ns, R, Ar ,Pc and Li; summer with Ap , Sp and Er ; winter presented by Er, Ma, Ar and Bu. All other species namely: Bp, Hn, Ma, Er, and Bu are always present during the study period (Figure 5).

Conclusion

The analysis of physicochemical parameters indicates that certain environmental factors are essential to the operation and evolution of the closed lagoon of the biological reserve of Massa. These parameters include nutrients (nitrate, orthophosphates), dissolved oxygen, pH, and temperature.

Rotifers were the dominant zooplankton group most of the study period in Massa lagoon. The studies of the distribution of zooplankton population in the three stations are characterized by the importance of rotifers that are better represented. The zooplankton is represented by a majority of smaller species of rotifers dominated in abundance and they contribute to a large share in the total zooplankton biomass. This illustrates their role and their importance in energy transfer. This rich zooplankton may be a source of important food for wading birds and sedentary abundant downstream. A better understanding of the structure, evolution and functioning of the aquatic and natural ecosystem is certainly a better preservation of biodiversity. The interactions between the components of biodiversity are fundamental to the understanding of how this ecosystem functions. Our study is still to be continued to better understand the structure,

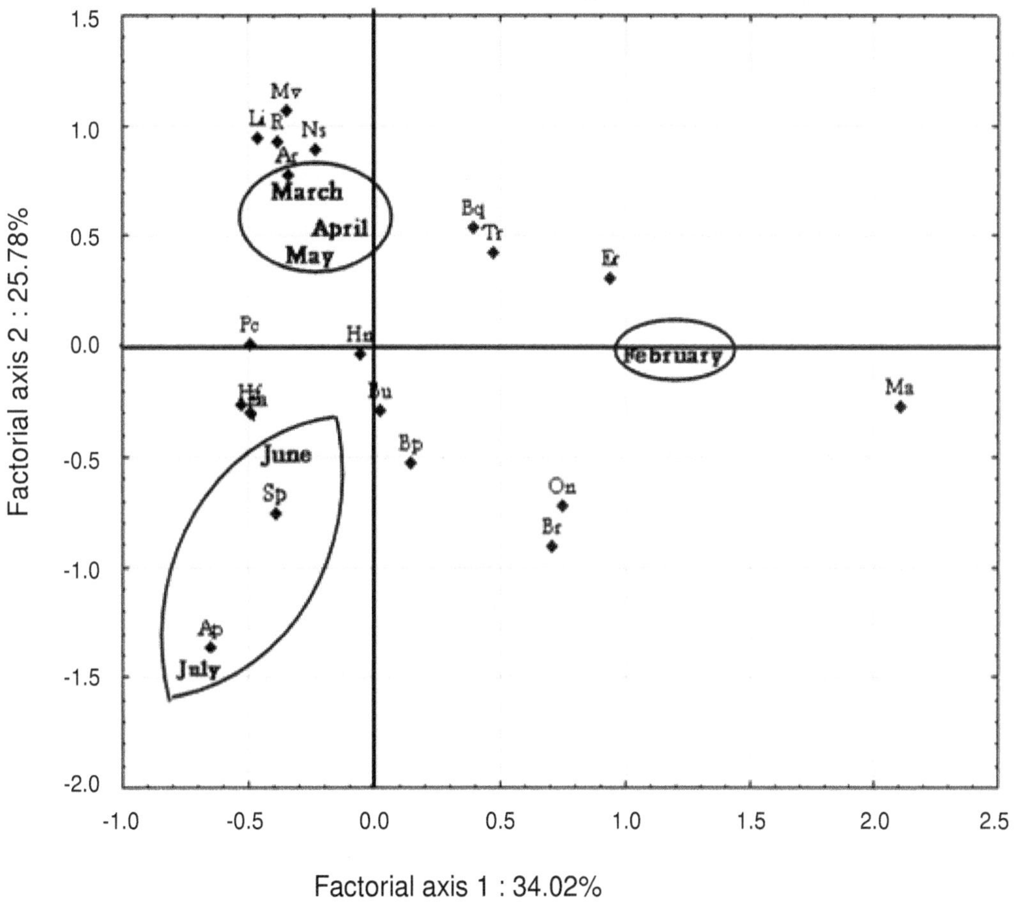

Figure 5. Plan [1-2] species and months of the correspondence analysis performed on the matrix composed of 48 species (rotifers and copepods) and 6 records (months).

functioning and ecology of zooplankton community of the biological reserve of Massa.

REFERENCES

AFNOR (1994). Recueil des normes françaises : Qualité de l'eau, Environnement, Paris.

Ait SH (1997). Organisation et structure des communautés zooplanctoniques de la retenue Youssef ben Tachfine (sud de Maroc). Thèse de 3 éme cycle, univ.Ibnou zohr, fac. Sc.Agadir., p. 200.

Amblard C, Boisson JC, Bourdier G, Fontvieille D, Gayte X, Sime-Ngandot T (1998). Écologie microbienne en milieu aquatique : des virus aux protozoaires.Rev. Se. Eau, n° spécial, pp. 145-162.

Amengal J (1987). Los Crustaceos de plancton de los embalses espagnoles. Oecol. Aquat. (Kindly provide page number).

Amoros C (1972). Contribution à l'écologie des étangs piscicoles de la Dourbes.

Amoros C (1984). Crustacés Cladocères., Univ. Claude-Bernard, Lyon I, Annales de l'Association Française de Limnologie, p. 63.

Aspe C, Pont P (1999). L'eau en représentation. Gestion de la qualité des milieux aquatiques et représentation sociale, Masson, Paris, pp. 1-101.

Ayadi H, Toumi N, Abid O, Medhioub K, Hammami M, Sime-Ngando T, Amblard C, et Sargos D (2002). Etude qualitative et quantitative des peuplements phyto-zooplanctoniques dans les bassins de la saline de Sfax, Tunisie, Rev. Sci. Eau., 15 : 1123-135.

Chaouti A (2005). Diversité taxonomique et structure de la macrofaune benthique des substrats meubles de la lagune de Smir. Travaux de l'Institut Scientifique, Rabat, série générale, 4: 33-42.

Dejen E, Vijverberg J, Nagelkerke LAJ, Sibbing FA (2004). Temporal and spatial distribution of microcrustacean zooplankton in relation to turbidity and other environmental factors in a large tropical lake (L. Tana, Ethiopia), Hydrobiologia, 513: 39–49.

Dussart B (1967). Les Copépodes des eaux continentales d'Europe occidentale. Tome I : Evolution des populations planctoniques d'Entomostracés. Thèse de 3eme cycle, Univ. Claude Bernard, Lyon, p. 59.

Dussart B (1969). Les Copépodes des eaux continentales d'Europe occidentale. Tome II.

Fernández-Rosado MJ, Lucena J (2001). Space–time heterogeneities of the zooplankton distribution in La Concepción reservoir (Istán, Málaga; Spain), Hydrobiologia, 455: 157-170.

Gannon JE, Stemberger RS (1978). Zooplankton (specially crustaceans and rotifers) as indicators of water quality. Trans. Amer. Micros. Soc., 97: 19-35.

Geraldes AM, Boavida MJ (2004). What factors affect the pelagic Cladocerans of the mesoeutrophic Azibo reservoir? Ann. Limnol. Int. J. Limnol., 40(2): 101–111.

Giraudel JL, Lek S (2000). A comparison of self-organizing map algorithm and some conventional statistical methods for ecological community ordination, Ecol. Model, 146(1-3): 329–339.

Kagalou H, Economidis G, Leonardos I, Papalouka C (2006). Assessment of Mediterranean shallow lentic ecosystem Lake

Pamvotis (Greece) using benthic community diversity: response to environmental parameters, Limnologica – Ecol. Manage. Inland, 36(4-8): 269-278.

Kirk KL, Gilbert JJ (1990). Suspended clay and the population dynamics of planktonic rotifers and cladocerans, Ecol., 71: 1741–1755.

Lair N, Taleb H, Marchant PR (1996).Horizontal distribution of the rotifer plankton of Lake Aydat, Aquatic Sci., 58: 3.

Lefebvre A, Guelorget O, Perthuisot JP, Dafir JE (1997). Evolution biogéologique de la lagune de Nador (Maroc) au cours de la période 1982-1993. Oceanologica Acta, 20(2): 371-385.

Loffler (1961) in Dussart (B. 1969). Les Copépodes des eaux continentales d'Europe occidentale. Tome II : Cyclopoïdes et Biologie quantitative. Boubée et Cie, Paris, p. 292.

Lorenzen CJ (1967). Determination of chlorophyll and pheopingments spectrophotometric equation, limnol. Oceangor., 12: 343-346.

Maemets A (1983). Rotifer as indicators of lake types in Estonia. Hydrobiologia, 104: 357-361.

Margalef R (1983). Limnologia.Omega, Barcelona.

Mishra SR, Saksena DS (1989). Industrial effluent pollution at Birla Nagar, Gwalior. Pollut. Res., 8: 76-86.

Orcutt JRD, Pace ML (1984). Seasonal dynamics of rotifer and crustacean zooplankton populations in a eutropic,monomictic lake with a note on rotifer sampling techniques. Hydrobiologia, 119: 73-80.

Pinnel-Alloul B, Masson S, Patoine A (1996). Control of zooplankton structure in southern Quebec lakes by natural and anthropic environmental factors in the wather shelds and lakes. Third Intr. Cong. limnol.-Oceanogr., Nante, France.

Pourriot R, Francez AJ (1986). Rotifères : Introduction pratique à la systématique des organismes des eaux continentales françaises, annales de l'association française de limnologie, p. 37.

Rabette C, Lair N (1998) Influence des facteurs abiotiques sur la sortie des sédiments de Cyclops vicinus et Chaoborus flavicans dans les zones sub-littorale et profonde d'un lac tempéré eutrophe .Annls Limnol., 34(3): 295-303.

Rodier J (1996).L'analyse de l'eau : eaux naturelles, eaux résiduaires, eaux de mer. Chimie, physico-chimie, bactériologie, 8e édition, Dunod, Paris, p. 1383.

Ruggiero AG, Solimini G, Carchini G (2004). Limnological aspect of an Apennine shallow lake, Ann. Limnol. Int. J. Limnol., 40(2): 89–99.

Ruttner-kilisko A (1977). Suggestions of biomass calculation of planktonic Rotifers., Arch. Hydrobiol. Ergebn. Limnol., 8: 71-76.

Sladecek V (1983). Rotifer as indicators of water quality.Hydrobiologia, 100: 169-201.

Velho LFM, Lansac-Tôha FA, Bonecker CC, Bini LM, Rossa DC (2001). The longitudinal distribution of Copepods in Corumbá Reservoir, State of Goiás, Brazil, Hydrobiologia. Pp. 385–391, 453–454.

Wetzel RG (2001). Limnology—Lake and River Ecosystems, third ed., Academic Press, p. 1006.

13

Diversity and distribution of ladybird beetles in District Dir Lower, Pakistan

Rahatullah[1], Faizul Haq[2]*, Sardar Azhar Mehmood[1], Kausar Saeed[1] and Shamsur Rehman[3]

[1]Department of Zoology, Hazara University, Mansehra, Pakistan.
[2]Department of Botany, Degree College, Battagram, Pakistan.
[3]Department of Science, High School Battamori, Battagram. Pakistan.

This paper communicates the diversity, distribution and host importance of ladybird beetles collected from District Dir Lower, Pakistan. A survey was conducted in the District Dir Lower over a period of two years. Specimens of coccinellid beetles were collected from Maidan, Jandool, Timergara and Adenzai. Identification of these beetles showed that 14 different species belonging to sub-families Coccinellinae, Chilocorinae, Scymninae and Epilachninae are present in the selected area. Out of 14 species from Dir Lower, two species *Hypraspis rahatiana* and *Cryptogonus faizihanum* were reported for the first time from Pakistan and is therefore, a new addition to Coccinellid fauna of Pakistan.

Key words: Dir Lower, Pakistan, ladybird beetles, fauna.

INTRODUCTION

Predaceous ladybird beetles (Coleoptera: Coccinellidae) (Latreille, 1807; Hodek and Honek, 1996) occur within the subfamilies *Chilocorinae*, *Coccinellinae, Coccidulinae*, *Scymninae*, *Sticholotidinae* whereas the *Epilachninae* are phytophagous. Although subfamilies of the Coccinellidae are more or less worldwide in distribution, many tribes within these subfamilies are restricted to particular biogeographical regions. About 6000 species of Coccinellidae are known worldwide (Vandenberg, 2000) with over 300 species known from the Indo-Pakistan sub-continent (Rahatullah et al., 2010). About 90% of coccinellid species are considered beneficial because of their predatory activity, mainly against homopteran insects and phytophagous mites injurious to various agricultural and forest plants. Some members of Epilachninae are however, serious pests of brinjal and cucurbits in Pakistan. Ladybird beetles undergo complete metamorphosis with distinct egg, larval, pupal and adult stages (Shah, 1985). Their life cycle is completed in one month depending upon location and temperature; two or three generations are generally produced in a year. Adults overwinter in sheltered locations such as tree holes and other natural hiding places (Majerus and Kearns, 1989). Gilani (1976) reported ten coccinellid species from the Faisalabad, Pakistan. Sasaji (1977) discussed the phylogenetic relationship among *Harmonia* species by the larval characters, between *H. axyridis* and *H. yedoeneis*. Inayatullah and Siddiqui (1978) studied the skeletal anatomy of the abdomen of *Coccinella septermpunctata* L., *Coccinella undecimpunctata* L. and discussed their taxonomic importance.

Khan et al. (2007) conducted survey of predatory Coccinellids in the Chitral District, Pakistan. Ghani and Mohyddin (1982) described seven species from Pakistan. Shah (1985) studied the Coccinellids of Peshawar valley and recorded 16 species along with geographical distribution and host plants. Din (2002) described predatory Coccinellid fauna of Chitral. Rafi et al. (2005) reported some species of predatory Coccinellids from Bagh, Sudhnuti and Poonch districts of Azad Jammu Kashmir. Keeping in view in importance of these useful predators, the present studies were conducted to enlist the species of Coccinellidae in Dir Lower.

The study area

Coccinellidae were collected from the Lower Dir area. District Dir is stretched over an area of 1,582 km^2 of Malakand, lying along the Afghanistan border between

*Corresponding author. E-mail: faizulhaq80@yahoo.com.

Chitral and Peshawar. Almost all coccinellid collection areas are administratively controlled by the Lower Dir District, the lower part of River Panjkora which raises high in the Hindu Kush at latitude 35.45 and joins the Swat River near Chakdara which is the main point of entry to the District (latitude 34.40). Dir is a mountainous area with peaks rising to 4876 m in the northeast and to 3048 m along the watersheds with Swat to the East and Afghanistan to the West (Anonymous, 1998; Rahatullah et al., 2010). Our objectives were: 1) document species of Coccinellidae found in the Dir Lower district, 2) record the distribution and seasonal occurrence of the species, 3) create awareness regarding Coccinellidae and their importance.

MATERIALS AND METHODS

The following strategy was adopted for exploring species diversity and elaborating distribution of the coccinellid species in the District Dir Lower.

Sample collection

To obtain in-depth knowledge of the general ecology of ladybird beetles, the area was visited frequently during the last quarter of 2007. It was more convincing to divide the study area into four sampling regions, namely,. Maidan, Jandool, Timergara and Adenzai tehsils (Table 1). Collection of beetles was done from different parts of these regions during 2008, from early spring to the autumn season. Each locality was visited frequently, bi-weekly, in general till the end of October 2008. A total of 224 plant species were studied for their host importance in different localities. The Coccinellid specimens were collected by hand netting and hand picking from each selected locality. Collected specimens were placed in a killing bottle and then pinned and placed in a collection box. Very small specimens were mounted on points. Each specimen was labeled noting the place of collection, date of collection, collector's name and host plant species.

Identification

The specimens were identified with the help of available literature and already identified specimens which are preserved in the insect Museum of KP Agricultural University, Peshawar. All the identified specimens were deposited in the Zoological Museum, Hazara University, Mansehra.

Description

Descriptions of the specimens were made on the visual observation and obviously differentiable traits.

RESULTS

A total of 510 specimens of coccinellid beetles were collected from four localities within the district. Identification of these beetles revealed 14 different species within 12 genera belonging to subfamilies Coccinellinae, Chilocorinae, Scymninae and Epilachninae were present in the area. Detailed descriptions of the specimens are provided as follows:

Bromoaid saturalis Fabricius

Distribution and seasonal occurrence

Brumoides saturalis was the most-frequently encountered species collected from each locality; twenty five specimens were collected from Maidan, Jandool and Timergara, and specimens of these species were abundant in September.

Host and importance

Fifty four plant species were studied for their host's importance in different localities. Among them, these species were abundantly found on *Pyrus malus* feeding on soft bodied nymphs of aphids.

Coccinella septempunctata Linnaeus

Distribution and seasonal occurrence

These species is widely distributed in the District and was found in all the localities. Thirty specimens were collected from all the selected localities. Highest number (10) was collected from Maidan, Namazkot. Individuals of these species were abundant from May to September. First specimens were collected on 1st April and last were collected on the 27th October.

Host and importance

Ninety three plant species were studied for their host's importance. Among them, these species were collected from *Zea mays*, *Solanum melongena, Lectuca sativa, Triticum aestivum, Morus alba* and *Brassica compestris,* feeding on *Aphis craccivora* Koch*, Aphis fabae,* Scopoli and *Aphis gossypii* Glover.

Coccinella transversalis Fabricius

Distribution and seasonal occurrence

Seven specimens of these species were collected from Maidan and Jandool. Five specimens were collected in June and two in July. These species appeared in June and was not available after July.

Host and importance

Thirty nine plant species were studied for their host's

Table 1. Total number of specimens collected from different localities.

Name of area	Total number of specimens
Maidan	255
Jandool	160
Timergara	50
Adenzai	45
Total	510

importance. Among them, these species were collected from *Glycine max*, *L. sativa*, *Hibiscus esculentus* and *Prunus domestica* and *Helianthus annus* feeding on *A. craccivora* Koch, *A. fabae* Scopoli and *A. gossypii* Glover.

Hypraspis rahatiana

Distribution and seasonal occurrence

Fifteen specimens of these species were collected from Maidan in July and August. These species were available in large number in August.

Host and importance

Sixty one plant species were studied for their host's importance. Among them, these species were collected from *H. annus* and *Z. mays,* feeding on *A. craccivora Koch, A. fabae* Scopoli and *A. gossypii* Glover.

Hippodamia varigata Goeze

Distribution and seasonal occurrence

Hippodamia varigata were recorded from all localities and 35 specimens were collected. Highest number of 10 was collected from Maidan and Jandool Smarbagh in months of April and May while one was collected in month of August from Timergara. Individuals of this species were abundant from April to June.

Host and importance

Seventy plant species were studied for their host's importance. Among lot of species, *H. varigata* were collected from *T. aestivum, Brassica compertris* and *Coriandrum sativum* feeding on *Aphidophagous* in nature.

Halyzia tschischerini Semenov

Distribution and seasonal occurrence

Twenty-five specimens of these species were collected from Maidan and Jandool. The species was abundant in October. The first specimen was collected in September.

Host and importance

Eighty one plant species were studied for their host's importance. Among lot of species, these species were collected from *Dilbergia sisso,* feeding on powdery mildew, also attacks on different species of Aphids.

Harmonia dimidata Fabricius

Distribution and seasonal occurrence

Ten specimens of these species were collected from Maidan, Jandool, Timergara and Balambat. They were present in large number in September. The first specimen was collected in August.

Host and importance

Fifty plant species were studied for their host's importance. Among lot of species, these species were collected from *Dilbergia sisso* and *Z. mays*; feeding on *Adelges sp.* and *A. craccivora* Koch.

Oenopia sauzeti Mulsant

Distribution and seasonal occurrence

These species were collected in Maidan and Jandool. Ten specimens were collected in April while two in August from Jandool. Individuals of this were available in large number in April.

Host and importance

These species were collected from *Z. mays* and an undetermined ornamental plant, feeding on *A. crasccivora* Koch, *A. fabae* Scopoli and *A. gossypii* Glover.

Illies confusa Timberlake

Distribution and seasonal occurrence

Thirty five specimens of these species were collected from two localities Timergara and Adenzai. The first specimen was collected in September. Individuals of this species were available in large number in October.

Host and importance

These species were collected form *D. sisso* and an

undetermined ornamental plant Alta; feeding on Aphidophagous in nature.

Menochilus sexmaculata Fabricius

Distribution and seasonal occurrence

The species is distributed in Maidan, Jandool, Timergara and Adenzai. Thirty specimens were collected from all the localities. Highest number collected was 6 in August and lowest number 2. Individuals of this species were available in large number from July to end of August in Maidan and Jandool, but in Adenzai the beetle was abundant in April.

Host and importance

Host plants of this species were *H. annus* and *Z. mays*, an undetermined ornamental plant, feeding on *Adelgids sp. Bemisia tabaci* and *Pyrilla perpusilla*.

Propylea leuteopustulata Mulsant

Distribution and seasonal occurrence

The species was recorded from all the localities and 15 specimens were collected. Highest number of 5 was collected from Maidan and Jandool in April and May. Lowest number collected was 1 in August from Maidan. Individuals of this species were abundant from April to June.

Host and importance

Thirty one plant species were studied for their host's importance. These species were mostly collected from *Triticum aestivum* and *B. compertris;* feeding on aphids, psyllids and whiteflies.

Cryptogonus nepalensis Bielawski

Distribution and seasonal occurrence

These species were collected from Maidan in June.

Host and importance

Seventy plant species were studied for their host's importance. Among lot of species, these species were collected from mot grass; it is in a predator of scale insects.

Cryptogonus faizihanum

Distribution and seasonal occurrence

Fifteen species were collected from Maidan and Jandool in August.

Host and importance

Seventy plant species were studied for their host's importance. These species were collected from mot grass; it is a predator of scale insects.

Epilachna sp. Linnaeus

Distribution and seasonal occurrence

Twelve specimens of these species were collected from two localities Jandool and Adenzai. These species were available in large numbers in July. The first specimen was collected in May and the last specimen was collected in July.

Host and importance

Host plants of this species were *Z. mays, T. aestivum* and *Solenum tuberosum.* It is phytophagus in nature.

DISCUSSION

During the present study, 14 species from 12 genera belonging to sub families Coccinellinae, Chilocorinae, Scyninae and Epilachninae were collected from the Dir Lower district. *B. saturalis* were collected from all localities. The species were most abundantly found on *P. malus* feeding on soft bodied nymphs of aphids. A total of 50 specimens of *Coccinellina septempunctata* L. were collected from all localities. This species was previously reported by Irshad (2001), (Tables 2 and 3) Rehman et al. (1960), Gilani (1976), Gilani and Mohyuddin (1982), Ahmad et al. (1999), Shah (1985) from Pakistan. It was collected from a large number of plants including *Z. mays*, *Hibiscus escolentus*, *S. melongena*, *L. sativa* and *G. max* etc. *Coccinellina transversalis* was also collected from Maidan and Jandool. It was reported by Gilani (1976), Shah (1985) and Irshad (2001). The species was collected from *G. max*, *L. sativa*, *H. esculentus* and *P. domestica*. Five specimens of *C. transversalis* were collected from *H. annus* in Maidan. *Hyperaspis rahatiana* was also collected from Maidan. This species were reported for the first time from Pakistan and is a new addition to *Coccinellidea fauna* of Pakistan. The host plants of the species were *Z. mays* and *H. annus. H. varigata* was reported with 12 specimens from Jandool,

Table 2. List of total number of species collected from different localities.

Maidan	Jandool	Timergara	Adenzai
Brumoides saturalis	*B. saturalis*	*B. saturalis*	*B. saturalis*
Coccinella septempunctata	*C. septempunctata*	*C. septempunctata*	*C. septempunctata*
Coccinella transversalis	*Coccinella transversalis*	-	-
Hyperaspis rahatiana	-	-	-
Hippodamia varigata	*H. varigata*	*H. varigata*	*H. varigata*
Halyzia tschitscherini	*H. tschitscherini*	-	-
Harmonia dimidata	*H. dimidata*	*H. dimidata*	*H. dimidata*
Oenopia sauzeti	*Oenopia sauzeti*	-	-
-	-	*Illeis confusa*	*Illeis confuse*
Menochilus sexmaculata	*M. sexmaculata*	*M. sexmaculata*	*M. sexmaculata*
Propylea leuteopustulata	*P. leuteopustulata*	*P. leuteopustulata*	*P. leuteopustulata*
Cryptogonous nepalensis	-	-	-
Cryptogonus faizihanum	*C. faizihanum*	-	-
-	*Epilachna* sp.	-	*Epilachna* sp.

Table 3. Coccinellids and their host plant relationships.

Name of species	Host plant	Season of collection
Brumoides saturalis	*Pyrus malus*	September
Coccinella septempunctata	*Z. mays*, *S. melongena*, *L. sativa*, *T. aestivum*, *M. alba* and *B. compestris*	April - October
Coccinella transversalis	*T. aestivum*, *H. annus*, *M. alba* and *B. compestris*	June - July
Hyperaspis rahatiana	*Z. mays* and *H. annus*	July-August
Hippodimia varigata	*T. aestivum*, *B. compestris* and *C. sativum.*	April- August
Halyzia tschitscherini	*D. sisso*	September-October
Harmonia dimidiate	*Z. mays* and *D. sisso*	August-September
Oenopia sauzeti	*Z. mays* and other ornamental plants	April- August
Illeis confuse	*D. sisso*	September-October
Menochilus sexmaculata	*H. annus*, *Z. mays* and other ornamental plants	April- August
Propylea leuteopustulata	*T. aestivum* and *B. compertris*	April- June
Cryptogonous nepalensis	Mot grass	June
Cryptogonus faizihanum	Mot grass	August
Epilachna sp.	*Z. mays*, *T. aestivum* and *S. tuberosum*	May-July

20 specimens from Maidan, 4 specimens from Timergara and 3 specimens from Adenzai. *H. variegata* was reported by Gilani (1976), Shah (1985) and Irshad (2001) from Pakistan. The species was collected from wide range of host plants including *T. aestivum*, *Brassica compestris* and *Coriandum sativum* feeding on *Aphidophagous* in nature. *Halyzia tschitscherini* was collected from Maidan and Jandool from *D. sisso* feeding on powdery mildew, also attacks on different species of Aphids. *Harmonia dimidata* were found in all localities. These species were collected from *D. sisso* and Zea mays; feeding on Adelges sp. and *A. craccivora* Koch. *Oenopia sauzeti* was collected from Maidan and Jandool. Host plants of these species were *Z. mays*, *H. annus*, an undetermined ornamental plant feeding on *A. crasccivora* Koch, *A. fabae* Scopoli and *A. gossypii* Glover. *Illeis cunfusa* was collected from Timergara and Adenzai. This species was collected from *D. sisso.*

Menochilus sexmaculata was recorded from all localities. Total 35 specimens were collected. Previously it was reported by Shah (1885) from Peshawar, Pakistan. It was collected from *H. annus*, *Z. mays*, *H. esculentus* and *P. domestica* feeding on Adelgids sp. *B. tabaci* and *P. perpusilla*. *P. leuteopustulata* were recorded from all localities. These species were collected from *Triticum aestivu* and *B. compertris*; feeding on aphids, psyllids and whiteflies. *Cryptogonus nepalunsis* was rarely distributed in Maidan and colleced from mot grass. *Cryptogonus faizihanum* was collected with two adults from Maidan and Jandool. The host plants of this species were mot grass. *C. faizihanum* was reported for the first time from Pakistan. *Epilachna sp.* was collected from Jandool and Adenzai. Species of subfamily Epilachininae were reported by Ghani and Mohyuddin (1982) from Pakistan. Shah (1985) reported species of the subfamily from Peshawar valley. These species were collected from *Z. mays, T. aestivum* and *S. tuberosum.* It is phytophagus in nature. Among the fourteen species recorded in the present study from Dir Lower, *H. rahatiana* and *C. faizihanum* species were reported for the first time from Pakistan (Table 2).

REFERENCES

Ahmad FK, Baig MU, Mustafa SG (1999). Spatial distribution and Phenology of adult *Chilocorus infernalis* on apple trees in Kashmir Pakistan. Sharhad J. Agric., 15: 343-346.

Anonymous (1998). Statsistics Section, Planning & Development Department, Govt. of State of Jammu & Kashmir, pp. 1-25.

Din S (2002). Distribution of Predatory Coccinellid (Coleoptera: Coccinellidae) beetles in District Chitral. Dept of Plant Protect, M.Sc. Thesis. NWFP Agric Univ Peshawar, Pak., p. 37.

Ghani MA, Mohyuddin (1982). Investigation on the natural enemies of *Eiplachna* sp. in Pakistan. Final Report, Project No. F.G. Pa-296. C.I.B.C. Rawalpindi. p. 31.

Gilani WA (1976). Studies on the predaceous Coccinellidai of Lyal Pur. Unpublished Thesis, M.sc. (Hons) Ento. Dept Agric. Univ Faisalabad. p. 81.

Inayatullah M, Siddiqu EM (1978). Comparative studies on the anatomy of the abdomen of *Coccinellae septerpunctala* L., *Coccinella underimpunctala* L. (Coccinellidae: Colleeptera), Pak. J. Zool., 10: 85-97.

Irshad M (2001). Distribution, host, Ecology and biotic potential of Coccinellids of Pakistan. Pak. J. Biol. Sci., 4: 1259-1263.

Khan I, Din S, Khalil SK, Rafi MA (2007). Survey of predatory Cocicnellids(Coleoptera: Coccinellidae) in the Chitral District, Pakistan. 6. J. Insect Sci., 7: 07.

Latreille PA (1807). Genera Crustaceorum et Insectorum Secundum Ordinem Naturalem in Familias Disposita, Iconobus Exemplisque Plurimus Explicata Paris, 3: 258.

Majerus M, Kearns PK (1989). "Lady Birds". University of Cambridge. Rechmond Publishing Co. Ltd. P.O. Box 963, Slough, SL 23 RS, England, pp. 1-101.

Mohyuddin AI (1992). A report of PARC-IIBC on Malakand Fruit and Vegetable Development Project, Swat, Pakistan, p. 50.

Rafi MA, Irshad M, Inyatullaha M (2005). Predatory Ladybird beetles of Pakistan. PARC/NWFP Agric .Univ. Roohani Art Press, Islamabad, Pakistan, p. 105.

Rahatullah, Ahmad H, Inayatullah M (2010). M. Phil. thesis on species diversity of coccinellidae of Dir valley.

Rehman MA, Ghani MA, Kazmi K (1960). Introduction of Sanjose Scale into pakistna. Publication of Common Wealth Institute of Biological Control Pakistan Station. Rawalpindi, pp. 165: 18.

Sasaji H (1977). Larval characters of the Asian species of the genus *Harmonia* Muls. (Coleoptera: Cccinellidae). Mem. Fac. Eduo. Fukui Unir. Ser II. Nat. Sci., 27: 1-18.

Shah ZM (1985). Ladybird Beetles (Cccinellidae: Coleoptera) of Peshawar region. M.sc (Hons) Thesis, Deptt, of Entomol NWFP Agric Univ. Peshawar (accepted), p. 109.

Vandenberg NJ (2000). Coccinellidae Latreille, 1807: 19.

Spatio-temporal distribution of drosophilids: A study at Jnanabharathi Campus, Bangalore, Karnataka, India

B. P. Harini* and D. S. Pranesh Sujaymeendra

Department of Zoology, Drosophila Culture Laboratory, Bangalore University,
Jnana Bharathi Campus, Bangalore 560 056, India.

Though studies associated with the geographic distribution of Drosophilidae in India have taken rapid strides in last few years, only a cursory survey has been undertaken in certain areas, whereas a vast area of the Indian subcontinent still awaits exploration. In view of this, an attempt has been made to record the Drosophilid fauna of Jnanabharathi habitat, which is a completely virgin field, to explore with a view to furnish a spatio-temporal distribution pattern of Drosophilid species. The present study has revealed that species diversity has enriched in undisturbed region where human habitat is poor. On the other hand, the diversity and distribution of the Drosophilids have been affected enormously where human habitat is frequently sensed. In addition to this seasonally related environmental factors have a direct or indirect effect on the population density and distribution pattern of *Drosophila* in space and time.

Key words: *Drosophila* fauna, biodiversity, distribution pattern, spatial, temperature.

INTRODUCTION

The Drosophilidae family encompasses 3,952 species distributed in 73 genera and two subfamilies (Steganinae and Drosophilinae). The subfamily Drosophilinae is the most diversified one and includes 3,240 species distributed across 44 genera, of which *Drosophila* is the most speciose, with 1,159 species recorded (Bächli, 2009). Some of these species are endemic, while others are cosmopolitan and may often disperse in association with human activities (Tidon-Sklorz and Sene, 1999). The *Drosophila* species are observed in essentially any environment, from the sea level to considerable altitudes, and in temperate as well as in equatorial zones (Throckmorton, 1975). However, Lachaise (1979) suggests that these species are subject to restrictions as regards the habitats they live in. Sporadic reports are available in regard to eco-distributional patterns, altitudinal and seasonal variations in South India (Hegde, 1979; Hegde et al., 1989, 1999, 2000, Guruprasad et al., 2010).

The Indian subcontinent, with its subtropical climate and varied physiographic conditions, including variable altitudes and luxuriant flora offers an abode for the rich and wide distribution of *Drosophila* fauna. During recent years, considerable data have been accumulated regarding faunal composition of Drosophilid species as a result of extensive field collections in different ecological habitat by Ayala (1970). The Drosophilid flies thus obtained have been utilized for various studies *viz.*, taxonomic, ecological, genetical, behavioral, and its distribution record. As a result more than 290 Drosophilid species have been reported so far from different ecogeographical areas in India (Gupta, 2005; Kumar et al., 2009) and most of them are new to the world of science. Studies on Drosophilid assemblages aim at offering a simplified explanation of the complex systems and circumstances that are repeatedly observed in different habitats. Thus the increased global habitat destruction, studies on the diversity of insects in various environments are highly important regarding the comprehension of biological assemblages and the devising of conservation strategies.

Other authors have also stressed the trend exhibited by certain species towards occupying inner forests or else some sort of open environment (Dobzhansky and Pavan, 1950). Studies that have evaluated Drosophilid assemblages in forest fragments demonstrate the clear segregation between the faunistic composition of the inner forest and the disturbed areas. Climatic variables

*Corresponding author. E-mail: bpharini@yahoo.co.in.

such as humidity, rainfall, temperature and incidence of sunlight, among others, are determining factors in the occurrence of Drosophilid species (Pavan, 1959). Similarly, biotic factors like the kind of vegetation that form natural gradients and changes associated to latitude, for example, are also important (Powell, 1997). Therefore, the composition and structure of a Drosophilid assemblage depends on the habitat in which it was established. The recognition of patterns at these organizational levels affords to propose hypotheses about the ecology and evolution of the groups studied (Begon et al., 1990).

MATERIALS AND METHODS

Drosophila flies were collected at Jnanabharathi Campus in the first week of the months of May, August, November, 2009 and February, 2010 at three different localities that is, in open field (human activity is more) termed Locality 1, forest edge (human activity scarce) termed Locality 2, and inner forest (absence of human activity) termed Locality 3. The method used to collect the flies was by net sweeping (Markow and O'Grady, 2006). The rotten mixed fruits were spread in the evening of the previous day. The fermented fruit was spread in each distant trap collection spots. Such bait as fermenting fruits retains its attractive odour for a long time. The collections were made early in the morning by sweeping in each traps at least three times and transferred to six quarter pint milk bottles filled with standard agar medium sprayed with yeast. The collected flies were brought to the laboratory, etherized, categorized, counted and species were identified under Leica Stereozoom Microscope. The males were studied as such but the individual females, which could not be identified, were isolated and allowed to breed in separate vials containing standard laboratory food medium. The progeny obtained from such single gravid females were used for species identification.

RESULTS

The field collection of Drosophilid from the Jnanabharathi Campus in different seasons have yielded a total of 3488 flies during year 2009 - 2010 (1609 males and 1879 females). The males were grouped on the basis of morphological characters and analyzed taxonomically on the basis of genital structure and sex comb. The taxonomic analysis on the basis of genital structure has helped us to identify several species of Drosophilids.

Distribution of different species of *Drosophila* and their numbers found during collections along with temperature and rainfall in Jnanbharathi Campus, Bangalore, Karnataka, India during 2009 - 2010. The data (Table 1) from the present survey has revealed 12 different species of *Drosophila* with no new species found distributed in different localities of JB Campus namely, *Drosophila melanogaste* > *Drosophila ananassae*> > *Drosophila bipectinata* > *Drosophila biarempis* > *Drosophila simulans* > *Drosophila kikkawai* > *Drosophila malerkotliana* > *Drosophila takashi* > *Drosophila jambulina* > *Drosophila nigra* > *Drosophila neonasuta* > *Drosophila nasuta* Of which, *D. ananassae* is more abundant than other 11 species in all the spots of the assessments. In addition to this, *D. nasuta* was not found to appear in the month of May and February in L-1 and L-2, but it was recorded in the month of August and November IN L-1 and L-2. Interestingly, very fewer number of *D. nasuta* was found in all the above assessed months in L-3 locality. Accordingly, *D. melanogaster* and *D. ananassae* are more adapted to the prevailing environment and *D. nasuta* is more sensitive to the temperature and the locality.

Figure 1 depicts the percent distribution of different species of *Drosophila* at different localities of the JB Campus in open field (L-1) is about 28.78%. At Forest edge that is, at L-2 location about 30.90% of flies were collected and subsequently at the Forest interior L-3 site 40.30% of flies were sampled. It clearly shows that the Drosophilid distributions are largely affected by the human habitat where disturbance gradient is frequent in L-1 and also at the edges of forest. The number of *Drosophila* species found was increasingly high at L-3 locality of the present study and least number (Figure 2).

The collection of flies' season wise (Figure 3a – c). It appears that the flies were found to be collected more in number in the month of August and November and lesser in the month of May and February. Thus the temperature has an impact on the development and the distribution of the flies. The sampling size varies with the season and temperature and it is as follows: November > August > May > February.

DISCUSSION

A better understanding of how different species are affected by current climates and why they sometimes respond differently to climate change is necessary for predicting future effects of climate change (Weatherhead, 2005). The *Drosophila* species are observed in essentially any environment, from the sea level to considerable altitudes, and in temperate as well as in equatorial zones (Throckmorton, 1975). However, Lachaise (1979) suggests that these species are subject to restrictions as regards the habitats they live in. Other authors have also stressed the trend exhibited by certain species towards occupying inner forests or else some sort of open environment (Dobzhansky and Pavan, 1950; Sene et al., 1980).

Studies that have evaluated Drosophilid assemblages in forest fragments demonstrate the clear segregation between the faunistic composition of the inner forest and the disturbed areas. In relation to these studies the present data also implies on the distribution of *Drosophila* flies depends on spatio-temporal factors. Out of 3488 flies recorded in JB Campus at different seasons of the year, about 28.78% of the flies (1004) were found in the highly disturbed niche. Whereas, 30.90% (1078) of flies in the niche edges and 40.30% (1406) of flies were

Table 1. Spatio-Temporal distribution pattern of different species of *Drosophila* at three different localities of JB Campus, Bangalore.

Locations	**Open field (human activity is more)- LI**				**Forest edge (human activity scarce)-LII**				**Inner forest (absence of human activity)- LIII**			
Months	**May**	**Aug**	**Nov**	**Feb**	**May**	**Aug**	**Nov**	**Feb**	**May**	**Aug**	**Nov**	**Feb**
Temperature (°C)	32	28	27	30	32	29	27	30	32	29	27	30
Rainfall (mm)	154	83.2	118.8	3.2	154	83.2	118.8	32	154	83.2	118.8	3.2
D. melanogaster	48	38	55	36	59	44	55	39	59	64	70	33
D. annanassae	44	66	69	33	47	75	69	35	57	85	85	45
D. bipectinata	45	44	47	36	49	52	47	40	51	62	67	40
D. biarempis	36	42	48	24	37	46	48	28	39	49	58	30
D. simulans	12	19	23	08	17	22	23	11	19	32	38	17
D. kikkawai	18	21	24	06	21	20	24	14	23	30	34	20
D. malerkotliana	12	19	22	09	15	25	22	05	19	29	32	12
D. takashi	04	09	10	00	08	11	10	03	07	17	19	07
D. jambulina	10	15	19	00	18	17	19	00	16	20	29	09
D .nigra	06	02	11	03	09	02	11	04	14	06	16	07
D.s.neonasuta	02	04	00	00	06	07	00	00	00	05	03	05
D. nasuta	00	02	03	00	00	00	08	00	03	08	07	09
Total	237	281	331	155	286	277	336	179	307	407	458	234

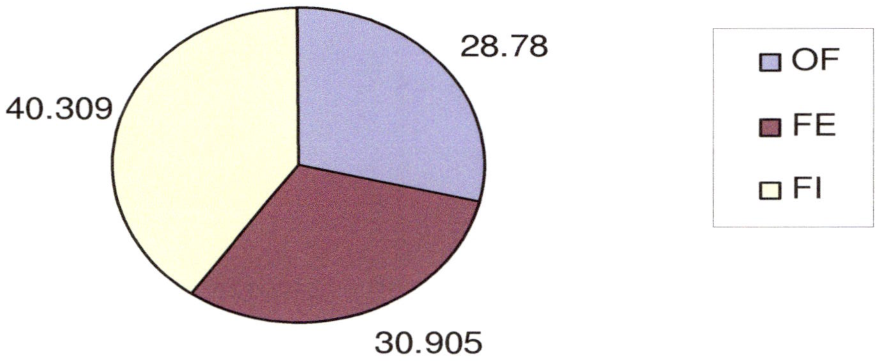

Figure 1. Percent distribution of *Drosophila* species at different localities (OF = Open field; FE = Forest edge; FI = Forest interior) of JB Campus.

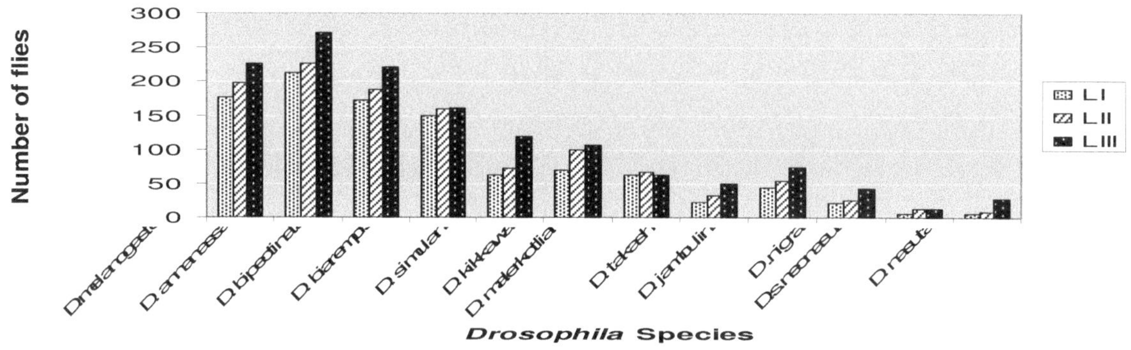

Figure 2. Number of *Drosophila* species at different localities (LI = Open field; LII = Forest edge; LIII = Forest interior) of JB Campus.

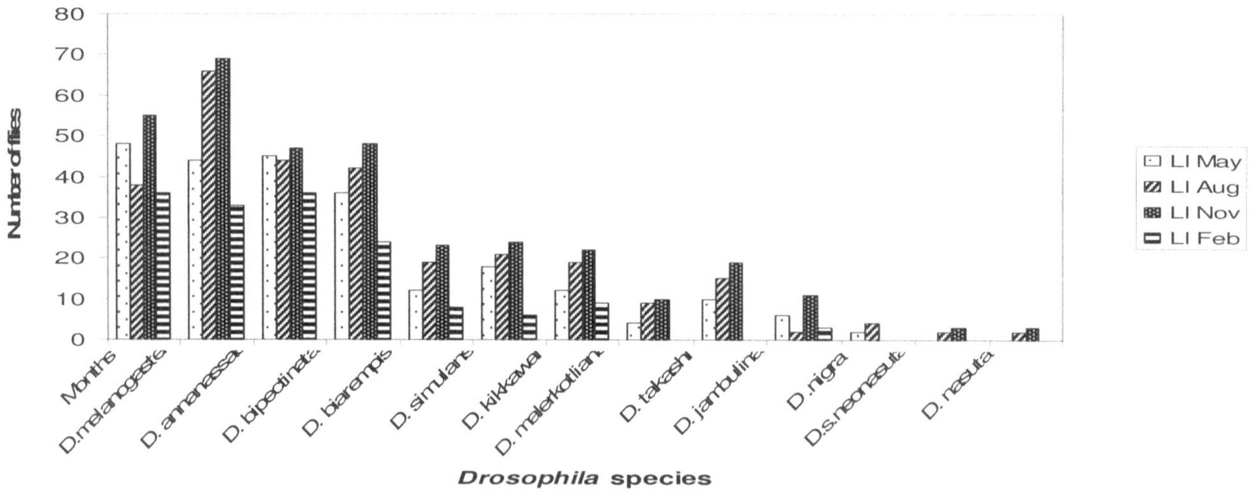

Figure 3a. Number of flies recorded in different months at LI locality at JB Campus.

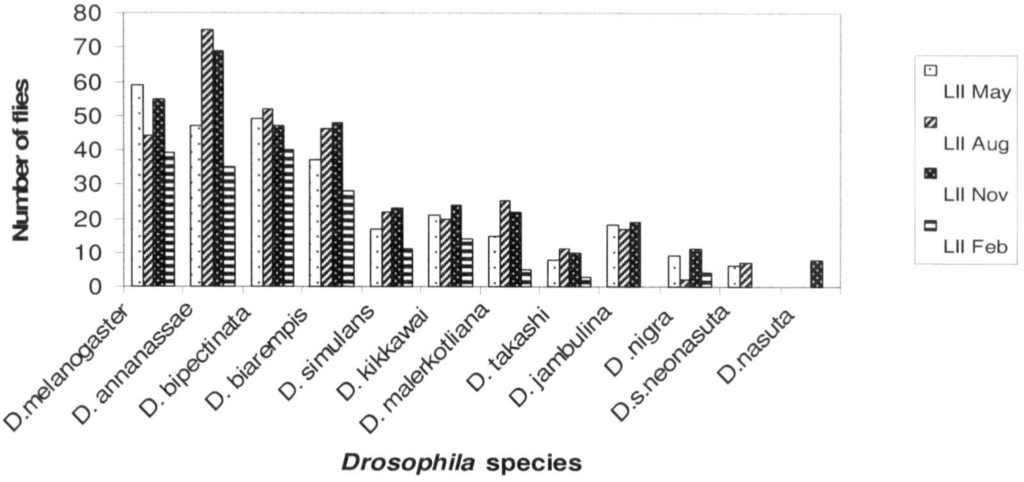

Figure 3b. Number of flies recorded in different months at LII locality at JB Campus.

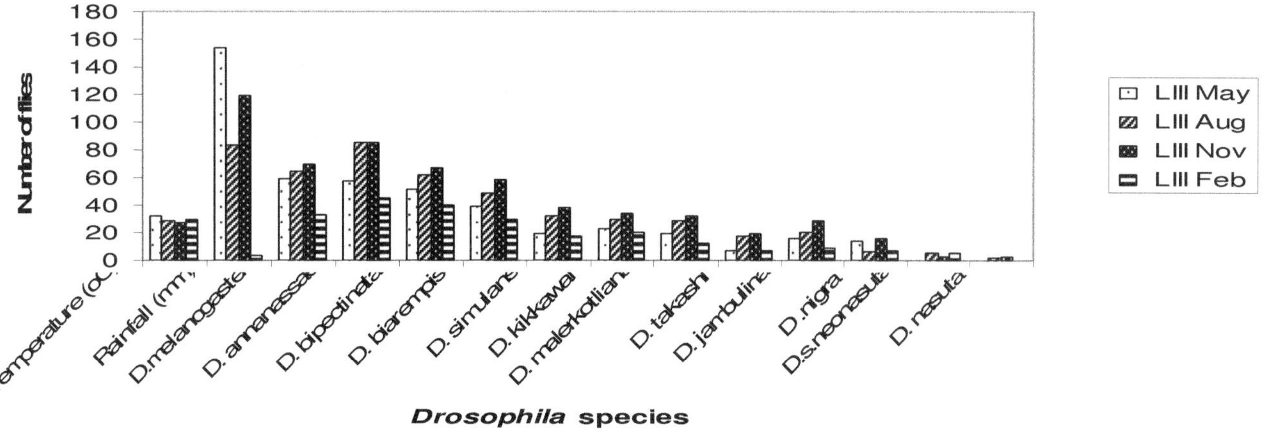

Figure 3c. Number of flies recorded in different months at LIII locality at JB Campus.

recorded and collected in the undisturbed area of the JB Campus (Figure 1). It is evident that, the flies are less sampled in the area were the human habitat is frequently seen. Species that are habitat specialists make up much of biodiversity, but the evolutionary factors that limit their distributions have rarely been considered, such species are likely to be constrained in their evolutionary responses to future climate changes (Kellermann et al., 2009).

Interestingly, it was also observed that the increase in temperature during the month of May and February the flies were less recorded, and with decreased temperature that is, during the month of November and August the collection was increasingly high. According to Hegde et al. (2000) the growth and size of the population depend on several environmental factors in addition to genetic structure. The fluctuation in population size of *Drosophila* through different seasons reflects the close relationship between populations. It should be emphasized that the months with higher species richness occur during the rainy season.

The present study also implies that the climatic variables such as humidity, rainfall, temperature are determining factors in the occurrence of Drosophilid species as suggested by (Pavan, 1959). Therefore, the composition and structure of a Drosophilid assemblage depends on the habitat in which it was established. Studies on Drosophilid assemblages aim at offering a simplified explanation of the complex systems and circumstances that are repeatedly observed in different habitats. The recognition of patterns at these organizational levels affords to propose hypotheses about the ecology and evolution of the groups studied (Begon et al., 1990). Greater divergence between a species pair involved divergent selection on both niche dimensions. The dimensionality of selection may complement genetic and geographic explanations for the degree of diversification in nature (Nosil and Sandoval, 2008).

Thereby the present study provides information on the spatial and temporal distribution of the Drosophilidae family in three environments (open field, forest edge and inner forest) along the four seasons during one year 2009 (May, August and Novenmber) to 2010 (February) in Jnanabharathi Campus at Bangalore, Karanataka, India. The study opines, diversity and distribution of the Drosophilids have been affected enormously where human habitat is frequently sensed. In addition to this seasonally related environmental factors have a direct or indirect effect on the population density of *Drosophila*. Thus the presence or absence of a species in an ecological niche, and its richness or abundance in that area is an indicator of both biological and ecological diversity of that ecosystem. In addition to physical and biotic factors, the topography and season also affect the animal distribution. *Drosophila* a representative system for population geneticists to understand genetic basis of ecological differentiation at the levels of population and species. *Drosophila* is ecologically a rather highly specialized but closely knit group which offers valuable opportunities for studies on organisms-environment relations.

ACKNOWLEDGEMENTS

The authors are thankful to UGC, New Delhi for providing financial support.

REFERENCES

Ayala FJ (1970). Competition, coexistence and evolution. In: Essays in Evolution and Genetics in Honour of Theodosius Dobzhansky (Hecht, M.K., and W.C. Steere, eds.). Appleton-Century-Crofts, New York): pp.121-158.

Bächli G (2009). TaxoDros: The database on Taxonomy of Drosophilidae.Electronic Database accessible at http://www.taxodros.unizh.ch.

Begon M, Harper JL, Townsend CR (1990), Individual,s Population,s and Communities. Ecology Blackwell Scientific Publications, Melbourne, Australia., p. 876.

Dobzhansky T, Pavan C (1950). Local and seasonal variations in relative frequencies of species of *Drosophila* in Brazil. J. Anim. Ecol., 19: 1-14.

Gupta JP, (2005), A monograph on Indian Drosophilidae. J. Sci. Res., 5: 1-252.

Guruprasad BR, Hegde SN, Krishna MS (2010). Seasonal and altitudinal changes in population density of 20 species of *Drosophila* in Chamundi hill. J. Insect Sci., 10: 1-12.

Hegde SN (1979). Studies on cyto taxonomy and genetics of a few members of the melanogaster species group of Drosophila (Ph. D. Thesis) University of Mysore, Mysore, India.

Hegde SN, Vasudev V, Krishna MS, Shakuntala V (1999). Description of two new species of montium subgroup of Drosophila (Diptera: Drosophilidae from South India). Entomon., 24: 149-156.

Hegde SN, Vasudev V, Shakuntala V, Krishna MS (2000). Drosophila fauna of Palani hills Tamilnadu, India. Droso. Inf. Ser., 81: 138.

Kellermann V, Van Heerwaarden B, Sgrò CM, Hoffmann AA (2009). Fundamental evolutionary limits in ecological traits drive *Drosophila* species distributions. Science, 325: 1244-1246.

Kumar R, Ajai K (2009). Studies on Drosophilidae (Diptera: Drosophilidae of Gujarat state in India. Dros. Inf. Serv., 92: 106.

Lachaise D, Pignal MC, Roualt J (1979). Yeast flora partitioning by Drosophilid species inhabiting a tropical African savanna of the ivory-coast (*Diptera*). Ann. Soc. Entomol., 15: 659–680.

Markow TA, O'Grady PM (2006). Drosophila : a guide to species identification and use. Elsevier Inc. Academic press pub., pp. 1-247.

Nosil P, Sandoval CP (2008) Ecological niche dimensionality and the evolutionary diversification of stick insects. PLoS One, 3: 1-11.

Pavan C (1959). Relationship between natural populations of Drosophila environment. Bolm. Fac Filos. Hist S. Paul, Biol. General, 11: 1-81.

Powell JR (1997). Progress and Prospects in Evolutionary Biology, The Drosophila Model (Powell, J.R., ed.), New York, Oxford University Press.

Sene F, Val FC, Vilela CR, Pereira MA (1980). Preliminary data on the geographical distribution of *Drosophila* species within morphoclimatic domains of Brazil. Pap. Avul. Dep. Zool. Sec. Agric., 33: 315-326.

Throckmorton LH (1975). In: Handbook of Genetics (King, R.C., ed.)Plenum Press, New York, pp. 421-467.

Tidon-Sklorz R, Sene FM (1999). Terrestrial invertebrates. Biodiversity of the State of São Paulo. Synthesis of Knowledge at the End of the Twentieth Century, FAPESP, São Paulo. p. 5.

Weatherhead PJ (2005). Effects of climatic variation on timing of nesting reproductive success and offspring sex ratios of red-winged black birds, Oecologia, 10: 1007.

15

Vegetation analysis of Montane forest of Western Ghats with special emphasis on RET species

Brilliant R.[1]*, Vincy Mary Varghese[2], Joby Paul[2] and Pradeepkumar A. P.[3]

[1]Department of Environmental Science, St. John's College, Anchal, India.
[2]School of Environmental Sciences, Mahatma Gandhi University, Kottayam, India.
[3]Department of Geology, University of Kerala, Kariavattom, India.

The paper provides a synoptic account of the floristic diversity of the montane forest of Western Ghats of India. Present study has revealed the occurrence of 286 species of plants belonging to 85 families. Among them, 88 rare, endangered and threatened (RET) species were recorded justifying the inclusion of the area in the Western Ghats Hotspot. The rapid growth in tourism has created more infrastructure, increased pollution, put unsustainable demands on local environments and generated adverse impacts on biodiversity. Damaging these natural resources is not only detrimental to the biodiversity of the world, but to the tourism industry itself. Good biodiversity management is good business for tourism companies. Through contributing to the conservation of biodiversity, tourism companies can improve the quality of the service or product provided, enhance their reputation, increase recognition, publicise themselves, save on costs and increase their income. Significantly can contribute towards making tourism sustainable and guaranteeing the livelihoods of people who work within or benefit from the tourism industry now and in the future. The Landuse/Land Cover (LULC) changes in the extent of different landuse categories during the period from 1969 to 2010 was carried out. The paper discusses the negative impacts of tourism on phytodiversity which pinpoints the importance for the conservation of grasslands by means of different diversity indices.

Key words: RET species, endemic flora, biodiversity indices.

INTRODUCTION

Life on Earth depends on the components of biological diversity. Biodiversity maintains essential ecosystem functions including primary production and decomposition. Humans are part of biodiversity, but their activities, deficient knowledge and ever-increasing demand on natural resources are among the main causes for the loss of diversity in the living world. Conservation of plant diversity assumes greater importance when the world is facing unprecedented loss of biological diversity. As per an estimate about 60,000 out of 2,87,655 species of plants known in the world are facing the threat of extinction. 11,824 species were evaluated for their threat status as per the revised 1994 IUCN Red List Categories; of these 8321 species are now on the IUCN Red List 2004 (IUCN, 2004). Shola forests are unique montane vegetation occupying temperate habitats in tropical latitude and are communities restricted to valley and depressions especially along the folds of hills.

The montane grasslands and adjacent evergreen tropical forests (locally called the shola forests) form a distinctive vegetation mosaic. The Western Ghats are a mountain range that runs north-south parallel to the western coastline of the Indian peninsula from 8° N to 22° N. The montane grasslands and adjacent forests face several threats largely due to increasing anthropogenic activities (Pramod et al., 1997). Once widespread across the high altitudes of Western Ghats, the grasslands are

*Corresponding author: E-mail:env.brilliant@gmail.com.

now highly fragmented and restricted to approximately 400 km (Karunakaran et al., 1998). Unlike the adjacent forests (Gadgil and Vartak, 1975; Chandran et al., 1998; Brown et al., 2006), the grasslands have not received much attention from biologists and conservationists. Landscape level analysis of vegetation, especially from the Vagamon grasslands of Kerala are particularly lacking.

Biodiversity hotspots are areas that support natural ecosystems that are largely intact and where native species and communities associated with these ecosystems are well represented. They are also areas with a high diversity of locally endemic species, which are species that are not found or are rarely found outside the hotspot. Emphasis was given mainly on the plants that are endemic (to India) and/or in the rare, endangered and threatened (RET) category according to IUCN (International Union for Conservation of nature and Natural Resources). A rare species is one with small population that is not presently endangered but is at risk, an endangered species is one, which is in danger of extinction throughout all or of a significant portion of its range and a threatened species is one, which is likely to become endangered in foreseeable future (IUCN, 1978; Bryde, 1979; Smith 1980; Nayar and Sastry, 1990).

Study area

Across the world, 25 hotspots have been identified on the basis of species endemism and degree of threat through habitat loss (Myer et. al., 2000). Out of these, two are confined to India sub-continent (that is, Western Ghats / SriLanka and Indo-Burma). The present study site falls within the Western Ghats.

Vagamon (9° 66' and 9° 73´ N latitude and 76° 86´ to 76° 98´ E longitude) has been identified as a biodiversity 'hotspot'. Urban development in the area is generally low; the climate is equitable, ranging from 32°C in summer to 16°C in winter. With an average annual rainfall of about 376 cm, Vagamon receives rain for about 200 days a year. The enchanting hill station dotted with tea gardens will soon be the venue of one of State's foremost eco-tourism projects. The Vagamon hills scattered at a height of about 1,100 m above the mean sea level with delightfully attractive meadows, open forests, waterfalls and valleys offer adventure enthusiasts and nature lover's endless opportunities. Introduction of tourism development activities may result in the deterioration of the ecology of the hills. Extensive developmental activities like road construction in hilly areas will undermine the stability of hill slopes and may cause hazards like landslides. Hence the concept of sustainable development becomes very pertinent for hilly regions, which is influenced to some extent by the climate of the plains; they contain the sources, the catchments and the watersheds of several major river systems, which flow down to the plains; they flourish in natural vegetation and mineral wealth.

The grassland ecosystems has been subjected to particularly intense pressure for the upcoming tourism activities, so current extinction rates of grassland species are expected to remain high or even increase. However, research on these grasslands is scanty and highly dispersed. The main aim of the paper was to consider current processes and future scenarios for grassland biodiversity. It also gives emphasis for the need for conservation of the grasslands and shoal forest.

MATERIALS AND METHODS

Remote sensing techniques are particularly suited for providing reliable, up-to-date and comprehensive data on land-use/land-cover. Apart from the field investigations, SOI Toposheets (1967) (58C/14) and IRS-P 6 LISS III data (Feb. 19, 2004) were used for preparing the land-use map of the area (Figure 2). Different thematic maps were prepared using map info version 5 and 7. The Landuse/Land Cover (LULC) changes in the extent of a,b land use categories during the period from 1967 to 2009 have been carried out.

The study was conducted during the year 2007 to 2010. Plots were laid put by stratified random approach. The plants were identified with the help of Flora of British India (Hooker, 1872-1897), Flora of Presidency of Madras (Gamble and Fischer, 1915-1936), Indian Medicinal Plants (Basu, 1975), Flora of Palghat (Vajravelu, 1990), Flora of Thiruvananthapuram (Mohanan and Henry, 1994), Flowering Plants of Thrissur Forest (Sasidharan and Sivarajan, 1996), 100 Red Listed Medicinal Plants of Conservation Concern in Southern India (Ravikumar and Ved, 2000) and other regional floras. The identification of endemic plants was done by way of standard books such as Endemic Plants of the Indian Region (Ahmedulla and Nayar, 1987), Red Data Book of Indian Plants (Nayar and Sastry, 1987, 1988, 1990), Hot Spots of Endemic Plants of India, Nepal and Bhutan (Nayar, 1996) and Flowering Plants of Kerala (Sasidharan, 2004).

Researchers have applied different indices in order to measure diversity (Krebs, 1998). In most studies about biodiversity, Alpha and Beta diversities have been considered (Pitkanen, 1998). Alpha diversity has been species diversity in a region (Whittaker, 1972) and Beta diversity mentions to the amount of changes of species in environmental gradient length (Whittaker, 1972; Wilson and Shmida, 1984). In the present article, alpha diversity and beta diversity have been studied and important indices of species diversity and evenness were calculated.

For vegetation study, qualitative and quantitative methods were employed. The qualitative method included the collection, identification and verification of plant components using standard flora and herbarium. The entire flora from these plots were counted and identified by following standard keys (Gamble and Fischer, 1915 - 1936). Final identification of materials was done after consulting relevant literature and expert judgment. The nomenclature of each species has been brought up to date as per rules given in the International Code of Botanical Nomenclature (1988).

For analyzing the vegetation quantitatively, quadrate method was employed. The size of the quadrate was determined by the species area curve method. Quadrates of different sizes such as 10 x 10 m for trees, 5 x 5 m for shrubs and climbers and 1 x 1 m for herbs and grasses were laid. The forest dominant communities such as the trees were given importance. The plant components were taxonomically identified, counted and measured. Girth at breast

Table 1. List of RET species.

Species	Vernacular name	Family	Distribution	Status
Actinodaphne campanulata Hook. f.		Lauraceae	Endemic to the Western Ghats	Endangered (IUCN, 2000).
Actinodaphne malabarica Balakr.	Kambilivirinji	Lauraceae	Endemic to Southern Western Ghats	Rare (Nayar, 1997)
Aeschynanthus perrottetii A. DC.		Gesneriaceae	Endemic to Western Ghats	
Allophylus concanicus Radlk.		Sapindaceae	Endemic to Western Ghats	Rare (Nayar and Sastry, 1988)
Amomum pterocarpum Thw.	Peralam	Zingiberaceae	Endemic to Western Ghats	Rare (Nayar, 1997)
Anaphalis lawii (Hook. f.) Gamble		Asteraceae	Endemic to Peninsular India	
Anaphyllum wightii Schott	Keerikizhangu	Araceae	Endemic to Southern Western Ghats	Threatened (Nayar, 1997)
Andrographis explicata (Clarke) Gamble		Acanthaceae	Endemic to Western Ghats	Rare (Nayar, 1997)
Arthraxon lanceolatus (Roxb.) Hochst.		Poaceae		Threatened
Artocarpus hirsutus Lam.	Angili	Moraceae	Endemic to Southern Western Ghats	Endangered
Asystasia gangetica (L.) Anders.	Upputhali	Acanthaceae	Endemic to Peninsular India, Sri Lanka, Arabia and Africa	
Begonia albo-coccinea Hook.		Begoniaceae	Endemic to Southern Western Ghats	Vulnerable (Nayar, 1997)
Begonia trichocarpa Dalz.		Begoniaceae	Endemic to Western Ghats	Vulnerable (Nayar, 1997)
Belosynapsis vivipara (Dalz.)		Commelinaceae	Endemic to Western Ghats	Vulnerable (Nayar, 1997)
Bentinckia condapanna Berry and Roxb.	Kantha-kamugu	Araceae	Endemic to Southern Western Ghats	Vulnerable (IUCN, 2000).
Boesenbergia pulcherrima (Wall.) O. Ktze.		Zingiberaceae	Endemic to Southern Western Ghats	Threatened (Nayar, 1997)
Calamus delessertianus Becc.	Chooral	Arecaceae	Endemic to Western Ghats	
Ceropegia maculata Bedd.		Asclepiadaceae		Possibly extinct
Christisonia tubulosa (Wight) Benth. ex Hook. f.		Orobanchaceae	Endemic to Southern Western Ghats	
Cinnamomum malabatrum (Burm. f.) Blume	Vayana	Lauraceae	Endemis to Southern Western Ghats	
Cinnamomum sulphuratum Nees	Kattu karuva	Lauraceae	Endemic to Western Ghats	
Crinum defixum Ker-Gawl.	Veluthapolathali	Amaryllidaceae	Endemic to India and Sri Lanka	
Crotalaria barbata Grah. Ex Wight and Arn.		Fabaceae	Endemic to Southern Western Ghats	Rare (Nayar, 1997)
Crotalaria grahamiana Wight and Arn.		Fabaceae	Endemic to Southern Western Ghats	Rare (Ahmedullah and Nayar, 1987)
Crotalaria obtecta Grah. ex Wight and Arn.		Fabaceae	Endemic to Southern Western Ghats	Rare
Cullenia exarillata Robyns	Kurunguplavu	Bombacaceae	Endemic to Southern Western Ghats	
Cyclea fissicalyx Dunn		Menispermaceae	Endemic to Southern Western Ghats	Rare and Threatened (Ahmedullah and Nayar, 1987)
Cymbopogon flexuosus (Nees ex Steud). Wats.	Chukkunari pullu	Poaceae	Endemic to India and South East Asia	
Cynanchum alatum Wight and Arn.		Asclepiadaceae	Endemic to Southern Western Ghats	Rare (Nayar, 1997)
Debregeasia ceylanica Hook. f.		Urticaceae		Threatened
Dendrobium haemoglossum Thw.		Orchidaceae		Threatened
Desmos viridiflorus (Bedd.) Safford		Annonaceae	Endemic to Southern Western Ghats	Endangered (Nayar, 1997)
Dictyospermum montanum Wight		Commelinaceae	Endemic to Peninsular India	

Table 1. Continued

Dictyospermum ovalifolium Wight		Commelinaceae	Endemic to Western Ghats	Rare (Nayar and Sastry, 1987)
Didymocarpus fischeri Gamble		Gesneriaceae	Endemic to Southern Western Ghats	Vulnerable (Nayar, 1997)
Ensete superbum (Roxb.) Cheesman.	Kalluvazha	Musaceae	Endemic to Peninsular India	
Exacum bicolor Roxb.		Gentianaceae	Endemic to Peninsular India	
Exacum courtallense Arn.		Gentianaceae	Endemic to Southern Western Ghats	Rare (Nayar, 1997)
Flueggea leucopyrus Willd.	Ambooripachila	Euphorbiaceae	Endemic to India, Sri Lanka and Myanmar	
Globba marantina L.		Zingiberaceae	Endemic to Peninsular India	
Glochidion bourdillonii Gamble		Euphorbiaceae	Endemic to Peninsular India	Vulnerable (IUCN, 2000).
Glycosmis macrocarpa Wight		Rutaceae	Endemic to South India and Sri Lanka	Rare (Nayar, 1997)
Goniothalamus wightii Hook. f.and Thoms.		Annonaceae	Endemic to Southern Western Ghats	
Grewia hirsuta Vahl.		Tiliaceae	Endemic to India, Sri Lanka and Bangladesh	Common
Hedyotis swertioides Hook. f.		Rubiaceae	Endemic to Southern Western Ghats	Rare (Nayar, 1997)
Hoya retusa Dalz.		Asclepiadaceae	Endemic to Western Ghats	Threatened
Impatiens acaulis Arn.		Balsaminaceae		Threatened (Nayar, 1997)
Impatiens cordata Wight.		Balsaminaceae	Endemic to Southern Western Ghats	
Impatiens herbicola Hook. f.		Balsaminaceae		Vulnerable (Nayar, 1997)
Impatiens parasitica Bedd.		Balsaminaceae	Endemic to Southern Western Ghats	
Litsea floribunda (Blume) Gamble	Pattuthali	Lauraceae	Endemic to Western Ghats	
Mallotus philippensis (Lam.) Meull. Arg.		Euphorbiaceae	Endemic to Peninsular India	
Medinilla beddomei Clarke		Marantaceae	Endemic to Southern Western Ghats	
Memecylon lawsonii Gamble		Melastomataceae	Endemic to Southern Western Ghats	Vulnerable (IUCN, 2000).
Mussaenda belilla Buch.-Ham.	Vellila	Rubiaceae	Endemic to Western Ghats	
Myristica malabarica Lamk.	Panampalaka	Myristicaceae	Endemic to Western Ghats	Vulnerable (IUCN, 2000).
Nothopegia racemosa (Dalz.) Ramam.	Naicheru	Anacardiaceae	Endemic to Western Ghats	
Ochlandra scriptoria (Dennst.) Fischer	Eera	Poaceae	Endemic to Western Ghats	
Ochlandra travancorica (Bedd.) Benth. ex Gamble	Eetta	Poaceae	Endemic to Western Ghats	
Osbeckia aspera (L.) Blume		Melastomataceae	Endemic to Southern Western Ghats	Possibly extinct (Nayar, 1997)
Pecteilis gigantea (Smith) Rafin.		Orchidaceae		Threatened
Peucedanum anamallayense Clarke		Apiaceae	Endemic to Southern Western Ghats	Endangered (Nayar, 1997)
Phaeanthus malabaricus Bedd.		Annonaceae	Endemic to Southern Western Ghats	Lower risk: near threatened (IUCN, 2000)
Phoenix sylvestris (L.) Roxb.		Arecaceae	Endemic to India, Sri Lanka, Himalaya and Myanmar	
Piper mullesua Buch.-Ham. ex D. zdon	Kattukurumulaku	Piperaceae	Endemic to Peninsular and north east India	
Premna glaberrima Wight		Verbenaceae	Endemic to Southern Western Ghats	Rare (Nayar, 1997)
Pterocarpus santalinus L. f.	Raktha Chandanam	Fabaceae	Endemic to India, Sri Lanka and Pakistan	
Pterospermum reticulatum Wight. and Arn.	Malyuram	Sterculiaceae	Endemic to Southern Western Ghats	Vulnerable (IUCN, 2000).

Table 1. Continued.

Rauvolfia micrantha Hook. f.		Apocynaceae	Endemic to Southern Western Ghats	Endangered (Nayar, 1997)
Rubus ellipticus Smith.		Rosaceae	Endemic to India, Sri Lanka and Myanmar	
Semecarpus auriculata Bedd.	Vellacheru	Anacardiaceae	Endemic to Southern Western Ghats	Lower risk: near threatened (IUCN, 2000)
Smithia venkobarowii Gamble		Fabaceae	Endemic to Southern Western Ghats	Possibly extinct (Nayar, 1997)
Solanum anguivi Lam.	Neelavazhutina	Solanaceae	Endemic to Peninsular India	
Sonerila sahyadrica Giri and Nayar		Melastomataceae	Endemic to Southern Western Ghats	Rare (Nayar, 1997)
Strobilanthes ciliatus Nees	Karimkurinji	Acanthaceae	Endemic to Peninsular India	
Syzygium cumini (L.) Skeels		Myrtaceae	Endemic to Southern Western Ghats	
Syzygium mundagam (Bourd.) Chitra	Kattuchampa	Myrtaceae	Endemic to Southern Western Ghats	Rare
Tabernaemontana heyneana Wall.	Kundalappala	Apocynaceae	Endemic to Southern Western Ghats	Lower risk: near threatened (IUCN, 2000)
Terminalia paniculata Roth.	Maruthu	Combretaceae	Endemic to Peninsular India	
Thunbergia mysorensis (Wight) Anders.		Acanthaceae	Endemic to Western Ghats	
Torenia bicolor Dalz.	Kakkapoovu	Scrophulariaceae	Endemic to Western Ghats	
Vanasushava pedata (Wight) Mukh. and Const.		Apiaceae	Endemic to Southern Western Ghats	Rare (Nayar, 1997)
Vanda tessellata (Roxb.) Hook. ex D. Don	Arathamaravazha	Euphorbiaceae	Endemic to India, Myanmar, China and Sri Lanka	
Vernonia multibracteata Gamble		Asteraceae	Endemic to Southern Western Ghats	Possibly extinct (Nayar, 1997)
Wrightia tinctoria (Roxb.)	Vetttupala	Apocynaceae	Endemic to India, Myanmar and Timor	
Xanthophyllum arnottianum Wight.	Maddaka	Xanthophyllaceae	Endemic to Western Ghats	
Xanthophyllum flavescens non Roxb.	Mottal	Xanthophyllaceae	Endemic to Western Ghats	
Xanthophyllum flavescens non Roxb.	Mottal	Xanthophyllaceae	Endemic to Western Ghats	

height (GBH), total height as well as the bole height were measured. For each vegetation type, the structural attributes and Importance value index (IVI) were computed as described in Muller-Dombois and Ellenberg (1978) and explained below. Importance value index is a statistical quantity, which gives the overall dominance of a species in the entire community. The importance value index (IVI) for the tree species was determined as the sum of the relative density, relative frequency and relative dominance (Curtis, 1959). These characters were analyzed by following the methods of Curtis and McIntosh (1950) and Misra (1968) (Table 1). Alpha and beta diversities for different species were computed.

Sum of basal cover of individual plants of a species will yield total stand basal cover of that species. Basal area = $(GBH)^2 / 4\pi$; where, GBH = $2\pi r$. Absolute value which is the simplest beta diversity is the total number of species that are unique between communities (Meffe et al., 2002). The Sorensen index is a very simple measure of beta diversity, ranging from a value of 0 where there is no species overlap between the communities, to a value of 1 when exactly the same species are found in both communities (Sorensen, 1948 in Looman and Campbell, 1960). The number of species per sample is a measure of richness. The more species present in a sample, the 'richer' the sample. Menhinick's richness index is the ratio of the number of taxa to the square root of sample size. Simpson's Index It measures the probability that two individuals randomly selected from a sample will belong to the same species. The tree species diversity was determined by using Shannon-Wiener information function (H′) (Shannon and Wiener, 1963). Berger-Parker dominance is simply the number of individuals in the dominant taxon relative to *n,* It is simply a measure of the numerical importance of the most abundant species. The Gini coefficient is a measure of the inequality of a distribution, a value of 0 expressing total equality and a value of 1 maximal inequality. The Gini coefficient can range from 0 to 1.

RESULTS

The total land use of the study area of Vagamon was 104.3 km^2. The various landuse patterns were classified as (i) open scrub, (ii) tea cultivation, (iii) flood plain, (iv) Kurishumala

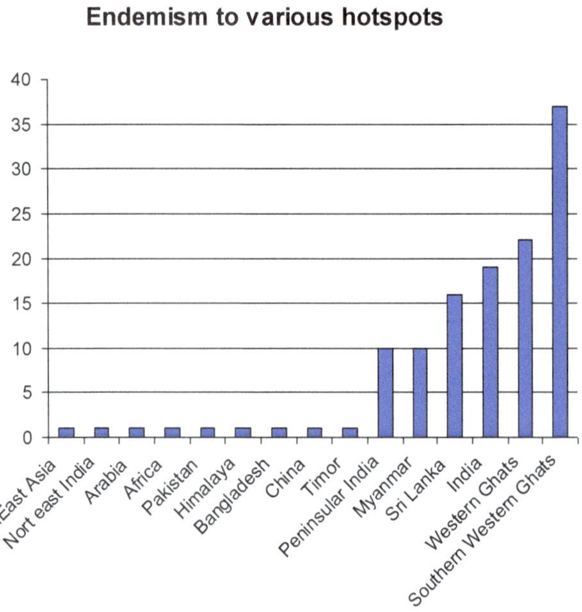

Figure 1. Endemism to various hotspots.

Ashramam, (v) green meadows, (vi) built up land, (vii) mixed crop with settlement, (viii) pine forest and (ix) eucalyptus.Majority of the land area consists of open scrub with the land area 76.48 km^2 in 1967 to 1969, which reduced to 71.49 km^2 in 2009. The cultivation of tea in the area commenced at the time of British rule. The tea plantation area in 1967 to 1969 was 25.02 km^2; but in 2009 it increased to 26.9 km^2. The flood plain area in 1967 to 1969 was 2.39 km^2; it was reduced to 2.08 km^2 in 2009 (Map 2). The land area of Kurishumala Ashramam was 0.08 km^2 and the area of green meadows in was 0.33 km^2. The built up land, mixed crop with settlement, pine forest and eucalyptus are the additional land use patterns after 1967 (Map 1). The built up land comprises an area of 2.187 km^2; the mixed crop with settlement 0.18 km^2; the pine forest covers 0.51 km^2 and eucalyptus area ranges up to 0.55 km^2.

Present study has revealed the occurrence of 286 species of plants belonging to 85 families. Among them, 88 endemic species were recorded justifying the inclusion of the area in the Western Ghats Hotspot (Table 1, 4). The most abundant species are from the families such as Asteraceae (19 spp.), Euphorbiaceae (14 spp.). Fabaceae (15 spp.) Rubiaceae (nine spp.), Orchidaceae (eight spp.) Araceae (nine spp.) Acanthaceae (9 spp.) and Poaceae (14 spp.). The quadrate analysis in 22 sites of the study area shows there are eight species of trees identified with *Memecylon* and *Cinnamomum* with great frequency of occurrence and *Syzygium cumini* with lesser frequency of occurrence. The percentage of Sorensen's similarity index is 100%, this shows the true grassland nature of the study area. Whittaker's Index is calculated as 1 and the total number of Common Species is

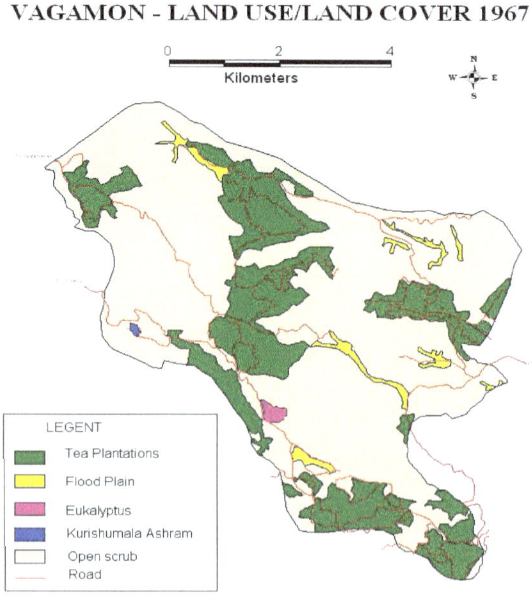

Figure 2a. Landuse/land cover of Vagamon, 1967 (Map 1).

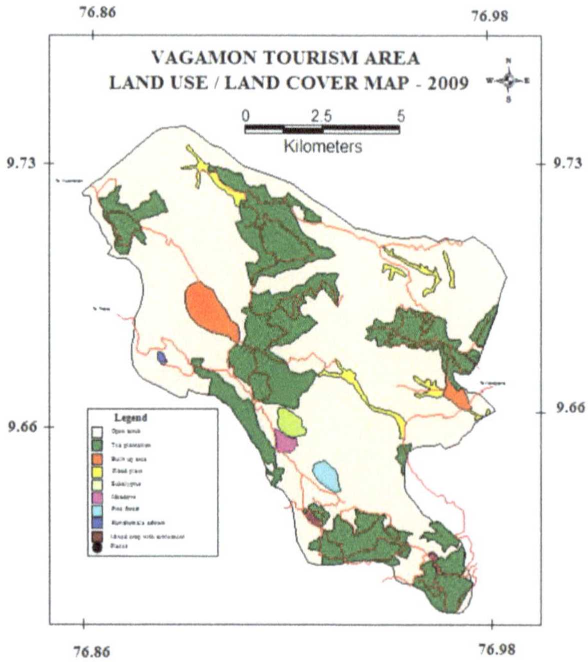

Figure 2b. Landuse/land cover of Vagamon, 2009.

calculated eight. Equitability is the evenness with which the individuals are spread out among the species in a community, the observed value in the study area is 0.7. 88 endemic species recorded as part of this study belong to 43 families (Table 3). Flora has been categorized according to IUCN (2004) (Table 2 and Figure 1).

Table 2. Categorization of plants according to IUCN (2004)

S/N	Category	Number
1	Endangered	5
2	Lower risk: near threatened	3
3	Possibly extinct	4
4	Rare	17
5	Threatened	8
6	Vulnerable	10

The quadrate analysis of the study area consists of 10 x 10, 5 x 5 and 1 x 1 m quadrates for the identification of trees, herbs, shrubs and seedlings and the grasses, respectively. Eight species of trees were identified. *Memecylon* and *Cinnamomum* had the highest frequency of occurrence followed by *Syzygium cumini* with lesser frequency of occurrence. High abundance is shown by *Maesa indica* and high relative density is shown by *S. cumini* and less by *M. indica*. High relative frequency is shown by *S. cumini* (12.6) *M. indica* (6.3), *Symplocos cochinchinensis* (6.3), *Olea dioica* (6.3), *Litsea* (6.3) and the tree belonging to Rubiacea (6.3). The high relative basal area is observed for *S. cumini* and it also shows high important value index (IVI).

The analysis of herbs, shrubs and seedlings and grasses show that high percentage is shown by *M. indica, C. odorata, Urena lobata*, and *Lantana camara*. The grass species with high percentage frequency include *Axonopus compresses, Justicia* spp., *Mimosa pudica* and *Ischaemum* sp. The density of the species *Pteridium aquilium, L. camara* and *M. pudica* is high compared to other observed species. *Pteridium aquilium, Symplocos* spp., *M. pudica* and *Ageratum conyzoides* also show high abundance. *Glochidion zeylanicum, Meliosma simplicifolia, Ficus hispida, Schefflera* spp. *Albizia, Desmodium triflorum, Ixora, Citrus, Flueggea, Cyclea, Pandanus, Plectranthus, Cymbopogon flexuosus, Desmodium heterophyllum, Alternanthera sessilis, Andrographis explicate, Vernonia cinerea* and *Chromolaena odorata* show high relative density. The relative frequency of herbs, shrubs, seedlings and grasses shows that *Vernonia, Clidemia hirta, G. zeylanicum, Meliosma simplicifolia, Cissampelos pareira, Solanum nguivi, Asclepias curassavica, Ficus hispida, Sida acuta, Triumfetta rhomboidea Pteris, Leucas ciliate, Smithia, Schefflera* sp., *Vernonia arborea, Xanthophyllum arnottianum, Albizia, D. triflorum, Andrographis, Anaphalis lawii, Blumea, Phoenix sylvestris, Psidium guajava, Jasminum* sps, *Osbeckia* spp., *Memecylon, Asparagus, Ixora, Citrus, Litsea, Flueggea, Cyclea, Pandanus, Plectranthus, Rauvolfia, Cymbopogon flexuosus, Desmodium heterophyllum, D. triflorum,Themeda sraes* and *Sporobulus* sp. are the relatively high frequency ones. *Smilax zeylanica, Vernonia, C. hirta, G. zeylanicum, Meliosma simplicifolia, Crotalaria, Cissampelos pareira, Solanum anguivi,Asclepias curassavica, Ficus hispida, Albizia, D. triflorum, Ixora, Citrus, Flueggea, Cyclea, Pandanus* and *Plectranthus* show high important value index (IVI).

Table 3. Number of families.

S. No.	Family	Number of occurrence
1	Acanthaceae	4
2	Amaryllidaceae	1
3	Anacardiaceae	2
4	Annonaceae	3
5	Apiaceae	2
6	Apocynaceae	3
7	Araceae	2
8	Arecaceae	2
9	Asclepiadaceae	3
10	Asteraceae	2
11	Balsaminaceae	4
12	Begoniaceae	2
13	Bombacaceae	1
14	Combretaceae	1
15	Commelinaceae	3
16	Euphorbiaceae	4
17	Fabaceae	5
18	Gentianaceae	2
19	Gesneriaceae	2
20	Lauraceae	5
21	Marantaceae	1
22	Melastomataceae	3
23	Menispermaceae	1
24	Moraceae	1
25	Musaceae	1
26	Myristicaceae	1
27	Myrtaceae	2
28	Orchidaceae	2
29	Orobanchaceae	1
30	Piperaceae	1
31	Poaceae	4
32	Rosaceae	1
33	Rubiaceae	2
34	Rutaceae	1
35	Sapindaceae	1
36	Scrophulariaceae	1
37	Solanaceae	1
38	Sterculiaceae	1
39	Tiliaceae	1
40	Urticaceae	1
41	Verbenaceae	1
42	Xanthophyllaceae	3
43	Zingiberaceae	3

Table 4. Quantitative analysis of floristic diversity of Vagamon.

S/N	Species	Frequency (%)	Density	Abundance	Relative density	Relative frequency	Relative basal area	IVI
Trees								
1	*Maesa indica*	40	6.6	16.5	0.28	6.3	0	6.58
2	*Symplocos cochinchinensis*	40	1.2	3	1.57	6.3	1.9	9.77
3	*Syzygium cumini*	20	0.2	1	9.42	12.6	4.5	26.52
4	*Olea dioica*	40	1	2.5	1.88	6.3	1.17	9.35
5	*Litsea*	40	0.6	1.5	3.14	6.3	2.21	11.65
6	*Cinnamomum*	60	0.6	1	3.14	4.2	2.2	9.54
7	*Memecylon*	80	0.8	1	2.35	3.1	1.77	7.22
8	*Rubiaceae tree*	40	0.4	1	4.71	6.3	3.21	14.22
Herbs, shrubs and seedlings								
1	*Maesa indica*	80	6.6	8.25	0.28	3.1	0	3.38
2	*Chromolaena odorata*	80	6.2	7.75	0.3	3.1	0	3.4
3	*Cinnamomum malabatrum*	40	1	2.5	1.88	6.3	0	8.18
4	*Pteridium aquilium*	40	47.8	119.5	0.39	6.3	0	6.69
5	*Stachytarpheta jamaicensis*	40	4	10	0.47	6.3	0	6.77
6	*Urena lobata*	80	1	1.25	1.88	3.1	0	4.98
7	*Actinodaphne malabarica*	40	1	2.5	1.88	6.3	0	8.18
8	*Mikania micrantha*	40	0.8	2	2.35	6.3	0	8.65
9	*Olea dioica*	40	1	2.5	1.88	6.3	0	8.18
10	*Symplocos*	40	12.6	31.5	0.59	6.3	0	6.89
11	*Smilax zeylanica*	40	0.6	1.5	3.1	6.3	0	9.4
12	*Cymbopogon flexuosus*	40	2.2	10.5	0.85	6.3	0	6.88
13	*Lantana camara*	80	24.8	31	0.75	3.1	0	3.85
14	*Vernonia*	20	1.4	7	1.34	12.6	0	13.94
15	*Clidemia hirta*	20	3	15	0.62	12.6	0	13.22
16	*Rubus ellipticus*	40	1.4	3.5	1.34	6.3	0	7.64
17	*Lobelia nicotianifolia*	40	2.4	6	0.31	6.3	0	6.61
18	*Glochidion zeylanicum*	20	0.2	1	9.42	12.6	0	22.02
19	*Meliosma simplicifolia*	20	0.2	1	9.42	12.6	0	22.02
20	*Crotalaria*	20	1.4	7	1.34	12.6	0	13.94
21	*Cissampelos pareira*	20	0.4	2	4.71	12.6	0	17.31
22	*Solanum anguivi*	20	0.6	3	3.14	12.6	0	15.74
23	*Asclepias curassavica*	20	0.4	2	4.71	12.6	0	17.31
24	*Ficus hispida*	20	0.2	1	9.42	12.6	0	22.02
25	*Sida acuta*	20	0.4	2	4.71	12.6	0	17.31
26	*Triumfetta rhomboidea*	20	0.6	3	3.14	12.6	0	15.74
27	*Pteris*	20	0.8	4	2.35	12.6	0	14.95
29	*Leucas ciliate*	20	0.6	3	3.14	12.6	0	15.74
30	*Smithia*	20	0.4	2	4.71	12.6	0	17.31
31	*Schefflera sp.*	20	0.2	1	9.42	12.6	0	22.02
32	*Vernonia arborea*	20	0.4	2	4.71	12.6	0	17.31
33	*Xanthophyllum arnottianum*	20	0.4	2	4.71	12.6	0	17.31
34	*Albizia*	20	0.2	1	9.42	12.6	0	22.02
35	*Desmodium triflorum*	20	0.2	1	9.42	12.6	0	22.02
36	*Andrographis*	20	0.4	2	4.71	12.6	0	17.31
37	*Anaphalis lawii*	20	0.4	2	4.71	12.6	0	17.31
38	*Blumea*	20	0.8	4	3.14	12.6	0	15.74
39	*Phoenix sylvestris*	20	1.6	8	1.17	12.6	0	13.77

Table 4. Continued.

40	*Psidium guajava*	20	1.4	7	1.34	12.6	0	13.94
41	*Jasminum* spp.	20	0.8	4	2.35	12.6	0	14.95
42	*Osbeckia*	20	1.2	6	1.57	12.6	0	14.17
43	*Memecylon*	20	0.8	4	2.35	12.6	0	14.95
44	*Asparagus*	20	0.4	2	4.71	12.6	0	17.31
45	*Ixora*	20	0.2	1	9.42	12.6	0	22.02
46	*Citrus*	20	0.2	1	9.42	12.6	0	22.02
47	*Litsea*	20	0.4	2	4.71	12.6	0	17.31
49	*Flueggea*	20	0.2	1	9.42	12.6	0	22.02
50	*Cyclea*	20	0.2	1	9.42	12.6	0	22.02
51	*Pandanus*	20	0.2	1	9.42	12.6	0	22.02
52	*Plectranthus*	20	0.2	1	9.42	12.6	0	22.02
53	*Rauvolfia*	20	0.4	2	4.71	12.6	0	17.31
Grasses								
1	*Axonopus compresses*	60						
2	*Cymbopogon flexuosus*	20	0.2	1	9.42	12.6	0	
3	*Urena lobata*	40	0.8	2	2.35	6.3	0	
4	*Desmodium heterophyllum*	20	0.2	1	9.42	12.6	0	
5	*Desmodium triflorum*	20	0.4	2	4.71	12.6	0	
6	*Justicia*	60	3.8	6.3	4.94	4.2	0	
7	*Mimosa pudica*	60	22.4	37.3	8.41	4.2	0	
8	*Themeda sraes*	20	0.4	2	4.71	1.26	0	
9	*Ischaemum sp.*	60				4.2	0	
10	Unknown grass	20				12.6	0	
11	*Sporobulus sp.*	40	1.4	3.5	13.4	6.3	0	
12	*Unknown grass*			1			0	
13	*Andrographis explicate*	20	0.2	1	9.42	12.6	0	
14	*Alternanthera sessilis*	20	0.2	9	9.42	12.6	0	
15	*Ageratum conyzoides*	20	1.8	16	1.04	12.6	0	
16	*Bidens pilosa*	20	3.2	2	5.88	12.6	0	
17	*Emilia sonchifolia*	20	0.4	1	4.71	12.6	0	
18	*Vernonia cinerea*	20	0.2		9.42	12.6	0	
19	*Mikania micrantha*						0	
20	*Elephantopus scaber*	40	7.8	4.75	2.45	6.3	0	
21	*Leucas ciliate*	40	0.8	2	2.35	6.3	0	
22	*Stachytarpheta*	40	0.4	1	4.71	6.3	0	
23	*Chromolaena odorata*	20	0.2	1	9.42	12.6	0	

DISCUSSION

Species diversity is generated by species interaction such as competition and niche diversification (Pianka, 1966; Bada, 1984) which is both greatly manifested in the tropics due to high humidity and temperature (Ojo and Ola-Adams, 1996). There are 57 tree species identified in the Vagamon area.

Simpson's diversity index

Simpson's Diversity Index is a measure of diversity which takes into account the number of species present, as well as the relative abundance of each species. As species richness and evenness increase, so does diversity. The value of Simpson's index ranges between 0 and 1. With this index, 1 represents infinite diversity and 0, no diversity (Simpson, 1949). The Simpsons index of the Vagamon area is calculated as 0.36 and the Dominance Index (D = 1 - Simpson) of Vagamon is 0.65 which shows moderate biodiversity in the area even though it is grassland.

Shannon index

Shannon index (entropy) is a diversity index, taking into

account the number of individuals as well as number of taxa. It varies from 0 for communities with only a single taxon to high values for communities with many taxa, each with few individuals.

H = -sum((n i /n)ln(n i /n)). Shannon-Wiener Index (log) is calculated as 2.09, Shannon-Wiener Index (ln) as 1.45 and the percentage of Shannon-Wiener Index (adjusted)* is 69.81%.

Menhinick's richness index

Menhinick's richness index is the ratio of the number of taxa to the square root of sample size. Menhinick Index of Vagamon is calculated as 1.06.

Margalef's richness index

Margale"s richness index is calculated by (S-1)/ln(n); where, S is the number of taxa, and n is the number of individuals. Margalef richness index for trees, grasses, herbs and shrubs are found to be 1.731, 3.539 and 7.975 respectively (Table 5, 6, 7)

Berger-Parker dominance

Berger-Parker dominance is simply the number of individuals in the dominant taxon relative to n. It is a simple measure of the numerical importance of the most abundant species. d = N max / N; where, N max is the number of individuals in the most abundant species, and N is the total number of individuals in the sample. The reciprocal of the index, 1/d, is often used, so that an increase in the value of the index accompanies an increase in diversity and a reduction in dominance. Berger-Parker dominance index of Vagamon is 0.58 and Inverted Berger-Parker dominance index is calculated as 1.73. It may also be expressed as its reciprocal, which has the advantage that low values indicate a low-diversity community and vice versa.

Absolute Beta value

Absolute beta value of Vagamon is calculated as 7.

Sorensen index

The Sorensen index is a very simple measure of beta diversity, ranging from a value of 0 where there is no species overlap between the communities, to a value of 1 when exactly the same species are found in both communities (Sorensen, 1948).

Sorensen's similarity index

Sorensen's similarity index is 1 and Sorensen's similarity index (%) is 100%, this shows the true grassland nature of Vagamon.

Whittaker's Index

Whittaker's Index is calculated as 1 and the total number of common species is calculated as 8.

Evenness

Equitability is the evenness with which the individuals are spread out among the species in a community, the observed value in the study area is 0.7.

Renyi entropy

The Rényi entropies are important in ecology and statistics as indices of diversity. The Rényi entropy, a generalization of Shannon entropy, is one of a family of functionals for quantifying the diversity, uncertainty or randomness of a system, the observed values of the study area shows higher diversity.

Gini index

The Gini coefficient is a measure of the inequality of a distribution; a value of 0 expressing total equality and a value of 1 indicate maximal inequality. The Gini coefficient can range from 0 to 1; it is sometimes multiplied by 100 to range between 0 and 100. A low Gini coefficient indicates a more equal distribution, with 0 corresponding to complete equality, the Gini coefficient in the Vagamon area is calculated as 2.23, which indicates higher equality of floral diversity in the grassland.

Total number of tree species identified in Vagamon is 56, with Simpson Index of 0.36, Shannon-Wiener Index (log):2.09 and Shannon-Wiener Index (adjusted) is 69.81%. The total number of herbs and shrubs identified in the Vagamon area by quantitative methods are 679 and the total species 53, the Simpson Index is 0.17, Dominance Index (D = 1 - Simpson) is calculated as 0.83. The index values consist of Shannon-Wiener index (log) 3.633, Menhinick index is 2.03, Sorensen's similarity index 1 and the adjusted Shannon-Wiener Index is the percent of the maximum possible diversity. The maximum is the log 2 ('total number of species'): 5.73 or ln('total number of species'): 3.97, depending on which formula is used. Total number of grasses is 501, Simpson index (∑(n i (n i -1)/(N(N-1)))): 0.17 and the dominance index (D = 1 - Simpson): 0.83. The different indexes clearly show that the area is rich in floristic diversity especially the patched shola in the hill valleys.

Many mountain ecosystems are host to higher species richness and levels of endemism than adjacent lowlands.

Table 5. Similarity index of trees of Vagamon.

Total number of tree species:	57	Total number of species:	8
Average population size:	7.125	Decimal accuracy:	4
Total Number of Regions:	1	Total number of region sets:	6
Alpha biodiversity			
Simpson Index ($\sum(n_i(n_i-1)/(N(N-1)))$):	0.3546	Alternate simpson index ($\sum((n/N))^2$):	0.366
Dominance Index (D = 1 - Simpson):	0.6454	Alternate dominance index (D = 1 - Simpson):	0.634
Reciprocal Simpson Index (1 / D):	2.82	Alternate Reciprocal Simpson Index (1 / D):	2.733
Shannon-Wiener Index (log):	2.094	Berger-Parker Dominance Index:	0.5789
Shannon-Wiener index (ln):	1.452	Inverted Berger-Parker dominance index:	1.727
Shannon-Wiener index (adjusted)*:	69.81%	Margalef richness index:	1.731
Menhinick index:	1.06	Rényi entropy/hill numbers (0,1,2,∞):	8, 4.283 2.733, ≈1.728
Buzas and Gibson's index:	0.5338	Gini coefficient:	2.234
Equitability index:	0.6981	ln() of Hill numbers (0,1,2,∞):	2.079, 1.455 1.005, ≈0.5471
Beta biodiversity			
Absolute beta value ((S_0-c)-(S_1-c)...):	7	Whittaker's index (S/alpha):	1
Sørensen's similarity index:	1	Alternate Whittaker's index (S/alpha-1):	0
Sørensen's similarity index (%):	100%	Number of common species:	8

Table 6. Similarity index of grasses of Vagamon.

Total number of organisms:	501	Total Number of Species:	23
Average population size:	21.78	Decimal Accuracy:	4
Total number of regions:	1	Total Number of Region Sets:	1
Alpha biodiversity			
Simpson index ($\sum(n_i(n_i-1)/(N(N-1)))$):	0.1656	Alternate Simpson index ($\sum((n/N))^2$):	0.1672
Dominance index (D = 1 - Simpson):	0.8344	Alternate dominance index (D = 1 - Simpson):	0.8328
Reciprocal Simpson index (1 / D):	6.04	Alternate reciprocal Simpson index (1 / D):	5.98
Shannon-Wiener index (log):	3.051	Berger-Parker dominance index:	0.2615
Shannon-Wiener index (ln):	2.115	Inverted Berger-Parker dominance index:	3.824
Shannon-Wiener index (adjusted)*:	67.44%	Margalef richness index:	3.539
Menhinick index:	1.028	Rényi entropy/hill numbers (0,1,2,∞):	23, 8.332 5.98, ≈∞
Buzas and Gibson's index:	0.3603	Gini Coefficient:	5.031
Equitability index:	0.6744	ln() of hill numbers (0,1,2,∞):	3.135, 2.12 1.788, ≈-INF
Beta biodiversity			
absolute beta value ((S_0-c)-(S_1-c)...):	22	Whittaker's index (S/alpha):	1
Sørensen's similarity index:	1	Alternate Whittaker's index (S/alpha-1):	0
Sørensen's similarity index (%):	100%	Number of common species:	23

Mountains at lower altitudes can support exceptional biodiversity, due to compression of a wide range of ecosystems into a relatively short distance. Endemism levels are often high, particularly on mountains at medium elevations in the tropics and warmer temperate zones. The environment of a region is negatively influenced by the increase of tourism, whereas the growth of tourism depends on the quality and characteristics of the environment (Bushell, 2002). Vagamon, a highland area coming under Western Ghats

Table 7. Similarity index of Herbs and shrubs of Vagamon.

Total number of organisms:	679	Total number of species:	53
Average population size:	12.81	Decimal accuracy:	4
Total number of regions:	1	Total number of region sets:	1
Alpha biodiversity			
Simpson index ($\sum(n_i(n_i-1)/(N(N-1)))$):	0.1723	Alternate Simpson index ($\sum((n/N))^2$):	0.1735
Dominance index (D = 1 - Simpson):	0.8277	Alternate dominance index (D = 1 - Simpson):	0.8265
Reciprocal Simpson index (1 / D):	5.805	Alternate reciprocal Simpson index (1 / D):	5.765
Shannon-Wiener index (log):	3.633	Berger-Parker dominance index:	0.352
Shannon-Wiener index (ln):	2.518	Inverted Berger-Parker dominance index:	2.841
Shannon-Wiener index (adjusted)*:	63.42%	Margalef Richness index:	7.975
Menhinick index:	2.034	Rényi Entropy/Hill numbers (0,1,2,∞):	53, 10 5.765, ≈∞
Buzas and Gibson's index:	0.234	Gini coeffificient:	9.395
Equitability index:	0.6342	ln() of hill numbers (0,1,2,∞):	3.97, 2.532 1.752, ≈-INF
Beta biodiversity			
Absolute beta value ((S_0-c)-(S_1-c)...):	52	Whittaker's index (S/alpha):	1
Sørensen's similarity index:	1	Alternate Whittaker's index (S/alpha-1):	0
Sørensen's similarity index (%):	100%	Number of common species:	53

is a biodiversity rich area with good number of endemic species includes flora and fauna; unsustainable tourism development in the area leads the disappearance of endemic species and results biodiversity loss.

REFERENCES

Ahmedulla M, Nayar MP (1987). Endemic Plants of the Indian Region. Botanical Survey of India, Kolkata.

Bada SO (1984). Growth patterns in an untreated tropical rainforest ecosystem. PhD thesis, University of Ibadan, Nigeria.

Basu BD (1975). Indian Medicinal Plants. International book Distributors, Dehradun, India. Vol. 1-4.

Brown N, Bhagwat S, Watkinson S (2006). Macrofungal diversity in fragmented and disturbed forests of the Western Ghats of India. J. Appl. Ecol. 43:11-17.

Bryde MB (1979). Information needed to use the Endangered Species Act for Plant Conservation, (Geographical Data Organization, Rare plant conservation, New York).

Bushell R (2002). Healthy Tourism: a new approach to achieve sustainable outcomes. Proceedings of Tourism Development, Community and Conservation: Shaping Ecotourism for the Third Millennium International. Conference. Jhansi, India. Feb, 2002.

Chandran MDS, Gadgil M, Hughes JD (1998). Sacred groves of Western Ghats of India. In: P. S. Ramakrishnan, K. G. Saxena and U. M. Chandrashekara (eds.). Conserving the Sacred for Biodiversity and Management. Science Publishers, Inc. New Hampshire. pp. 211-232.

Curtis JT, McIntosh RP (1950). The interrelations of certain analytic and synthetic phytosociological characters. Ecology 31:434-455.

Curtis JT (1959). The Vegetation of Wisconsin, An Ordination of Plant Communities. University Wisconsin Press, Madison. Wisconsin.

Gadgil M, Vartak VD (1975). Sacred groves of India – A plea for continued conservation. J. Bombay Nat. Hist. Society 72:314-320.

JS, Fisher CEC (1936). Flora of the presidency of Madras. London. (Reprinted 1957), Calcutta.

Hooker JD (1897). The flora of British India London: L. Reeve and Co., 7:4.

IUCN (1978). Plant Red Data Book, (Royal Botanic Garden, Kew, England).

IUCN (2000). Plant Red Data Book, (Royal Botanic Garden, Kew, England).

IUCN Red List of Threatened Species (2004). A Global Species Assessment, The IUCN Species Survival Commission.

Karunakaran PV, Rawat GS, Uniya VK (1998). Ecology and Conservation of the Grasslands of Eravikulam National Park, Western Ghats. Wildlife Institute of India, Dehra Dun, India.

Krebs JC (1998). Ecological Methodology. Addison Wesley Longman Inc., ISBN: 13: 978-0321021731, 620 pp.

Looman J, Campbell JB (1960). Adaptation of Sorensen's K (1948) for estimating unit affinities in prairie vegetation. Ecology 41(3):409-416.

Meffe GK, Nielsen LA, Knight RL, Schenborn DA (2002). Ecosystem management: Adaptive, community-based conservation. Washington, D. C., U.S.A: Island Press.

Myer N, Russell A, Mittermelert C, Mittermelert G, Gustavo AB, Da Fonseca, Kents J (2000). Biodiversity hotspots for conservation priorities. Nature 24:853-858.

Misra R (1968). Ecology Work Book. Oxford and IBH Publishing Company, Calcutta.

Mohanan M, Henry AN (1994). Flora of Thiruvananthapuram, Kerala. Botanical Survey of India. Calcutta.

Mueller-Dombois D, Ellenberg H (1978). Aims and Methods of Vegetation Ecology. John Wiley and Sons, New York.

Nayar MP, Sastry ARK (1987, 1988, 1990). Red Data Book of Indian Plants, eds. 1-3. Botanical Survey of India, Howrah (Calcutta), India.

Nayar MP, Sastry ARK (1990). Red Data Book of Indian Plants, eds. 1-3. Botanical Survey of India, Howrah (Calcutta), India.

Ojo LO, Ola-Adams BA (1996). Measurement of tree diversity in the Nigerian rain forest. Biodivers. Conserv. 5:1253-1270.

Pianka ER (1966). Latitudinal gradients in species diversity: a review of concepts. Am. Nat. 100:33-46.

Pitkanen S (1998). The use of diversity indices to assess the diversity of vegetation in managed boreal forests. For. Ecol. Manag. 112:121-137.

Pramod P, Daniels RJR, Joshi NV, Gadgil M (1997). Evaluating bird communities of Western Ghats to plan for a biodiversity friendly development. Curr. Sci. 73:156-162.

Ravikumar K, Ved DK (2000). Illustrated Field Guide to 100 Red-Listed

Medicinal Plants of Conservation Concern in Southern India, FRLHT, Bangalore.
Sasidharan N, Sivarajan V (1996). Flowering Plants of Thrissur Forest. Sci. Pub., Jodhpur.
Sasidharan N (2004). Flowering plants of Kerala Kerala Forest Research Institute, Peechi, Thrissur.
Shannon CE, Weiner W (1963). The Mathematical theory of communication. University of Illinois Press. Urhana 117.
Simpson EM (1949). Measurement of diversity. Nature 163:688.
Smith EV (1980). Laws and information needs for listing plants. Rhodora 82:193.
Sorenson T (1948). A method of establishing groups of equal amplitude in plant sociology based on similarity of species content. Detkong Danske Vidensk Selk Boil. Skr. Copenhagen) 5:1-34.
Stafleu FA, Cowan RS (1988). International Code of Botanical Nomenclature (Saint Louis Code).
Vajravelu E (1990). Flora of Palghat District, including Silent Valley national park, Kerala (Calcutta: Botanical Survey of India) pp. 245-260.
Whittaker RH (1972). Evolution and measurement of species diversity. Taxon 21:213-251.
Wilson MV, Shamida A (1984). Measuring beta diversity with presence – absence data. J. Ecol. 72:1055-1064.

Comparative study of herb layer diversity in lower Dachigam National Park, Kashmir Himalaya, India

S. A. Shameem[1]*, P. Soni[2] and G. A. Bhat[3]

[1]Division of Environmental Sciences, SKUAST-Kashmir, Srinagar (J and K), Forest Research Institute University, Dehradun, Uttarakhand, India.
[2]Ecology and Environment Division, Forest Research Institute Dehradun, Uttarakhand, India.
[3]Postgraduate Department of Environmental Sciences/CORD, University of Kashmir, Srinagar (J and K), India.

The present study was conducted at two different ecosystems that is, site I (pastureland) and site II (forest) in the lower Dachigam National Park of Kashmir, Himalaya. The pasture site is located outside the National Park and is under grazing whereas forest site is located inside the National Park and is protected. The study was done on seasonal basis and the average results revealed comparatively more or equal values of diversity (H') for both sites (site I = 2.435 and site II = 2.395) while dominance index showed higher value at site I (average = 0.147). The richness index (average = 3.842) and equability index (average = 0.90) both showed higher value at site II. Seasonal trend of Shannon diversity (site I = 3.03, site II = 2.87), richness index (site I = 3.70, site II = 5.83) and evenness or equability index (0.94, site I and II) depicted highest value during summer season whereas lowest variation in Shannon diversity and richness index was observed in winter season at both sites. However, dominance index was recorded lowest in summer season at both sites (site I = 0.06 and site II = 0.07) hence inversely related to diversity (H'). The frequently occurred dominant species during prominent seasons based on importance value (IV) were *Cynodon dactylon*, *Salvia moorcroftiana* and *Thymus serphyllum* at site I and *Fragaria nubicola*, *Galinsoga parviflora*, *Stipa sibirica* and *Viola indica* at site II. The abundance to frequency ratio (A/F) indicated most of the species performed contagious pattern of distribution.

Key words: Biodiversity, community structure, seasons, species, grazing.

INTRODUCTION

The global biodiversity crisis has given rise to a growing concern at the prospect of a rapidly accelerating loss of species, population, domesticated varieties, medicinal herbs and natural habitats. Recent estimates suggest that more than half of the habitable surface of the planet has already been significantly altered by the human activity (Hannah and Bowles, 1995) and we are on the verge of mass extinction of the species (Wilson, 1985). Conservation biologists warn that 25% of all species could become extinct during the next 20 to 30 years. The cause for the loss of species is numerous but the most important is the loss and fragmentation of natural habitats. Biological diversity implies the variety of living organisms and includes diversity within species, between species and of ecosystems and the ecological processes of which they are a part (Gaston and Spicer, 2004). Spices diversity is considered to be one of the key parameters characterizing ecosystems and a key component of ecosystem functioning (Hutchinson, 1959; Schulze and Mooney, 1994; Larsson, 2001; Loreau et al., 2002; Scherer-Lorenzen et al., 2005). Globally, biodiversity is changing at an unprecedented rate as a complex response to several human-induced changes (Vitousek, 1994; Hooper et al., 2005). These changes in biodiversity cause concern for ethical, economical, ecological, and aesthetic reasons, but they also have a strong potential to alter ecosystem services such as the prevention of soil erosion and maintenance of hydrologic cycles, and ecosystem goods, like tourism and recreation. Beyond the ecosystem services, biodiversity influences many ecosystems properties such as productivity, decomposition rates, nutrient cycling,

*Corresponding author. E-mail: sa78@rediffmail.com.

resistance and resilience to perturbations (Loreau et al., 2001). Moreover, a high biodiversity is seen as an insurance against a decline in ecosystem services, and should therefore be preserved (Yachi and Loreau, 1999).

The structure of plant as well as animal communities in many natural ecosystems is largely influenced by the disturbances, frequently occurring in the system naturally or due to anthropogenic activities (Armesto and Pickett, 1985; Bennett and Adams, 2004; Eldered and Doak, 2006; Kwit and Platt, 2003). In many of these systems, disturbances change overall community structure (Shaforth et al., 2002; Sousa, 1979) which in turn can ultimately affect community and population dynamics. In other words, the global environmental degradation has been severely occurred and has been introduced as one of the main environmental troubles worldwide. Studying vegetation and various environmental factors (e.g. physiographic, climate, soil, etc.), the community stability and the factors correlation with the vegetation can be reached, which is crucial in terms of forest communities development and rehabilitation (Basiri, 2003). Grazing areas have become less and less productive resulting from over stocking of livestock. Conflicts over the use of land have increased due to increased demand for land by different sectors of the economy. Of particular concern are the conflicts among cultivators, livestock keepers, wildlife conservationists, individual land users and governments due to encroachment of humans into the protected areas (Hoare, 1999; Campbell et al., 2003; Western, 1976; Wells and Brandon, 1992). Forests are the primary source to rejuvenate productivity of land through recycling of nutrients, which make physico-chemical conditions of the soils favourable for plant growth (Bargali et al., 1998). Due to increasing human population, the biotic pressure on native forest is inevitable. The uncontrolled lopping and felling of trees for fuel wood, leaf fodder, burning of ground vegetation, livestock grazing and harvesting of ground vegetation for forage are some of the factors responsible for exploitation of forests (Bargali et al., 1998).

The herbaceous layer composition is changing continuously in space and time due to multitude of factors such as grazing, fire, and rainfall which differs in intensity and duration. Kashmir Himalaya, due to its rich repository of vegetation has attracted naturalists and botanists for more than two centuries (Dar et al., 2001). Numerous studies dealing with diverse aspects of vegetation from different areas of the region have been carried out from time to time (Stewart, 1982; Dar et al., 2001). The general vegetation of Dachigam has been dealt in detail by Singh and Kachroo (1976). They have recognized a number of vegetational types based on habitat, form and density of dominant species, though the vegetation patterns are controlled by such factors as habitat, slope, exposure to sunlight and altitude, besides biotic factors. Anthropogenic disturbances in forests followed by livestock grazing in pasture lands adversely affected the composition of herbaceous vegetation, it is therefore imperative to conserve the herbaceous vegetation of the two selected sites. In this context, the present study was therefore conducted to assess the seasonal variations in phytodiversity and distribution pattern of herbaceous vegetation in the two selected sites.

METHODOLOGY

Study area

Dachigam National park is located between 34°.04′-34°.11′ N - latitude and 74° 54′-75° 09′ E-longitude is nearly about 20 km away from Srinagar city of Kashmir Valley with an undulating mountain valley topographic system. The entire area of the park is distinguishable into two sectors upper and lower Dachigam which is spread over an area of 141 km^2. The present study was confined to the lower Dachigam National park conducted on seasonal basis at two different ecosystems viz.; site-I (pastureland falls within the catchment of Dachigam but located outside the official boundary of the Park) and site-II (forest located inside the official boundary of the Park).

Vegetation analysis

To study the community composition and other phytosociological characteristics of the herbaceous vegetation at two selected sites, thorough field surveys were conducted during four prominent seasons Spring (March-May), Summer (June-August), Autumn (September-November) and Winter (December-February). Phytosociological attributes of plant species were studied by randomly laying 25 quadrats of 1×1 m^2 size at each site (Sharma et al., 1983; Rajvanshi et al., 1987). Specimen of plant species encountered at each site during the study period were collected and herbarium was prepared and identified at Centre of Plant Taxonomy, Botany Department University of Kashmir, Forest Research Institute Dehradun, Uttarakhand.

Data analysis

The vegetation data recorded was quantitatively analysed for density, frequency and abundance following Curtis and McIntosh (1950). The relative values of these indices were determined as per Phillips (1959). These values were summed up to get importance value index (IVI) of individual species (Curtis, 1959). The ratio of abundance to frequency (A/F) for different species was determined by eliciting the distribution pattern. This ratio has indicated regular (<0.025), random (0.025 - 0.05) and contagious (>0.05) distribution (Curtis and Cotton, 1956). Plant diversity in the two study sites were evaluated using the following indices:

Diversity index (H') Shannon-Weaver (1949)

$$H' = -\sum_{i=1}^{s} p_i \mathrm{Ln} p_i$$

H′ is Shannon-Wiener's diversity index, S is the total number of species (richness) and pi is the proportion of individuals in the ith species (Pi= ni/N, ni is the number of individuals in the ith species and N is the total number of individuals) (Shannon-Weaver, 1949).

Simpson Index (Simpson, 1949):

$$D = \Sigma\, pi^2$$

Evenness Index (Pielou, 1966),

$$J' = \frac{H'}{LnS}$$

Richness Index (Margalef, 1958)

$$R = \frac{S-1}{Ln\ (N)}$$

RESULTS

Diversity indices

During the study period total number of herbaceous species reported during prominent seasons was 48 at site I (pastureland) and 41 at site II (forest) (Figure 1). The seasonal break-up of species recorded at both sites showed maximum species occurrence during spring and summer season (site I, spring = 28; summer = 25), (site II, spring = 18; summer = 21). During autumn and winter season species number at both sites showed overall a declined trend (site I, autumn = 11; winter = 10) and (site II, autumn = 14; winter = 8). However, out of the total plant species encountered at both sites during different seasons results of 22 highly dominant species based on importance value (IV) are depicted in Figure 2 and elaborated below as season wise and site wise. Dominant species based on importance value (IV) during spring season at site I were *Bothriochloa pertusa* (39.51), *C. dactylon* (37.29), *S. moorcroftiana* (35.30), *Organum vulgare* (22.13), *Tulipa stellata* (21.66), *S. sibirica* (13.45) *Ranunculus* sp. (17.56) and *T. serphyllum* (15.79) whereas at site II dominant species recorded were *F. nubicola* (74.17), *G. parviflora* (34.15), *S. sibirica* (25.53), *Dropteris flixmass* (24.37) and *Poa* sp (23.86) in the same season. During summer season *T. serphyllum* (34.01), *C. dactylon* (27.66), *S. moorcroftiana* (26), *Oxalis corniculata* (22.56) and *Plantago lanceolata* (20.26) showed higher dominance at site I and *Arthraxon prinoides* (37.42), *Setaria viridis* (32.15), *F. nubicola* (31.93) *Viola indica* (21.18) and *Stipa sibirica* (19.46) depicted high dominance at site II during the same season.

S. moorcroftiana (140.97), *T. serphyllum* (40.60), *C. dactylon* (12.10), *P. lanceolata* (18.57) and *S. sibirica* (18.10) are the species showing high dominance at site I in autumn season. Site II was dominated by *Poa annua* (101.35), *F. nubicola* (53.34), *Impatiens grandiflora* (20.45), *Galinsago parviflora* (22.98) and *Viola indica* (32.13) during this season.

High dominance in winter season at site I was shown by *S. moorcroftiana* (IVI = 108.97), *C. dactylon* (22.27), *S. sibirica* (26.22), *Artemisia* sp. (20.31) followed by *T. serphyllum* (41.59) and *P. lanceolata* (57.56). At site II during this season high dominance was reported by *Poa angustifolia* (IVI = 55.88), *F. nubicola* (43.93), *Poa annua* (37.14), *Stellaria media* (30.18) followed by S. sibirica (30.37). Comparatively S. moorcroftina, T. serphyllum and *C. dactylon* remained as dominant species during different seasons throughout the study period at site I whereas *F. nubicola*, *G. parviflora*, *S. sibirica* and *Viola indica* were dominant. Different diversity indices recorded at both sites are presented in Figure 3. The perusal of the data revealed diversity index (H'), evenness index and richness index obtained maximum value during summer season at both sites (site I = 3.03, site II = 2.87), evenness index (0.94 each at site I and II) and richness index (site I = 3.70, site II = 5.83). However, minimum value of diversity index (H') was recorded in winter (1.80 = site I) and autumn season (2.04 = site II). Moreover, evenness index showed lowest value during autumn season at both sites (0.73-site I and 0.85- site II). Species richness was recorded lowest at both sites during winter season (1.87-site I and 2.47-site II). Dominance index depicted inverse trend to that of Shannon diversity with lowest values reported in summer season at both sites (site I = 0.06 and site II = 0.07) and highest during autumn at both sites (0.25-site I and 0.18-site II) hence inversely related to diversity (H'). In general diversity showed an increasing trend from spring to summer and thereafter a declined trend was observed at both sites.

Distribution pattern

The abundance to frequency ratio (A/F) indicated most of the species at both sites performed contagious pattern of distribution. A seasonal picture of both sites noted most of the species as contagious>random>regular distribution during different seasons. At site I about 26.09 (spring) to 57.15% species in winter season showed contagious distribution followed by 39.14% (summer) to 56.53% species in winter as randomly distributed (Figure 4). Seasonal trend indicated that during spring season at site II maximum (50%) species fall in random distribution followed by contagious (44.45%) and only one species (*S. media*) fall in regular distribution (5.56%). In summer season (47.62%) species showed contagious distribution followed by regular (28.58%) and random (23.81%). During autumn and winter season maximum species (72.73, 70%) showed contagious distribution followed by random (27.28%, 30%) during the same season. Regular distribution of species was absent during autumn and winter season at site II (Figure 5).

DISCUSSION

Diversity is considered to be an outcome of evaluation of species in a bio-geographic region. It is considered to be synthetic measure of the structure, complexity and stability of a community (Hubble and Foster, 1983). It is a combination of two factors: the number of species present, referred to as species richness and the distribution of individuals among species, referred to as species evenness or equability. Species diversity therefore, refers to the variation that exists among the

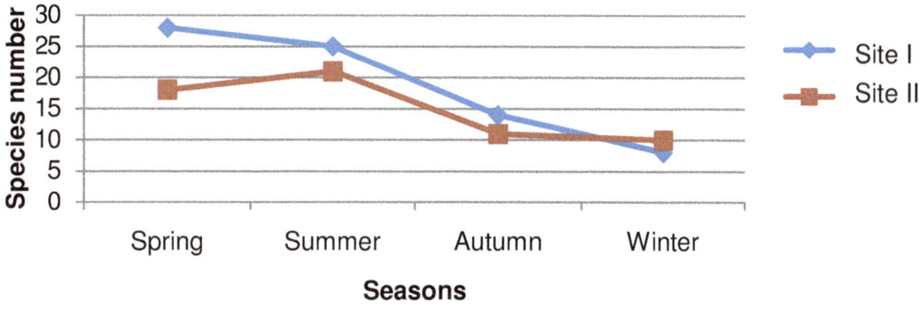

Figure 1. Species recorded at two sites during different seasons.

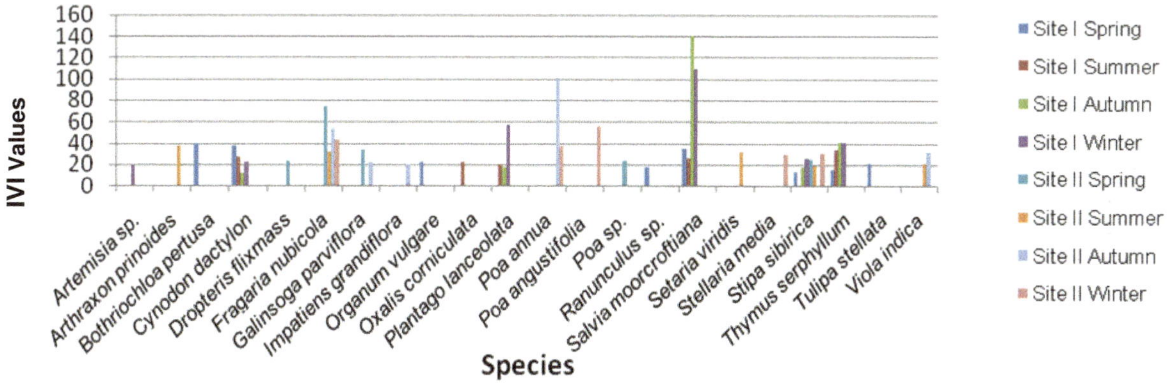

Figure 2. Twenty two highly dominant herbaceous plant species recorded at two sites during different seasons.

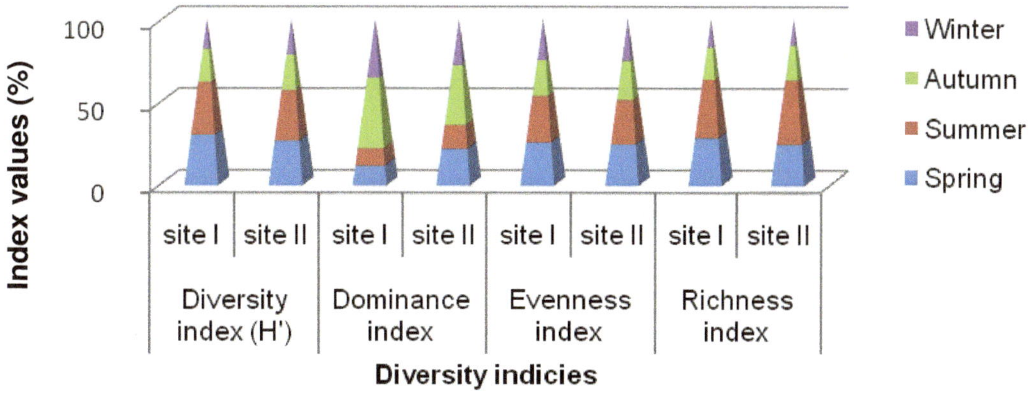

Figure 3. Diversity estimates of the herbaceous vegetation at site I and site II using different diversity indices.

different life forms. In the present study general structure of vegetation at both sites depicted an increasing trend in species number particularly during spring and summer seasons (site I, spring = 28; summer = 25) and (site II, spring = 18; summer = 21). The reason for their maximum occurrence during the two seasons could be due to the availability of moisture present in the form of rains and other environmental factors. Alhassan et al. (2006) during their study period reported similar factors responsible for the variation in species number and diversity. The sequence of observations mirrored to present study was also mentioned by Sharma and Upadhyay (2002). The species diversity in the present study ranged from 1.80 (winter) to 3.03 (summer) at pasture site and 2.04 (autumn) to 2.87 (summer) at forest site. An increasing trend in species diversity was observed

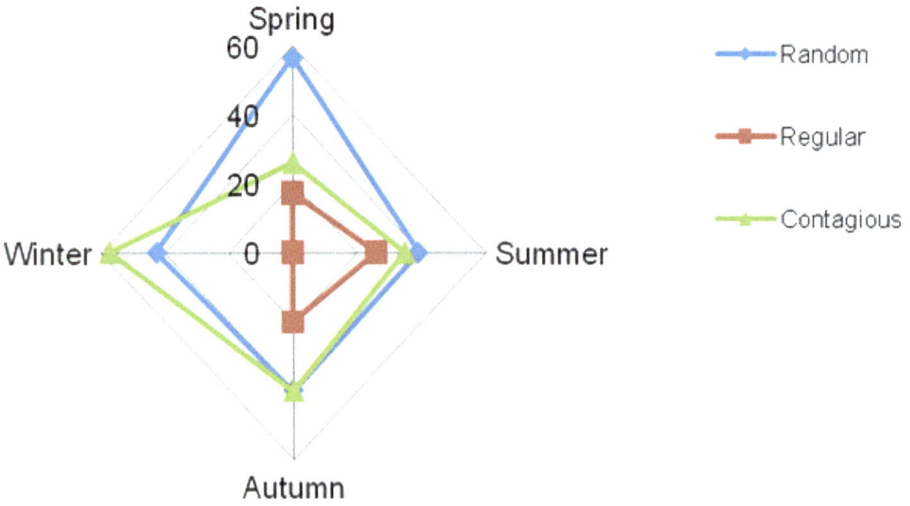

Figure 4. Distribution pattern (%) of herbaceous vegetation at site I during different seasons.

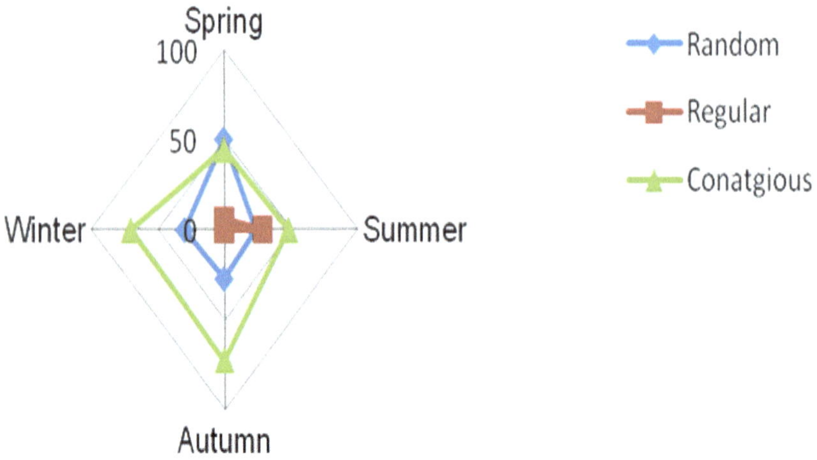

Figure 5. Distribution pattern (%) of herbaceous vegetation at site II during different seasons.

from spring onwards which declined with the commencement of autumn and winter seasons at both sites. This character is attributed to the fact that during spring and summer seasons, new species goes on sprouting depending upon the root / seed stock in the soil and thereby adding to species in total resulted more diversity. During autumn and winter season the rate of sprouting of root/seed stock is diminished and species number declined owing to adverse climatic conditions (Shadangi and Nath, 2005). The lower diversity during autumn and winter at both sites may also be due to lower rate of evolution and diversification of communities (Fischer, 1960; Simpson, 1964) and severity in environment (Connel and Oris, 1964). Comparatively, results of Shannon diversity at both sites fall within the range of the study carried out by Kiss et al. (2004); Yadav and Gupta (2007). Lalfakawma et al. (2009) and Kharkwal et al. (2004) are in the same view while comparing the results of Shannon-diversity in concurrent to the present study. However, highest species diversity during summer season at site I might be due to the moderate level of grazing or anthropogenic disturbances and invasion of new species. Connell (1978) and Decocq et al. (2004) also reported species diversity highest in intermediate disturbance ecosystem than in undisturbed systems. Many other studies mentioned similar results pertaining to the present study emphasizing moderate level of grazing promoted species diversity (Rikhari et al., 1993; Singh et al., 2003). Pandey and Singh (1985) in their study on disturbed ecosystem of Kumaon Himalaya are in the same agreement that species diversity increased in disturbed ecosystem due to moderate disturbances. However, other suggestions made by Lubchenco (1978) and Huston (1979) considered it as a positive force that might increase species diversity in the community by preventing competitive exclusion by dominant species. Kakinuma and Takatsuki (2008) investigated the change in plant communities by grazing in northern Mongolia and observed that species diversity and biomass of forbs decreased with increasing grazing intensity. Highest trend in species diversity during summer season at site II could be due to various environmental and climatic factors. Concentration of dominance at site I ranged from 0.06 (summer) to 0.25 (autumn) at site I and 0.07 (summer) to 0.18 (autumn) at site II hence inversely related to diversity index (H'). Compatible results of inverse relationship between diversity and dominance were also reported by Kharkwal et al. (2004). The lower value of dominance at forest site showed that dominance of herb layer plants is shared by many species. The Pielou's indices at both sites were around 0.0825 (site I) and 0.90 (site II) on average, indicating low dominance and more or regular distribution of plant species at both sites. Site I during autumn/winter and site II during winter season was at lower side in species richness which could be due to dry environmental conditions and also due to slow growth rate, to a maximum during spring/ summer (site I), spring/summer and winter season at site II which could be due to favourable climatic conditions. Abdullah at al. (2009) in their study also mentioned climatic factors as a reason that influenced the distribution of species in certain habitats. High importance value (IV) of a species indicated its dominance and ecological success, its good power of regeneration and greater ecological amplitude. It does vary with the season. The reason why certain species grow together in a particular environment is usually because they have similar requirements for existence in terms of environmental factors such as light, temperature, water and soil nutrients and drainage etc. They may also share the ability to tolerate the activities of animals and humans such as grazing, burning, cutting or trampling (Wood et al., 1994). In accordance to our results for site I *S. moorcroftiana* and *T. serphyllum* showed maximum importance value (IV) during autumn and winter season indicating its dominance due to environmental suitability and ability of the species against grazing during the two seasons. However, their dominance at a particular site could be due to the availability of optimum conditions for their growth. Favourable observations in support of results achieved for site I was also reported by (Kukshal et al. 2009) based on seasonal changes in the IVI value of species that makes them dominant during different seasons. The growing dominance of non-palatable and other species at pasture site is probably an indication of adaption against herbivory and adverse climatic conditions. Bhandari et al. (1999) while working in pasturelands of Garhwal Himalaya reported same trend in their results as concurred in the present study. However, at site II maximum IVI was shared by *Fragaria nubicola, Galinsoga parviflora, Stipa sibirica* and *Viola indica* during most of the season. Their dominance during a particular season can be well correlated with the study conducted by (Kukshal et al. 2009). Moreover, high IVI value by any individual species indicated that most of the available resource are being utilized by that species and left over are being trapped by another species as the competitors and the associates. This could be the reason why IVI was reported always highest by few species during autumn than rest of the seasons. Other reason for their dominance during autumn and winter season could be as the rate of sprouting of root/seed stock is diminished and the species number declined owing to adverse climatic conditions. It is generally argued that each individual species depends on some set of other species for its continued existence and the species have co-evolved in the ecosystem on which they depend (Paine, 1966). The loss of natural associations may be the probable reason for supporting low number of species (Walker, 1992). It is to be mentioned that distribution of niche space or availability of resource was equally distributed among all species that showed maximum dominance during autumn season at site II. Whereas at site I only 2-3 species occupied more niche space than other species during a particular season.

The nature of plant community at a place is determined by the species that grow and develop in such environment (Bliss, 1962). Difference in the species composition from site to site is mostly due to micro-environmental changes (Mishra et al., 1997). The pattern of distribution depends both on physico-chemical natures of the environment as well as on the biological peculiarities of the organisms themselves. Abundance and frequency ratio (A/F) ratio was used to assess the distribution pattern of species. It reveals that most of the species were contagiously distributed whereas as regular distribution was reported almost negligible during most seasons.

The study conducted by Shadangi and Nath (2005) reported maximum species in contagious distribution. Ilorkar and Khatri (2003) investigated herb layer species in contagious distribution followed by random. Dominance of contagious distribution may be due to the fact that the majority of species reproduce vegetatively in addition to their sexuality. Odum (1971) described that in natural conditions contagious distribution is most common type of distribution and is performed due to small but significant variation in environmental conditions while random distribution is found only in very uniform environment. Contagious distribution in natural vegetation has been reported by Greig-Smith (1957); Kershaw (1973); Singh and Yadava (1974) hence compatibly favours the results achieved in the present study. However, observations indicated that contagious distribution in vegetation (as recorded for both sites) was due to multitude factors and the vegetative reproduction may not be the only reason (Kershaw, 1973; Saxena and

Singh, 1982).

Conclusion

The study concluded that the improvement in vegetation cover is possible through regular monitoring of livestock grazing and human disturbance at site I and biotic interference at site II. Furthermore, new sites need to be explored for seasonal grazing of livestock as an alternate followed by protection for few years that certainly will allow the vegetation to regenerate which is constantly under threat due to various factors at site I. However, increasing biotic inference at site II within the national park need urgent attention and the human activities for fuel fodder collection, harvesting of medicinal herbs, burning of ground vegetation and livestock grazing requires sustainable control measures. It is also suggested that new sites within the national park must be allowed for future use so that constantly increasing pressure in the present selected site can be reduced which ultimately will be helpful in the enhancement of species diversity at this site. It is further recommended that species with lower IVIs need priority measures for protection and those with higher IVIs need monitoring effort in order to maintain diversity in the selected sites during different seasons.

ACKNOWLEDGMENTS

The authors are grateful to the Director, Forest Research Institute University Dehradun, India for allowing them to conduct the Research work in the Dachigam National Park. Director, Centre of research for development (CORD)/P. G. Department of Environmental Science, University of Kashmir is highly acknowledged for providing laboratory facility. Thanks are also due to Chief Wildlife Warden, J and K Government for allowing us to use the National Park for research purpose.

REFERENCES

Abdullah MB, Sanusi SS, Abdul SD, Sawa FBJ (2009). An Assessment of the Herbaceous Species Vegetation of Yankari Game Reserve, Bauchi, Nigeria. Am-Eur. J. Agric. Environ. Sci., 6(1): 20-25.

Alhassan AB, Chiroma AM, Kundiri AM (2006). Properties and classification of soils of Kajimaram oasis of Northeast Nigeria. Int. J. Agric. Biol., 8: 256–261.

Armesto JJ, Pickett STA (1985). Experiments on disturbance in old field plant communities: Impact on species richness and abundance. Ecology, 66: 230-240.

Bargali K, Bargali SS, Singh RP (1998). Ecological relationship of Bidens biternata and Galingsonga ciliata in open and closed habitats. Ind. J. Ecol., 25(2): 107-113.

Basiri R (2003). Ecolgogical study on Quercus libani site by analyzing environmental features in Marivan. Ph. D Thesis University of Tarbiat Modarres.

Bennett LT, Adams A (2004). Assessment of ecological effects due to forest harvesting: approaches and statistical issues. J. Appl. Ecol., 41: 585-598.

Bhandari BS, Nautiyal DC, Gaur RD (1999). Structural attributes and productivity potential of an alpine pastures of Garhwal Himalaya. J. Ind. Bot. Soc., 78: 321-329.

Bliss LC (1962). Rosine and lipid contents in alpine Tundra plants. Ecology, 43:753-757.

Campbell D, Gichohi H, Reid R, Mwangi A, Chege L, Sawin T (2003). Interactions between People and Wildlife in Southeast Kajiado District, Kenya. LUCID Working paper No. 18. Int. Livestock Res. Institute. Nairobi.

Connel JH, Oris E (1964). The ecological regulation of species diversity. Am. Nat., 48: 399-414.

Connell JH (1978). Diversity in tropical forests and coral reefs. Science, 1999: 1302-1310.

Curtis JT, Cotton G (1956). Plant Ecology Workbook: Laboratory Field Reference Manual. Burgess Publishing Co., Minnesota. p. 193.

Curtis JT, McIntosh RP (1950). The interrelation of certain analysis and systematic phytosociological characters. Ecology, 31: 434-455.

Curtis JT (1959). The vegetation of Wisconsin: An ordination of plant communities. University of Wisconsin Press Madison, Wisconsin. p. 657.

Dar GH, Bhagat RC, Khan MA (2001). Biodiversity of Kashmir Himalaya. Valley Bok House, Srinagar, J & K, India.

Decocq G, Aubert M, Dupont F, Alard D, Saguez R, Wattez-Franger A, Foucault B. DE, Delelis-Dusollier A, Bardat J (2004). Plant diversity in a managed temperate deciduous forest: understorey response to two silvicultural systems. J. Appl. Ecol., 41: 1065-1079.

Eldered BD, Doak DF (2006). Comparing the direct and community-mediated effects of disturbance on plant population dynamics: flooding, herbivory and Mimulus guttatus. J. Ecol., 94: 656-669.

Fischer AG (1960). Latitudinal variation in organic diversity. Evolution, 14: 64-81.

Gaston KJ, Spicer JL (2004). Biodiversity: an introduction. 2nd ed. Oxford, Blackwell Science. p. 191.

Greigh-Smith P (1957). Quantitative plant ecology, 2nd ed. Butterworth, London.

Hannah L, Bowels I (1995). Global priorities. Biol. Sci., 45:122-132.

Hoare RE (1999). Determinants of human-elephant conflict in a land use mosaic. J. Appl. Ecol., 36: 689–7.

Hooper DU, Chapi FS, Ewel JJ, Hector A, Inchausti P, Lavorel S, Lawton JH, Lodge D, Loreau M, Naeem M, Schmid SB, Setala H, Symstad AJ, Vandermeer J, Wardle DA (2005). Effects of biodiversity on ecosystem functioning: A consensus of current knowledge. Ecol. Monogr., 75: 3-35.

Hubble SP, Foster RB (1983). Diversity of canopy trees in a neo-tropical forest and implications to conservation. Tropical Rain Forest: Ecol. Manage., (Sutton L, T. C.

Huston M (1979). A general hypothesis of species diversity. Am. Nat., 113: 81-101.

Hutchinson GE (1959). Homage to Santa Rosalia, or why are there so many kinds of animals? Am. Nat., 93: 145–159.

Ilorkar VM, Khatri PK (2003). Phytosociological Study of Navegaon National Park, Maharashtra. Ind. For., 129(3): 377-387.

Kakinuma K, Takatsuki S (2008). Change in plant communities by grazing in north Mongolia and land assessment of herders. 21st. International Grassland Congress and 8th International Rangeland Congress in Hohhot, China. pp. 493-493.

Kershaw KA (1973). Quantitative and Dynamic Plant Ecology. 3rd Edn, ELBS and Edward Arnold Ltd., London.

Kharwal GP, Mehrotra RYS, Pangtey YPS (2004). Comparative study of herb layer diversity in pine forest stands at different altitudes of central Himalaya. Appl. Ecol. Environ. Res., 2(2): 15-24.

Kiss T, Sipos G, Bodis K, Barta K (2004). Community composition, species diversity, and secondary succession in grazed and ungrazed alpine meadows of the west Himalaya, India. Int. J. Field work Stud., 2: 1.

Kukshal S, Nautiyal BP, Anthwal A, Sharma A, Bhat AB (2009). Phytosociological investigation and life form pattern of grazing lands under pine canopy in temperate zone, Northwest Himalaya, India. Res. J. Bot., 4: 55-69.

Kwit C, Platt WJ (2003). Disturbance history influences regeneration of non-pioneer under storey trees. Ecology, 84(10): 2575-2578.

Lalfakawma-Sahoo UK, Roy S, Vanlalhriatpuia K, Vanalalhluna PC (2009). Community Composition and Tree Population Structure in undisturbed and Disturbed Tropical Semi-Evergreen Forest Stands of North-East India. Appl. Ecol. Environ. Res., 7(4): 303-318.

Larsson TB (eds) (2001). Biodiversity evaluation tools for European forests. Ecol. Bull., 50: 1–236.

Loreau M, Naeem S, Inchausti P, Bengtsson J, Grime JP, Hector A, Hooper DU, Huston MA, Raffaelli D, Schmid B, Tilman D, Wardle D (2001). Biodiversity and ecosystem functioning: current knowledge and future challenges. Science, 294: 804-808.

Loreau M, Naeem S, Inchausti P (eds). (2002). Biodiversity and ecosystem functioning. Oxford, University Press. p. 294.

Lubchenco JL (1978). Plant species diversity in a marine inter-tidal community: importance of herbivore food preference and algae competative abilities. Am. Nat., 112: 23-29.

Margalef R (1958). Perspective in ecological theory. University of Chicago Press, Chicago.

Mishra D, Mishra TK, Banerjee SK (1997). Comparative phytosociological and soil physic-chemical aspects between managed and unmanaged lateritic land. Ann. For., 5(1):16-25.

Odum EP (1971). Fundamentals of Ecology. Saunders Co., Philadelphia.

Paine RT (1966). Food-web complexity and species diversity. Am. Nat., 100:65-75.

Pandey AN, Singh JS (1985). Mechanism of ecosystem recovery: A case study from Kumaun Himalaya. Recl. Reveg. Res., 3: 271-292.

Philips EA (1959). Methods of vegetation study. Henry Holt and Co.Inc; New York. p. 318.

Pielou EC (1966). Species diversity and pattern diversity in the study of ecological succession. J. Theor. Biol., 10: 370-383.

Rajvanshi R, Kumar V, Bachpari W, Rajgopal K, Raj SFH (1987). Herbaceous under growth in some forest habitats in Nilgiris. Ind. For., 113(9): 599-608.

Rikhari HC, Negi GCS, Ram J, Singh SP (1993). Human-induced secondary succession in alpine meadow of Central Himalaya, India. Art. Alp. Res., 25: 8-14.

Saxena AK, Singh JS (1982). A phytosociological analysis of woody species in forest communities of a part of Kumaon Himalaya. Vegetation, 50: 3-22.

Scherer-Lorenzen M, Korner C, Schulze ED (2005): Forest diversity and function:Temperate and boreal systems. Berlin, Springer. p. 399.

Schulze ED, Mooney HA (eds), (1994): Biodiversity and ecosystem function. Berlin, Springer. p. 525.

Shadangi DK, Nath V (2005). Impact of seasons on ground flora under plantation and natural forest in Amarkantak. Ind. For., 131(2). 240-250.

Shaforth PB, Stromberg JC, Patten DT (2002). Riparian vegetation response to altered disturbance and stress regimes. Ecol. Appl., 12: 107-123.

Shannon CE, Weaver W (1949). The Mathematical Theory of communication. University Illinois Press, Urbana.

Sharma KP, Upadhyaya BP (2002). Phytosociology, primary production and nutrient tetention in herbaceous vegetation of the forestry arboretum on the Aravalli hills at Jaipur. Trop. Ecol., 325-335.

Sharma SK,Georga M, Prasad KG (1983). Forest vegetation survey and classification with special reference to South India. 1. Vegetation survey and quadrate analysis. Ind. For., 109(6): 384-394.

Simpson EH (1949). Measurement of diversity. Nature, 163: 688.

Simpson GG (1964). Species diversity of North American recent mammals. Syst. Zool., 13: 57-73.

Singh G, Kachroo P (1976). Forest Flora of Srinagar and Plants of Neighborhood. B. Singh and M.P. Singh, Dehradun.

Singh JS, Yadava PS (1974). Seasonal variation in composition, plant biomass, and net primary productivity of tropical grassland at Kurukshetra, India. Ecol. Monogr., 44: 351-376.

Singh YP, Kumar A, Rai JPN (2003). Species diversity as related to grazing pressure in alpine meadows of Nanda Devi Biosphere Reserve. Pro. Nat. Seminar on Biodiversity conservation and management, EPCO, Bhopal., pp.147-153.

Sousa W (1979). Disturbance in marine intertidal boulder fields: the monequillibrium maintenance of species diversity. Ecology, 60: 1225-1239.

Stewart RR (1982). History and exploration of plants in Pakistan and adjoining Areas (Nasir E, Ali SI eds.) Pan Graphics, Islamabad, Pakistan.

Vitousek PM (1994). Beyond global warming: ecology and global change. Ecology, 75: 1861-1876.

Walker BH (1992). Biodiversity and ecological redundancy. Conserv. Biol., 6: 18-23.

Wells M, Brandon K (1992). People and Parks: Linking Protected Area Management and Local Communities. The World Bank, Washington, DC.

Western D (1976). "Amboseli; Integration of people, land and wildlife seeks to end the conflicts which threaten this national park." Parks, 1:1-4.

Wilson EO (1985). The biological diversity crisis. Bio-science, 35; 700-7006.

Wood J, Low AB, Donaldson JS, Rebelo AG (1994). Threats to Plant Species Diversity through Urbanization and Habitat Fragmentation in the Cape Metropolitan Area, South Africa, pp: 259-274. In: Huntely BJ (Ed.) Strelitzia 1, Botanical diversity in South Africa. Proceedings of a Conference on the Conservation and Utilization of Southern African Botanical Diversity. Cape Town, September 1993. Pretoria: National Botanical Institute.

Yachi S, Loreau M (1999). Biodiversity and ecosystem productivity in a fluctuating environment: The insurance hypothesis. Proc. Nat. Acad. Sci., USA. 96: 1463-1468.

Yadav AS, Gupta SK (2007). Effect of micro-environment and human disturbance on the diversity of herbaceous species in Sariska Tiger Project. Trop. Ecol., 48(1): 125-128.

Role of participatory research on natural resource management: A case of Karnali Watershed Area, Nepal

Kabi Prasad Pokhrel

Geography, Research Division, TU, Kirtipur, Tribhuvan University, Nepal. E-mail: drkabipokhrel@gmail.com

This paper discusses the development trends of participatory research in Nepal with reference to natural resource management, taking example from Karnali watershed area. The issues discussed in this paper clearly indicate that there is urgent need to promote the participatory action research on environment and development for the proper links between population, development, technological implication and institutional strengths, using a micro level conservation and development model that suits local environments. The author opines that both government organizations and academic institutions have to join hands with an effort to promote the participatory research culture in Karnali watershed area for sustainable natural resource management.

Key words: Scientific research, relaxed rapport, extractive survey, eco-development, integrated conservation and development planning, village level model.

INTRODUCTION

Over the decades, there have been drastic changes in modes of research and learning. Shifts from logical-positivism paradigms (quantitative) to phenomenological paradigms (qualitative) have been witnessed in research and development practices since 1980s. These shifts move now from top-down to bottom-up and from extractive survey questionnaires to participatory appraisal, in which more and more activities for innovation, management and development of available resources are done by local people rather than outsiders. It is seen that various methods and approaches of participatory research evolved from here, with local people learning these methods themselves in order to map out their conditions in planning and action. An important method of participatory research is rapid rural appraisal (RRA), which had a widespread in the 1980s. Its further evolution, which was met in the 1990s, into participatory rural appraisal (PRA) has grew fast and promoted participatory action research (PAR), participatory learning action, focus group discussion and perception study for development dialogue with a view to enable local people share, enhance and analyze their knowledge of life and conditions to plan and to act. Thus, the participatory research in natural resource management and development appeared since 1990s with several sources, reasons and traditions; however, the term activist participatory research was introduced in the 1970s by the work and inspiration of Paulo Freire in his book Pedagogy of the Oppressed (Rhoades, 1990; Chambers, 1995). Historically, it evolves continuously quicker and closer communication, transfers and sharing of activist participatory research, agro-ecosystem analysis, applied anthropology, field research and rapid rural appraisal. Particularly, field research on farming systems and conservation of natural and environmental resources by geographers, agricultural economics or biological scientists have contributed in shaping the present form of the participatory research in a complex structure.

LOCATION OF THE KARNALI WATERSHED AREA

The Karnali- Watershed Basin (KWB) lies in the western part of the Mid Western Development Region of Nepal, which is about 500 km west from the capital city, Katmandu, of the country. It is located between 27°42' N

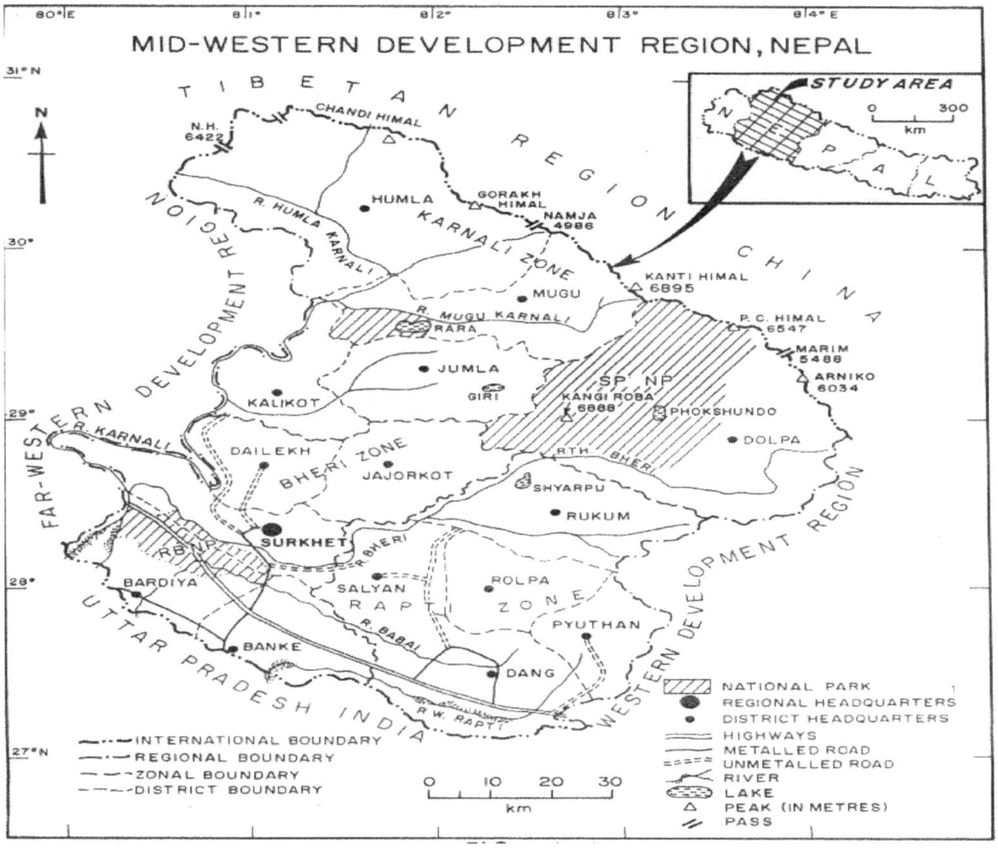

Figure 1. Map of Karnali watershed area.

to 30°27' North latitude and 81°E to 83°45' East longitude, extending from Dhaulagiri Himal in the east to Bayas-Rishi Himal in the west. The total spatial cover of the area is 42,559 sq km, which is about 29.02% of the country. Its average length is 176.59 km from the east to west with non-uniform width of 240 km from the north to south (MOFSC, 2010). Hence, the north-south extension of the basin makes great variations between ecological regions namely: mountain, hill and *tarai* in regard to relief features, climatic characteristics, and distribution pattern of resources. However, about 80% of the land is covered by rugged terrain. The elevation ranges from 129 m in the southern plain area to 7,000 m height in the northern mountain (CBS, 2007). It borders the Western Development Region in the east, Far-Western Development Region in the west, Uttar Pradesh of India in the south, and the Tibetan Region of China in the north (Figure 1).

REVIEW OF PARTICIPATORY RESEARCH ACTIVITIES

Researchers and scientists believe that participatory research is a legitimate and useful method of scientific research in natural and social sciences by rapid assessment procedures like conversation, observation, informal interviews and focus group discussion of realities. Chamber (1989) pointed out many reasons to the origin of the rapid rural appraisal that further evolved into participatory rural appraisal and participatory action research. Among them, the first reason was dissatisfaction with anti-poverty biases of rural development by urban-based professionals. These biases were spatial (visit to nearby cities, on roadsides, and to the villages' centers with neglected peripheries), a project that gave special attention to the donor desires rather than people's demands, a person meeting men more than women, and elites more than the poor and disadvantaged groups, seasonal visits to sites in favorable weather, and diplomacy, that is, where the outsiders do not wish to cause offence by asking to meet poor people in bad conditions. The next one is disillusionment with the normal processes of the questionnaire survey and their results. This is as a result of the fact that the conventional approach to research tended to be long drawn-out, tedious, difficult to administer, a nightmare to process and write up, inaccurate and unreliable data obtained, and difficult to use in wider perspectives. Another important preseason was more positive and was in favor of this study to make it a more cost-effective method of learning. It was developed by the growing recognition of the development professional to discern the fact that rural people are knowledgeable on many issues which touched their lives and which are recognized as the indigenous technical knowledge (ITK) to tap it more effectively, as a source of information for analysis and use by outsiders.

In the mid of the 1990s, it became a more popular model of research and development in Asia, Africa and South America in both the GOs and NGOs sectors. It also became the main theme of field research for academic institutions and it spread internationally. Many international NGOs contributed to shape its present form, which is important for need based development activities. Among them, International Institute for Environment and Development (IIED) London, Institute of Development Studies (IDS) UK, World Bank, Action Aid and Win rock International have played significant role to promote the participatory research. Therefore, it is a simple method, with set of approaches for scientific inquiry that enable local people conduct their own analysis and often plan and take action.

Principles of participatory research

To carry out a good and practicable participatory research, some important performances are required by the practitioners, and they are briefly discussed here.

A reversal of learning: To learn from rural people directly, on site and face-to-face, thereby gaining from local physical, technical and social knowledge.

Learning rapidly and progressively: With conscious exploration, the flexible use of the methods of opportunism, improvisation, iteration and cross checking do not follow a blueprint program, but the ability to adapt in any learning process.

Offsetting biases: Rural development is all about relaxing and not rushing, listening and not lecturing, probing instead of passing on to the next issues, being unimposing instead of important, and seeking out the deprived people and learning their concerns and priorities.

Optimizing trade-off: Relating the cost of learning to the useful truth of information, with trade-off between quantity, relevance, accuracy and timeliness. This includes the principles of optimal ignorance - knowing what is not worth knowing, and of appropriate imprecision - not measuring more than needed. As Keynes is reputed to have said; it is better to be approximately right than to be precisely wrong.

Triangulating: Using a range of methods, types of information, investigators and disciplines to cross-check.

Seeking diversity: This has been expressed in terms of seeking variability rather than average and has been described as the principle of maximum diversity or maximizing the diversity and richness of information. This can involve sampling in a non-statistical sense. It goes beyond the cross checking of triangulation. Defined broadly, it deliberately looks for notices and investigates contradictions, anomalies and differences.

Facilitating: Rural people are allowed to facilitate investigation, analysis, presentation and learning by themselves, so that they present and own the outcomes, and as well learn from it. This has been expressed as 'handing over the stick'. This often entails an outsider starting a process and then sitting back or walking away, and not interviewing or interrupting.

Self-critical awareness and responsibility: Facilitators continuously examine their behavior, and try to do better. This includes embracing error-welcoming error as an opportunity to learn to do better and using one's own best judgment at all times, that is, accepting personal responsibility rather than vesting it in a manual or a rigid set of rules.

All these principles are behavioral, since they are applied in practice by people doing things. Thus, the quality and accuracy of participatory research in environment and development area depend on the moral responsibility and technical efficiency of researchers.

In the new millennium, it is more applicable in integrated conservation and development planning with respect to climate change adaptation and humanitarian issues. Thus, it requires picking up an appropriate theme and area of research that has in the priority of the country issues that suite the interdisciplinary research. However, funding institutions do not provide grants to such research in general due to the tradition of providing grant based on donor desires rather than local needs and demands. It further demands that the old concept of research and fund granting traditions should be modified to make it more applicable for use in research outputs into socio-economic development with due consideration of natural resources conservation and development policies of the country, which could be carried out with in-depth investigation, critical inquiry and exact assessment in seeking facts through collective efforts. Conducting such research in an environment with natural resource appraisal and development, self searching methods of inquiry that promote the culture of self-research with participatory mapping and modeling of what is on ground are used. It also requires that trend analysis, livelihood analysis, analysis of difference and participatory planning and budgeting should be done.

Innovation of participatory research

Participatory research has broken new ground of research

and methods of rediscoveries that encourage people to improvise the spirit of play. Various discoveries and innovations of participatory research have existed since the last three decades, and are briefly reviewed.

Villagers' knowledge and capabilities: The first discovery is that villagers have a greater capacity to map, model, estimate, rank and score their surrounding resources than outsiders. Findings of the participatory research show that rural people have more extensive and detailed mental maps than urban people and that when given the right conditions and materials, they can express this visibly on ground or on paper as three dimensional models (that is, watersheds). Rural people have now noticed social details using local materials such as seeds, manure, vegetables; land resource, water resource, market and other information. It evinces that having the right methods and materials, villagers can show themselves capable of generating and analyzing information far beyond normal professional expectations. It is evident that rural people are able to use largely independent culture of literacy to reflect and rank the problems and opportunities as they perceive them, and express their preferences for improving their farming system, managing and using common property resources for better livelihoods and for development actions in their communities.

Relaxed rapport: The second discovery (the relaxed rapport between outsiders and rural people) can be established early in the process, which is the key to facilitating participation. To measure trust and minimize predisposing conditions for inquiry, it promotes the culture of honest and accurate sharing of detailed knowledge and values between outsiders and villagers.

Diagramming and visual sharing: The discovery is the power of diagramming and visual sharing information. It helps to transfer the world of the person interviewed into paper and to change the information into public goods from personal to unverified. With visual sharing of map, the model, diagram or unit can be used for ranking, scoring, counting, or quantification, and all who are present can see, point, discuss, manipulate and alter physical objects or representation. Triangulation takes place with people crosschecking and correcting each other. Learning becomes more progressive and information is visible by the public, checked, verified, amended, added to and owned by the participants. Through participatory mapping and modeling, villagers can draw and model their resources with location, deciding what to include, debate, add and modify in the details. In shared diagramming information, seasonal changes dimensions, such as rainfall, agricultural production, labor, income, indebtedness, food security and migration trends can be shown in paper or other local materials. In the media, villagers commend and alter the fact of whether they are literate or not with confidence.

Sequences: The next popular discovery is the sequences of participatory methods like participatory mapping, social mapping and matrix scoring or ranking, which have more striking combinations and sequences that present new and complementary information on socio-cultural, natural resources and livelihood improvement aspects. In many areas, participatory resource map leads to transect planning, in which villagers made the map act as guides for outsiders. It further leads to the identification and discussion of problems and opportunities and then leads to listing and ranking options. A village social map provides up-to-date household information which leads to discussion, negotiation and reconciliation of priorities. Thus, the power of such sequences is four fold. First, the commitment of participation increases, before it further makes the action more likely, spontaneous and sustainable. Secondly, sequences triangulate and reveal errors of omissions in earlier presentations. Thirdly, the different activities interact cumulatively, and each activity add a dimension to the quality of details and enrich others, so that when taken together the whole become more than the sum of the parts. Fourthly, all concerned learn through the process, through people sharing what they know, through observation and through analysis. Therefore, participatory methods fit well with flexible learning process approach which is even more open-ended and adaptable to the villagers. Thus, it enables them use their own categories and criteria to generate their own agenda, and to assess and indicate their own priorities.

Training and reorientation for outsiders: The fifth innovation of participatory mapping, which is the initial training and orientation given to outsiders, can be conducted in practice in a short span of time. Face-to-face field experience is the main thrust of participatory research. Hence, the principles of optimal unpreparedness are applied. Usually, it is best to start sooner; as such, it seems safe or sensible for newcomers not to wait, but to start, stumble, self-correct and then share. Through participatory methods, villagers are encouraged to map, participate and plan for the training of outsiders to facilitate changes in perceptions and action, listening and not lecturing, learning progressively, embracing error and being critically self-aware of participation. Thus, the reversing role is taught to villagers in order for them to perform village tasks. For some outsiders, especially those who have had a strictly formal professional training, this can threaten trauma. They deserve sympathetic understanding, and no significant change may take place. For some though, a new range of possibilities and sense of freedom to experiment and innovation opens up. Thus, there is no need for the villagers to be trained in all methods. The methods can be tried, improvised and adopted subsequently, and new ones can be invented.

Sharing and spread: The last and most important discovery is sharing the culture and spread of technologies through participatory research. Participatory research is recognized now to have three foundations: methods, behavior and attitudes, and sharing. Previously, methods appeared to be the most important than behavior and attitudes of the outsiders, but now sharing is the most important for spreading of experience and mutual learning. To share and exchange methods and experience, interchanges of staff are efficient; as such, staff of one organization spending time with staff of other organizations can enhance the creativity and inventiveness of villagers to come into play. In such ways, innovations can be continuously stimulated. Thus, all forms of sharing-trainings, trainers, ideas, experiences, methods and innovations take place.

Practical application of participatory research

Participatory research methods, approaches and techniques have been widely used for appraisal, analysis and research in many subject areas such as agro ecosystems, natural resources, environment, technology and innovation, health and nutrition, family system research and extension, pastoralism marketing, disaster risk reduction, organizational assessment, socio-cultural and economic conditions. Participatory appraisal and planning, participatory implementation, monitoring and evaluation, and training and orientation for outsiders and villagers are the major processes that have been used in development and resource management areas. However, most of the application of the participatory research can be listed out in four categories.

Natural resources management: Participatory research is more popular and it uses widely acceptable techniques for watershed planning and management including: rapid catchment assessment, soil and water conservation, degraded forest assessment, nurseries and planting, identification of trees' uses, rural energy assessment, green enterprise development, wildlife reservation, and village resource management plans.

Agriculture: The next application is farmer participatory farming system research done by farmers in addressing crop farming, livestock farming, irrigation system, and market network for their products.

Equity and inclusive development: Another application focuses on a research on policy level to sustain the growth momentum in order to ensure inclusive development, where benefits of economic development are to percolate to the poor and under privileged with an objective to boot economic growth along with augmenting employment opportunities, alleviating poverty and ensuring regional equity and prosperity.

Health and nutrition: Health and nutrition monitoring is the fourth important application of participatory research which focuses on reproductive health, disease problem ranking, healthcare providers and costs, planning health projects, food security, nutrition assessment and monitoring, and water and sanitation assessment, planning and location.

All these applications are positive evidences that are scaled and used by various organizations and individuals, who have been engaged in training and appraisal, to respond to demand and their own sense of priorities and experiences. The use of monitoring and evaluation was found to be limited and academic researchers have been slow to recognize what was happening in practical application of participatory research. These methods have usually proved more reliable in information flow arguments and discussions taking place among the villagers, outsiders and between both villagers and outsiders. However, the success and failure much depend on the behaviors and attitudes of the outsiders whether or not they have the time, patience and will to get closer to reality. Any way, reversals of modes like from closed to open, from verbal to visual, from counting to comparing, and reversals of dominance as from extracting to empowering, reserve to rapport and from tedium to fun are some important changes in the research that was brought out by participatory research. In Nepal and other developing countries, like India and the middle east countries, it proved more useful to carry out the research and development activities for conservation of environmental and natural resources and livelihood improvement of rural communities. Besides these, policy research and change, personal behavior, attitudes and learning, farming system research, quality assurance, inclusive growth and empowerment, and substituting for survey are potential fields to participatory research in Nepal for achieving flexibility by empowering people, learning to love change, becoming obsessed with listening, deferring to the front line, and building system for a world turned upside down. Therefore, we have challenged the locals to further develop and disseminate approaches and methods to promote their own analysis and make their own needs and priorities known to scientists.

Usage of participatory methods in Karnali watershed area

Since 1990s, participatory approaches and methods of research and development have been widely used in many parts of the country. Particularly, this has been done in rural resources development and management practices through community based organizations and users' groups at community level, and it has become more popular in the world as the model of participatory development, such as community forestry and water

Table 1. Perceived challenges for ensuring people's participation.

Challenges	Key informants	
	No.	%
Lack of ownership feeling	12	19.57
Low level of public awareness	9	14.28
Unclear policy	7	11.11
Growing unemployment	9	14.28
Lack of technical assistance	7	11.11
Lack of social mobilization	7	11.11
Lack of need basis plan	11	17.46
Total	63	100

Source: Field Survey (2010).

resource management through users' groups. Many gainful evidences indicate that rural communities have benefited from using participatory methods to development for improving their farming system, conserving their available resources and preparing their own plan based on their own choices. However, it has little effects in Karnali watershed area. There are many reasons behind this which are discussed here with resolutions in brief.

Low level of people's participation

The Onward Fifth Five Year Plan (1975 to 1980) was initiated with the central objective of institutionalization of people's initiative in all development plans. The Decentralization Act of 1982 was enacted in the mid 1980s. However, in practice, particularly, in the Karnali watershed area, it is often seen that both the government and non-governmental organizations, involved in the local resource development, did not promote people's participation, rather the local people were kept away from taking active part in development. In the most cases, local people had not been given the choice to take decision for their development. Even though the financial authority was kept by the Central Government, the role of government was still found as the body in-charge of development and welfare of the local people. Further, there is growing number of middlemen who have bred corruption and mismanagement of available resources. Thus, only a small number of elite people are the beneficiaries of the development activities. The local people (real beneficiaries) are always kept far from the mainstream of development. It is evinced that the drawbacks of top-down development approach have largely been attributed to promote the people's participation in local level development. Sundaram (1997) has nicely noted that the government mechanism in Asian countries, like Nepal and India, hardly want to enhance the people's participation in the real term. It may be seen from Table 1 that the unclear government policy towards the participatory development, owing to the lack of strong sense of ownership feeling of local people on the ongoing development programs, was observed as the main challenges for increasing people's participation in the area. The skewed distribution of resources has also been hampering people's participation and the major bulk of resources are to be occupied by a small number of elites. About 11% of key informants have noticed the lack of appropriate social mobilization policies where people are educated, organized, motivated and enabled to undertake social enquiry and analysis for understanding their life situation and taking decisions and actions to change it for their well-being. As a result, 14% of the key informants further noted the dominant role of resource rich family on community-based organizations, which is further responsible for mass unemployment, and low level of local participation. Moreover, people's involvement has been considered to volunteer a form in the implementation stage of local development in the region. Thus, it is the Central government authority that takes almost all development decisions. Consequently, the local needs have not been taken seriously at the time of programming. All these make the people to be less willing to participate in development activities in Karnali watershed area. Among the various issues, the most remarkable was observed to be technological drawbacks. The present use of technology for local resource mobilization is not appropriate and adequate. The level of participation has also been observed to vary among the ethnic groups based on resource base and priorities. Therefore, an appropriate need-based and time specific strategic intervention in various stages is of utmost importance for the mobilization of local resources. On the whole, the imposition development from the top level without due consultancy from the village level failed to activate people's participation in local resource mobilization. So, there is a need to look at the opinion and idea of the local people in utilizing the local resources effectively at the local level. It shows that people's participation is found as the most effective social measure for empowerment of local communities and the only way by which backwardness of the area can be eradicated (Pokhrel, 2004). It further permits to note that genuine public participation of people at the centre of development is the hallmark of eco-development. This is so, because it is an endogenous development process in which local people themselves assume the responsibility to manage their resources, define their needs, goals and aspiration, and to make decision affecting their well being and sustainable development. People's participation is the process by which the rural poor would be able to identify their needs, share in design, implementation and evaluation of the participatory action for resource development (FAO, 1990). Some important activities that have been conducted by local communities for the conservation and management of natural and livelihood resources in Karnali watershed area are presented in the following photographs (Figures 1, 2a, 2b, 2c and 2d).

Figure 2a. Briefing of project activities with authors at the field office of the Local NGO.

Figure 2d. Discussion with beneficiary women of organic farming.

Figure 2b. Meeting with beneficiary farmers of Jumla.

Figure 2c. Discussion with INRM user groups in Surkhet valley.

Under such circumstances, the collective efforts of locals become self-organizing and self-sustaining in unleashing the potential of the local community and in meeting their inner urges and preferences with regard to development through self-reliance. The gravity of such development is to be self-generated, self-organized and self-sustained at the local level (Gultung, 1980; Charles, 1987; Bajpai, 1998). Thus, the insight development of the locals is most important. In doing so, there is an immediate need of resource knowledge to the local people which is further possible through people's resource right (Joshi, 2002). Of course, to ensure people's involvement in eco-development activities of the concerned communities, they must have a sense of belonging to the local and natural resources that is the main process of gaining power by real actors. Therefore, efforts are needed to promote the capabilities of local and indigenous institutions for making their functions more productive. This is possible only by strengthening such community institutions through providing sufficient autonomy as part of the organization involved in resource development. So, if local communities are to become principal actors in promoting their social well-being, they must be able to give vision and focus to their endeavors. It is indicative that devolution of power from the top to bottom (that is, grassroots) for capacity building with generation, application and diffusion of knowledge through a unified approach social change at the grassroots level could be achieved, and the resource poor could participate directly in efforts to improve their own well being. Furthermore, their participation must be substantive and creative in order to enhance the development of material resources and technologies with regard to serving and matching its real needs. It is clear that eco-development can only change the potential beneficiaries into functional groups. Thus, government authority should be given a priority for ushering in people's involvement. Moreover, it needs to be a strong support of finance, policy and technical inputs from the government to local institutions for enabling their capacity with regard to resource mobilization. Most importantly, government authorities need to establish broad eco-technological strategies in partnership with key stakeholders. Evidences suggest that in the national

Table 2. Need of people's participation for eco-development.

Perceived needs	**Key informants**	
	No.	**%**
Need basis/Bottom up planning	19	30.15
Ownership feeling	10	15.87
Coordination between outsider and insiders	6	9.52
Verification of implementing program	5	7.93
Development of indigenous knowledge	6	9.52
Organized/Collective Efforts	8	12.60
Institutionalization	9	14.28
Total	63	100

Source: Field Survey (2010).

level, local resources user groups and self-managed enterprises are emerging and functioning successfully. It proves that such grassroots organizations have great potential in those regions like the Karnali watershed area where environmental, as well as development challenges are immense. For this purpose, local people (that is, primary beneficiaries) need to form their own organization different from that of the primary local unit for mobilizing resources to acquire ownership of resource utilization and management overtime and assume full responsibility on decision making and implementation of community based resource development activities (Bavuskar, 1999; Sethi, 2003; World Bank, 2003). To achieve this, first, efforts need to be made to access the attitudes, behaviors and priorities of the locals with deep understanding of the group dynamics and heterogeneous characters of the society. Secondly, steps should be taken to identify and support formal and informal groups and institution. So, special attention must be given to explore the resource and resource rights. It is necessary thus to ensure eco-education of people to make them able to understand their needs and problems. Thirdly, investment has to be made in institutional capacity development which will further enhance the internal capabilities of local units. Such institutions should be considered as the form of public services. Therefore, local authorities must be strengthened to secure common benefits through accepting mutual obligation on resource mobilization. It shows that people's involvement in resource development is necessarily desirable for a need base bottom-up planning, strong feeling of ownership on local resources, development of indigenous technology and promotion of collective action of the grass-roots people in the region. It can further be noted that coordination among and between the line agencies who are involved in community based resource development, correction and verification of their implementation procedures, and institutionalization of such community based organization are also possible at the micro level by adopting eco-development strategy (Table 2).

Thus, people's participation is essential to replace a relationship based on dominance and competition with that based on reciprocity and collaboration, and to promote solidarity and unity of purpose among all members of the community. However, several challenges are ahead in ensuring people's participation in the study region. In most cases, the mechanisms involved for this purpose have failed to reach the grassroots level. The possible reasons might be over dependence on foreign aid and central controlled programs. Therefore, need base and resource base planning with right advocacy skill development and technical dissemination should be emphasized on government policy. At the moment, local efforts should be institutionalized to empower themselves with regard to income generation, social change, environment management and ecological balance of community life in the region. Emphasis should also be given to human resource development in order to promote the indigenous resource management system. Such indigenous systems have significant role on conservation of forest and other natural resources. For example, community forest user groups have proved themselves to be more responsible and reliable than the government officials. So, the participation of user groups and others who have rights in decision-making process about utilization and conservation of local and natural resources is what is needed now with respect to the region's sustainable development. Therefore, cultural, social and ecological dimensions should be prioritized in the implication process of integrated eco-development strategies in Karnali watershed area, because the gravity of cultural values, social norms and people's perception is much more on the control mechanism that acts for resource utilization and management. In short, the current emphasis should be on policies for strengthening indigenous resource management systems and institutionalizing such systems in order to succeed in the integrated eco-development model at the village level.

ECO-DEVELOPMENT PRACTICES

It is true that in Karnali region, the traditional farming system might have some little effect on ecological stability, so long as the population of man and his animals do not exceed the carrying capacity of the environment, but when they do, there has been consequent pressure on land, less suitable for farming, depletion of forest resources, erosion, and environmental destruction particularly of the marginal and sensitive areas. Lack of ecological consideration in development activities has also been intensified by environmental assaults and the life support systems have undergone diminishment. Thus, there is need to modify current land use policies that can make the region available for ecologically sound and sustainable agro-farming with man-environment kinship needs to be restored in order to link the socio-economic interests of the local people with the sustainable management of natural resources for environmental stability and ecological security. Therefore, shift in priorities, change in implementation methods, modification in the planning process, and examination in resource allocation patterns are needed for maximizing the economic returns of development without ecological destruction. Due to the wide range of topography, climatic elements, soil characteristics, and socio-economic condition, the Karnali watershed area presents very special ecological features having more than 75% of terrain rugged and steep, and the unit cost of infrastructure development is high, while returns are relatively low. The ever growing human demands, faulty method of cultivation, and development patterns have brought considerable damage to fragile eco-systems in the area. In this regard, an integrated approach of development and conservation by involving local people in all processes and steppes could be effective to promote the quality of life of the people in the area, because eco-development refers to sustainable development in which all development activities are performed in such a way that regional ecological balance can be maintained. Practically, environmental planning is done for regional (Meso) and rural (Micro) level development with a rational use of the resources and application of technological styles (Bhati, 1985; Guleria, 1987; Peet, 1989). Through the principal objectives of the development strategies, protection of the available resources, regeneration of resources and development of human knowledge could enhance the quality of environment and the economy of the area. Furthermore, through social development with respect to maximizing the productivity of both the natural and human resources, and through preventive planning and development policy, many adverse ecological consequences could be avoided and the interest of the local people to environment management could be enhanced. Finally, people will feel comfortable with their environment and process of development. Therefore, such development approach would make people adopt the integrated conservation and development in which people can decide their real goals and objectives for the proper utilization of local and natural resources. At present, it is shaping into people's movement, and the active people's participation is envisaged from the grassroots rather than from the remote control exercised by the corporate executives. Consequently, a fair hope that people can decide what their priorities of work are for conservation and development is expected. Thus, it is necessary to give top priority to the livelihood of the locals for designing such development strategies. This can be achieved through social planning norms that promote an organized collective effort of local people at the community level. All these are possible on a resource-base planning unit which may be a watershed unit, be it macro or micro, but more than the political boundary. It also enables the integration of environment management and socio-economic development by adopting watershed management approach at the village level. It shows that the holistic vision of the proposed development strategy is essential for the integration of the environment into development practice with regards to rational utilization of local resources and minimization of the environmental hazards.

WATERSHED MANAGEMENT

It is clear that the plans of the administrative units and central government failed to address the real needs of the area and the utilization of regional resources in Karnali watershed area. This is mainly due to top-down planning for local development and less interest of local people towards the government launched program. Further selection of the planning unit which is based on political boundary is also the factor responsible for ignoring the ecological bases of development. Thus, a socially desirable, economically viable and ecologically sound development approach is required to promote harmonious relationship between development and ecology. It is possible, to adopt only the watershed management strategy that prove to be an appropriate development path for the area. Pokhrel (2009) has noted that watershed management strategy should be the core for development activities, and there is need to popularize it in Himalayan region. Some other case studies have also shown that the integrated watershed development approach is more practicable and it evolves a compact technical, economical, ecological and geographical unit of planning for resource utilization and conservation (GOI, 1988; Grewel et al., 2001; Narayan, 1993). Thus, in Karnali watershed area, watershed management approach could play a significant role in increasing the productivity of the total resources. Watershed is a very appropriate spatial unit for examining the physical aspects of Karnali and its tributary rivers with

Table 3. Number and area of watersheds.

Basins	Catchment area (km^2)	Watersheds		Sub watersheds		Micro watersheds	
		No.	Area (km^2) per watershed	No.	Area (km^2)	No.	Area (km^2)
Karnali	42,890	7	6,127	15	2,859	90	476.55
Babai	3,270	2	1,635	6	545	27	121.11
Rapti	6,500	5	1,300	10	1,083	42	240.71
Total	43,260	14	3,090	31	1,395	159	272

Source: Sharma (1977); Field Survey (2010).

their size, which may vary from small to some large catchment areas. So, watershed is a logical planning unit for evaluating the biophysical linkages of upstream and downstream activities of the basin. It is only the sixth five-year plan introduced as an integrated development approach in Nepal as the synonymous of soil and water conservation, but the failure of large scale development programs and the realization of the importance of interlink ages between resource productivity and sustained increase in rural income have led to the shift in watershed management approach with the prime focus of integrating environment and development in regional level. In operational level, it made efforts to bring all concerned institutions into single window system through interdisciplinary approach. Recently, it is more anthropo-eco-centric and community based rather than technology based, thereby leading empowerment and self-reliance of the primary stakeholders as an entry point for outsiders to understand local environment, and for making development plan in a comprehensive way. However, evidences show that macro watershed management programs in Nepal have some remarkable failures, in that many watersheds are still being degraded. Particularly, in the Karnali watershed area, large scale programs have rarely been successful in conservation and utilization of local and natural resources with people's participation. The degraded watersheds are further worsening due to over intervention of outside technical teams (Shah, 2000). Thus, social mobilization is needed to enhance the success of watershed development approach for the strong feeling of belongingness of local people towards the ongoing development and conservation programs. This is possible only by adopting micro-watershed development approach that can provide an appropriate path way for eco-development. Further, it would also be helpful in devising location specific solutions that may emerge by resolving activities that are conflicting (Head reach-tail end beneficiaries), and build-on complementary activities (forestation and soil conservation in hill and mountain regions enables better productivity in the *Tarai* region). Thus, emphasis should be shifted from large to small watershed areas in which both management of environment and socio-economic development can be integrated (Agarwal, 2003). Therefore, an effort is made here to delineate estimated watersheds, sub watersheds, and micro watersheds in the region. Based on the drainage pattern, transport network, population statistics and relief features, the whole Karnali-West Rapti basin (43,260 km^2) was divided into four hierarchical orders: resource regions, macro watersheds, sub watersheds and micro watershed. So, 14 macro watersheds, 31 sub watersheds and 159 micro watersheds were identified as the unit of eco-development planning in the area (Table 3). These micro watershed units can further be sub-divided into small and mini watershed units varying in size (50 km^2) for different land use purposes. Therefore, it is the micro watershed development strategy that could solve the problem of the five 'es': employment, economy, ecology, expert and equity, and would then be ushered in eco-development with a sharp focus on resource conservation and poverty alleviation in the region. Furthermore, the institutional and technical co-ordination amongst the involved agencies would also be strengthened in the region (Figure 3).

As the concept of integrated eco-development refers to sustainable resource development, it requires local initiations for integrating the environmental issues into socio-economic development. This development approach can make a substantial contribution to resource management at the micro level incorporating biophysical, technological, social, political and other variables which account for resource utilization and conservation. Further, such type of development approach maps the complex connection, direct-indirect effects, feedback loops, and reciprocal relations and effects of environmental and human resources to each other. Experiences have shown that local institutions in essence, community resource user groups are so effective in conservation and development of natural resources at regional and national levels in the country. Even in many parts of the Karnali watershed area, especially in the hills, such user groups have also exhibited extra-talents in forest resource protection, utilization and management. Thus, it is strongly felt that a micro level resource utilization and conservation model is necessary in the precise and synthetic form that would increase the productivity of environmental resources in a sustainable manner. Such grassroots model can as an aspect of bottom-up strategy call for mass action at the micro level (that is, village level). So, the kingpin of the natural resource

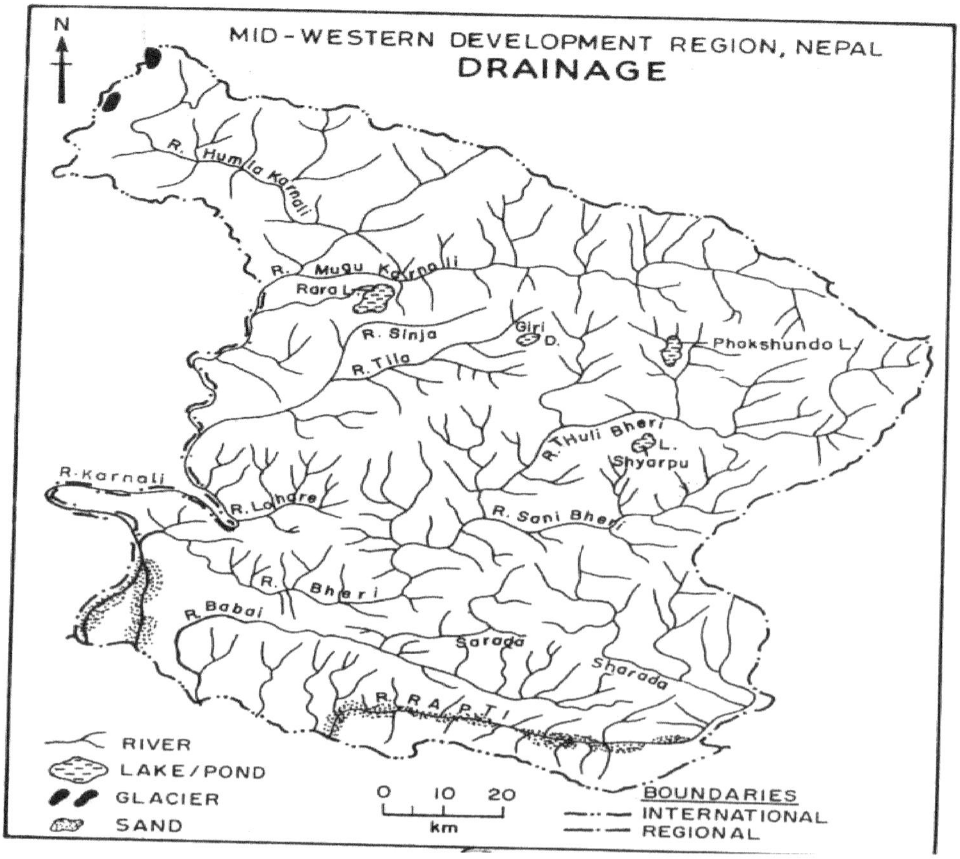

Figure 3. Watershed Management in Karnali watershed Area.

development strategy of Karnali watershed area is self-reliance on resource mobilization by using indigenous knowledge and experiences. Thus, the right mix of traditional knowledge would be a winning combination of resource development. The central thrust of participatory development has to integrate development and conservation, government authorities and local beneficiaries, and coordination of the involved institutions between the decision making and implementation level and that of the micro watershed level. Further, such adaptive model of resource management focuses on environmental feedback to shape policy. Thus, the model explicitly considers social learning and institutional development. Under such provision, large numbers of community based organizations would have evolved out of the region, and would be able to develop locally suitable technology, appropriate policy feedback and strategic clarity for resource development and conservation. Better information system, participatory monitoring and systematic implementation of resource mobilization will also evolve through local initiations. Finally, a specific technique would be modeled against the top-down tendencies for resource based regional development planning in Karnali watershed area and in Nepal in general. For this purpose, it is necessary to build-up local capacity for the total reforms of institutions in order to change the environmental policies in different orders. The principal concepts and components of this grassroots development model are:

(i) Formulating policy, strategy and programming the project based on the resource knowledge of the local people and an application of a suitable technology. Thus, it has to have a local base including people, traditions and indigenous institutions.
(ii) Implementation of resource development activities through people's involvement to strengthen local organizations and to bring about positive change in the local environment.
(iii) Participatory monitoring system with a view to avoid previous drawbacks of resource development planning at the micro level.

Thus, such innovative model can eventually lead to a district, as well as village level conservation and development plans. Further, such model could successively imply that the local organizations have to control the local and natural resources, adopt a sustainable pattern of development and avail appropriate technologies in subsequently distributing local resources amongst beneficiaries on an equitable basis. Therefore, government authorities need to develop suitable methodology and guideline in order to ensure people's participation and facilitate local organizations for the process of institutionalization in the proper way. Similarly, to provide some financial support in the form of seed money is also needed. To build up local capabilities by imparting training and techniques at the local level is also essential from the government side. Hence, it should be emphasized that the catalytic role of government is to galvanize the people to initiate resource development actions according to their priorities at self reliance. In this regard, district development committee (DDC) and village development committee (VDC) as the authority of government need to make a framework, in which this model may function and evolve. To achieve this, there is need to modify outside intervention for the dynamics of local institutions. Actors, like non-government organizations, should collaborate with resource owned communities to develop the integrated micro level model, and explore and exhibit relevant alternatives that would stimulate the development of new resource management institutions (Saravanan, 2001).

CONCLUSION

It is evident that such participatory resource-based research and development model enables both the local people and collaborating authorities to learn about each other (their goals, knowledge and techniques) and finally make development efforts that are socially desirable and ecologically suitable. All these imply far-reaching changes and cooperation among and between the beneficiaries with respect to up-grading of skills and adaptation of new technologies for community based resource utilization and management in Karnali watershed area.

REFERENCES

Agarwal SK (2003). Integrated Watershed Development: Technology based Model for Rural Development, Kurukshetra. 51(12): 49-57.

Bajpai PK (1998). People's Participation in Development: A Critical Analysis, in S.P. Srivastava (ed).The Development Debate: Critical Perspectives. Rawat Publishers, Jaipur.

Baviskar A (1999). Participating in eco-development: The case of the Great Himalayan National Park. In Jeffery R, Sundar N (eds.). A New Moral Economy for India's Forests? Discourses of Community Participation, Sage Publication, New Delhi.

Bhati JP, Swarup R (1985). Why Eco-development Planning Fails in the Himalayas: Some Problems and Policy Issues. In Singh TV, Kaur J (ed.) Integrated Mountain Development, Himalayan Books, New Delhi.

Chambers R (1995). Rural Appraisal: Rapid Relaxed and Participatory. In Amitava M (eds.) Participatory Rural Appraisal Methods and Applications in Rural Planning, Vikash Publishing House, New Delhi.

Chambers R (1989). Farmer First: Farmer Innovation and Agricultural Research, Intermediate Technology Publications, London.

Charles KJ (1997). What is Development? In George KK, Nair VS Jacob J (eds). Economics, Development and the Quest for Alternatives. Concept Publishing Company, New Delhi.

FAO (1990) Farming Systems Development, Food and Agriculture Organization of the United Nations, Rome.

Gultung J (1980). Self-Reliance – A Strategy for Development, Bogle Liovertune Publications, London.

GOI (1988). Integrated Watershed Development Project (Hills): An Analytical Overview and Future Strategies.National Workshop on Watershed Approach to Rain fed Farming: Strategies for Integrated Watershed Development for Hill Regions, Rain fed Farming System, Government of India, Vigyan Bhawan, N Delhi.

Grewel AS, Jai TC (2001). Poverty Alleviation and Resource Conservation through Integrated Watershed Management in a Fragile Foot-Hill Ecosystem. In Stott DE,. Mohtar RH, Steinhardit GC (eds) Sustaining the Global Farm. USDA-ARS National Soil Erosion Research Laboratory, Purdue.

Guleria AS (1987). A Strategy for Eco-Development in Himalayan Region: Watershed Approach in Pangtey YPS (ed) Western Himalaya: Environment, Problems and Development, Gyanodaya Prakashan, Nainital.

Joshi AP (2002). Resource Rights – Community Domain, Kurukshetra, 50(7): 12-14.

Narayan VD (1993). Watershed Management: Concept and Framework in Soil and Water Conservation Research, Indian Council of Agricultural Research, New Delhi.

Peet R (1989). New Models of Environment and Resources. In Peet R, Thrift N (ed) New Models in Geography, Unwin Hyman Ltd., London.

Pokhrel KP (2009b). Climate Change Adaptation: Strategy for Ecological Restoration and Livelihood Security, VOICE, 1(1): 5-12.

Pokhrel KP (2004). Local Efforts on Resource Conservation and Eco-development: A Case Study of Mid Western Development Region Nepal, Unpublished. Ph.D. Thesis, BHU, India.

Pokhrel KP (2009a). Environment and Development Practice in Nepal: A Geographical Analysis, EDRC, Banke, Nepalgunj.

Rhoades RE (1990). The Coming Revolution in Methods for Rural Development Research, Users' Perspective Network (UPWARD), International Potato Centre (CIP), Manila.

Saravanan VS (2001). Institutionalizing Community Based Watershed Management in India: Institutional Elements for Sustainability National Council of Applied Economic Research (NCAER), New Delhi.

Sethi S (2003). Stakeholders' Participation in Local Level Planning and Development, Kurushetra, 53(3): 26-30.

Shah P (2000). Participatory Watershed Management in India: The experience of the Aga Khan Rural Support Program. In Ian S, John T (ed.) Beyond Farmer First-Rural People's Knowledge, Agricultural Research and Extension Practice, Intermediate Technology Publications Ltd., London.

Sundaram KV (1997). Development from Within: Rethinking on an Alternative Development Path, in George KK, Nair VS, Jacob J (ed.) Economics Development and the Quest for Alternatives, Concept Publishing Company, New Delhi.

World Bank (2003). World Development Report, Sustainable Development in a Dynamic World, Transforming Institutions, Growth, and Quality of Life, The World Bank and Oxford University Press, New York.

Among and within population variation in growth dynamics and floral sex ratios in *Inula racemosa;* a critically endangered medicinal herb of N. W. Himalayas

Peerzada Arshid Shabir , Irshad Ahmad Nawchoo and Aijaz Ahmad Wani

Economic Botany and Reproductive Biology Laboratory, Department of Botany, University of Kashmir, Srinagar-190006. J&K, India.

The purpose of the present study was to determine the extent to which different morphological and reproductive characteristics of a critically endangered, Himalayan herb (*Inula racemosa*) varied along an altitudinal gradient. To this end, seven populations growing at different altitudes (1595 - 2800 m asl) across the Kashmir Himalaya were assessed. ANOVA expressed significant differences (P=0.05) in all growth parameters as a function of altitude. The highest difference between the populations was apparent in the stem height and in all populations patterns in the foliar and floral number and dimensions closely followed those in plant height. In four out of the seven populations, significant variation in floret number per head was detected among plants within all seven populations. Only 19-34% of variation was found within plants and in contrast 38-72% variation was found among plants. As compared to total ovule number per capitulum the species show low percentage of reproductive success (59-72% seed set). However, increasing altitude resulted in a decrease in relative and absolute allocation of biomass to reproductive structures in the form of decreasing dry weight of individual capitula as well as in the form of absolute declining of total number of capitula produced. There was also a trend for increased relative allocation to below-ground rhizome with increasing altitude, even though altitude did not affect absolute allocation to below-ground and vegetative structures.

Key words: Altitude, floral sex ratio, growth, *Inula racemosa,* phenotypic variability, reproductive success.

INTRODUCTION

Plants grow in a modular manner and are indeterminate in their growth which allows them to be highly plastic in both morphology and reproduction. Moreover, because buds act as the germ, any mutations acquired will be passed along through subsequent cell divisions. It is this mode of growth that allows the large degree of morphological plasticity, characteristic of plant growth, resulting in wide intraspecific differences in many features of plant architecture (Silvertown and Lovett Doust, 1993).

Whether this variation is genetic or environ-mental can be debated, however it is most certainly mediated physiologically (Waller, 1988). However, plant growth pattern influences a number of important traits that affect fitness, including pollination and dispersal (Waller, 1988). It can also provide phylogenetic infor-mation (Walton and Hufford, 1994). Moreover, plant growth and allocation patterns are influenced by a complex suite of potentially-interacting factors, providing a mechanism by which, over the long term, adaptive associations among external factors and the various morphological traits can occur. However, morphological variation is not necessarily due to an evolved response and precise causal mechanisms for architectural variation is yet to be established for most species (Waller, 1988).

Altitudinal system presents strong, multidimensional environmental gradients (Korner, 2003) and plants at different positions along these gradients face distinct growth conditions. Each of these factors may influence plants directly and also indirectly by altering plant response to other factors. Different altitudinal ranges are therefore excellent model systems for studying plant responses to environmental variation over small geographical scales. Elevational gradients are thus convenient scenarios to investigate such responses and key traits for plant functioning vary in a clinal fashion along elevation gradients. For instance, the reduction in overall plant size is the most conspicuous structural alteration observed along elevational gradients (Korner et al., 1989).

Flower size does not seem to be affected in a proportional way and therefore alpine plant species are often believed to invest relatively more in reproductive structures (Fabbro and Korner, 2004). For perennial plants absolute allocation to reproduction usually increases with size (Clauss and Aarssen, 1994). However, the proportional reproductive allocation, that is, reproductive effort has been found to increase (Aarssen and Taylor, 1992), decrease (Waite and Hutchings, 1982) or be independent of plant size (Pickering, 1994). Such varied effects of plant size on reproductive effort may be genetically or environmentally induced (Schmid and weiner, 1993) and/or result from different morphologyical/developmental constraints (Geber, 1990).

In this study, we used four years of field observations to explore within and among population variation and/or adaptive plasticity in phenotypic traits and reproductive success of perennial herb *Inula racemosa* Hook.f. (Asteraceae). The species has a limited distribution in Hindukush Himalayan region across Afghanistan, Pakistan, India and Nepal and is primarily found in open temperate meadows and forest margins of North-western Himalayan mountainous regions between 2100 and 2800 m in altitude. The species is known for its potent medicinal properties (Wani et al., 2006), is facing ruthless over-exploitation and is at the verge of extinction. Once being considered as critically endangered, the species is now drawing favourable attention in various conservation programmes.

The present study aimed to present a comparative study of within and among-population variation in various morphological and reproductive traits. A specific aim of this research project was to quantify and contribute to the body of knowledge of the phenotypic variability and allocation pattern for an individual species and its interactions with its environment. To achieve these aims, data were collected from naturally growing populations across a wide geographic range. Specifically, we search for trends in trait variation reflecting the changes along a steep altitudinal gradient. We focus on variation in growth morphology and architectural traits, reproductive success and changes in allocation patterns and their relation to the environmental conditions along the altitudinal gradient. This paper also presents an attempt to make a formal quantitative analysis of the floral composition in populations of *I. racemosa* growing over the wide range of different altitudinal gradients. Here we reported observations of total floret numbers and percentage female ray and hermaphrodite disc florets per capitulum and the variation within and among individuals within different studied populations of *I. racemosa.*

MATERIALS AND METHODS

Study species

I. racemosa Hook. f. (Asteraceae); endemic to North West Himalaya and its adjoining areas is a long lived herbaceous perennial species found in moderately-moist, shady and sloppy habitats of North Western Himalayas at an altitudinal range of 2100 to 2800 m-asl. Plants produce multiple stems each year, about 1-2 m tall. Mature flowering individuals are about 60to 200 cm tall with a terminal raceme of yellow flowers visited by bumblebees (Shabir et al., 2013). Mean date of flowering (first flower) is about 15 July in our study site. Flowers are arranged in terminal discoid capitula of 5-8 cm in diameter, with 5-30 capitula per stem. Fruit ripening (small achene) occur in September.

Study sites

Six natural populations of *I. racemosa* were studied for four consecutive years (2009-2012). Study populations were located along an altitudinal gradient representing a wide range of geographical and environmental conditions across the Kashmir Himalaya (Table 1). The geographical co-ordinates of these sites were recorded using geographical positioning system (GPS). Because of the difficult access to the wild populations, one additional transplanted population at KUBG was maintained to have a ready access to the germplasm and to acclimatize the plants at low altitude conditions. Comprehensive field notes on habit, habitat, growth dynamics, floral sex ratio and reproductive success were recorded.

Plant morphological characterization

In order to address the overall morphology of the species and to capture the variability in floral and vegetative traits within and among populations of *I. racemosa*, a series of traits for 30 individuals per population were measured at peak flowering time. The different readily identified characters were chosen for morphological measurements which include: plant height, leaves number, foliar dimensions, floral density and floral dimensions (Table 2). Differences among populations in various morphological features were analysed by one-way ANOVA using SPSS.

Floral sex ratios and reproductive success

Heads at peak flowering from all the selected populations of *I. racemosa* were collected during all the four seasons and each

Table 1. Salient features of selected sites for the collection of *I. racemosa*.

Study site	Altitude (m-asl)	Latitude and longitude	Climatic zone	Habitat	Threat factor
Gulmarg	2650	34°03'N 74^{0}22'E	Temperate	Moist open slope	Grazing, extraction and habitat destruction
Drung	2150	34°4' N 74°35'E	Temperate	Moist open slope	Grazing, extraction and habitat destruction
Harwan	1800	34°9' N 74°54'E	Temperate	Moist open slope	Grazing with partial landslides
Sonamerg	2800	34°20'N 75°20'E	Temperate	Moist open field	Grazing and habitat destruction
Dawar Guraz	2580	34°63'N 74°83'E	Temperate	Roadside	Heavy land slides
Kanzalwan	2580	34°63'N 74°83'E	Temperate	Roadside	Heavy land slides
KUBG^{+}	1595	34°5'N 74°35'E	Temperate	Moist open field	Nil

Table 2. Morphological variations recorded among seven different populations of of *I. racemosa*

Attribute	Gulmarg	Drung	Harwan	Sonamerg	Dawer	Kanzalwan	KUBG	P value
Plant height	164.85±16.06	189.43±17.34	181.19±14.78	161.25±14.21	166.02±12.41	165.29±18.33	216.81±18.26	0.000
Leaf no.	15.21±2.09	17.41±2.31	17.99±3.22	15.26±3.29	15.33±2.88	15.41±2.01	19.61±2.41	0.000
Max. leaf length (cm)	57.61±7.41	57.14±8.32	66.52±6.59	56.29±7.22	52.21±8.29	53.64±7.16	76.22±6.29	0.001
Max. leaf breadth (cm)	14.36±2.21	12.67±1.86	15.29±2.11	13.34±2.01	14.55±1.99	14.92±1.85	19.41±3.61	0.000
Capitulum number	14.81±2.85	16.51±2.87	18.29±3.21	14.55±2.02	15.52±2.25	16.01±1.87	21.31±2.79	0.005
Capitulum diameter (cm)	7.49±1.62	7.54±1.41	6.92±2.03	7.89±1.88	7.82±1.38	7.09±1.19	7.63±1.82	0.001
Ray floret number	55.06±6.57	55.11±4.45	54.24±5.41	55.92±6.07	55.61±4.99	54.99±5.33	57.61±3.67	0.005
Disc floret number	415.21±23.26	416.61±26.52	431.45±31.29	445.92±37.34	401.66±23.11	429.24±29.66	451.91±35.88	0.005
Achene number (per capitulum)	342.34 ±8.38	317.67 ±11.76	321.24±12.22	338.29±10.51	313.72±17.64	326.77±12.28	305.67±15.43	0.000
Achene length (mm)	3.11±0.31	3.15±0.64	3.62±0.91	3.94±0.23	3.66±0.35	3.54±0.22	3.91±0.42	0.000
Achene width (mm)	0.41±0.01	0.44±0.01	0.48±0.11	0.49±0.02	0.45±0.04	0.47±0.02	0.49±0.01	0.001

collection involved selecting plants at regular intervals (1-2 m, depending on the population sampled) along a transect. We sampled 20 plants in each population and two heads were collected arbitrarily from scattered positions on each plant. In total, we collected 280 flower heads per season. The heads were dissected under a bifocal magnifying glass and the number of ray and disc flowers per capitulum was counted to examine the proportion of variation in the ray and disc flowers within and among plants as well as among populations. In late September mature fruiting heads were similarly collected from all the selected populations. On each head we determined the percent seed set, expressed as the ratio of developed achenes to the total number of florets per capitulum. Developed seeds could easily be distinguished from undeveloped ones because the former are black and hard while the later are soft and empty. The data were analysed using ANOVA, followed by calculations of the percentage variation at each level (Sokal and Rohlf, 1995).

Reproductive allocation

To determine resource allocation pattern, 15 individuals of each population were dug up and separated into rhizomes, stems, leaves and inflorescence. The flower fraction contains all flowers including their involucres, whereas stems and flower stalks form the stem fraction and the leaves with their petioles included form the leaf fraction. In cases of clonal, multistemmed individuals, the sampling unit was a ramet, also referred to as flowering shoot. All plant material was dried to a constant mass in an oven at 60°C and weighed. Reproductive effort was calculated from dry weight of the different parts of individual plants, that is, dry weight or biomass allocated to reproductive and vegetative structures (Abrahamson and Gadgil, 1973).

$$\text{R.E} = \frac{\text{Dry weight of the inflorescence}}{\text{Total dry weight of the above ground tisssues}} \times 100$$

RESULTS

Growth attributes

I. racemosa is a perennial ascending hairy herb with 100-200cm grooved orthotropic stem and rhizome that grows 5-10 cm below the soil surface. Leaves are winged petiolate with an elliptical-lanceolate lamina (average length 30-50 cm) and having villous lower surface. In all reproductive plants varying number of 5-30 inflorescences per plant emerge per growing season. Inflorescence is capitula which are 4-8cm in diameter. Involucral bracts are 4-5 seriate. Marginal ray florets are ligulate, pistillate and 2-3 toothed while central florets are tubular, perfect and 5 toothed.

The populations of *I. racemosa* analysed in this field survey varied considerably in their morphological features showing an admirable phenotypic variation both at inter and intra-population level (Table 2). There were highly significant differences among populations in stem height, leaf dimensions, capitulum size and capitulum number (Table 2). Based on the means of the thirty plants per population, the strongest difference between the populations was apparent in the shoot size which was almost 1.5 times as high in Harwan (189.43±17.34; P=0.03) as in Sonamerg (161.25±14.21; P=0.001) with an average decrease of 2.74% between 1800 and 2800 m-asl (Table 2). The other growth parameters that followed the similar trend as shown by the plant height across the study sites include leaf number, leaf dimensions and capitula number. Plants flowered for a period of 8-10 weeks and the highest percentage of flowering plants and the largest mean number of capitula per plant were reached between late-July and early-August. Log (capitulum number) per individual also decreased significantly and substantially with altitude (by an average of 14% between 1800 m and 2800 m a.s.l.) with a mean capitulum number ranging from 14.55±2.02 to 21.31±2.79, (P=0.05) across populations. This decline in capitulum number might partly be related to decreasing plant size with altitude, since log (capitulum number) is correlated with stem height across all species and regions (r = 0.61, 0.001). Contrary to patterns among natural populations, KUBG plants were almost 2 times larger than plants in the natural populations. In addition, plants from the cultivated population also show the highest values of foliar and floral dimensions (Table 2).

Floral sex ratio and reproductive success

A significant reduction in both total floret number and ray floret number per capitulum occurred with increased altitude (Table 2), however, the reduction in ray floret number was minor relative to total floret number. In four out of seven populations, significant variation in floret number per head was detected, and among plants within all seven populations (Table 3). Only 19-34% of variation was found within plants and in contrast 38-72% variation was found among plants. The added variance component within plants in the proportion of disc florerts was significant for only transplanted KUBG population while a significant variation in this variable did exist among different sampled populations, and 36-54% of total variation occurred among populations (Table 3). In all sampled populations, most of the variation in total floret number was among plants (58-77%). Percentage seed set per capitulum also varied along altitudinal gradient and lowest seed set was observed in transplanted KUBG population (59.94%). A significant variation was observed in achene number per head among the plants within all the seven populations (Table 2).

Reproductive statistics for the seven studied populations of *I. racemosa* (Table 3), which includes the estimation of number of ovules developing per capitulum and the percentage of reproductive success showed that the species showed 59-72% of seed set. Multiplying the mean number of achenes per capitulum by the mean number of capitula per stem gives the estimate of seed-reproductive capacity per stem as shown in Table 2. The number of capitula per stem varied with overall plant height, from zero on small stems to over 30, with a mean of 14-30 across populations. In addition, the number of ramets per rhizome varied from one to seven, and thus represents a great plasticity in numbers of capitula per stem and thus the reproductive success.

Table 3. Distribution of variation in total flower number and percentage of disc flowers and achene number per head among and within populations and within plants.

Population	No. of heads per plant	No. of plants per population	Percentage of variation in flower number per head		Percentage of variation in disc flowers per head		Percentage of variation in achene number per head	
			Within plants	Among plants	Within plants	Among plants	Within plants	Among plants
Gulmarg	5	10	24.9	45.8**	18.3	32.2*	53.1	49.2***
Drung	5	10	34.6***	57.3**	26.1	48.8*	44.2*	38.4**
Harwan	5	10	19.2	38.6***	16.4	34.5**	54.8**	64.2**
Sonamerg	5	10	28.7**	61.5***	30.8	56.9***	45.9	38.8*
Qanzalwan	5	10	26.8**	45.3*	44.1	37.5**	32.8**	43.6*
Dawer	5	10	22.1	52.6**	27.9	45.6*	38.9***	49.7***
KUBG	5	10	31.3**	72.3*	52.7**	56.1***	78.4*	65.2**

Reproductive allocation at flowering

For the studied species, plant size varied across altitudes, the total resource pool thus varied over several orders of magnitude and a significant difference was observed in total above ground dry weight biomass, dry weight of different vegetative structures and the total reproductive effort among the plants of studied populations, growing at different altitudes. Allocation of biomass to different organs was size dependent for plants in all sites (Figure 1) and at lower altitudes, mean plant biomass was approximately 1.5 to 2 times higher than at higher altitude. Plant size decreased with increasing altitude ($r = 0.91$). However, increasing altitude resulted in a decrease in relative and absolute allocation of biomass to reproductive structures in the form of decreasing dry weight of individual capitula ($r=0.81$) as well as in the form of absolute declining of total number of capitula produced ($r=0.79$). There was also increase in relative allocation to below-ground rhizome with increasing altitude, even though altitude did not affect absolute allocation to below-ground and vegetative structures. On average, flowering individuals of high altitude, Sonamerg population weighed 191.47 ± 18.0 g, while the KUBG population growing at lower altitude had mean individual/ramet weights of 244.85 ± 38.24 g, that is, were 1.5 times heavier. However, mean flower mass fraction (flower mass/total shoot mass) was larger in higher altitude Sonamerg (0.19) population than in lowland populations (0.21 in KUBG). The stem mass fraction differed in the opposite direction and was almost two times lower at high (0.37, $p \leq 0.01$) than at low altitude (0.58, $p \leq 0.04$). The leaf mass fraction of shoots did not differ significantly between altitudes. In terms of biomass, reproductive effort increased significantly with altitude (26.93 in Gulmarg, Figure 1) while at lower altitude populations (20.85 in KUBG, Figure 1), reproductive effort varied inversely with plant size. Further, the percentage allocation to reproductive parts, that is, reproductive effort was more towards maleness (0.32±0.21 to 0.27±0.09) than femaleness (0.16±0.43 to 0.15±0.07), which suggests an outbreeding nature of the species.

DISCUSSION

In this study, we found distinct directional trends in growth morphology, architecture and fitness-related traits across different populations of *I. racemosa* growing along a steep altitudinal gradient. A great degree of phenotypic variability was observed and on an average, plant traits decreased linearly with increasing altitude of population origin. Plants growing along an altitudinal gradient exhibit differences in their height which generally became expressed by a shortening of their stems at high elevations. Decreasing plant size as an adaptation to increasing altitude is a well-known phenomenon (Baret et al., 2004; Korner, 2003; Willis and Hulme, 2002). It results from a slower growth rate as a response to harsher growing conditions and shorter season at high altitude (Bennington and McGraw, 1995; Grime, 1979). This decrease in plant height along with altitudinal elevation may prove advantageous for the species as the stem shortening allows plants to avoid the damaging effects of the strong winds blowing at high altitudes and to improve photosynthetic conditions by keeping the leaves closer to the warmer soil surfaces (Korner and Cochrane, 1983).

The main effects of altitude on leaf and capitulum number might partly be related to decreasing plant size with altitude. The maximum

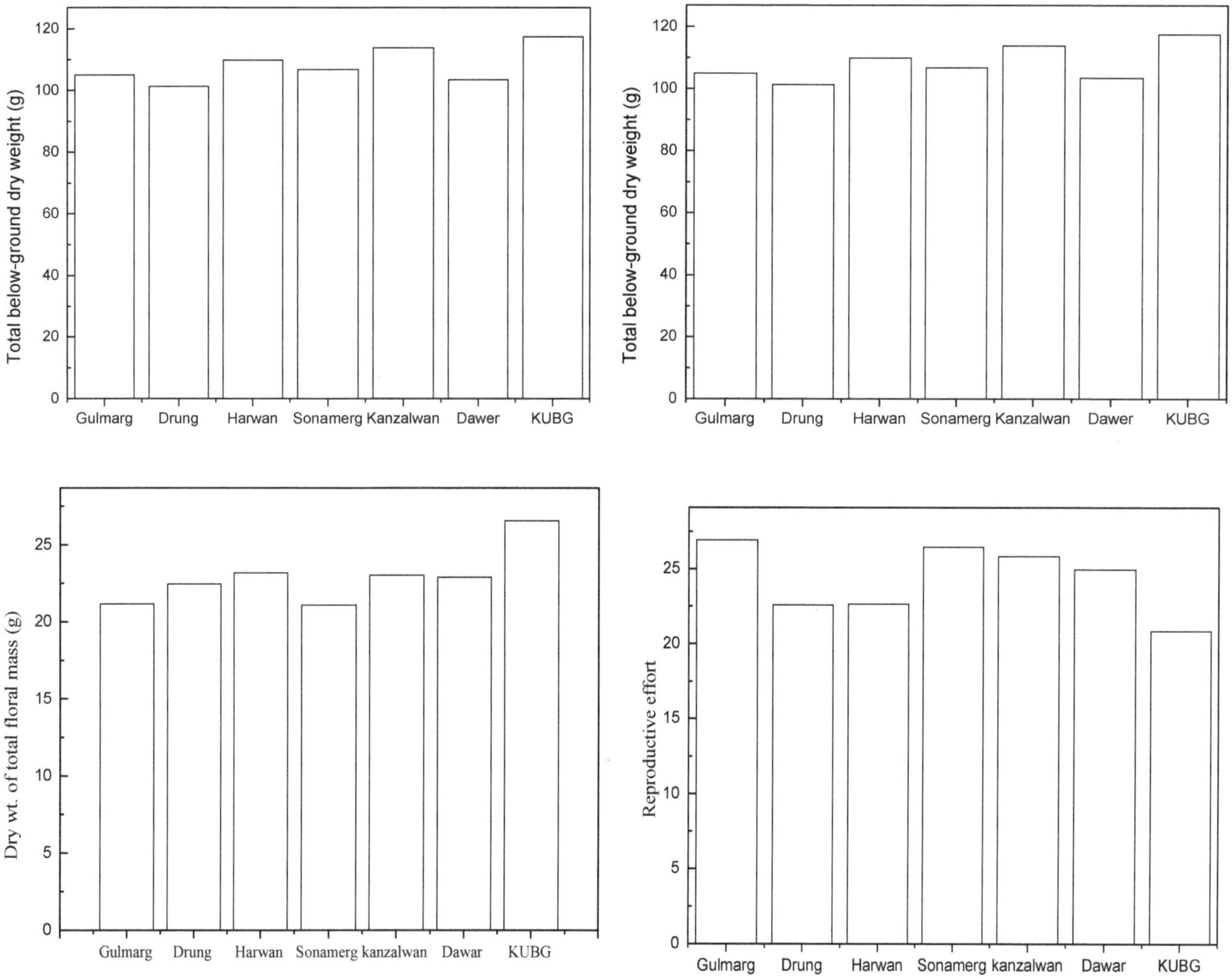

Figure 1. (a) Total above-ground dry weight, (b) total below ground dry weight, (c) dry weight of total floral mass, (d) reproductive effort.

mean difference for capitula number was observed between the transplanted population of KUBG and the population of Sonamerg. These findings suggest that plants growing at lower altitudes are distinctly more vigorous, much taller and bearing larger number of leaves and inflorescences as compared to plants growing at higher altitudes as also previously revealed by Johnson and Cook (1968) and Hickman (1975). Also, the relationships between plant size and reproduction, that is, increased allocation to reproduction (number of capitula) with plant size; is in conformity with the observations of Ohlson (1988), Aarssen and Taylor (1992), Kudo (1993) and Clauss and Aarssen (1994).

In the seven populations of *I. racemosa* considered in this study, variation in total flower number per head was often evident. However, the floral sex ratio remained within narrow limits despite much variation (interpopulation) in floret count across populations. Specifically, large plants bore significantly more flowers per head than small plants in all studied populations. Of the seven studied populations, significant variation in total floret number was found in four populations and it accounted for a significant amount of the variation in proportion of ray and disc florets across populations.

Knowledge of the resource allocation pattern of plants is beneficial to understand the life history strategy operative in the species. Of the total resource pool, reproductive allocation is often determined by plant size (Lacey, 1986; Samson and Werk, 1986; Saulinier and Reekie, 1995) and thus reflects the reproductive potential of a species. Studied populations of *I. racemosa* significantly differed in proportional biomass partitioning, with high altitude plants having low biomass allocation to stem, leaves and inflorescences and high to belowground rhizome. At lowest altitude, plants dry mass partitioning averaged 27% for rhizome (P=0.0001), 19% for stems

(P=0.008), 16% for leaves (P=0.089) and 13% for inflorescences (P=0.002). At the highest altitude population, these values averaged 29% for rhizome (P=0.0001), 18% for stems (P=005), 17% for leaves (P=024) and 11% for inflorescences (P=0.0001). The study revealed that the flower size did not seem to be affected in a proportional way and high altitude Dawar population allocated more of their percentage aboveground biomass to floral structures than lowland KUBG population, while the shoot mass fraction allocated to stems decreased in high altitude populations.

Therefore, alpine plant species are often believed to invest relatively more in reproductive structures. This suggests that alpine plants have enough resources to allocate more proportional aboveground biomass to flowers. It also implies a clear prioritizing of reproduction over growth at high altitude. These results suggest that environmental selective forces could result in variation of reproductive strategies in the species (Silverton and Lovett-Doust, 1993). Studies of other plant species have also shown that reproductive effort decreases significantly with increasing plant size (Klinkhamer et al., 1990). In addition, sub-division of reproductive resources among floral parts e.g. stamens and carpels are also variable with more allocation towards male (0.27±0.09 to 0.32±0.21) than carpels (0.15±0.07 to 0.16±0.43). Such increased strategies in the biomass allocation to male reproductive structures suggest an outbreeding nature of the species (Huang et al., 2004).

In summary, analysis of the growth responses of *I. racemosa* along the altitudinal gradient indicates significant effects on several characters measured ($P < 0.001$). The altitude showed positive correlation with belowground biomass ($r = 0.84$) and negatively correlated with the plant height ($r = -0.91$) number of leaves, number of capitula and above ground biomass ($r = -0.59$). Negative correlation of all morphological parameters to the increasing altitudes indicates better growth performance of this species at lower elevations.

ACKNOWLEDGEMENT

The first author is highly grateful to Council of Scientific and Industrial Research, (CSIR) for providing financial assistance as JRF.

REFERENCES

Aarssen LW, Taylor DR (1992). Fecundity allocation in herbaceous plants. Oikos 65:225-232.

Abrahamson WG, Gadgil MD (1973). Growth form and reproductive effort in goldenrods (*Solidago*, Compositae). American Naturalist. 107:651-661.

Baret S, Maurice S, Le Bourgeois T, Strasberg D (2004). Altitudinal variation in fertility and vegetative growth in the invasive plant *Rubus alceifolius* (Rosaceae), on Reunion Island. Plant Ecol. 172:265-273.

Bennington CC, Mc Graw JB (1995). Natural-selection and ecotypic differentiation in *Impatiens pallida*. Ecol. Monogr. 65:303-323.

Clauss MJ, Aarssen LW (1994). Phenotypic plasticity of size- fecundity relationships in *Arabidopsis thaliana*. J. Ecol. 82:447-455.

Fabbro T, Körner CH (2004). Altitudinal differences in flower traits and reproductive allocation. Flora 199:70-81.

Geber M (1990). The cost of meristem limitation in *Polygonum arenastrum*: Negative genetic correlations between fecundity and growth. Evolution. 44:799-819.

Grime JP (1979). *Plant Strategies and Vegetation Processes.* New York: John Wiley and Sons.

Hickman J (1975). Environmental unpredictability and plastic energy allocation strategies in the annual *Polygonum cascadense* (Polygonaceae). J. Ecol. 63:689-701.

Huang SQ, Tang LL, Yu Q, Guo YH (2004). Temporal floral sex allocation in protogynous *Aquilegia yabeana* contrasts with protandrous species: support for the mating environment hypothesis. Evolution 58:1131-1134.

Johnson MP, Cook SA (1968). Clutch size in butter cups. American Naturalist. 102:405-411.

Klinkhamer PGL, de Jong TJ, Meelis E (1990). How to test for size dependent reproductive effort. American Naturalist 135:291-300.

Korner C (2003). Alpine plant life, 2nd edition. Springer, Heidelberg, Germany.

Korner C, Cochrane PM (1983). Stomatal responses and water relations of *Eucalyptus pauciflora* in summer along an elevational gradient. Oecologia 66:443-455.

Korner C, Neumayer M, Menendez-Riedl SP, Smeets-Scheel A (1989). Functional morphology of mountain plants. Flora 182:353-383.

Kudo G (1993). Size dependent resource allocation pattern and gender variation of *Anemone debilis* Fisch. Plant Species Biol. 8:29-34.

Lacey EP (1986). Onset of reproduction in plants: size-verses age-dependency. Trends in Ecol. Evol. 1:72-75.

Lovett-Doust J, Lovett Doust L (1989). Reproductive Ecology of Plants: Patterns and Strategies. Oxford University Press, Oxford. pp. 98-124.

Ohlson M (1988). Size dependent reproductive effort in three populations of *Saxifraga hirculus* in Sweden. J. Ecol. 76:1007-1016.

Pickering CM (1994). Size-dependent reproduction in the Australian alpine *Ranunculus*. Aust. J. Ecol. 19:336-344.

Samson DA, Werk KS (1986). Size dependent effects in the analysis of reproductive effort in plants. American Naturalist 127:667-680.

Schmid B, Weiner J (1993). Plastic relationships between reproductive and vegetative mass in *Solidago altissima*. Evolution 47:61-74.

Shabir PA, Nawchoo IA, Wani AA (2013). Floral phenology, secondary pollen presentation and pollination mechanism in *Inula racemosa*. J. Threatened Taxa. 5(10):4498-4503.

Silverton JW, Lovett-Doust J (1993). *Introduction to Plant Population Biology.* Blackwell Scientific Publications, Oxford 141-157.

Silvertown JW, Loven Doust J (1993). Introduction to plant population biology. Blackwell Scientific Publications, Oxford.

Sokal RR, Rohlf FJ (1995). Biometry. W. H. Freeman, New York, NY.

Waite S, Hutchings MJ (1982). Plastic energy allocation patterns in *Plantago coronopus*. Oikos 38:333-342.

Waller DM (1988). Plant morphology and reproduction. In Plant Reproductive Ecology: Patterns and Strategies (J. Lovett Doust and L. Lovett Doust, eds.), Chapter 10. Oxford University Press, New York, NY.

Walton GB, Hufford L (1994). Shoot architecture and evolution of *Dicentra cucullaria* (Papaveraceae. fuarioideae). Inter. J. Plant Sci. 155(5):553-568.

Wani PA, Ganaie KA, Nawchoo IA, Wafai BA (2006). Phenological episodes and reproductive strategies of *Inula racemosa* (Asteraceae) - a critically endangered medicinal herb of North West Himalaya. Int. J. Bot. 2(4):388-394.

Willis SG, Hulme PE (2004). Environmental severity and variation in the reproductive traits of *Impatiens glandulifera*. Funct. Ecol. 18:887-898.

Ecology and biodiversity in Pangong Tso (lake) and its inlet stream in Ladakh, India

F. A. Bhat[2*], A. R. Yousuf[1], Aftab A.[1], Arshid, J.[1], M. D. Mahdi[1] and M. H. Balkhi[2]

[1]Centre of Research for Development (CORD), University of Kashmir, Srinagar, India- 190 006.
[2]Faculty of Fisheries, Sher-e-Kashmir University of Agricultural Sciences and Technology of Kashmir, Srinagar, India -190 006

Pangong Tso is a land locked lake situated in eastern part of Ladakh (Indian Tibet), at an altitude of 4,266 m A.S.L. and remains frozen for about three months during winter. There is no outlet to the lake and loss of water is only through evapotranspiration. The lake was found to be highly alkaline (pH ≤ 9.0) with high conductivity (≥ 1639 μS) and nutrients (NH4-N ≥ 54 μg/L; NO_3-N ≥ 299 μg/L; TP ≥ 464 μg/L) as compared to its inlet Cheshul stream. The progression of the cations in the lake was Mg^{++} > Na^+ > K^+> Ca^{++} while as in the stream it was Ca^{++} > Mg^{++} > Na^+ > K^+. Mg^{++} content in the lake was higher than the Ca^{++} and Na^+ due to its more solubility. The lake belonged to cold monomictic group of lakes. Twenty-three identified and some unidentified taxa of phytoplankton were recorded in the lake while as the periphytic community in the stream was represented by 34 taxa. Zooplankton in the lake was represented by only red coloured copepods, *Diaptomus* spp. and their larvae. No fish was observed in the lake while as in the stream three fishes *Schizopygopsis stoliczkae, Triplophysa stoliczkae* and *Triplophysa gracilis* were recorded. The low biodiversity in the lake was found due to high salinity and harsh environmental conditions prevailing in the lake. The inorganic nitrogen in the lake was recorded in very low concentration, Phosphate, on the other hand was present in very high concentrations. It seemed that instead of phosphate, nitrogen may act as a limiting factor for the aquatic community in the lake.

Key words: Pangong Tso, Ladakh, phytoplankton, *Diaptomus* spp., biodiversity, Cheshul stream.

INTRODUCTION

The Himalaya is home to numerous lakes and wetlands of many shapes and sizes, which make it a unique ecosystem that fulfils an important function in the overall water cycle of the basins. High altitude lakes of Himalaya located above timber line represent a relatively common ecosystem in mountain ranges in general; however, they remain less intensely studied than lowland lakes, mainly because of their remoteness and the short summer open-water period. Nevertheless, high altitude lakes are sensitive reference systems of global climatic change and other human impacts (Schmidt and Psenner, 1992). In fact, although remote high altitude lakes are in general protected from direct human impacts, in the last few decades they have been increasingly affected by airborne contaminants, such as acids and nutrients (Marchetto et al., 1995; Rogora et al., 2006), organic pollutants and heavy metals (Carrera et al., 2002). Due to the extreme environmental conditions (low temperature, strong radiations, mostly low buffering capacity and low nutrient level) these ecosystems have a relatively simple food web and react more rapidly and more sensitively to environmental changes than other lakes (Psenner, 2002). Even minor impacts are able to significantly affect the physical and chemical properties of soft water high altitude lakes, to induce changes in species composition and abundance of the biota and to cause accumulation of trace substances in higher trophic organisms (Hofer et al., 2001).

In spite of the socio-economic and ecological importance of high mountain lakes, better knowledge of

*Corresponding author. E-mail: fabhat_fb@yahoo.com.

several ecological aspects (especially regarding species distribution patterns and biogeography, diversity and functional interaction among the different components of the food web) is needed for better understanding of their relationships with the environmental variables. These wetlands and lakes have received little attention so far in terms of their limnology, diversity, conservation and water management, but they are becoming increasingly important due to the possible consequences of the global climate change. Conservation of these unique systems needs special efforts and some of these lakes and wetlands are trans - boundary extending into more than one country, while in other places the catchments are shared by more than one country.

Ladakh is a dry land of rugged mountains and open plains in the north of the main Himalaya and south of Karokarum mountains, the two mountain ranges which enclave the Ladakh. The region is peculiar in having a number of salt water lakes, which receive snow, melt water from the catchment but lack an outlet. These include Pangong Tso, Tso Moriri, Tso Khar and Tso Rul and many more. Most of the Ladakh waters, including these peculiar saltwater habitats, have so far received very little attention from scientific community except for few short reports such as Hutchinson (1933, 1936, 1937); Mir and Suri (1975); Gopal et al. (2002), Rashid and Pandit (2005) and Bhat (2009). Pangong Tso is a long, narrow, brackish lake spanning the Indian/ Chinese border, in a valley in the upper drainage basin of the Indus river, at the east end of the Karakoram Range. Only the westernmost one-third of the lake lies in Indian territory. The lake has been suggested as a Ramsar site due to its biological, cultural and geological values (Chatterjee et al., 2002). It is the largest and most brackish wetland in the cold desert ecosystem of the Trans-Himalaya. It is with this background that the present Lake and one of its important streams (Cheshul stream) was analyzed for various physico-chemical and biological characteristics in order to generate a base line data. The data obtained during the present study is presented in the present communication.

STUDYAREA

Pangong Tso is one of the trans-boundary lakes, major part of which, about two third lies in Tibet (China) while its remaining part is located in Ladakh (Jammu and Kashmir, India). The Lake is situated in eastern Ladakh with a catchment area of about 2000 km^2. It is about 134 km long, bisected by the international border between India and Tibet (China). Lake is having a maximum width of 8 km in Indian side and the western bank of the lake which is more than 50 km in length in Indian side is accessible through a rough road, on which Jeep Safaris, trucks etc. are able to move. The lake remains cut off from the rest of the Ladakh for about six months (winter months) as the road leading to lake across the Changla pass (Alt. > 5300 m ASL) remains closed for more than six months. The lake remains frozen for more than three months from the last week of December to first week of April.

Four sampling sites on the basis of length of the lake in Indian side were selected during the study period (during October, 2004, September, 2005 and September, 2006), one located in the Cheshul stream (RS) and three in the Pangong Tso (LS 1 to LS 3) (Figure 1). The geographical coordinates of selected sites are given in Table 1.

MATERIALS AND METHODS

The surface water samples were collected in one litre polyethylene bottles after 11 a.m. at all sites during the period of investigation. Temperature, pH and conductivity were recorded on the spot using Hg thermometer, digital pH meter-1011E and portable conductivity meter - DB104 model respectively. Water samples were collected in 250 ml glass bottles for estimation of dissolved oxygen following the modified Winkler's methods (APHA, 1998). Collected water samples were brought to the laboratory for immediate estimation of total hardness, total alkalinity, chloride, calcium, sodium, potassium, magnesium, total phosphate phosphorus, ammonium nitrogen, nitrate nitrogen following the Mackereth et al. (1978), CSIR Pretoria (1974) and APHA (1998).

For the quantitative and qualitative estimation of phytoplankton and zooplankton 20 L of the lake water was sieved through a plankton net made of nylobolt (mesh size 64 μm) and the filtered samples were then preserved in 4% formalin solution. Periphytic community was collected by scratching 1 cm^2 of the substratum (bottom stones) in inlet stream only. The scratched material was preserved in 4% formalin solution. Enumeration of the samples was done by taking 1 ml of the sub-sample in Sedgwick rafter chamber in triplicate and counting its entire contents under microscope to obtain the statistical accuracy after their identification with the help of standard taxonomical works (Edmondson, 1959; Heurek, 1896; Randhawa, 1959; Pal et al., 1962). The results have been expressed as units per litre for phytoplankton, individuals per litre for zooplankton and taxa per cm^2 for periphytic algae.

Species diversity index was calculated by using the Shannon diversity index (1963):

$$H' = - \Sigma \, pi \log^2 pi$$

Where H′ = Shannon diversity index; pi = the importance of probability of each species (*ni*/N), N = total no. of individuals in "S" species and ni = number of individuals in *i*th species.

Similarity analysis between the sites was calculated by using Jaccard similarity index (J):

$$J = (100 \times c) / (a + b - c)$$

Where 'a' is the number of species in community A, 'b' is number of species in community B, while 'c' is the number of species common to both communities.

RESULTS AND DISCUSSION

The physico-chemical features of the lake are given in Table 2. In the Lake, water temperature and air temperature were found close to one another due to the standing feature of the water. In the stream the water

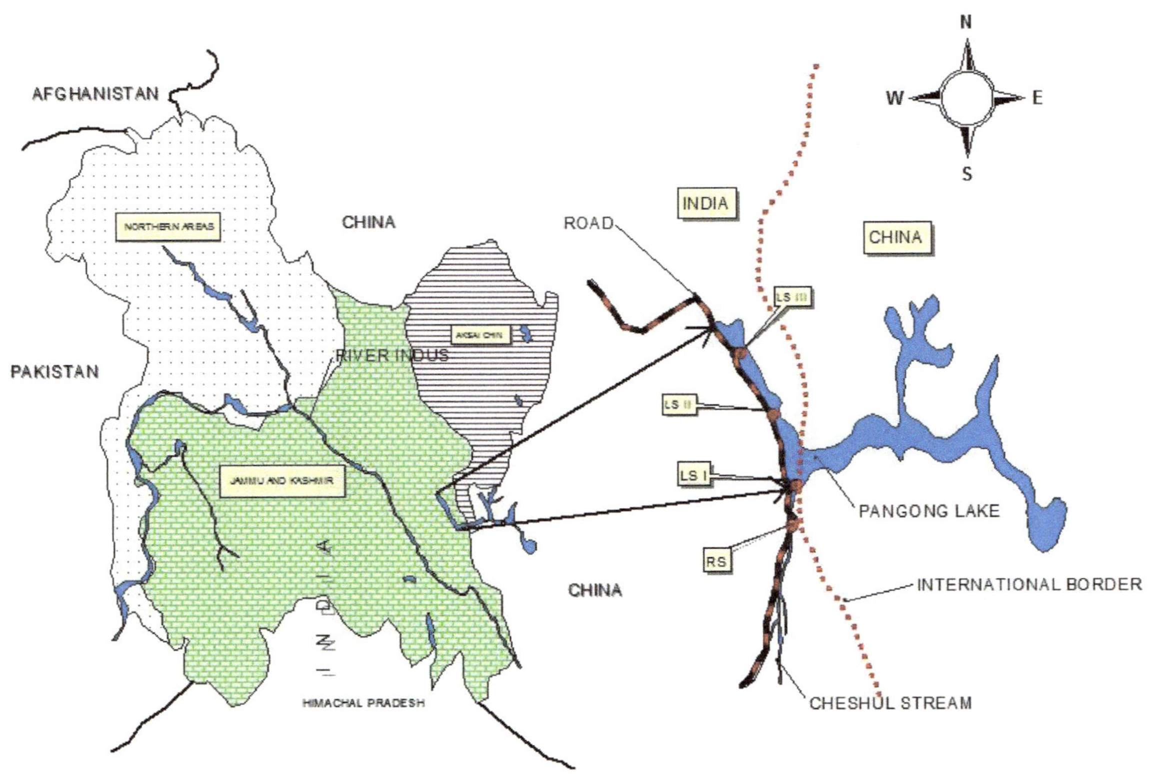

Figure 1. Location map of Pangong Tso Lake showing Study sites.

Table 1. The geographical coordinates of selected sites.

	RS	LS I	LS II	LS III
Latitude	33° 44' 16.3" N	33° 45' 06.3" N	33° 50' 32.4" N	33° 50' 50.5" N
Longitude	78° 38' 12.1" E	78° 38' 08.1" E	78° 34' 19.4" E	78° 39' 11.5" E

temperature showed much variation with respect to the air temperature, due to the running nature of the water (Hynes, 1979; Bhat, 2003). Dissolved oxygen in the lake was present with a mean value of 6.17 mg/L as compared to stream which was having a mean value of 8.67 mg/L. There was not much variation in dissolved oxygen in different sites of the lake and the little variation seemed due to mixing of atmospheric air by high waves

Table 2. Physico-chemical characteristics of Pangong Lake (LS I – LS III) and Cheshul Stream (RS I).

	RS	LS I	LS II	LS III
Parameter	**Range(Mean ± SD)**	**Range (Mean ± SD)**	**Range(Mean ± SD)**	**Range (Mean ± SD)**
Air temp.(°C)	11 – 26(21±8)	11 – 28(22 ± 10)	10 – 28(22±10)	9 – 27(21±10)
Water temp.(°C)	8 – 12(10±2)	12 – 16(14 ± 2)	13 - 16(14±2)	12 – 15(14±2)
pH	8.42 – 8.44(8.43 ±0.01)	8.74 – 9.05(8.89 ± 0.16)	8.92 – 9.08(9.00±0.08)	8.76 – 9.11(8.98±0.19)
D. O_2 (mg/L)	8 – 9(9 ± 1)	6 – 6.5(6 ± 0)	6 – 6(6±0)	6 – 6.5(6±0)
Conductivity (µS)	140 – 162(153 ± 12)	1524 – 1577(1550 ± 27)	1582 – 1621(1598±21)	1621 – 1639(1631±9)
Total alkalinity (mg/L)	240 – 268(257 ± 15)	1500 – 1866(1663 ± 186)	1640 – 2302(1909±348)	1700 – 2196 (1902±260)
TDS (ppm)	60 – 82(73 ± 12)	410 – 431(420 ± 11)	370 – 534(438±86)	452 – 564(524±62)
Total hardness (mg/L)	340 – 388(365 ± 24)	1988 – 2140(2049 ± 80)	2122 – 2420(2234±162)	2198 – 2344(2251±81)
Calcium (mg/L)	26 – 28(27 ± 1)	26 – 38(33 ± 6)	22 – 32(28±5)	21 – 30(27±5)
Magnesium (mg/L)	14 – 18(15 ± 2)	404 – 437(418 ± 17)	456 – 480(470±12)	488 – 520(509±18)
Chloride (mg/L)	14 – 16(15 ± 1)	1500 – 1700(1626 ± 109)	1600 – 1700(1638±54)	1500 – 1530(1517±16)
Sodium (mg/L)	6 – 11(9 ± 3)	182 – 264(230 ± 43)	227 – 246(239±10)	218 – 260(249±27)
Potassium (mg/L)	2 – 4(3 ± 1)	118 – 132(127 ± 8)	138 – 144(141±3)	98 – 140(125±24)
Nitrate (µg/ L)	260 – 288(275 ± 14)	244 – 310(281 ± 340	266 – 344(315±42)	251 – 330(302±44)
Ammonia (µg/ L)	42 – 56(48 ± 7)	55 – 68(61 ± 7)	47 – 64(56±9)	44 – 46(45±1)
Total phosphate phosphorus (µg/ L)	140 – 166(156 ± 14)	390 – 451(420 ± 31)	370 – 584(465±109)	452 – 580(507±66)

and due to little temperature difference of air and water. In the study area there are always howling wind afternoon and it sometimes even becomes difficult to collect the samples. However, LS II showed low DO values as compared to other two sites, the reason being that this site is located in the centre of the lake in the Indian territory with no inlet in its vicinity as compared to other two sites.

The pH of the stream ranged from 8.42 to 8.44 while as that of the lake ranged from 8.74 to 9.11. The lake was found devoid of any macrophyte, even on its littoral sides and the photosynthetic rate does not seem to be responsible for the higher pH value, instead the geological features of the catchment as well as that of basin and the evaporation pattern of water from the lake surface seem to be responsible for the high pH. Odour et al. (2003) have related the loss of solutes through underground seepage to low pH in Lake Baringo (Kenya). Ebel and Koski (1969) and Hannon et al. (1979) reported that the higher total alkalinity in the bottom waters is related to the release of carbonates from sediments and their subsequent convergence to the bicarbonates by carbonic acid is formed due to decomposition of organic matter. However, in the present lake due to negligible quantities of organic matter present (absence of macrophytes), decomposition does not seem to have much influence on the release of carbonates from sediments and their subsequent convergence to the bicarbonates is insignificant.

According to Hutchinson (1967) carbonates appear in water only when pH exceeds 8.34 and with increase in the concentration of carbonates, pH value also shows an increasing trend. This seems to be true for the present lake as well as its tributaries. Total alkalinity, which is taken as temporary hardness of $CaCO_3$ mg/L in the lake on the average was 1825 mg/L (1988 to 2420 mg/L). Freiser and Fernando (1966) stated that when total alkalinity is higher than HCO_3^- system, $CO_3^=$ system prevails and pH is usually found on the higher alkaline side. This is true for the present system as CO_2 was altogether absent in all the sampling sites.

Conductivity in the lake ranged from a minimum of 1524 µS at LS I during October 2004 to a maximum of 1639 µS at LS III during September,

2006 and the mean concentration was 1593 μS. Zutshi et al. (1980) reported that the conductivity of a water body decreases with the increase in altitude but in the present lake such phenomenon was not found to prevail, though the lake is located at a very high altitude. Compared to Kashmir Himalayan lakes it was observed that the values are quiet high which can be related to its non open type lake, which over the geological period of time has concentrated much salts mostly due to evaporation of water (Zutshi et al., 1980; Yousuf and Qadri, 1981; Zutshi et al., 1984). Kiplagat et al. (1999) asserted that conductivity and alkalinity are strongly influenced by the nature of the incoming water and its periodicity. It has also been observed that the geological character of rocks and subsoils not only in the catchment area but that of the lake basin (Hutchinson, 1937, 1967) as well influences the conductivity of the lake water. The snowmelt water of the stream (RS) of the Pangong Tso recorded generally very low conductivity, the average value being 153μS, however, the conductivity of the lake water (mean = 1593 μS) was much higher than that of the incoming water, which may be due to the landlocked nature of the lake which over the geological time period has accumulated the salts mostly due to evapotranspiration.

Water source in high altitude Himalayan lakes is mostly snow melt and rain water, which is usually low in ionic content. As this surface flow reaches a lake, the latter inherits the water type of these inflow waters before salts are deposited in the lake (Mianping, 1997). Total hardness and TDS in the lake and stream were found with a mean value of 2178 and 462 mg/L and 365 and 73 mg/L respectively. Total hardness in the lake was not only due to the calcium and magnesium but other salts contributed substantially to it. Calcium content in the lake was less in concentration as compared to magnesium and the mean value of calcium in the lake and stream was 29 and 27 mg/L respectively. The mean magnesium concentration in the lake and stream was 466 and 15 mg/L respectively. Sodium concentration was comparatively higher as compared to potassium in the lake as well as stream. Sodium and potassium on average in the lake and stream were present with a mean value of 239 and 131 mg/L and 9 and 3 mg/L respectively. The cation progression in the lake was as; $Mg^{++} > Na^{+} > K^{+} > Ca^{++}$. Comparison of the data clearly revealed that Na^{+} and K^{+} recorded a similar pattern in the lake water as well as tributaries but the concentration of Ca^{++} and Mg^{++} in the inflows and lake water showed a reverse trend, thus the calcium concentration in the lake was lower than magnesium although in incoming stream calcium was always higher than magnesium. Zutshi et al. (1980) have reported that in Kashmir lakes calcium is generally the dominant cation because of the predominance of lime rich rocks in the catchments areas. But in the present study the predominance of magnesium over calcium may be explained on the grounds that calcium, which is brought in large quantities in the incoming water, gets quickly precipitated as calcium carbonate in the highly alkaline lake water and therefore settles at the bottom. Magnesium salts, on the other hand, being more soluble in water, get concentrated in the water column due to the evaporation of water from the lake surface (Hutchinson, 1933; Mianping, 1997).

Chloride like that of magnesium was too high in the main lake. Mean concentration of chloride in the lake and stream was 1594 and 15 mg/L respectively. The high chloride content in the lake may be contributed to the land locked nature of the lake. Nitrate - nitrogen, Ammonium– nitrogen and Total phosphate concentration in the lake were recorded with a mean concentration of 299, 54 and 464 μg/L respectively. Their concentrations in the stream were 275, 48 and 156 μg/L respectively. Ammonium – nitrogen, Nitrate - nitrogen and total phosphate which are good indicators of anthropogenic pressures (Zutshi et al., 1984) but the human pressure in the vicinity of the lake being negligible but still these were higher in the lake. Vass (1980) and Zutshi et al. (1980) reported the distribution of the main plant nutrients like nitrate, ammonia and phosphate to be governed by physical factors such as input source to the lake (catchment characteristics), changes in mixing depth during stratification and complete mixing during isothermal conditions and by biological processes such as algal photosynthesis. The nitrogenous compounds in the water bodies are derived to an appreciable degree from the atmosphere, whereas ammonia is the chief product of decomposition of plant and animal proteins (Schmittou et al., 1988; Sharma, 2000). As the decomposition of plant and animal proteins is negligible in the lake, catchment characteristics (geological features) and atmosphere and closed nature of the lake seem to be responsible for their higher concentration.

Bhat (2009) while working on Tso Moriri, a similar land locked lake in Ladakh, with same altitude and a depth of 72 m, reported that after the melt of ice cover during summer the maximum surface water temperature reached up to 19°C while the maximum bottom temperature during the same month was 8°C and after words the temperature again started to decrease till the lake gets frozen during late December. The same condition seems to prevail in the present lake also and thus one can easily keep the lake Pangong under the cold monomictic group of lakes (Hutchinson, 1937; 1967; Bhat, 2009).

Biological measurements provide direct information on the condition of groups of biota resident in the water resources and also on the conditions of the resource. They address management issues, more directly and can provide a more sensitive time-integrated assessment of a water body than physical or chemical variables directly. In the lake a total of 23 taxa of phytoplankton and some unidentified taxa were recorded, which belonged to four classes. Out of 23 taxa, 17 belonged to

Bacillariophyceae, 3 to Chlorophyceae, 2 to Cyanophyceae and 1 to Xanthophyceae (Table 3). The dominant genera among the phytoplankton species in the lake were: *Gomphonema germinatum, Navicula* spp. *Achnanthes ovalis, Nitzschia* spp., *Diatoma elongatum, Caloneis* spp., *Fragilaria* spp., *Schaeroplea annulina, Spirulina major* and *Arachnochloris minor* and these were found throughout the lake. The distribution and density of other taxa varied significantly from site to site. Some of the species were found site specific; *Achnanthes ovalis, Amphora* spp., *Schroederia anchora, Lyngbya* spp. were recorded only at LS I while as *Cyclotella* spp. was present at LS III only. LS II and LS III showed almost the same pattern of taxa formation having 14 taxa and 12 taxa respectively while as the LS I showed comparatively maximum (23 taxa) number of taxa (Figure 2). The LS II and LS III showed the maximum Jaccard similarity index of (73.33) followed by LS I and LS III (70.57) (Table 4). In the lake on the average, Bacillariophyceae contributed 71% to the total taxa followed by Chlorophyceae (13%), Cyanophyceae (8%), Xanthophyceae (4%) and unidentified taxa (4%) (Figure 3).

In the stream (RS), the density and diversity of periphyton was higher and a total of 34 taxa were recorded, which belonged to five classes. Bacillariophyceae in the stream was represented by 22 taxa, Chlorophyceae by 7 taxa, Cyanophyceae by 2 taxa, Xanthophyceae by one 1 taxa and Rodophyceae by 2 taxa (Figure 2). On the average in the stream Bacillariophyceae formed 65% of the total taxa while as Chlorophyceae, Cyanophyceae, Xanthophyceae and Rodophyceae formed 21, 6, 3 and 7% of the total taxa respectively (Figure 4). Twenty two (22) taxa recorded in the stream where present in the lake also, though planktonic in nature. The dominant taxa of Bacillariophyceaea in the stream were *Cymbella tumida, Diatoma anceps, Eunotia* spp., *Liomophora anglica, Meriodon* spp., *Stauroneis* spp., *Tabellaria fenestrate,* the dominant taxa of Chlorophyceae were *Closteriopsis irregular, Desmidium aptogonum, Hormidium subtile, Tetraedron spp. Zygnema spp.*, Cyanophyceae was represented by *Lyngbya* spp. and *Spirulina major.* Rodophyceae was present in the stream only and was represented by two taxa, *Thorea ramosissima* and *Batrachospermum boryanum*. The Jaccard similarity index between the stream site (RS) and lake sites was low as compared among the 3 sites of the lake. However, the lake site which was near to the mouth of the stream, that is, LS I showed maximum similarity of 50 with the RS , while as the LS II and LS III showed a similarity of 33.33 and 31.42 respectively with the RS I. The maximum Shannon diversity of algal taxa in the lake was recorded at LS I (mean, H´ = 1.282) followed by LS II (mean, H´ = 0.967) and LS III (mean, H´ = 0.928) Figure 5. The average Shannon Diversity Index value of lake was 1.059 while that of the stream site was 2.434. Zooplankton in the lake was represented by red coloured adult *Diaptomus* species and their larval stage Nauplius larvae. Their density varied between the sites and the dominance pattern of zooplankton in the studied sites was; Site LS I (mouth of the Cheshul stream) > LS II > LS III. The more density of zooplankton in the studies sites was recorded during the September, 2005. No fish was recorded from the lake; however, the fishes were present in the stream. Three species i.e. *Schizopygopsis stoliczkae, Triplophysa stoliczkae* and *Triplophysa gracilis* were recorded from the stream.

The low biodiversity in the lake is considered due to its high salinity and harsh environmental conditions prevailing in the lake as the lake remains frozen for about more than three months (Bhat, 2009). The biotic life during the frozen period is expected to have decreased further as it is only during the warmer period the phytoplankton are being reported to be dominant in such water bodies. Khan (2003) and Bhat (2009) reported that physico–chemical parameters like light, temperature and nutrient concentration, lake currents, and grazing by zooplankton control the phytoplankton growth and reproduction. Williams (1992) and Kipriyanova et al. (2007) reported that the diversity and density of phytoplankton decline with increase in salinity. Similar observations were also found in the present lake as the maximum density and diversity was recorded near the mouth of the stream, where due to dilution of the stream water, salinity is low as compared to other sites, which is also confirmed by the high Jaccard similarity index between the mouth site (LS I) and the stream site (RS) as compared to other lake sites. Tuchman et al. (1998) and Alocer and Hammer (1998) also found species richness in the saline lake ecosystems of Mexico to be decreasing with increase in salinity.

The extreme conditions which influence the Pangong Tso are reflected in all groups of life forms mostly in zooplankton. One of the effects of harsh environmental conditions is also the absence of fishes from the lake. Hansson et al. (2007) reported on the effect of UV radiation on the zooplankton at high altitudes and found that to save the body from any damage zooplankton commonly synthesizes photo protective compounds that function either as sun-screens or as scavengers of photo-produced radicals. Due to strong ultraviolet radiations all the species have developed the red colour due to pigmentation of carotenoids and melanin in their body surface to overcome the harmful effects of ultraviolet radiations. This phenomenon was exhibited by all the taxa (adult as well as juveniles) in the Lake and due to the pigmentation the individuals appeared reddish in colour, especially during bright hot day. The fishes as were recorded only from the inlet stream of the lake and were absent from the main lake, it is possible that in the past the fishes might have been present in the lake but over the geological period of time as the salinity of lake increased due to evaporation, these might have ventured into the inlet stream and have adapted themselves well

Table 3. Distribution pattern of periphyton, phytoplankton, zooplankton and fishes in Cheshul stream (RS) and Pangong Lake (LS I – LS III).

Serial No.		RS	LS I	LS II	LS III
A	Algal taxa	Periphyton	Phytoplankton		
	Bacillariophyceae				
1	*Achnanthes ovalis*	++++	++	-	-
2	*Amphora sp.*	+++	+	-	-
3	*Caloneis sp.*	++	+	+	+
4	*Closterium* spp.	+	+	+	-
5	*Cyclotella sp.*	++	-	-	+
6	*Cymbella tumida*	+	-	-	-
7	*Diatoma elongatum*	+++	++	+	+
8	*Diatoma anceps*	+	-	-	-
9	*Diatoma hiemale*	+	+	-	-
10	*Epithemia zebra*	++	+	+	-
11	*Eunotia* spp.	+	-	-	-
12	*Eupodiscus argus*	-	+	+	-
13	*Fragillaria sp.*	+++	+	+	+
14	*Gomphonema herculeanum*	++	+	-	-
15	*Gomphonema germinatum*	+++	++	+	+
16	*Liomophora anglica*	+	-	-	-
17	*Meriodon* spp.	+	-	-	-
18	*Navicula* spp.	+++	+++	++	++
19	*Navicula radiosa*	+	+	+	+
20	*Nitzschia sp.*	++	++	+	+
21	*Nitzschia lanceolata*	-	+	-	-
22	*Stauroneis* spp.	+	-	-	-
23	*Synedra* spp.	++	+	-	-
24	*Tabellaria fenestrata*	+	-	-	-
	Chlorophyceae				
25	*Closteriopsis irregulare*	+	-	-	-
26	*Desmidium aptogonum*	++	+	-	-
27	*Hormidium subtile*	+	-	-	-
28	*Schaeroplea annulina*	++	+	+	+
29	*Schroederia anchora*	++	+	-	-
30	*Tetraedron* spp.	+	-	-	-
31	*Zygnema* spp.	++	-	-	-
	Cyanophyceae				
32	*Lyngbya* spp.	+	+	-	-
33	*Spirulina major*	+	+	+	+
	Xanthophyceae				
34	*Arachnochloris minor*	+	+	+	+
	Rodophyceae				
35	*Thorea ramosissima*	+	-	-	-
36	*Batrachospermum boryanum*	++	-	-	-
37	Un identified algae	-	+	+	+
B	**Zooplankton**				

Table 3. Contd.

1	*Diaptomus* spp.	-	+++	++	-
2	Nauplius larvae	-	++	+	+
C	**Fishes**				
1	*Schizopygopsis stoliczkae*	++	-	-	-
2	*Triplophysa stoliczkae*	+++	-	-	-
3	*Triplophysa gracilis*	++	-	-	-

++++= Very dense, +++ = Dense, ++ = moderate, + = low - = absent .

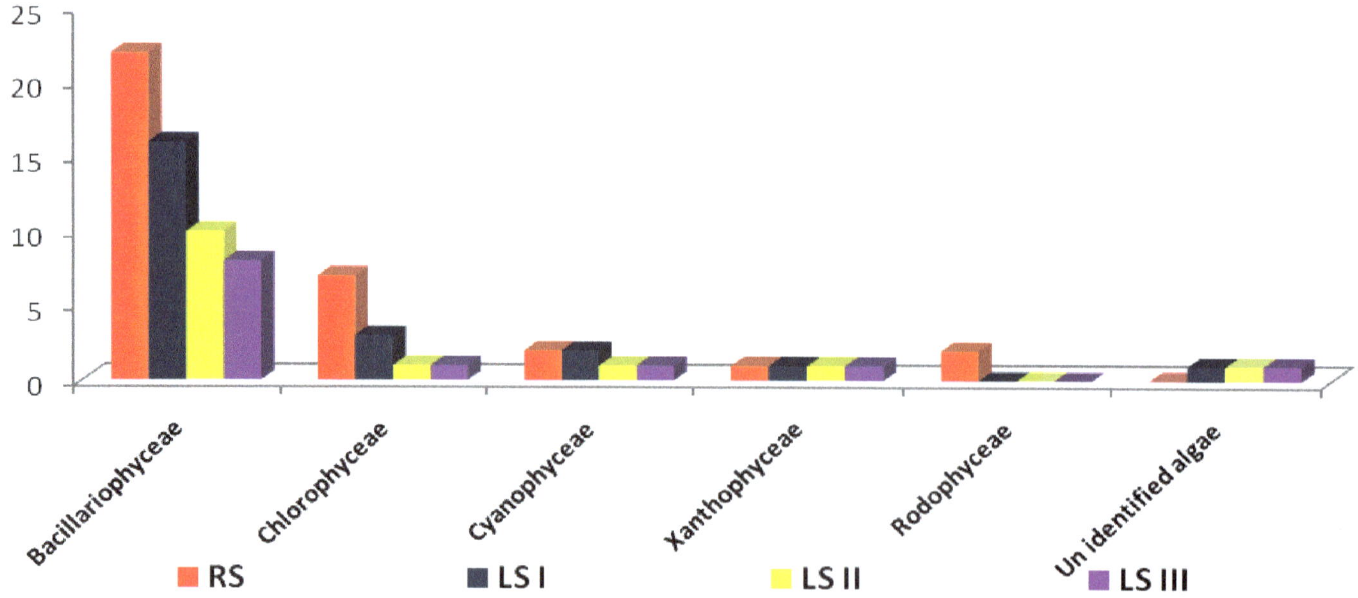

Figure 2. Dominance of different taxa to different algal classes at different sites.

Table 4. Jaccard similarity index (J) of algal taxa between the four sites.

	RS I	LS I	LS II	LS III
RS I		50	33.33	31.42
LS I			60.89	70.57
LS II				73.33
LS III				

there. The species recorded in the Pangong Lake are also reported from the other parts of the Himalaya, mostly from the fresh water lakes of Kashmir valley (Wanganeo and Wanganeo, 1991) as well from other parts of the world (Gunkel et al., 2002 and Tolotti et al., 2006) and some have also been reported from the saline lakes (Hodgson et al., 2001). This clearly points to the fact that only eurytopic species that have the capability of resisting wide range of fluctuations in environmental factors have been able to colonize this extreme ecosystem of the cold desert of Ladakh.

The low biodiversity in the lake is considered due to its high salinity and harsh environmental condition as the diversity and density of phytoplankton decline with increase in salinity as the maximum density and diversity was recorded near the mouth of the stream. Due to the extreme conditions and due to the effect of UV radiation at high altitudes, zooplankton to save the body from any

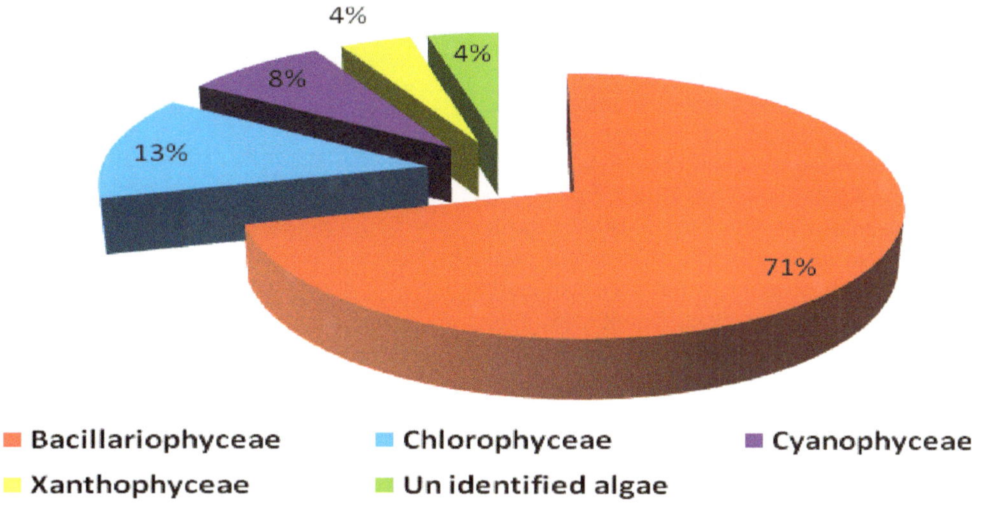

Figure 3. Mean percentage contribution of different classes to total phytoplankton taxa in the lake.

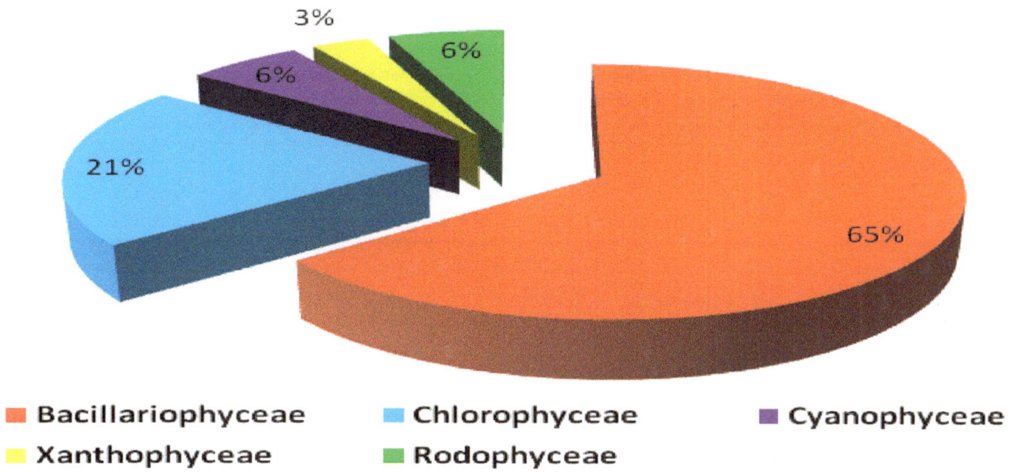

Figure 4. Mean percentage contribution of different classes to total periphyton taxa in the stream.

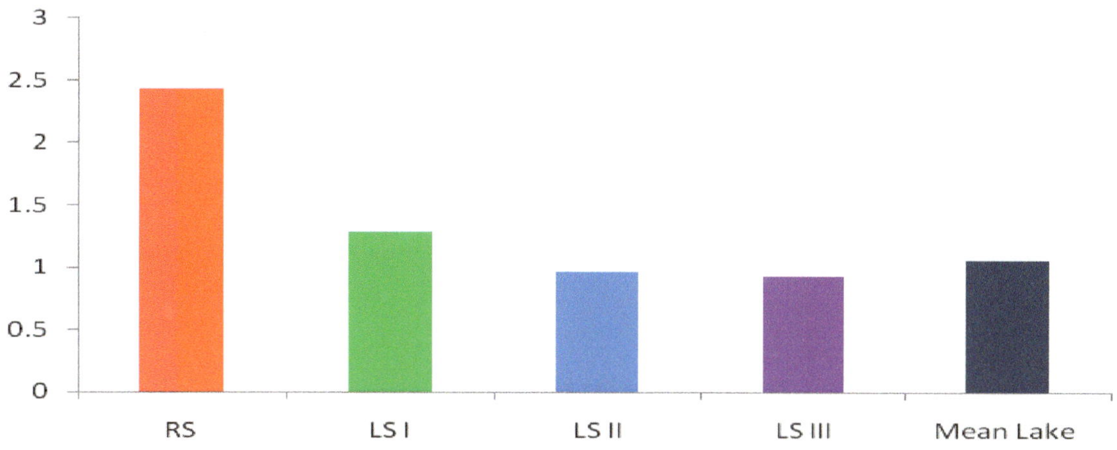

Figure 5. Mean Shannon diversity Index in the stream and different sites in Pangong Lake.

damage, commonly synthesize photo protective compounds that function either as sun-screens or as scavengers of photo-produced radicals. Due to strong ultraviolet radiations all the species have developed the red colour due to pigmentation of carotenoids and melanin in their body surface to overcome the harmful effects of ultraviolet radiations. The fishes as were recorded only from the inlet stream of the lake and were absent from the main lake. It is possible that in the past the fishes might have been present in the lake but over the geological period of time as the salinity of lake increased due to evaporation, these might have ventured into the inlet stream and have adapted themselves well there.

ACKNOWLEDGEMENTS

Thanks to the National Hydro Power Corporation, Delhi and Indian Space Research organization (S.A.C.), Ahmadabad for their financial assistance for EIA / EMP of Nimoo – Bazgoo Hydroelectric Electric Project, Nimoo, Leh and Biodiversity of Lake Tso Moriri and Tso Khar respectively. It was during these project tenures that our teams were able to visit the Pangong Lake. Thanks to the all research scholars of the limnology laboratory for their help one way or other way.

REFERENCES

Alocer J, Hammer UT (1998). Saline lake ecosystems of Mexico. Aquatic Ecosystem Health Manage., 1: 291-315.

APHA (1998). Standard Methods for the examination of water and wastewater. 20th ed; APHA, AWWA, WEF, Washington, DC.

Bhat FA (2003). Ecology of Schizothorax Heckel in Lidder River, Kashmir. M. Phil Dissertation, University of Kashmir.

Bhat FA (2009). Limnological characterization of Lake Tso Moriri, Ladakh. Ph.D. Thesis, University of Kashmir.

Carrera G, Fernandez P, Grimalt JO, Ventura M, Camarero L, Catalan J, Nickus U, Thies H, Psenner R (2002). Atmospheric deposition of organochlorine compounds to remote high mountain lakes of Europe. Environ. Sci. Technol., 36: 2581-2588.

Chatterjee A, Chandan P, Gautam P, Droz BH (2002) High Altitude Wetlands of Ladakh: A Conservation Initiative. WWF-India, New Delhi. p. 38.

CSIR (1974). Analytical Guide (Laboratory Techniques) CSIR, Pretoria, South Africa.

Ebel WJ, Koski CH (1969). Physical and chemical limnology of Brownlee reservoir, 1962-1964. Fish. Bull., 67(2): 295-336.

Edmondson WT (1959). Freshwater Biology, John Wiley, NY.

Freiser H, Fernando O (1966). Ionic equilibria in analytical chemistry. New York, Wiley and Sons.

Gopal B, Chauhan M, Zutshi DP (2002). High altitude lakes of Ladakh: changes since the Yale North India Expedition. Verh. Internat. Verein. Limnol., 28: 519-523.

Gunkel G, Casallas J (2002). Limnology of an equatorial high mountain lake – Lago San Pablo, Ecuador: the significance of deep diurnal mixing for lake productivity. Limnologica, 32: 33-43.

Hannon HH, Fuchs IR, Whitenberg DC (1979). Spatial and temporal patterns of temperature, alkalinity, dissolved oxygen and conductivity in oligomesotrophic deep storage reservoir in Central Texas. Hydrobiologia, 60: 209-221.

Hansson L, Hylander S, Sommaruga R (2007). Escape from UV threats in zooplankton: A cocktail of behavior and protective pigmentation. Ecology, 88(8): 1932-1939.

Heurek HV (1896). A treatise on Diatomaceae. William Wisley and Son, Essex street, strand, WC.

Hodgson DA, Vyverman W, Sabbe K (2001). Limnology and biology of saline lakes in the Rauer Islands, eastern Antartica. Antarctic Science, 13(3): 255-270.

Hofer R, Lackner R, Kargl J, Thaler B, Tait D, Bonetti L, Vistocco R, Flaim G (2001). Organo-chlorine and metal accumulation in fish (Phoxinus phoxinus) along a northsouth transect in the Alps. Water Air and Soil Pollution. 125: 189–200.

Hutchinson GE (1933). Limnological studies at high altitudes in Ladakh. Nature, 132-136.

Hutchinson GE (1936).The Clear Mirror: A Pattern of Life in Goa and in Indian Tibet, Cambridge University.

Hutchinson GE (1937). Limnological studies in Indian Tibet. Int. Hydrobiol., 35: 134 177.

Hutchinson GE (1967). A Treatise on Limnology Vol. I, Geography, physics and Chemistry of lakes. John Wiley and Sons New York, U.S.A.

Hynes HBN (1979). The Ecology of Running Waters. Liverpool University Press.

Khan TA (2003) Biology of four saline lakes in western Victoria, Australia. II. Biological Parameters. Limnologica, 33: 327–339.

Kiplagat K, Njuguna SG, Francis MM, Krienitz L (1999). The physico-chemical conditions of Turkwel gorge reservoir, a new man-made lake in Northern Kenya. Limnologica, 29: 377–392.

Kipriyanova LM, Yermolaeva NI, Bezmaternykh MM, Dvurechenskaya SY, Mitrofanova EY (2007). Changes in the biota of Chany Lake along a salinity gradient. Hydrobiologia, 576: 83 – 93.

Mackereth FJH, Haron J, Talling JF (1978). Water analysis. Fresh Wat. Biol. Assoc. Sct. Publ., p. 36.

Marchetto A, Mosello R, Psenner R, Bendetta G, Boggero A, Tait D, Tartari GA (1995). Factors affecting water chemistry of alpine lakes. Aqua. Sci., 55: 81–89.

Mianping Z (1997). An Introduction to Saline Lakes on the Qinghai-Tibet Plateau. Kluwer, Dordrecht, p. 294.

Mir AM, Suri BL (1975). Phytoplankton of Ladakh. Geobios., 2(6): 199-200.

Pal DP, Kundu BC, Sudarlingam PVS, Venkataraman GS (1962). Charophyta, ICAR, Publ. New Delhi. pp. 0–130.

Psenner R (2002). Alpine waters in the interplay of global change: complex links - simple effects. In: Global Environmental Change in Alpine Region, New Horizons in Enironmental Economics (Steininger KW, Weck-Hannemann H eds.). Edward Eldgar, Cheltenham, UK, pp. 271.

Randhawa MS (1959). Zygnemaceae. ICAR Publ. New Delhi. pp. 91-437.

Rashid H, Pandit AK (2005). Periphytic algal community in relation to the physico – chemical features of seven water bodies of Ladakh region, J & K. J. Res. Dev., 6: 71 – 80.

Rogora M, Mosello R, Arisci S, Brizzio MC, Barbieri A, Balestrini R, Waldner P, Schmitt M, Stahli M, Thimonier A, Kalina M, Puxbaum H, Nickus U, Ulrich E, Probst A (2006). An overview of atmospheric deposition chemistry over the Alps: present status and long-term trends. Hydrobiologia, 562: 17-40.

Schmidt R, Psenner R (1992). Climate changes and anthropogenic impacts as causes for pH fluctuation in remote alpine lakes. Documenta Ist. Ital. Idrobiol. 32: 31-57.

Shannon CE, Weiner W (1963). The mathematical theory of communication. University of Illinois Press, Urbana. p. 117.

Sharma AP (2000). Manual on Fishery Limnology. G. B. Pant, University of Agric. and Tech. Pantnagar.

Tolotti M, Manca M, Angeli N, Morabito G, Thaler B, Rott E, Stuchlik E (2006). Phytoplankton and zooplankton associations in a set of Alpine high altitude lakes: geographic distribution and ecology. Hydrobiologia, 562: 99–122.

Tuchman ML, Stoermer EF, Carney HJ (1984). Effect of increased salinity on the Diatom assemblage in Fonda Lake, Michgam. Hydrobiologia, 109: 179 – 188.

Vass KK (1980). On the trophic status and conservation of Kashmir lakes. Hydrobiologia, 68(1): 9-15.

Wanganeo A, Wanganeo R (1991). Algal population of Kashmir

Himalaya. Arch. Hydrobiol., 121(2): 219-233.
Williams WD (1992). The biological status of Lake Corangamite and other lakes in western Victoria. A report to the Department of Conservation and Environment, Colac, Victoria, Australia.
Yousuf AR, Qadri MY (1981). Ecology of *Diaphanosoma brachyurum* Lieven (cladocera:Crustacea) in Lake Manasbal, Kashmir. J. Indian Inst. Sci., 63(C): 35–42.
Zutshi DP, Subla BA, Khan MA, Wanganeo A (1980). Comparative limnology of nine lakes of Jammu and Kashmir Himalayas. Hydrobiologia, 72: 101–112.
Zutshi DP, Vishin N, Subla BA (1984). Nutrient status and plankton Dynamics of a Perennial Pond. Proc. Indian Nat.. Sci. Acad., 50(6): 577-581.

Impact of land use types on population structure and extent of bark and foliage harvest of *Afzelia africana* and *Pterocarpus erinaceus* in Eastern Burkina Faso

Blandine Marie Ivette Nacoulma[1]*, Salifou Traoré[1], Karen Hahn[2] and Adjima Thiombiano[1]

[1]Department of Plant Biology and Physiology, Laboratory of Plant Biology and Ecology, University of Ouagadougou, UFR-SVT, 09 BP 848, Ouagadougou 09, Burkina Faso.
[2]Department of Ecology and Geobotany, Institute for Ecology, Evolution and Diversity, J. W. Goethe University, Siesmayerstraße 70, 60323 Frankfurt am Main, Germany.

In the West African Sudanian regions, people depend on natural products, especially on highly valued species as source of income, fuel wood, food, medicine, fodder for livestock etc. However, land-use management coupled with unsustainable uses of highly valued trees might jeopardize the long-term viability of some species' populations. Thus, we compared the population structures of two trees, *Afzelia africana* and *Pterocarpus erinaceus* and the extent of bark and foliage harvesting within two contrasting land-use types using a random stratified design with 45 replication plots for each species. For both species, population structures were stable in the protected area whereas they showed a declining structure in the agroforestry parklands with lower densities of seedlings and adults as well as a total lack of saplings and young mature trees. In addition, both species were over-exploited. More individuals of *A. africana* and *P. erinaceus* were harvested with a weak to severe intensity in the parklands, while only few individuals were harvested in the protected area, with a higher proportion of weak to medium intensity. To ensure conservation of these highly valued species, participatory introduction of juveniles and sensitization for seedling protection are required in the agroforestry parklands.

Key words: Population structure, W National Park, agroforestry parklands, pruning, debarking, West Africa.

INTRODUCTION

In the Sudanian zone people depend heavily on the natural resources and on the services provided by ecosystems and agro-ecological systems (Devineau et al., 2009). Indeed, the majority derives their livelihood from agriculture combined with extensive livestock breeding and the use of a variety of natural products. Among these services, non-timber forest products (NTFPs) play an important role (Lykke, 1998; Cunningham, 2001; Belem et al., 2007). Therefore, the increase of population in these recent years and agricultural intensification (notably cotton cultivation) combined with climate change result in an increasing pressure on natural resources. This ecological crisis is particularly accentuated in West Africa leading to habitat degradation and loss of biodiversity. For example, more than 14.5% of savanna habitats were lost in the peripheral areas of the W-Arly-Pendjari (WAP) protected area complex in the period 1984 to 2002. During this period, the potential capacity of the area to conserve species richness decreased considerably (Clerici et al., 2007).

Afzelia africana (Sm.) or African mahogany (Caesalpiniaceae) and *Pterocarpus erinaceus* Poir or African rosewood (Papillionaceae) are highly valued

*Corresponding author. E-mail: nblandine@gmail.com or nac_marivet@yahoo.fr.

species throughout the Sudanian zone and particularly in the area around the WAP complex (Sieglstetter and Wittig, 2002; Krohmer et al., 2006; Koadima, 2008). Both are known for the good quality of the wood for timber, fuel and charcoal. *A. africana* and *P. erinaceus* also provide medicinal products. The barks, leaves and roots are used by local people in the treatment of diarrhoea, cough, mental illness, gastrointestinal disorders, general pain relief and trypasonomiasis in domestic animals (Arbonnier, 2002; Akinpelu et al., 2008; Ouédraogo-Koné et al., 2008). Leaves are important fodder and nitrogen sources for animals during the dry season in the agrosylvopastoral system (Petit and Mallet, 2001; Ouédraogo-Koné et al., 2008).

Due to the importance of these species, especially for medicinal and fodder purposes, they are very often subject to bark and foliage harvesting in Eastern Burkina Faso, which could affect the long-term viability of their populations. According to Peter (1994) and Delvaux et al. (2009) harvesting of vegetative structures (like stems, leaves and barks) significantly threatens the survival of plant populations. Due to these threats *A. africana* was classified as "Endangered" in the IUCN red list and in the national red list species data base of threatened species in Burkina Faso (Paré et al., 2009). Consequently these species need conservation efforts, which should be based on a detailed knowledge about their current population status.

Two main land-use types characterize the West African ecosystems: protected areas controlled for *in situ* biological conservation goals and agroforestry parklands, exposed to human impact *via* land clearing for agriculture, livestock grazing and/or harvesting of forest products. Thus, studying how strong land-use affects population structures of highly valued species might help to develop effective conservation strategies of the given species in accordance to the needs of local population. Especially understanding species population dynamics under harvesting impact is important for improved conservation. The use of population structures as a tool to investigate the demographic health of harvested populations can be the basis for strong management decisions if it is combined with information related to species specific growth rate, spatial distribution (Fandohan et al., 2010a), patterns of use and harvest (Gouwakinnou et al., 2009; Schumann et al., 2010).

Previous studies on *A. africana* and *P. erinaceus* have focused primarily on regeneration and seedling survival constraints (Bationo et al., 2001; Ouédraogo et al., 2006), characterization of nutritive value, chemical composition (Ouédraogo-Koné et al., 2008) and medicinal properties (Akinpelu et al., 2008). Sinsin et al. (2004) used the dendrometric characteristics as indicators of pressure of *A. africana* to assess the dynamic changes in trees between climatic zones. None of the studies assessed the relationships between bark and leave harvestings and population structures under land-use impact for conservation purposes. Our study will especially emphasize the effect of land-use types on *A. africana* and *P. erinaceus* populations by comparing their populations in a National Park and its surrounding agroforestry parklands under harvesting impact. Thus, it will provide a profound knowledge and information basis for developing management strategies for a more sustainable use and conservation of these species.

The aim of this study is to (i) assess the population structure of *Afzelia africana* and *Pterocarpus erinaceus* by comparing populations of two land-use types: parklands *versus* a protected area and (ii) identify the harvesting patterns of *Afzelia africana* and *Pterocarpus erinaceus* products. We draw on our results to provide recommendations for better conservation strategies for the species populations in the West African savannas.

MATERIALS AND METHODS

Study site

The study was undertaken in two different land-use types namely protected and agroforestry parklands located in Eastern Burkina Faso. The protected area comprises the W National Park (WNP) and its adjacent hunting zones. WNP is a component of a transfrontalier biosphere reserve (1 023 000 ha) shared between three countries: Benin, Burkina Faso and Niger (Figure 1). The protected area is managed through prescribed annual early fires (October or November) (Clerici et al., 2007). Grazing and fuel wood extractions are prohibited inside, while exploitation of baobab fruits and straw by neighbouring local communities is authorized and regulated. The vegetation is composed of a mosaic of various types of woodland and riparian forests where *Afzelia africana* and *Pterocarpus erinaceus* characterize two main vegetation communities namely the *Afzelia africana-Tamarindus indica* and the *Isoberlinia doka-Pterocarpus erinaceus* associations (Mahamane, 2005).

The agroforestry parklands are characterized by farming systems with alternating cycles of cultivation and fallows. Also heavy and extensive livestock grazing, uncontrolled fires as well as wood and NTFP collection for daily subsistences (legume, fodder, medicinal products, domestic uses like traditional hives and pestles, etc.) takes place (Belem et al., 2007). *Vitellaria paradoxa* Gaertn. C. F., *Parkia biglobosa* (Jacq.) R.Br. ex G.Don, *Piliostigma thonningii* (Schumach.) Milne-Redh., *Guiera senegalensis* J.F. Gmel., *Combretum collinum* Fresen., *Pterocarpus erinaceus* and *Afzelia africana* trees are commonly found in this area.

The average annual rainfall in the study area ranges from 750 to 950 mm and the dry season lasts 6 to 7 months which corresponds to the Sudanian phytogeographical sector (Fontès and Guinko, 1995). The main soil types are luvisols, lixisols and leptosols. Human population density bordering the WNP in Burkina Faso is about 16 inhabitants per km^2 (INSD, 2007). The largest ethnic groups are Gourmantché, Fulani and Djerma who are farmers and cattle-breeders. Livestock density (mainly cattle, sheep and goats) in the province Tapoa is about 50.28 per km^2 (ENEC, 2003).

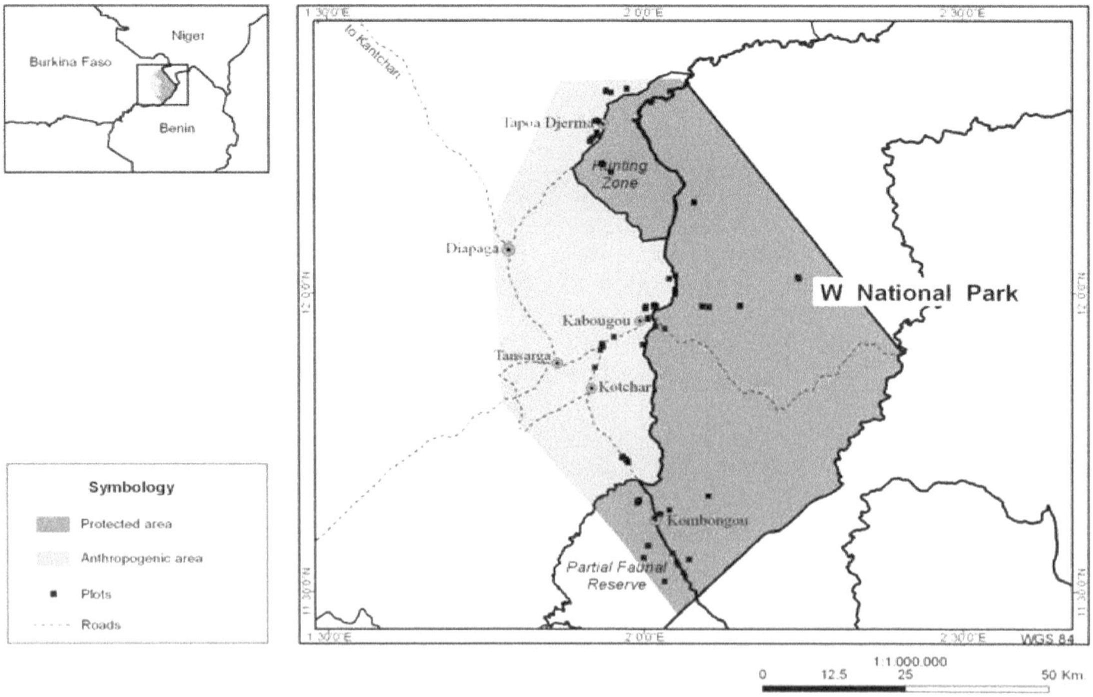

Figure 1. Location of the study area with the position of the plots (UTM zone 31 North, WGS 84).

Table 1. Description of the four levels of intensity of harvesting based on Cunningham (2001).

Percentage of harvesting	Intensity of harvesting
Individual without damage (without branch neither cut nor mutilated trunk)	Null
Individual with 0–25% of their crown pruned or trunk debarked	Weak
Individual with 25–50% of their crown pruned or trunk debarked	Medium
Iindividual with >50 % of their crown pruned or trunk debarked	Severe

Studied species

Afzelia africana is a fire-tolerant tree species (Stark, 1986) occurring in sub-Saharan Africa between the Guinean and the Sudano-guinean bioclimatic zone. The species is mostly found on deep sandy and alluvious soils. While *Pterocarpus erinaceus* is widely distributed in West African savanna regions on diverse types of soil. The later occurs in the Sudano-guinean bioclimatic zone (Arbonnier, 2002). Both species colonise woodland savannas and dry or open forests. They attain a height up to 25 to 35 m (*A. africana)* and 8 to 15 m (*P. erinaceus).* In Burkina Faso *A. africana* is found between the latitude 12 to 13°N and *P. erinaceus* below the latitude 13°N (Terrible, 1984).

Sampling and data collection

Trees and other sapling populations of *A. africana* and *P. erinaceus* were sampled following a random stratified (protected and parklands) design with 45 replication plots (30 x 30 m) for each species. Individuals with a diameter between 5 to 10 cm at breast height (dbh) were considered as saplings, whereas those with dbh >10 cm were recorded as trees. Seedling populations (dbh <5 cm) were sampled in one random subplot (10 x 10 m) within each plot. The parameters measured for each individual of tree and sapling populations were dbh, height, the percentage of debarked trunk and pruned branches. The adapted scale of human pressure on trees from Cunningham (2001) is used to estimate separately for each individual the percentage of trunk debarking as well as the percentage of pruned branches. Four levels of disturbance pressure were thus considered as shown in Table 1. All these data were coupled with informal talks and interviews with farmers in order to know the reasons of harvests.

Data analysis

To assess the impact of land-use on the two high value species we compared variables of population structure (density, height, dbh and the height/dbh ratio) following development stages between the land-use types using the Mann-Whitney non parametric test since the data were not normally distributed.

Table 2. Comparison of seedling, sapling and tree densities (mean individuals ha-1 ± SE) of *A. africana* and *P. erinaceus* in protected and parklands.

Species	Diameter by developmental stage	Protected area	Parklands	Z score	p-value
	Seedlings (0-5)	2466.67 ± 369.62	4.44 ± 4.44	Z = -8.048	p = 0.000
A. africana	Saplings (5-10)	3.46 ± 1.05	0	Z = -3.516	p = 0.000
	Trees (>10)	52.10 ± 4.16	21.48 ± 1.95	Z = -6.091	p = 0.000
	Seedlings (0-5)	244.44 ± 101.98	6.67 ± 6.67	Z = -3.280	p = 0.001
P. erinaceus	Saplings (5-10)	3.95 ± 1.28	0	Z = -3.514	p = 0.000
	Trees (>10)	43.46 ± 3.70	20.25 ± 1.94	Z = -5.312	p = 0.000

The significance of land use differences were tested using a Wilcoxon matched pairs test; z values and significance level are given.

The proportions of trees ranging by bark and foliage harvests intensities were calculated for the two land-use types; and the comparison of the extent between land-use types was assessed using Kruskal-Wallis test.

To test if the proportion of trees harvested (pruned or debarked) is dbh or species dependent we used the Fisher's exact test (two-tailed) instead of a chi-square test, because some cells of our contingency table had frequencies less than five.

The dbh was used as a measure of tree size, since this parameter is strongly correlated with the height (Spermann's rho = 0.670, p = 0.000 and rho = 0.583, p = 0.000 respectively for *A. africana* and *P. erinaceus*).

The trend of *A. africana* and *P. erinaceus* populations in both land-use types was analyzed through size class diameters (SCD) using the method proposed by Condit et al. (1998) and Lykke (1998). Ten diameter classes were used: [0 to 5], [5 to10], [10 to 20], [20 to 30], [30 to 45], [45 to 60], [60 to 70], [70 to 80], [80 to 90] and [90 to 100 cm]. Square linear regression was calculated with size class median as the independent variable and the average number of individuals in that class (Ni) as the dependant variable. The size class variable was logarithmically-transformed (ln) and the average number of individuals was transformed by ln (Ni+1), because some classes were without individuals. SCD slopes were used as indicators of population structure and their interpretation was based on the type of SCD described by Everard et al. (1995) and Lykke (1998). All statistical analyses were performed by PASW Statistic 18.0.0 (IBM).

RESULTS

Impact of land-use on species population structures

The seedlings, saplings and trees of *A. africana* and *P. erinaceus* differed significantly between the two sites (Table 2).

For both species stem density of adult trees and density of seedlings was significantly higher in the protected area compared to the parklands. Saplings and adults densities were two to three times higher in the protected area than in the parklands. However, for both species, the densities of saplings were very low in the protected area and totally absent in the parklands. Comparing the mean dbh between the land-use systems, no significant difference was observed for both species. However the species showed a significantly higher mean height in the protected area than in the parklands (Table 3). Moreover, the height/dbh ratio was significantly higher in the protected area compared to the parklands for both species.

Impact of land-use on species diameter (dbh) class distribution

Based on slope values and regression coefficients, the structure of the species populations showed different trends according to the two land-use types (Table 4).

In the protected area the population structure of *A. africana* and *P. erinaceus* showed strong negative slopes (respectively -1.30 and -0.95) and high coefficients of SCD regressions (0.73 and 0.85) with a typical inverse-J-shaped distribution. In contrast, the population structure of *A. africana* and *P. erinaceus* in the parklands deviated from J-shaped distribution (Figures 2a and 2b). In this area, slopes and regression coefficient of SCD were not significant. Moreover, these slopes were flat (respectively -0.088 and -0.141).

Furthermore, regardless of the overall population structure, young mature trees and some of the large size-classes were totally absent for both species in the sample plots of parklands. For *A. africana* individuals with a dbh of 5 to 10, 15 to 20 and 70 to 80 cm only occurred in the protected area. Similarly, *P. erinaceus* individuals with a dbh of 5 to 15 cm and 55 to 60 cm were only found (Figure 2).

Extent and degree of bark and foliage harvest

The extent of trees bark and foliage harvests of *A. africana* and *P. erinaceus* differed significantly between land-use types (Figures 3 and 4). In the parklands 95.51 ± 2.21% of individuals of *A. africana* and 96.34 ± 2.09% of individuals of *P. erinaceus* were pruned from weak

Table 3. Comparison of mean dbh (cm), height (m) and ratio (m) (mean ± SE) of *A. africana* and *P. erinaceus* within protected and parklands.

Species	Parameters measured	Protected area	Parklands	Z score	p-value
	Dbh (cm)	36.81 ± 1.19	40.42 ± 1.43	Z = -1.628	p = 0.104
A. africana	Height (m)	9.22 ± 0.19	7.19 ± 0.23	Z = -5.757	p = 0.000
	height/dbh Ratio (m)	30.17 ± 0.95	18.34 ± 0.47	Z = -9.068	p = 0.000
	Dbh (cm)	28.56 ± 0.94	30.76 ± 1.17	Z = -1.299	p = 0.194
P. erinaceus	Height (m)	8.71 ± 0.25	6.11 ± 0.22	Z = -6.119	p = 0.000
	height/dbh Ratio (m)	34.32 ± 1.06	20.68 ± 0.66	Z = -8.552	p = 0.000

The significance of land use differences were tested using a Wilcoxon matched pairs test; z values and significance level are given.

Table 4. Regression slope and coefficients values of *A. africana* and *P. erinaceus* following the land use regime.

Land-se types	Regression parameters	*A. africana*	*P. erinaceus*
	Slope	- 1.30	- 0.95
Protected area	r^2	0.73	0.85
	p	0.002	0.001
	Slope	- 0.09	- 0.14
Parklands	r^2	0.13	0.20
	p	0.301	0.266

The significant level is 0.05.

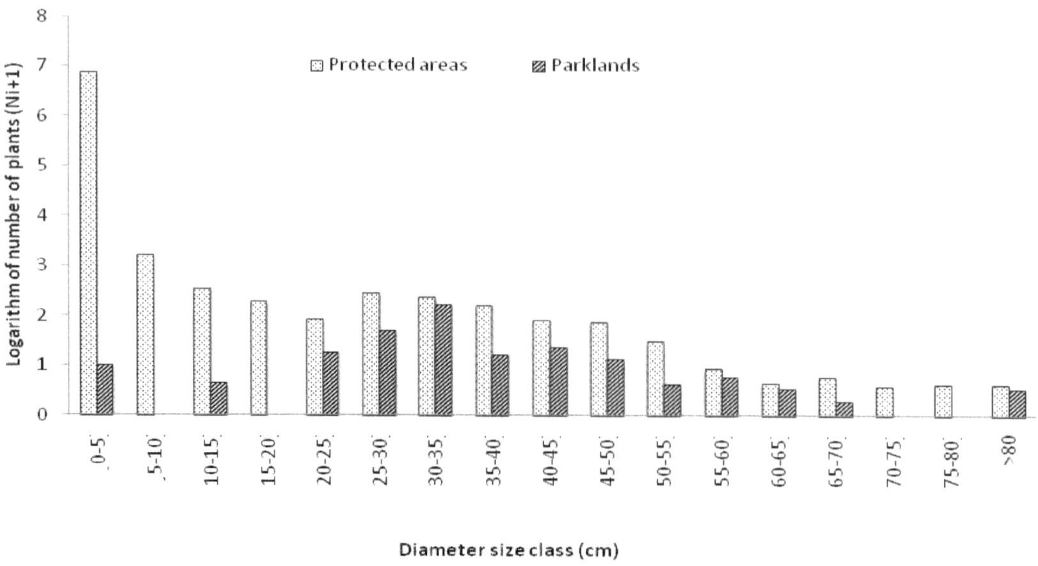

Figure 2a. Dbh class distribution of *A. africana* in the protected and parklands.

(between 0 to 25%) to severe intensity (>50%), while only 0.99 ± 0.76 and 0.74 ± 0.42% of individuals of *A. africana* and *P. erinaceus* respectively were not pruned (Figures 3a and 4a). Similarly, 83.15 ± 3.99 and 87.80 ± 3.64% individuals of *A. africana* and *P. erinaceus* respectively were debarked following the same levels. However, more

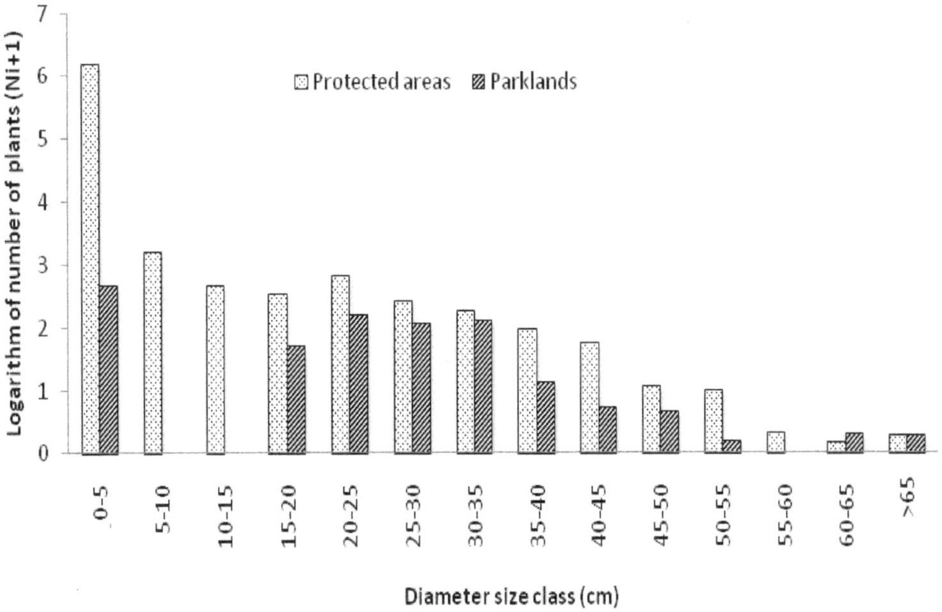

Figure 2b. Dbh class distribution of *P. erinaceus* in the protected and parklands.

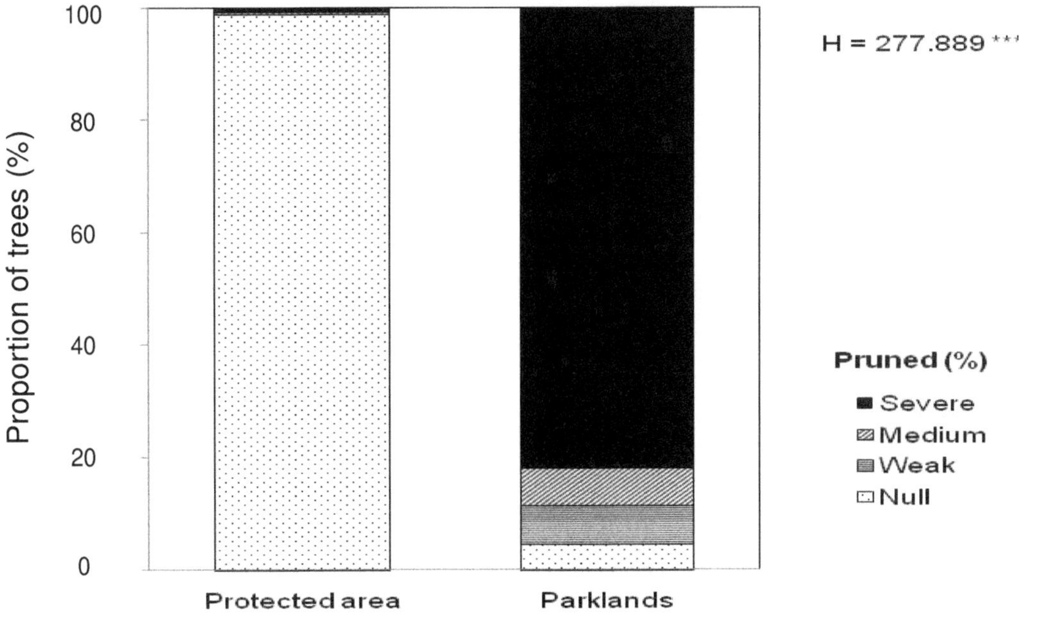

Figure 3a. Proportion of *A. africana* individuals foliage harvest within protected and parklands. *** p<0.001.

individuals were not debarked compared to pruning (3.70 ± 1.12 and 2.96 ± 0.74% for *A. africana* and *P. erinaceus* respectively).

In contrast, in the protected area 99.80 ± 0.001% of *A. africana* and 97.7 ± 0.01% individuals of *P. erinaceus* were not pruned, whereas, only 1.33 ± 0.76% individuals of *A. africana* were pruned at severe intensity and 4.17 ± 1.62% individuals of *P. erinaceus* were pruned at weak and severe intensity (Figures 3a and 4a). Similarly, 97.70 ± 0.001% of individuals of *A. africana* and 95.40 ± 0.01% of individuals of *P. erinaceus* were not debarked, while 1.77 ± 0.88% of *A. africana* were debarked at weak

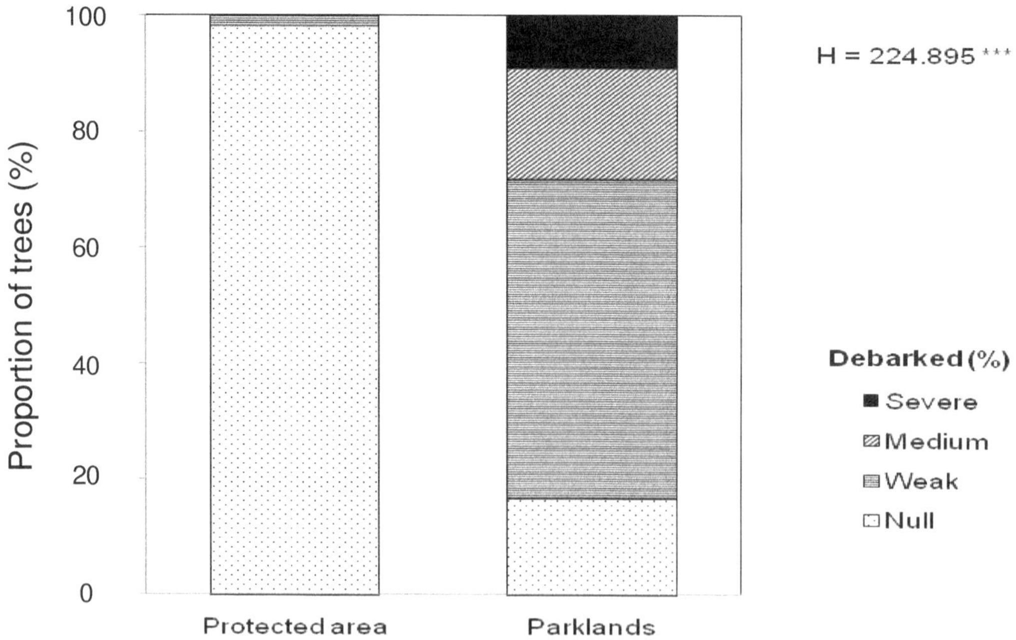

Figure 3b. Proportion of *A. africana* individuals bark harvest within protected and parklands. *** p<0.001.

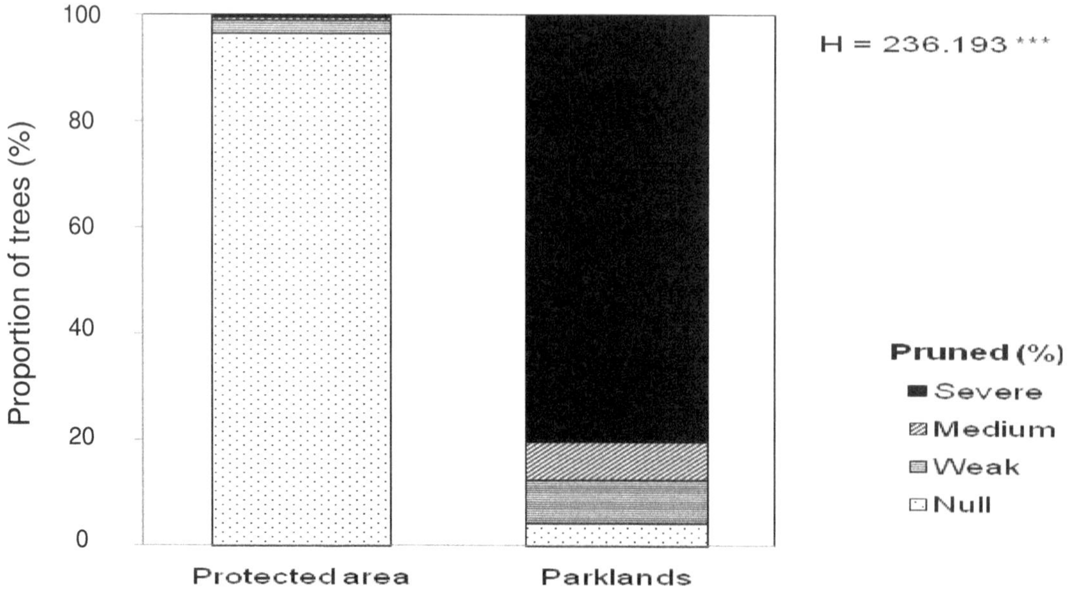

Figure 4a. Proportion of *P. erinaceus* individuals foliage harvest within protected and parklands. *** p<0.001.

intensity and 8.33 ± 2.13% of individuals and *P. erinaceus* were debarked at weak and medium intensity (Figures 3b and 4b).

In addition following the harvesting type (pruning or debarking), the severe intensities occurred with pruning type (88.15% of species) compared to bark harvest (50.36% of species). However, the proportion of tree foliage harvest was similar within the two species (Fisher's Exact: p = 0.99). Conversely, the proportion of trees debarked differed significantly between the species (Fisher's Exact: p = 0.0389). Higher proportions of *P. erinaceus* trees (74.07 ± 15.39%) were severely

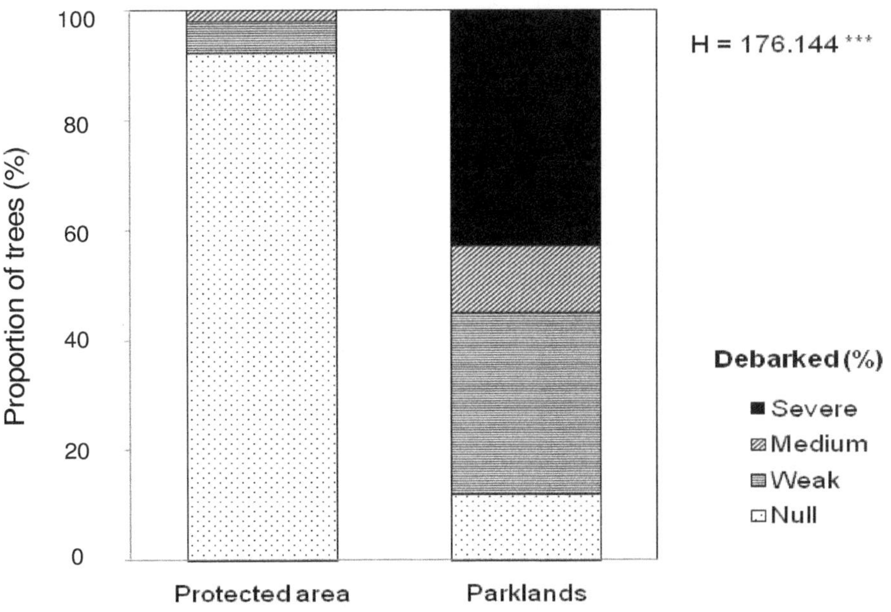

Figure 4b. Proportion of *P. erinaceus* individuals bark harvest within protected and parklands. *** p<0.001.

debarked compared to *A. africana* (9.88 ± 4.77%). Furthermore, following the dbh size class distribution, the intensity of pruning (Fisher's Exact: p = 0.3132 for *A. africana* and p = 0.7866 for *P. erinaceus*) and debarking (Fisher's Exact: p = 0.7484 for *A. africana* and p = 0.1993 for *P. erinaceus*) did not vary significantly within the size classes (small, medium and large).

DISCUSSION

Impact of land use on species population structures

The differences in species population structures observed between the protected and the parklands are principally related to the land-use types. In the protected area, the inverse J- shape of the curve indicated a stable population structure, with abundant individuals in small diameter size classes and a progressive decrease of individuals in upper size classes. In opposite, the shape of the populations of both species in the parklands indicated an unstable population structure suggesting important perturbations affecting population structures. This might be due to the combined effects of human impact such as land clearing for agriculture, livestock grazing and/or harvesting of fodder and medicinal products. Thus, the absence of protection measures led to the unstable structure. In regard to seedlings recruitment, herbivore pressures through seeds and seedlings consumption (Bationo et al., 2001; Ouédraogo-Koné et al., 2008) and trampling (Onana and Duvineau, 2002) could hinder the regeneration process. Similarly, land clearing and tree density reduction for agricultural purposes may be detrimental to seedling establishment and, hence threaten the species (Fandohan et al., 2010b). The shoots are cut during the rainy season in order to reduce their adverse effects (shade, moisture competition) on crops (Gijsbers et al., 1994) and also on the more preferred fruit-trees like *Vitellaria paradoxa* (Djossa et al., 2008). These findings are consistent with recent studies on other Sudanian trees (*Acacia* spp, *Sclerocarya birrea* (A. Rich.), *Tamarindus indica* L., *Adansonia digitata* L.) which have shown that land-use types and habitats effect differently the population structures of highly valued trees (Traoré et al., 2008; Gouwakinnou et al., 2009; Fandohan et al., 2010b; Schumann et al., 2010). Protection might efficiently provide better conditions for seedling regeneration, favour the recruitment and conserve trees densities (Figures 2 and 3 and Table 2).

However, the observed low number of seedlings in this area could also be due to other biotic and abiotic stresses (Bognounou et al., 2009), such as shade. This emphasizes the partial positive effect of protected areas in conserving highly valued plant species (Fandohan et al., 2010a). In addition, the differences in tree heights between the sites could be explained by high intra- and inter-specific competitions (Shackleton, 2002) for light (Scholes and Archer, 1997). This might lead to faster growth in height than in diameter in the protected area, where vegetation stands are denser.

The non-significant difference of mean diameter between the two sites may be due to the effect of the higher number of individuals in small size class in the protected area. Moreover, the high densities of seedlings and the low density of saplings found in the protected area indicated the high ability of species to regenerate. This confirmed the assumption that regeneration potential of *A. africana* and *P. erinaceus* species exists but the main challenge is the establishment as well as the development of saplings in land-use areas (Gijsbers et al., 1994; Bationo et al., 2001; Ouédraogo et al., 2006) (Figures 2 and 3).

Furthermore, the absence of the larger diameter size classes in the sample plots of the agroforestry parklands indicate other types of pressure such as commercial loggings besides bark and foliage harvests. Especially, large trunks of *A. africana* are used to make large mortars (personal observation) which are sold on the local markets.

To summarize, the stable population structures of *Afzelia africana* and *Pterocarpus erinaceus* observed in the protected area was due to better regeneration conditions and less disturbance of the species in this site because of the protection status. In the parklands, the gap created by the lack of regeneration and the total absence of saplings in combination with traditional agricultural practices, over grazing, seed predators and commercial loggings lead to a declining structure. This result corroborated the findings of Ouédraogo et al. (2006) for the same species and Lykke (1998) for other exploited species in West African savannas.

Extent and degree of harvests

The high proportions of harvested *A. africana* and *P. erinaceus* trees show that these species are used and appreciated particularly by helders of livestock (Petit and Maillet, 2001; Ouédraogo-Koné et al., 2008) and by local populations for the use as medicine (Akinpelu et al., 2008; Delvaux et al., 2009). This emphasizes their role as high value species in the Sudanian zone.

Nevertheless, the designation as protected area could be the best option to protect highly valued species in the over-exploitation context. The higher proportions of severe harvests intensity and the non-size dependant harvests conditions observed in the parklands, coupled with the highest extent of non-harvested individuals, and to the lowest proportion of severe harvestings in the protected area support this assumption. This shows the importance of protected area in preserving the highly valued species.

Furthermore, the non-size class specific harvest conditions (even the smallest diameter sizes) on the parklands, shows that harvesters seem to have no preference about size class. In contrast to Gaoue et al. (2007) on *Khaya senegalensis* and Schumann et al. (2010) on baobab, neither pruning nor debarking on *A. africana* and *P. erinaceus* was size specific. The higher pressures to which species are face might explain this result comparing to the others. Indeed, for species which is highly exploited for its bark, debarked trees of all sizes were found (Stewart, 2003). We also noticed that *A. africana* was subjected to harvest for magico-religious rituals (personal observation) by local people, especially Gourmantché community. For this ethnic group, some *A. africana* trees were sacred, so they were regularly debarked for sacrifices in order to pray to God for good health and protection of families. The sacred trees were debarked and the wound caused by this harvest was recovered with blood and feathers. Moreover, the higher pruning pressure is due to the higher density of livestock in the WNP area, especially in the dry season. Highest livestock densities may be explained by the transhumance activity (Convers et al., 2007) and by the positive effect of cotton culture, which allows farmers to spend money gained for livestock buying (Petit and Maillet, 2001). This result also supported the fact that species were more involved for pruning than debarking (Figures 4 and 5).

In addition, two main reasons might explain the different debarking rates between the two species in the parklands: the higher demand for *P. erinaceus* bark and the fact that *A. africana* does not recover well from debarking (Delvaux et al., 2009). However, according to later authors, harvesting disrupts the tree physiological functioning and stimulated a higher hormonal activity in order to close the wound quickly.

To conclude, the assessment of the population structure and the harvesting patterns of the highly valued species *Afzelia africana* and *Pterocarpus erinaceus* in two different land-use areas showed that both species are severely over-exploited in the parklands due to bark and foliage harvests combined with agricultural practices. However, both species were more affected by pruning than by debarking. To ensure more sustainable use of these highly valued species further research as well as participatory actions with local communities is required: In a participatory way introduction of juveniles and sensitization for seedlings protection should be carried out in the agroforestry parklands. For this purpose, best tree individuals for seed production for the nursery (seedlings production for planting) should be selected and protected for current and future use. The impact of bark and foliage harvests on both species fruit production should be assessed in order to provide indications on the harvesting tolerance.

ACKNOWLEDGEMENTS

The work was funded by sustainable use of natural vegetation in West Africa (SUN) (EU FP6 031685) project. We thank the Eastern environmental region of Burkina Faso, all foresters and people of the Tapoa Province for their collaboration. We also thank all technicians and scouts for field work assistance. Finally, we thank Richard Byer for linguistic improvements.

REFERENCES

Akinpelu DA, Aiyegoro OA, Okoh AI (2008). *In vitro* antimicrobial and phytochemical properties of crude extract of stem bark of *Afzelia africana* (Smith). Afr. J. Biotechnol., 7: 3665-3670.

Arbonnier M (2002). Trees, shrubs and lianas of dryland West Africa. CIRAD-MNHN-IUCN.

Bationo BA, Ouédraogo SJ, Guinko S (2001). Seed longevity and stress survival of seedlings of Afzelia africana Sm. in savannah woodland in Burkina Faso. Ann. For. Sci., 58: 69-75.

Belem B, Nacoulma BMI, Gbangou R, Kambou S, Hansen HH, Gausset Q, Lund S, Raebild A, Ouedraogo M, Lompo D, Theilade I, Boussim IJ (2007). Use of Non Wood Forest Products by local people bordering the "Parc National Kaboré Tambi", Burkina Faso. J. Transdisciplinary Environ. Stud., 6: 1-21.

Bognounou F, Thiombiano A, Savadogo P, Boussim IJ, Odén PC, Guinko S (2009). Woody vegetation structure and composition at four sites along a latitudinal gradient in Western Burkina Faso. Wood Trop. For., 300: 29-45.

Clerici N, Bodini A, Eva H, Gregoire JM, Dulieu D, Paolini C (2007). Increased solation of two Biosphere Reserves and surrounding protected areas (WAP ecological complex, West Africa). J. Nat. Conserv., 15: 26-40.

Condit R, Sukumar R, Stephen H, Foster RB (1998). Predicting population trends from size distributions: a direct test in a tropical tree community. Am. Nat., 152: 495-509.

Convers A, Chaibou I, Bino A, Dulieu D (2007). The management of transhumance in the area of influence of the regional park of the Program ECOPAS w: a "project approach" to development of the periphery of the park. Vertigo, hors serie, 4: 1-9.

Cunningham AB (2001). Applied Ethnobotany, People, Wild Plant Use and Conservation.WWF. Earthscan Publications Ltd, London, U.K.

Delvaux C, Sinsin B, Darchambeau F, Van Damme P (2009). Recovery from bark harvesting of 12 medicinal tree species in Benin, West Africa. J. Appl. Ecol., 46: 703–712.

Devineau JL, Fournier A, Nignan S (2009). "Ordinary biodiversity" in western Burkina Faso (West Africa): what vegetation do the state forests conserve? Biodvers. Conserv., 18: 2075-2099.

Djossa BA, Fahr J, Wiegand T, Ayihouenou BE, Kalko EK, Sinsin BA (2008). Land use impact on *Vitellaria paradoxa* C.F. Gaerten. Stand structure and distribution patterns: a comparison of Biosphere Reserve of Pendjari in Atacora district in Benin. Agrofor. Syst., 72: 205–220.

ENEC (2003). Statistics pets Tapoa Province, Burkina Faso.

Everard D, Midgley JJ, van Wyk GF (1995). Dynamics of some forests in KwaZulu-Natal, South Africa, based on ordinations and size class distributions. South Afr. J. Bot., 61: 283-292.

Fandohan AB, Assogbadjo AE, Glele Kakaï RL, Sinsin B, Van Damme P (2010b). Impact of habitat type on the conservation status of tamarind (*Tamarindus indica* L.) populations in the W National Park of Benin. Fruits, 65: 11–19.

Fandohan B, Assogbadjo AE, Glèlè Kakaï RL, Sinsin B (2010a). Effectiveness of a protected areas network in the conservation of Tamarindus indica (Leguminosea–Caesalpinioideae) in Benin. Afr. J. Ecol., doi: 10.1111/j.1365-2028.2010.01228.x.

Fontès J, Guinko S (1995). Vegetation map and land use in Burkina Faso. Explanatory note: French Ministry of Cooperation: Campus project (88313101).

Gaoue OG, Ticktin T (2007). Patterns of harvesting foliage and bark from the multipurpose tree *Khaya senegalensis* in Benin: variation across ecological regions and its impacts on population structure. Biodvers. Conserv., 137: 424–436.

Gijsbers HJM, Kessler JJ, Knerel MK (1994). Dynamics and Natural regeneration of woody species in farmed parklands in the Sahel region (Province of Passoré, Burkina Faso). For. Ecol. Manage., 64: 1-12.

Gouwakinnou GN, Kindomihou V, Assogbadjo AE, Sinsin B (2009). Population structure and abundance of *Sclerocarya birrea* (A. Rich) Hochst subsp. *birrea* in two contrasting land-use systems in Benin. Int. J. Biodvers. Conserv., 1: 194-201.

INSD (2007). General Census of Population and Housing in 2006 in Burkina Faso. Ministry of Territorial Administration, Burkina Faso, p. 51.

Koadima M (2008). Inventory of commercial timber species w Park and adjacent land in Burkina Faso and population status of three species has great socio-economic value. Master thesis, University of Ouagadougou.

Krohmer J, Hahn-Hadjali K, Wittig R (2006). Use of wild populations by Fulani in Burkina Faso and Benin. Studies on the flora and vegetation of Burkina Faso, 10: 33-48.

Lykke AM (1998). Assessment of species composition change in savanna vegetation by means of woody plants' size class distributions and local information. Biodver. Conserv., 7: 1261-1275.

Mahamane A (2005). Studies floristic, phytosociological and phytogeographical vegetation Regional Park W Niger. State Thesis in Agronomic Sciences and Biological Engineering, Free University of Brussels, Belgium.

Onana J, Devineau JL (2002). Afzelia africana Smith ex Persoon in North Cameroon. Current status of populations and pasture use. J. Ani. Husbandry Vet. Med. Trop. Countries, 55: 39-45.

Ouédraogo A, Thiombiano A, Hahn-Hadjali K, Guinko S (2006). Diagnosing the state of degradation of stands of four woody species in Sudanian zone of Burkina Faso. Drought, 17: 485-491.

Ouédraogo-Koné S, Kaboré-Zoungrana CY, Ledin I (2008). Important characteristics of some browse species in an agrosilvopastoral system in West Africa. Agrofor. Syst., 74: 213-221.

Paré S, Tigabu M, Savadogo P, Odén PC, Ouadba JM (2009). Does designation of protected areas ensure conservation of tree diversity in the Sudanian dry forest of Burkina Faso? Afr. J. Ecol., 48: 347-360.

Peters CM (1994). Sustainable Harvest of Non - Timber Plant Resources in Tropical Moist Forest: An Ecological Primer. USAID Biodiversity Support Programme, Washington DC.

Petit S, Mallet B (2001). The pruning of fodder trees: detail of a pastoral practice. Trop. Woods For., 4: 35-45.

Scholes RJ, Archer SR (1997). Trees-grass interactions in savannas. Annu. Rev. Ecol. Syst., 58: 517-544.

Schumann K, Wittig R, Thiombiano A, Beckerc U, Hahn K (2010). Impact of land-use type and bark- and leaf-harvesting on population structure and fruit production of the baobab tree (*Adansonia digitata* L.) in a semi-arid savanna, West Africa. For. Ecol. Manage., 260: 2035-2044.

Shackleton C (2002). Nearest-neighbour analysis and the prevelance of woody plant competition in South African savannas. Plant Ecol., 158: 65–76.

Sieglstetter R, Wittig R (2002). The use of wild wood and its effect on vegetation in the region Atakora (North-West Benin). Studies on the flora and vegetation of Burkina Faso, 7: 23-30.

Sinsin B, Eyog-matig O. Assogbadjo AE, Gaoue OG, Sinadouwirou T (2004). Dendrometric characteristics as indicators of pressure of *Afzelia africana* Sm. dynamic changes in trees found in different climatic zones of Benin. Biodvers. Conserv., 13: 1555–1570.

Stark MA (1986). Relationship between fire and basal scarring on *Afzelia africana* in Benoue Nationa Park, Cameroon. Afr. J. Ecol., 24: 263-271.

Stewart K (2003). The African cherry (*Prunus africana*): Can lessons be learned from an over-exploited medicinal tree? J. Ethnopharmacol., 89: 3-13.
Terrible M (1984). Test on ecology and sociology of trees and shrubs of the Upper Volta. Bookstore Savannah: Bobo Dioulasso, Burkina Faso, p. 257.
Traoré S, Kaboré O, Millogo-Rasolodimby J, Thiombiano L, Guinko S (2008). Impact of protected areas and anthropogenic land use on regeneration of Acacia woodland in Eastern Burkina Faso. Flora et Vegetatio Sudano-Sambesica, 11: 17-24.

21

Diversity and distribution of amphibians in the Kakum National Park and its surroundings

K. A. Monney, M. L. Darkey and K. B. Dakwa*

Department of Entomology and Wildlife, School of Biological Sciences, University of Cape Coast, Cape Coast, Ghana.

This study investigated amphibian distribution and diversity in Kakum National Park and its surrounding communities. Sampling was carried out in lowland swampy areas and upland zones of both locations; thus totalling four main habitat types as areas of survey. A total of 1434 specimens belonging to four Families comprising 12 species of anurans were recorded. The 12 species were irregularly distributed in the four habitat types. The Arthroleptidae and Ranidae were localized with low levels of abundance and richness while *Phrynobatrachus alleni* was widespread. There were no significant differences in both diversity and distribution between the reserve and its surrounding communities. This is a probable indication that the current anthropogenic disturbances off-reserve may have little or no effect on the amphibians; or may affect them in both the reserve and off-reserve together in the same way. However, the park's upland forests favoured amphibian distribution and diversity most and the results support the hypothesis that elevation may affect the distribution and diversity of some species of amphibians. Nevertheless, the results of the study calls for a critical consideration for efficient amphibian conservation in both in- and off-reserve of the study area as the number of species recorded is lower than expected.

Key words: Kakum National Park, lowland, upland, amphibian distribution, diversity, anurans, surrounding communities, decline, extinction, herpetofauna, environs.

INTRODUCTION

Currently, there is expression of great concern about extinction of amphibians globally (Reid and Zippel, 2008); as one in three amphibian species is threatened with extinction (Norris, 2007). This concern is justified on the grounds that worldwide, amphibians are crucially important in their ecosystems. The general ecological importance of amphibians lies in their being predators; acting as primary and secondary carnivores on insects some of which are crop pests or disease vectors (Behangana, 2004). The known important roles amphibians play in the food webs of most biological communities cannot therefore be overemphasized. For example a detailed study of upland streams in the Central Panamanian Highlands revealed that ecological effects of amphibian decline could not just be taken for granted. The effects included changes in algal community structure and primary production, altered organic matter dynamics, negative impacts on aquatic insect predators, such as snakes and reduced energy transfer between streams and associated riparian habitats (Ranvestel et al., 2004). The probable causes of the amphibian decline are established to be multi-faceted involving various complex combinations of obvious ones such as habitat destruction, habitat fragmentation or loss, pollution (industrial, agricultural and pharmaceutical) over-harvesting and invasive species (Beebee and Griffiths, 2005; Moore and Church, 2008). A rather non-traditional cause is realized to be the growing ecological impacts of climate change (Araújo et al., 2006). The causative factors are even suspected to be acting in synergy in some instances (Blaustein and Kiesecher, 2002; Davidson and Knapp, 2007). Quite recently, a disease

*Corresponding author. E-mail: dakb92@yahoo.com

caused by a chytrid fungus, *Batrachochytrium dendrobatidis* has been implicated as the factor most commonly associated with enigmatic declines and catastrophic extinctions of amphibian populations and species (Moore and Church, 2008).

Even though habitat loss may evidently remain the most significant threat overall (impacting 90% of those species currently considered threatened), it is believed that amphibians are more sensitive than other organisms to environmental deterioration and that the decimation of amphibians is a warning sign of an ever increasing poisoned environment. It is observed that many small mammals and herpetofauna species generally have relatively short generation time and a quick response to habitat and microclimatic variations within forest fragments (Cain et al., 2007). Thus, amphibians as biomonitors can be used to measure this feature (Wasonga et al., 2006). An alteration of microclimatic conditions is also thought to be of major importance to plants and animals generally in fragmented forests (Harper et al., 2005). The global extinction threats of amphibians have generated calls for proactive conservation activities (Stuart et al., 2004) and has even prompted a 5-year strategic plan by the Amphibian Conservation Action Plan (ACAP) in 11-pronged thematic areas. The strategy has designed a network of conservation sites for amphibians taking into consideration freshwater resources and associated terrestrial landscapes, climate change, biodiversity loss, amphibian declines and captive programmes (Moore and Church, 2008). The 11 thematic areas also translate into 4 phases of activities (Moore and Church, 2008). In the ensuing activities, it appears there is a favour for breeding programmes; but Pounds et al. (2007) for example have expressed reservations about the utility of breeding in zoos in favour of field oriented conservation programmes for amphibians. The World Zoo and Aquarium Conservation Strategy (WZACS, 2005) of the World Association of Zoos and Aquariums (WAZA) have suggested a compromise strategy through all the foregoing complexities and are trying to bridge the gap between conservation activities in nature (*in situ*) and work in zoos and aquariums (*ex situ*). Their line of action demonstrates potential for fully integrated and sophisticated approaches that are truly complementary.

It is in the light of concerns about worldwide amphibian declines that the need has been expressed about gathering base-line and long-term information. Base-line information on amphibians would be needed for at least most future monitoring and conservation efforts.

Proposals of WZACS emphasize the urgent need for more studies of the ecological impacts of amphibian declines and extinctions and there has been a call for accumulation of data on amphibian population biology and habitat viability in the wild. Furthermore, to ensure the appropriate husbandry conditions and management protocols in zoos, it cannot be denied that there also has to be sufficient knowledge of population biology and habitat viability in the wild (Reid and Zippel, 2008). However, such information about the natural history and conservation status of a very large number of tropical species is lacking in mainland Africa (Halliday, 2008). As it is in West Africa generally, the herpetofauna of Ghana remains largely uninvestigated and ecological research of amphibians linked to conservation activities has generally lagged behind in West Africa. The high biological significance of protected areas, the escalating human exploitation and the negative impact that human exploitations have on wildlife all suggest that there is the need for a detailed herpetofaunal investigation to document information we can gather now before some disappear. Data on the number of species of amphibians found in any one locality in Ghana remain unexplored. Since few detailed studies on the herpetofauna of Ghana have been carried out, this study was undertaken mainly to contribute to the scanty literature on Ghana's herpetofauna and to provide background data against which analyses such as amphibian declines, among other parameters, will be measured in future. Such information is vital for conservation priorities and the study focuses on diversity of amphibians in Kakum National Park (KNP) and its surroundings as a baseline for analyzing trends of changes and predictability, as well as setting up recovery goals in future as the need arises. The two main null hypotheses tested were that, there is no significant difference between the species diversity of amphibians in the Kakum National Park and its surrounding communities; and elevation does not affect the distribution of amphibians in the Kakum National Park.

Study area

The investigation was carried out in the 210 km^2 Kakum National Park (KNP) and its surrounding communities. The KNP is a protected area but part of the irregular remnant of the fast dwindling Ghana's portion of the moist evergreen Upper Guinea Forest and dominated by *Celtis zenkeri* and *Triplochiton scleroxylon* (Hall and Swaine, 1976) although there are about 100 tree species. The KNP spans the Twifo Praso, Assin and Abura districts of the Central Region of Ghana (1°30' - 1°51' W; 5°20'- 5° 40' N) and is sited about 165 km west of Accra, the national capital. The general climatic conditions of the country characterized by bimodal rainfall and two dry seasons (Durand and Skubich, 1982) prevail in the park. A heavy rainy season from April to July is followed by a light dry season from August to September. A light rainy season from October to early December is then followed by a heavy dry season from December to March (Kouadio et al., 2008). The fauna may concentrate in and around the few water spots available in the park during the dry harmattan from December to March; that is the heavy dry season. The average annual rainfall is about

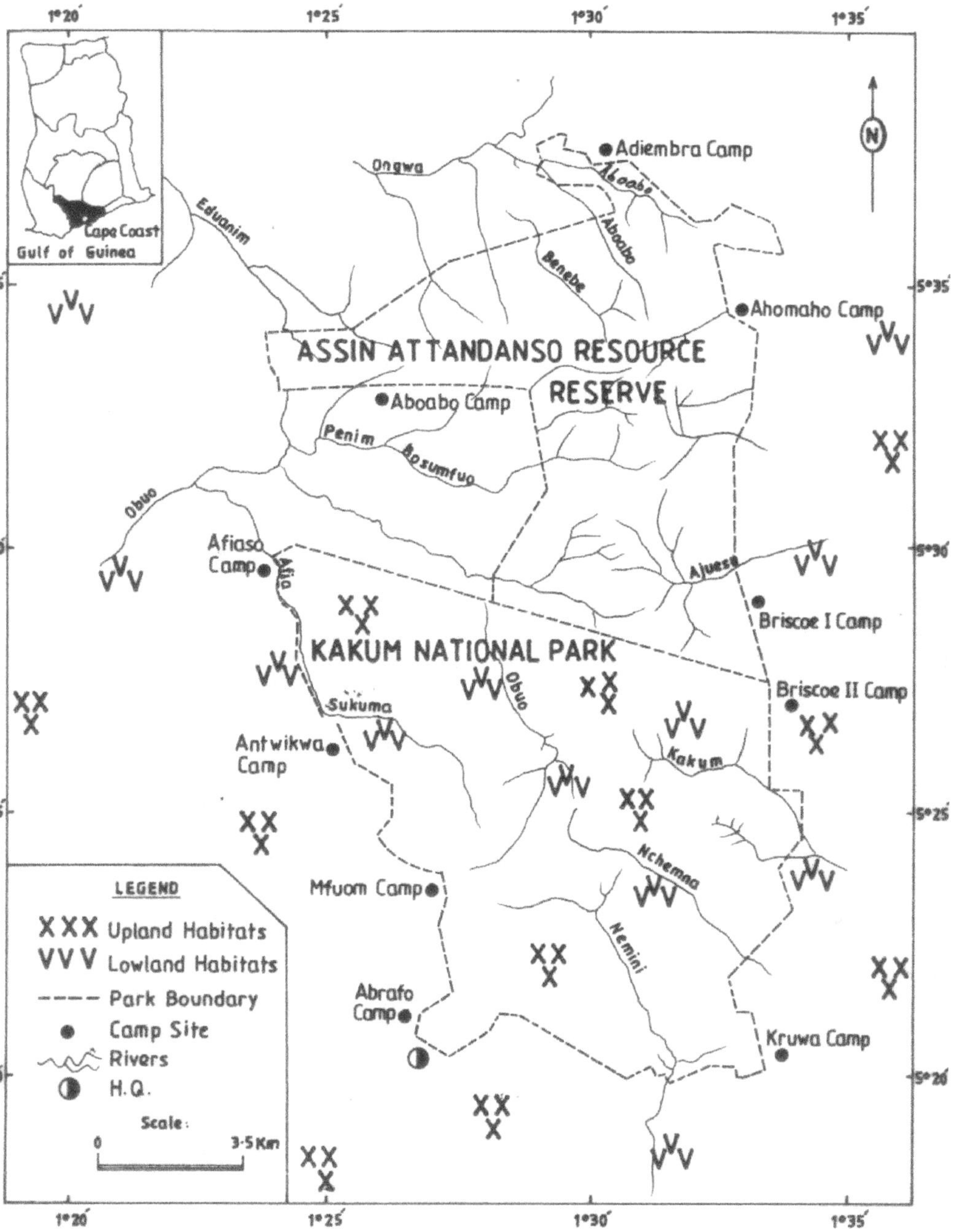

Figure 1. Map of study area.

1600 mm (Forestry Commission, 2007).

The park protects the headwaters of permanent rivers such as Kakum, Obuo and Nemini and rivulets like Ajuesu which may dry up in the heavy dry season (Figure 1). The average relative humidity is about 80% throughout the year while temperature ranges from 18.2 to 32.1°C. The terrain is flat to slightly undulating with an elevation of between 15 to 250 m above sea level (asl) (Forestry Commission, 2007). Most of the elevations occur in the south-western portion of the park. Light south westerly winds blow over the area almost throughout the year. The park is fringed by about 50 communities; six of which are designated by the Wildlife Division of Ghana as camps where park guards are stationed. The majority of the inhabitants in the communities engage in subsistence agriculture with poor farming methods such as slash and burn and small land holdings sometimes right down to the park's boundary. On the other hand, the park itself is increasingly being developed to a holiday resort offering great potential for both tourists and holidaymakers locally, nationally and internationally.

METHODOLOGY

The study area was classified into two land use types and sampled during the period of October 2008 to March 2009. The two land use types were the protected area i.e. the reserve and its recreation area; and the other was the surrounding communities, i.e. the human settlements with their adjacent farmlands. The farmlands consisted mainly of cash crops like cocoa, orange or oil palm and food crops such as cassava, cocoyam, plantain and vegetables (pepper, tomato, garden egg). Some parts of the human settlements consisted of permanently dry pond beds. Each land use type was further sub-divided into lowland which consisted of swampy areas adjacent to watercourses and elevations of dry terrain. We considered the lowland as terrain from sea level up to 25 m asl and elevations as 25-250 m asl. Thus four habitat types were studied, namely in-reserve lowland, in-reserve upland, off-reserve lowland and off-reserve upland. In each land use type sixteen 100x200 m zones were randomly selected out of about twenty five demarcated plots (Figure 1). Visual Encounter Surveys (VES) as defined by Crump and Scott (1994) was the detection technique used in all the habitats because our main aim was to maximize the man hour available. For each sampling, a team of at least six people were employed to search thoroughly for about one hour (at least 6 members were available out of 10 crew for 6 man hours/plot) in cracks and crevices, under deadwood, under leaf litter and around stagnant water bodies. There were 3-5 plot searches/day. Anuran calls were followed to locate male specimens early in the morning at about 05.30 GMT on about 10 occasions after it had rained in the night. Some incidental observations and opportunistic encounters (in which case amphibians encountered in the study area outside search times) were also encouraged, though these yielded no different results.

All individuals collected were identified directly in the field after Rödel (2007) and then released. Initial toe clipping to avoid double counting was discontinued because no specimen was caught twice even in adjacent sampling sites and it was feared that the clipping might introduce infection. Also, mark-recapture techniques were irrelevant for the purposes of this study. The intensive search method adopted favoured diurnal sampling and produced better results than nocturnal sampling and so the latter was discontinued after some time. Also, October which falls in the minor rainy season in Ghana was very wet with heavy rains during the study period and therefore both dry and wet conditions which normally influence amphibian abundance and distribution were adequately represented during the period of study. Voucher specimens were killed with chlorobutanol solution and preserved in 10% formaldehyde and kept in the Department of Entomology and Wildlife Museum, University of Cape Coast. It is obvious that some human activities have effects on amphibian diversity, distribution and abundance. While such activities like farming and spraying chemicals of any kind is not allowed in the reserve, the study sought to know the extent of such activities in the off-reserve sites of the study area through interviews with people in the communities studied.

A one-way ANOVA at 0.05 alpha level was used to test whether there was any significant difference in the means of population of the amphibians in the four habitat types. Variance among the populations in the four habitats was tested with Levene's test for homogeneity, and a Mann-Whitney U test was conducted to evaluate population distribution between swampy lowland and upland habitats in the reserve. Also, Simpson's and Shannon-Weiner's indices were calculated for diversity t-test by the use of SPSS to compare the diversity of the amphibians in the four habitat types.

RESULTS

A total of 1434 individual anurans were recorded during the study. These comprised 12 species, which belong to 5 genera of 4 families (Figures 2 and 3). Apart from the family Ranidae which occurred in off-reserve only, the others, namely: Arthroleptidae, Bufonidae, and Petropedetidae occurred in both in and off-reserves of the study area with Petropedetidae recording the highest encounter rate (Figure 3). The highest number of 963 amphibians was recorded in the upland habitats within the reserve as compared to the least of 67 in lowland swampy habitats in the reserve. The upland and lowland swampy habitats off-reserve recorded 237 and 167 amphibians respectively (Table 1). None of the four habitat types of the study area recorded all the 12 species but results of the off reserve upland habitat was most represented with 11 species contrasted with the least of 6 species in off-reserve swampy lowland habitats (Table 1).

Thus, amphibians are more fairly distributed in the upland habitats within the reserve than other habitat types (Figure 2). *Phrynobatrachus alleni* recorded the highest mean abundance of 113.8 (S.D.=165.6) though *Phrynobatrachus calcaratus, Phrynobatrachus. liberiensis* and *Phrynobatrachus poecilonotus* also occurred in fairly high abundance and the other species occurring at very low abundance in the study area, the least of 1.0 (S.D. = 1.4) being *Phrynobatrachus bibroni* (Figure 4). The data indicate clumped distribution of the anurans in almost all the habitats sampled. The means of amphibian population that occurred in all the four habitat types differed significantly ($F=3.71$, $p=0.02$). Furthermore, homogeneity test indicated a significant difference between the populations of amphibians in all the four habitat types. This therefore supports the hypothesis that amphibian species population differs for each of the four habitat types of amphibian distribution in the study area. Evaluation of the amphibian population distribution revealed an insignificant difference between swampy lowland in the reserve and upland habitats in the reserve that is, $U=49.5$, $p=0.20$; between upland habitats in the reserve and upland habitats off-reserve, $U=60.5$, $p=0.52$; and between lowland habitats in the reserve and off-reserve lowland habitats, $U=61$, $p=0.54$. In contrast, there was significant difference between the populations of amphibians in the off-reserve upland habitats and off-reserve lowland habitats ($U=35.50$, $p=0.04$).

Various diversity indices have been provided (Table 1) and diversity t-test conducted between lowland and upland habitats in the reserve indicated a significant difference between them as $t = -3.29$, $p=0.001$. Similarly, a significant difference was found in the diversity between off-reserve lowland and off-reserve upland habitats as t=-

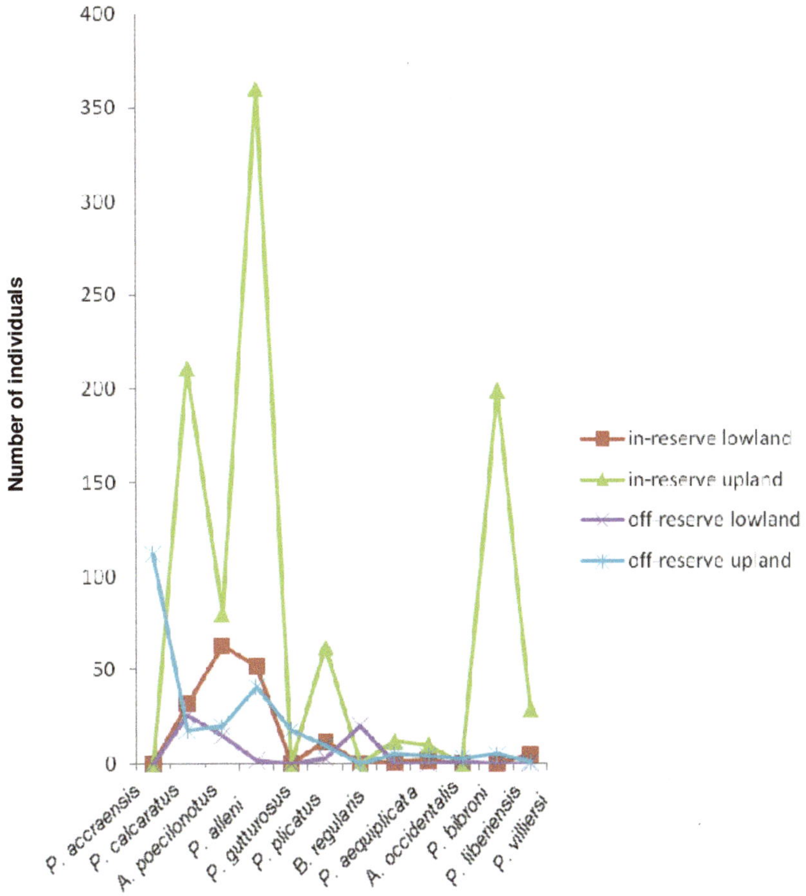

Figure 2. Distribution of amphibians in the study area.

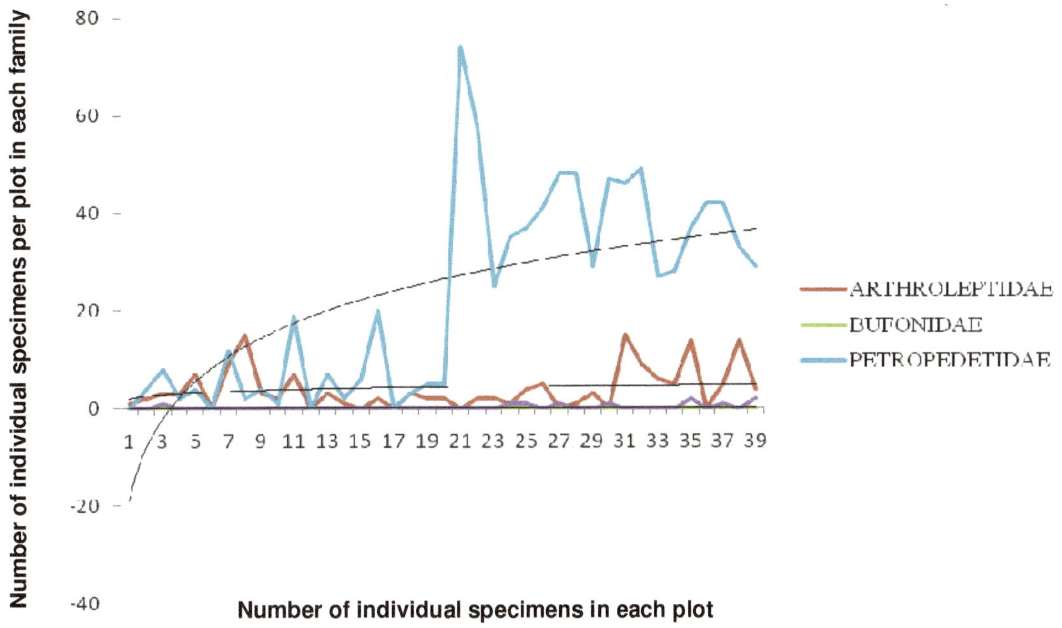

a

Figure 3. a) Encounter curve of amphibians in the reserve.

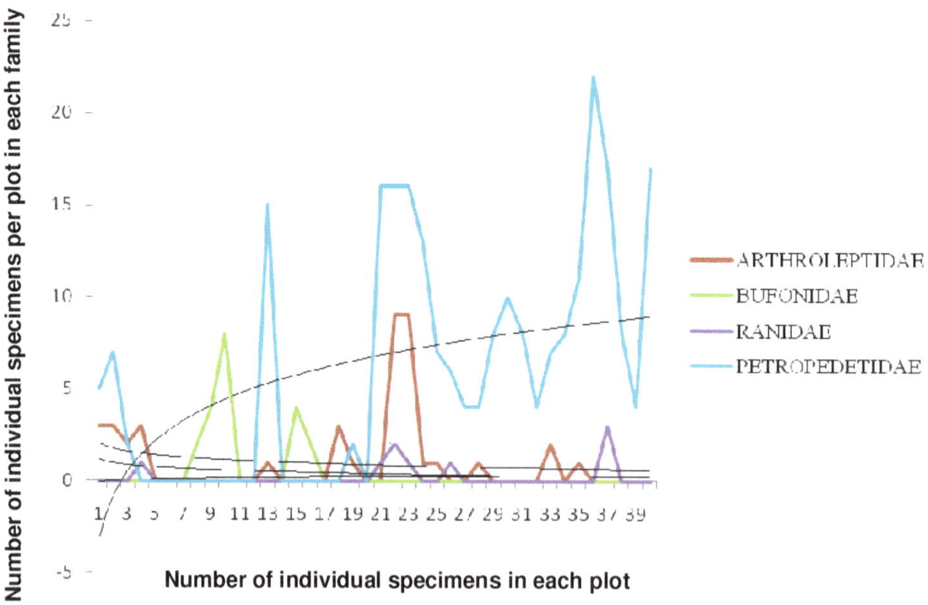

b

Figure 3. b) Encounter curve of amphibians in the off-reserve area.

Table 1. Diversity indices of the amphibians in the study area.

Parameter	In-reserve lowland	In-reserve upland	Off-reserve lowland	Off-reserve upland
Taxon (species)	7	8	6	11
Individuals	167	963	67	237
Dominance	0.2822	0.2427	0.2929	0.2717
Shannon-Weiner index (H)	1.426	1.617	1.37	1.709
Simpson's index (D)	0.7178	0.7573	0.7071	0.7283
Equitability/evenness	0.7326	0.7777	0.7646	0.7128

3.12, p=0.002. However, there was no significant difference in diversity between in- and off-reserve lowland habitats (t = 0.72, p=0.474) as well as between in- and off-reserve upland habitats (t = -0.91, p=0.366). These therefore support the hypothesis that diversity of species of amphibians could be altered by elevation. Figure 5 displays the species percentage composition at the four habitat types and only four species namely, *Arthroleptis poecilonotus, P. alleni, P. calcaratus, and P. plicatus* occurred in all the habitat types studied at different population levels. *Bufo regularis* and *Phrynobatrachus accraensis* occurred abundantly in only a single habitat type while the remaining six species occurred in two or three of the four habitat types at different population levels. Casual observation of the landscape of the study area pointed to a trend, allowing a few inconsistencies, cycling (Hanlin et al., 2000), as indicators of ecosystem health (Gibbons et al., 2000) and compose an important that the uplands were more of a closed canopy forest type. Interviews with farmers in the communities indicated frequent and indiscriminate use of agrochemicals, especially on vegetables and on cocoa farms off-reserve.

DISCUSSION

Wildlife inventories form the foundation for selecting priority sites and thereby help to identify priority species (Daily et al., 2003; Pineda and Halffter, 2004) and the relevance has become even more paramount because of the alarming rate of species extinction (Santos-Barrera et al., 2008). Amphibians hold vital positions in forest and aquatic food webs; they are also important for nutrient portion of the vertebrate biomass (Hutchens and DePerno, 2009). Rather unfortunately the lack of

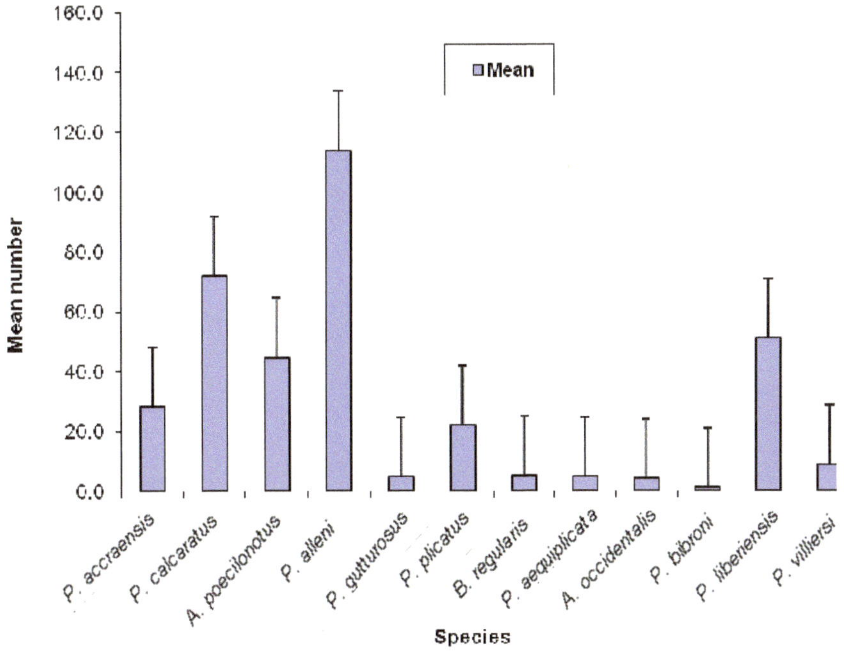

Figure 4. Mean abundance of amphibians in the study area.

systematic inventories precludes the solid evaluation of the possible trends documented.

Amphibians particularly have not attracted much attention in Africa and most information in that regard is conjectural. For example, it is estimated that there are about 221 species of amphibians and reptiles in Ghana (MES, 2002) and so this report may be the first attempt to make an inventory about amphibians for a specific location in a certain part of Ghana. The survey may represent the best natural diversity, abundance and distribution of vertebrates in the KNP since human predation for consumption and trade in amphibians is not known in Ghana. The results give a high conservation value to the amphibians of the KNP and its surrounding communities as reported for parts of Africa where similar studies have been undertaken (Behangana, 2004; Behangana and Arusi, 2004; Wasonga et al., 2006). It is noteworthy that the species present in the reserve are also represented in the surrounding communities.

Little or virtually nothing is known about the effects of pesticides or other agrotoxic chemicals, climate change or invasive species on amphibians in the study area; even though there is indiscriminate use of agrochemicals especially on vegetables and on cocoa farms off-reserve. However, the results reveal in some ways that such chemicals may have little or no effect on the amphibians considering that the distribution and diversity of these animals in and off reserve show no significant difference and that some other factors may account for the low diversity in both in- and off-reserve areas. While it is agreeable that forestry practices can negatively affect local habitat quality for selected herpetofaunal species, at least in the short-term (deMaynadier and Hunter, 1995), or for a long-term (Petranka et al., 1993) the results of this study did not show any significant difference in distribution, diversity or abundance of the amphibians between the intact habitats in the reserve and the cultivated and logged off-reserve habitats. Scott (1983) observed that even when extensive deforestation and developing land for agriculture have occurred in south western Costa Rica, a high diversity of historically known herpetofauna was still present. It is therefore not too clear why only 12 species were recorded for a total of 1434 individuals after an intensive and extensive search in the study area whereas about 76 species of amphibians are believed to be housed in the African rain forest (Poynton, 1999; IUCN and NatureServe, 2004). About 17 valid species of *Phrynobatrachus* alone are recognized in West African forests (Schiotz, 1964; Rödel and Ernst, 2002; Rödel et al., 2005) but only 7 species were encountered in this study area.

In a study carried out in the relatively undisturbed rain-forest tract of Kalakad-Mundanthurai Tiger Reserve, of the Western Ghats, India, a total of 509 individuals belonging to 20 species were captured (Karthikeyan et al., 2008). But also in the temperate region, Strojny and Hunter (2010) reported amphibian representation of only 12 species for 9,069 individuals over 1,540 ha in the Penobscot Experimental Forest (PEF) in Penobscot County, Maine, USA. Thus, expectation for more species than listed cannot be substantiated by mere logic, but habitat conditions in the study area favour high

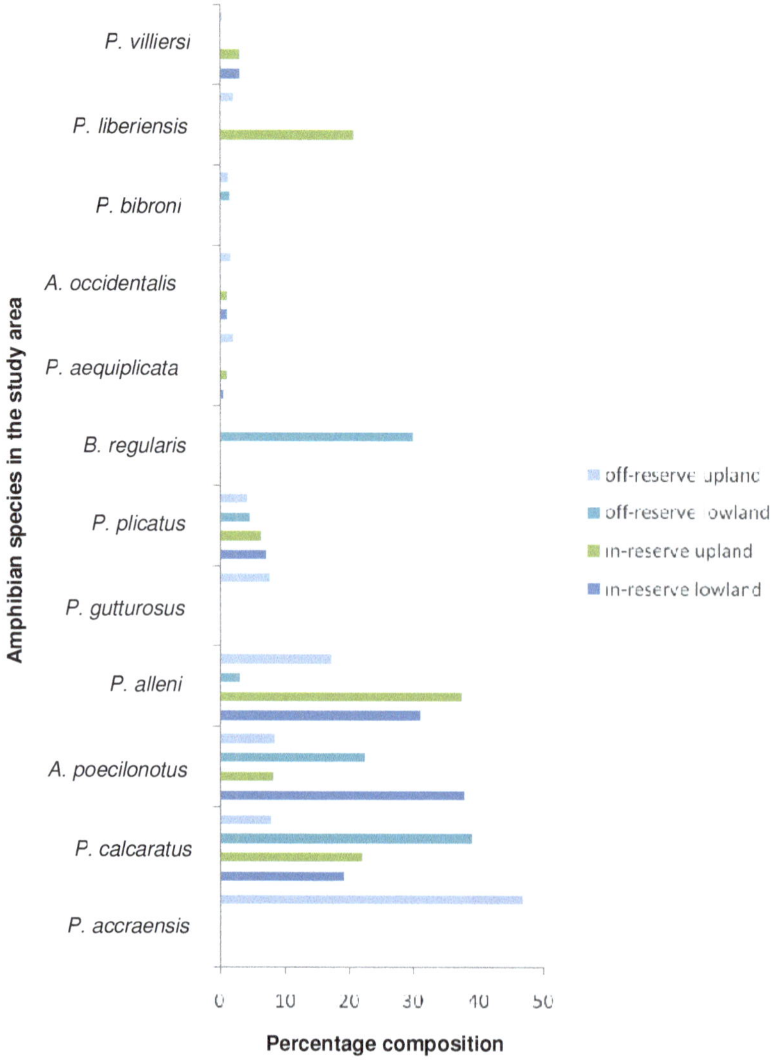

Figure 5. Percentage composition of various species of amphibians in the study area.

amphibian diversity. The study reveals that amphibian species population differs for each of the four habitat types of amphibian distribution in the study area and that uplands are more favourable habitats than lowlands for amphibians in KNP. Thus landscape is a major factor responsible for the distribution and diversity of amphibians in the study area. It is possible that the closed canopies of the uplands provided a more favourable condition for the anurans.

It is possible also that the animals have evolved to avoid predation by occupying upland habitats as their predators have learnt to associate them with lowland swampy areas over the years. The study did not take into account the effect of dispersal of juvenile amphibians on the distribution in the study area. Further studies are likely to show that the presence of juveniles contributed to the results in the lowland habitats as they dispersed to the uplands from the lowland habitats. Many authors such as Alford and Richards (1999) and Houlahan et al. (2000) have reported that amphibians are in decline worldwide, although the reasons for this decline have not been agreed upon (Collins and Storfer, 2003). The situation at the KNP and its surrounding communities predicts loss of species over time and this raises legitimate caution by reason of the fact that causes of species disappearances have not been studied and are unknown in this area.

Conclusion

We conclude that the number of species revealed by this study is low and lends credence to the fear that there is a global decline in amphibian diversity and express the urgent need to map out effective strategy for the conservation of amphibians worldwide. We report that any strategy adopted should include investigation into any possible cause of amphibian decline in the study area and also recognize the importance of uplands in- and off- reserves as a good habitat for amphibian survival.

REFERENCES

Alford RA, Richards SJ (1999). Global amphibian declines: a problem in applied Ecol. Annl. Rev. Ecol. Syst., 30: 133-165

Araújo MB, Thuiller W, Pearson RG (2006). Climate warming and the decline of amphibians and reptiles in Europe. J. Biogeogr., 33: 1712-1729.

Beebee TJC, Griffiths RA (2005). The amphibian decline crisis: a watershed in Conservation biology? Biol. Conserv., 125: 271-285.

Behangana M (2004). The diversity and status of amphibians and reptiles in the Kyoga Lake Basin. Afr. J. Ecol., 42 (Suppl. 1): 51-56.

Behangana M, Arusi J (2004). The distribution and diversity of amphibian fauna of Lake Nabugabo and surrounding areas. Afr. J. Ecol., 42 (Suppl.1): 6-13.

Blaustein AR, Kiesecher JM (2002). Complexity in conservation: lessons from the Global decline of amphibian populations. Ecol. Lett., 5: 597-608

Cain ML, Damman H, Lue RA, Yoon CK, Morel R (2007). Discover Biology. 3rd Edition. New York, W. W. Norton & Company

Collins JP, Storfer A (2003). Global amphibian declines: sorting the hypotheses. Diversity Distrib., 9: 89-98

Crump ML, Scott Jr NJ (1994). Visual Encounter Surveys, In Heyer WR Donnelly MA McDiarmid RW Hayek LC Foster MS (Eds) Measuring and Monitoring Biological Diversity, Standard Methods for Amphibians, Smithsonian Institution, Washington DC, USA. pp. 84-92.

Daily GC, Ceballos G, Pacheco J, Suzán G, Sánchez-Azofeifa A (2003). Countryside Biogeography of Neotropical Mammals: Conservation Opportunities in Agricultural landscape of Costa Rica. Conserv. Biol., 17: 1-11

Davidson C, Knapp RA (2007). Multiple stressors and amphibian declines: dual impacts of pesticides and fish on yellow-legged frogs. Ecol. Appl., 17: 587-597.

De Maynadier PG, Hunter ML (1995). The relationship between forest management and amphibian ecology: a review of the North American literature. Environ. Rev., 3: 230-261.

Durand JR, Skubich M (1982). Lagoons of Ivory Coast. Aquaculture, 27: 211-250

Forestry Commission (2007). Monitoring staff deployment, patrol effort, illegal activity and wildlife trends to facilitate adaptive and performance management in nine protected areas in Ghana. Second Evaluation, Accra, p. 55.

Gibbons JW, Scott DE Ryan TJ, Buhlmann KA, Tuberville TD, Metts BS, Greene JL,Mills T, Leiden Y, Poppy S, Winne CT (2000). The global decline of Reptiles, déjà vu amphibians. Bioscience, 50: 655-666.

Hall JB, Swaine MD (1976). Distribution and ecology of vascular plants in a tropical rainforest: forest vegetation in Ghana. Junk Publishers, The Hague, Netherlands.

Halliday TR (2008). Why amphibians are important. International Zoo Yearbook, 42: 7-14

Hanlin HG, Martin FD, Wike LD, Bennett SH (2000). Terrestrial activity, abundance and species richness of amphibians in managed forests in South Carolina. American Midland Naturalist, 143: 70-83

Harper KA, Macdonald SE, Burton PJ, Chen J, Brosofske KD, Saunders SC, Euskirchen ES, Roberts D, Jaiteh MS, Esseen PA (2005). Edge influence on forest structure and composition in fragmented landscapes. Conserv. Biol., 19: 768-782.

Houlahan JE, Findlay CS, Schmidt BR, Meyer AH, Kuzman SL (2000). Quantitative evidence for global amphibian population declines. Nature, 404:752–755.

Hutchens S, DePerno C (2009). Measuring species diversity to determine land-use effects on reptile and amphibian assemblages. Amphibia-Reptilia, 30: 81-88.

IUCN, Conservation International, & NatureServe 2004. Global amphibian assessment. Available from http:// www.globalamphibians.org. Retrieved 23-11-2010.

Karthikeyan V, Kumar A, Noon BR, Chellam R (2008). Density and Diversity of Forest Floor Anurans in the Rain Forests of Southern Western Ghats, India. Herpetologica, 64 (2): 207-215.

Kouadio KN, Diomande D, Ouattara A, Koné YJM, Gourène G (2008). Taxonomic diversity and structure of benthic macroinvertebrates in Aby Lagoon (Ivory Coast, West Aftrica) Pak. J. Biol. Sci., 11: 2224-2230.

MES (2002). Ministry of Environment and Science. National Biodiversity Strategy for Ghana. Accra, Government of Ghana.

Moore RD, Church DR (2008). Implementing the amphibian conservation action plan. International Zoo Yearbook, 42: 15-23.

Norris S (2007). Ghosts in the midst:coming to terms with amphibian extinctions. BioScience, 57: 311-316.

Petranka JW, Eldridge ME, Haley KE (1993). Effect of timber harvest on southern Apalachian salamanders. Conserv. Biol., 7: 363-370.

Pineda E, Halffter G (2004). Species diversity and habitat fragmentation: frogs in a tropical montane landscape in Mexico. Biol. Conserv., 117: 499-508.

Pounds JA, Carnavel AC, Puschendorf R, Haddad CFB, Masters KL (2007). Action on amphibian extinctions: going beyond the reductive. Science, (E-letter, 28/08/2007) http://www.science-mag.org/cgi/eletters/313/5783/48#9870.Retrieved20/09/2009.

Poynton JC (1999). Distribution of amphibians in sub-Saharan Africa, Madagascar, and Seychelles. In: Duellman W. (Ed) Patterns of distribution of amphibians, a global perspective. John Hopkins University Press, Baltimore pp. 483-539.

Ranvestel AW, Lips KR, Pringle CM, Whiles MR, Bixby RJ (2004). Neotropical tadpoles influence stream benthos: evidence for the ecological consequences of decline in amphibian populations. Freshwater Biology, 49: 274-285.

Reid GM, Zippel KC (2008). Can zoos and aquariums ensure the survival of amphibians in the 21st century? International Zoo Yearbook, 42: 1-6

Rödel MO (2007). Herpetofauna of West Africa savanna. Frankfurt am Main (Edition Chimaira)

Ródel MO, Ernst R (2002). A new *Phrynobatrachus* from the Upper Guinean rain forest, West Africa including a description of a new reproductive mode for the genus. J. Herpetol., 36: 561-571.

Rödel MO, Gil M, Agyei AC, Leache AD, Diaz RE, Fujita MK, Ernst R (2005). The amphibians of the forested parts of south western Ghana. Salamandra, 41: 107-127.

Santos-Barrera G, Pacheco J, Mendoza-Quijano F, Bolaños F, Cháves G, Daily GC, Errlich PR, Ceballos G (2008). Diversity, natural history and conservation of amphibians and reptiles from the San Vito region, southwestern Costa Rica. Rev. Biol. Trop., 56(2), 755-778

Schiotz A (1964). A preliminary list of amphibians collected in Ghana. Videnskabelige Meddelelser fra Dansk Naturhistorisk Forening, 127: 1-17.

Scott NJ Jr. (1983). The abundance and diversity of the herpetofauna of tropical forest litter. Biotropica, 8: 41-58.

Strojny CA, Hunter ML Jr. (2010). Relative abundance of amphibians in forest canopy gaps of natural origin versus. Timber harvest origin. Ani. Biodivers. Conserv., 33(1): 1–13.

Stuart SN, Chanson JS, Cox A, Young BE, Rodrigues ASL, Fischman L, Waller RW (2004). Status and trends of amphibian declines and extinctions worldwide. Science, 306: 1783-1786.

Wasonga DV, Bekele A, Lötters S, Balakrishnan M (2006). Amphibian abundance and diversity in Meru National Park, Kenya. Afri. J. Ecol., 45: 55-61.

WZACS (2005) Building a future for wildlife: the world zoo and aquarium Conservation strategy. Bern; WAZA. http://www.waza.org/conservation/wzacs.php Retrieved 30 -10-2010.

Diversity of lichens in Kollihills of Tamil Nadu, India

Shyam Kumar R.[1,4*], Thajuddin N.[2] and Upreti D. K.[3]

[1]Department of Biotechnology, Kamaraj College of Engineering and Technology, Virudhunagar, 626 001, India.
[2]Department of Microbiology, Bharathidasan University, Trichirapalli, 620 024, India.
[3]Lichenology Laboratory, National Botanical Research Institute, Lucknow, 226 001, India.
[4]Department of Botany and Microbiology, College of Science, King Saud University, Riyadh, KSA, Saudi Arabia.

An enumeration of 48 species belonging to 23 genera and 12 families of lichens from Kollihills, Namakkal District of Tamil Nadu is provided. Species of the lichen genera *Heterodermia*, *Parmotrema* and *Pertusaria* dominates the area.

Key words: Lichens, Kollihills, Easternghats, India.

INTRODUCTION

The symbiotic association between an alga and a fungus has resulted in a new life form called lichen. It is present in a wide range of habitats throughout the world and dominates terrestrial ecosystems (about 8%). 20,000 species of lichens have been reported globally. The Indian subcontinent has 2,450 species of lichens, of which India alone has about 2040 species (Awasthi, 2000). The Eastern Ghats of India is a broken chain of hills that extends from Orissa to Tamil Nadu and because of such topography; the hilly terrain and the surrounding planes are densely populated. Accessibility to the forest is rather easy. The forests in the Eastern Ghats are the most affected, compared to the Western Ghats and Himalayas, as they are experiencing heavy pressure on all sides (Jha et al., 2000) (human activities, exploitation of forest resources etc.). Though several studies have been carried out in Western Ghats and The Himalayas with respect to forest conservation and change detection, no reports are available about Eastern Ghats on the above aspects. Although Eastern Ghats are in a highly degraded stage (Jayakumar et al., 2002), it is absolutely essential to evaluate and reevaluate the biodiversity and species richness of all higher and lower plants and animals. The diversity of higher plant flora and fauna has been studied in great detail since European colonization in India. Although, there has been substantial research in terms of medicinal and ecosystem values, lower plants such as lichens, bryophytes, pteredophytes, algae and fungi have not received the adequate attention even in Western and Eastern Ghats. Thus to get an idea about the diversity of lichens in the Eastern Ghats of Tamil Nadu, a random survey of lichens in this area has been started in the year 2000 to 2001, and the enumeration of lichens from the Kollihills is presented in the present communication.

MATERIALS AND METHODS

Collection area

Kollihills (Kollimalai) lies between 11°10'5" -11°30' 00" N latitude and 78°15' 00" E longitude. It is situated in the Namakkal district of TamilNadu above the river Cauvery, covering an area of about 503 km^2. Physiographically, it is a hilly region with altitude ranging from 180 m at the foothill to 1415 m at the plateau. The slope of this region varies from gentle to very steep. Geologically, the study area occupied by the hill is highly undulating, cut by a network of streams and most of them are semi–perennial and seasonal flowing in all directions, but mostly in the eastern and southeastern directions and ultimately draining into Ayyar river. The forest types located in Kollihills are:

(1) Dry evergreen, (2) Semi evergreen (3) Dry deciduous

*Corresponding author. E-mail: kingshyam2003@yahoo.co.in

Table 1. The lichen flora of Kollihills.

S/No.	Name of lichen	Family	Collection area								
			1	2	3	4	5	6	7	8	9
1	*Canoparmelia ecaperata* (Müll. Arg.) Elix and Hale.	Parmeliaceae			+	+		+	+	+	
2	*Cetraria* sp.	Parmeliaceae	+		+	+					
3	*Cladonia cartilaginea* Müll. Arg.	Cladoniaceae				+					+
4	*Collema auriforme*(With.) Coppins and J.R. Laundon.	Collemataceae				+			+	+	
5	*Collema rugosum* Kremp.	Collemataceae					+			+	
6	*Cryptothecia dispersa* Makhija and Patw.	Arthroniaceae	+			+					
7	*Diplotoma proximatum* (Magnusson) Szat. ex D.D. Awasthi	Physciaceae			+						
8	*Dirinaria confluens* (Fr.) D.D. Awasthi	Physciaceae			+			+			
9	*Dirinaria consimilis* (Stirt.) D.D. Awasthi.	Physciaceae			+				+		
10	*Heterodermia angustiloba* (Müll. Arg.) D.D. Awasthi	Physciaceae			+		+				
11	*Heterodermia cosmosa* (Eschw.) Follm and Redon	Physciaceae			+		+				
12	*Heterodermia dendritica* (Pers.) Poelt	Physciaceae		+						+	
13	*Heterodermia diademata* (Taylor) D.D. Awasthi	Physciaceae			+				+		
14	*Heterodermia dissecta var koyana (* Kurok) J.C. Wei	Physciaceae					+				
15	*Heterodermia hypocaesia* (Yasuda) D.D. Awasthi	Physciaceae		+				+			
16	*Heterodermia incana* (Stirt.) D.D. Awasthi	Physciaceae				+			+		
17	*Heterodermia isidiophora* (Vainio) D.D. Awasthi.	Physciaceae				+				+	
18	*Heterodermia* boryi *(*Fée) Kr.P. Singh and S.R. Singh	Physciaceae				+		+		+	+
19	*Heterodermia speciosa* ((Wulfen) Trevis.	Physciaceae						+		+	
20	*Lecanora* sp.	Lecanoraceae	+	+							
21	*Lepraria* sp.	Imperfect fungi	+	+		+		+			
22	*Leptogium javanicum* Mont.	Collemataceae				+			+	+	
23	*Myriotrema microporum* (Mont.) Hale	Thelotremataceae		+							
24	*Ocellularia diacida* Hale	Thelotremataceae		+			+				
25	*Parmotrema austrosinense* (Zahlbr.) Hale	Parmeliaceae		+					+		
26	*Parmotrema crinitum* (Ach.) Choisy.	Parmeliaceae						+			+
27	*Parmotrema nilgherrense* (Nyl.) Hale.	Parmeliaceae			+		+		+		+
28	*Parmotrema praesorediosum* (Nyl.) Hale	Parmeliaceae						+	+	+	
29	*Parmotrema tinctorum* (Nyl.) Hale	Parmeliaceae		+		+				+	
30	*Pertusaria granulata* (Ach.) Müll. Arg.	Pertusariaceae			+						
31	*Pertusaria* sp. (Ach.)	Pertusariaceae			+						+
32	*Pertusaria leucosorodes* Nyl.	Pertusariaceae	+	+							
33	*Pertusaria melastomella* Nyl.	Pertusariaceae		+							
34	*Pertusaria pseudococcodes* Müll. Arg.	Pertusariaceae				+		+			
35	*Pertusaria* sp. (Taylor)	Pertusariaceae					+		+		

Table 1. Continued.

			1	2	3	4	5	6	7	8	9
36	*Phaeophyscia* sp.	Physciaceae	+	+			+		+		
37	*Phyllopsora corallina* (Eschw.) Müll. Arg	Biatoraceae		+							
38	*Physcia integrata* Nyl.	Physciaceae			+						
39	*Physcia tribacoides* Nyl.	Physciaceae					+				+
40	*Phyllopsora parvifolia* (Pers.) Müll. Arg.	Biatoraceae				+					
41	*Pseudocephallaria arggraceae* (Bory.) Vain..	Lobariaceae								+	
42	*Ramalina pacifica* Asahina	Ramalinaceae								+	+
43	*Rimelia clavulifera* (Räsänen) Kurok.	Parmeliaceae								+	+
44	*Rimelia reticulata* (Taylor) Hale and Fletcher	Parmeliaceae								+	+
45	*Teloschistes flavicans* (Sw.) Norman	Teloschistaceae			+		+	+		+	+
46	*Tephormela atra* (Huds.) Hafellner	Lecanoraceae		+							
47	*Usnea stigmatoides* G Awasthi	Parmeliaceae								+	+
48	*Usnea undulata* Strit.	Parmeliaceae								+	+
Total			6	13	13	13	10	10	11	17	12

1. Kottar – 600 m; 2. Kovilur – 1000 m; 3. Arapaleeswarar temple area – 1148 m; 4. Kulivalavu area – 1300 m; 5. Vasalur area – 1300 m; 6. Peramparappapatti – 1316 m; 7. Chemmedu – 1325 m; 8. Cholakkadu – 1400 m; 9. Sengarai – 1415 m+ = indicates the presence.

(4) Tropical riverine (5) Southern thorn (6) Euphorbia scrub and (7) Man made forests.

Samples were collected randomly from different altitude across the hills: Kottar – 600 m, Kovilur – 1000 m, Arapaleeswarar temple area – 1148 m, Kulivalavu area – 1300 m, Vasalur area – 1300 m, Peramparappapatti – 1316 m, Chemmedu – 1325 m, Cholakkadu – 1400 m, Sengarai – 1415 m. As showed in Table 1.

The present study is based on intensive collection of lichens made from January 2000 to February 2001 at various altitudes and across the Kollihills. Lichens were collected along with substratum using sharp knife. Much care was taken to collect the specimen without any damage to the thallus margin. Fruticose lichens were collected with their holdfasts intact on the substratum. The specimens were cleaned by careful removal of debris, sun dried and deposited in the herbaria of Department of Microbiology, Bharathidasan University, Trichirapalli, Tamil Nadu.

The morphological features were studied using trinocular zoom dissection microscope (Meiji optics, Japan). Anatomy of the samples was studied with the help of a compound microscope. The sections were obtained with a microtome, at a thickness up to 15 µm. The K, C, KC P tests and TLC analysis (Culberson and Kristinsson, 1970; Culberson, 1972; Santesson, 1973; Awasthi, 1988, 1991), which are important for identification of chemical substance in lichen, were made on thallus cortex and medulla.

RESULTS

The Kollihills exhibit good growth of almost all the growth forms of lichens. Twenty seven species of foliose lichens dominates over the three fruticose and eighteen crustose and squamulose lichens. Lichen family Physciaceae dominates the area with 16 species followed by the members of Parmeliaceae and Pertusariacea. The cyanophycean lichens exhibit their poor growth as represented by three species only. Among the different genera *Heterodermia* with 10 species dominates the area.

DISCUSSION AND CONCLUSION

Most of the localities situated above an altitude of 1000 m exhibit rich diversity of lichens. The Cholakkadu area at an altitude of 1400 m exhibit occurrence of 17 species has the maximum diversity of lichens followed by Kovilur, Arapaleswarar temple area, Kulivalavu, Vasalur, Peramparappapatti, Chemmedu, and Sengarai with 10 to 13 lichen species. The only locality at an altitude of 600 m, (Kottar) has the poor growth of lichens represented by six species of crustose lichen genera *Pertusaria* and *Cryptothecia.* The probable reason for poor diversity of lichens in the Kottar area may be due to a number of factors influencing the abundance and richness of lichens, such as temperature, moisture, altitude, forest type and Human activities. Among the various places surveyed, the Cholakkadu area has thick moist forest vegetation which provide suitable habitat for colonization of a number of lichen groups on different substrates. The lichen flora of Kollihills exhibit similarity with the lichen flora of the nearby areas of Tamil Nadu, such as Megamali Wild life Sanctuary, Kambam district (Nayaka et al., 2001) and Shervaroy hills (Hariharan et al., 2004) the members of the lichen family Physciaceae and Parmeliaceae dominates the Kollihills too.

ACKNOWLEDGEMENTS

The authors are thankful to the Director, National Botanical Research Institute, Lucknow, for providing laboratory facilities to work for identification of lichen samples. One of the authors (R.Shyam Kumar) is thankful to the Management, Thanthai Hans Roever College, Perambalur, Tamil Nadu, India for providing facilities to work in the department of Microbiology.

REFERENCES

Awasthi DD (1991). A key to the microlichens of India, Nepal and Sri Lanka. Biblioth. Lichenol., 40: 1–337.

Awasthi DD (1988). A key to the macrolichens of India and Nepal. J. Hattori Bot. Lab., 65: 207–302.

Awasthi, D.D. (2000). A Hand Book of Lichens. Shiva Offset Press, Lucknow, India, pp. 87-92.

Culberson CF, Kristinsson H (1970). A standardized method for the identification of lichen products. J. Chromatogr., 46: 85-93.

Culberson CF (1972). Improved conditions and new data for the identification of lichen products by a standardized thin – layer chromatographic method. J. Chromatogr., 72: 113-125.

Hariharan GN, Krishnamurthy KV, Upreti DK (2003). Lichens of Shervaroy hills of Eastern Ghats, India. Phytotaxonomy, 3: 1–23.

Jayakumar S, Arokiasamy ID, John BS (2002). Conserving forest in the Eastern Ghats through remote sensing and GIS- A case study in Kollihills. Curr. Sci., 82: 1259–1267.

Jha CS, Dutt CBS, Bawa KS (2000). Deforestation and land use changes in Western Ghats, India Cur. Sci., 79(2): 231–237.

Nayaka S, Upreti DK, Divakar PK (2001). Distribution and diversity of lichens in Meghamalai Wildlife Sanctury, Kambam district, Tamil Nadu, India. Biol. Memoirs., 27(2): 51–58.

Santesson J (1973). Identification and isolation of lichen substances. In: Ahmadjian, V, Hale ME (eds), The lichens. Academic press. London, 633-652.

Vegetation dynamics of an old-growth lowland tropical rainforest in North-east India: Species composition and stand heterogeneity

P. Deb[1,3] and R. C. Sundriyal[1,2]*

[1]G.B. Pant Institute of Himalayan Environment and Development, Kosi-Katarmal, Almora, Uttaranchal, 263643 India.
[2]Herbal Research and Development Institute, Gopeshwar (Chamoli) - 246401, Uttarakhand, India.
[3]Centre for Biodiversity and Natural Resources Conservation, Assam University, Silchar, India.

The North East India forms an important hotspot of biodiversity by virtue of its unique biogeographical situation and healthier growing conditions. The region has an area of 13302.80 km^2 under the protected area network (PAN), which is not yet systematically inventorised. In this paper, we investigated Namdapha National Park that comprises 14% of the total PAN of the region and forms the largest remnants of the tropical dipterocarp forest. We examined land use pattern of Namdapha National Park and studied the vegetation dynamics of lowland tropical rainforest communities in the buffer zone area. The study revealed a total of eight cover classes and four vegetation categories with three forest types, namely, Stand I: Altingia-mixed species, Stand II: Shorea-Dipterocarp (both are tropical wet evergreen forests), and Stand III: Albizia (riverine tropical semi-evergreen forest) types. These three stands were systematically studied for general physiognomy, vegetation stratification, species area curve, plant structure and stand heterogeneity (species richness, diversity and dominance), tree girth class structure, and dominance-diversity relationship. These three stands collectively comprised 198 species (including 135 tree species), thus exhibited 34% of total floral and 50% of total tree diversity of the Namdapha National park, which is significant. The tropical wet evergreen forests showed five distinct vegetation strata. The species area curve depicted 123 and 19 tree species ha^{-1} in tropical wet evergreen and riverine semi-evergreen forests, respectively. Higher frequencies of lauraceae, diperocarpaceae, Euphorbiaces, Fagaceae, Leguminosae, Meliaceae and Magnoliaceae species clearly show that the forest had all attributes of tropical rainforests. The Stands I, II and III showed densities of 418, 390, and 245 trees ha^{-1}, and total basal area of 45.47, 49.68 and 18.33 m^2 ha^{-1}, respectively, which is well comparable with other similar forests. Species dominance and rarity analysis revealed that 63, 35, and 45% of the species were rare at local level in Stands I, II and III respectively, however such species contributed highly to forest stand diversity. The species number and density declined with increasing tree-girth sizes that denotes an evolving population with old growth trees. Comparisons of the floristic data with other similar stands have been discussed.

Key words: Rainforest, tropical evergreen and semi-evergreen forest, protected area network, tree diversity, species area curve, dominance-diversity curve, forest management.

INTRODUCTION

Tropical forests occupy 7% of the earth's area, about half of the world's forest cover and 65% of global biodiversity (FRA, 2002). These forests, however, are extremely susceptible to biotic pressure (Zent and Zent, 2004). Therefore, nature reserves and protected areas have high implications for conservation of genetic resources of the world (McNeely and Nees, 1996). At present protected areas (PAs) form the only repositories of biological wealth and refugia for native plants, animals and microorganisms that act as an outdoor laboratory for harboring rich biodiversity of a given region (Myers et al.,

*Corresponding author. E-mail: pannadeb@rediffmail.com or rcsundriyalrc@yahoo.com.

2000). In Asia as well as in other tropical regions the primary forests have been the foremost victim of anthropogenic pressure to the extent that most areas are either replaced by secondary vegetation or denuded completely (Menon et al., 2001). Such loss is witnessed at a time when knowledge on their structure and dynamics still considered inadequate (Romero-Duque et al., 2007) and may alter the habitats, and spatial and temporal variation in species richness, composition and productivity (Zent and Zent, 2004).

In India the tropical forests occupy 84% of the total forest cover (637293 km^2), which is 19.39% of the total geographic area. The tropical wet evergreen forest extends up to 15010 km^2, thus covers 10% of the tropical forest cover of the country (IIRS, 2002). Such forests face serious threats because of widespread landuse changes, leading to hamper species survival (Menon et al., 2001). This emphasizes the need to conserve biodiversity rich sites by bringing more area under conservation network, at the same time update our knowledge on species distribution, floristic composition, ecosystem diversity and plant structure protected areas (Rodgers et al., 2000). To protect biodiversity rich areas of the country, a protected area network (PAN) programme initiated by bringing large number of habitats and ecosystems under PAN in the Himalaya as well as other parts of India (Kothari et al., 1989; Rodgers et al., 2000). Unfortunately, most of the protected areas are not well investigated with relation to species structure and stand heterogeneity, though such information forms the basis for conservation and management along with further scientific investigations on such stands.

The northeast India is a global hotspot of biodiversity because of its geographical position, climatic conditions and altitudinal variations (Myers et al., 2000). The region has 49 protected areas (11 national parks and 38 wildlife sanctuaries) covering a land area of 13936.80 km^2, which comprises 5.46% of the total area of the northeast (Rodgers et al., 2000). Namdapha National Park, Arunachal Pradesh comprised the largest area among all (14.24% of total PAN of the northeast) with extremely diverse vegetation and habitat types (Chauhan et al., 1996; Ghosh, 1987; Deb and Sundriyal 2008). Although there are a few studies on biodiversity characterization of selected vegetation types in the northeast India (Rao et al., 1997; Khan et al., 1986, 1987; Barik et al., 1996). However such information is highly limiting for protected areas, which otherwise has significant implications for forest management and biodiversity conservation of the region (Proctor et al., 1998; Nath et al., 2005). A thorough understanding of dynamics of the forest stand with relation to species composition is essential to maintain forest structure and productivity, and conserve plant diversity (Oliver and Larsons, 1990; Philips, 1994; Bhat et al., 2000). In this study, we sampled vegetation of Namdapha National Park and specifically addressed the following questions: Which are the major landuses/covers in the buffer zone area? Which are the tree and shrub species distribution pattern along with the floral diversity, tree structure, species diversity and dominance, and how the plant-family contributes to the stand diversity in different forests? Finally, we discussed the conservation implications of our findings to provide a strong basis to frame more targeted management strategy for the area.

Study area

Namdapha National Park (27°23'30"- 27°39'40" N to 96°15'2" - 96°58'33" E longitude) located in the Changlang district of Arunachal Pradesh state, northeast India (Deb and Sundriyal, 2008). It comprised an area of 1985.25 km^2 with 177.43 km^2 in buffer zone and 1807.82 km^2 in the core zone (Figure 1). The park shares southern and eastern boundaries with Myanmar, and the northern boundary with the Kamlang wildlife sanctuary of Lohit district of the state. The park area falls under the Eastern Himalaya (2D) biogeographic province, which covers the Paleartic and the Indo-Malayan (Oriental) realms (Rodgers et al., 2000). It is wedged between the Dapha Bum range of Mishmee Hills, an outspur at the tail end of North Eastern Himalaya, and the Patkai range with an elevational variance of 200 to 4571 m above sea level. General topography of the park is rugged with steep hills and narrow valleys intersected by several streams. Geologically the park is of recent origin and owes its formation to the upheaval of the Himalaya in Pleiocene period of the tertiary age (Chauhan et al., 1996). The area exhibits tropical climate, it receives an annual rainfall of 2500 to 3000 mm, and the temperature and relative humidity remains high throughout the season (Deb and Sundriyal, 2008). The present study was confined to the buffer zone area, which mainly comprised the lowland tropical wet evergreen and semi-evergreen forest types (Kaul and Haridasan, 1987). The core zone of the park is mostly inaccessible except for the river banks and mainly visited during the biennial Tiger Census. The soil of the forests was sandy-loam and acidic in nature (Table 1). The nitrogen, phosphorus and carbon content of the soil recorded higher under lowland tropical evergreen forests sites than and semi-evergreen (riverine) forest (Table 1). The former site also maintained a good humus and litter depth though it was <2 cm in riverine forest (Deb, 2006).

METHODS

Land use-cover of buffer zone

The buffer zone area was delineated using the maps prepared by the office of the Field Director, Namdapha National Park, Department of Environment and Forests, Government of Arunachal Pradesh, which was traced on the Survey of India topo sheets. To assess landuse-cover of buffer zone area we used satellite images

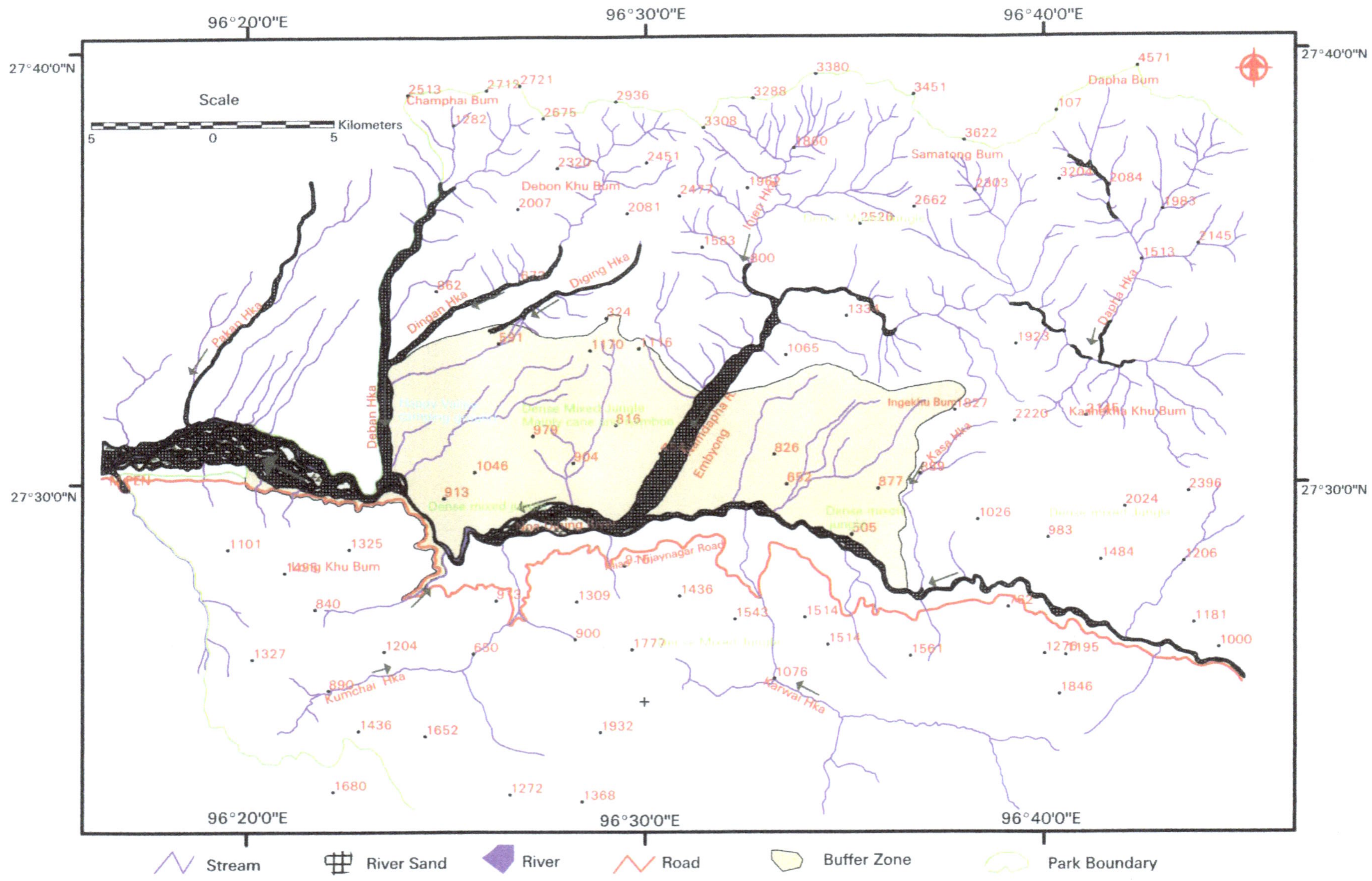

Figure 1. Digitized map of the buffer zone of Namdapha National Park in the northeast India.

Table 1. Soil characters of the study sites comprising *Altingia*-mixed species, *Shorea-Dipterocarp*, and *Albizia* stands in Namdapha National Park.

Parameters	*Altingia*- mixed species stand	*Shorea-Dipterocarp* stand	*Albizia* stand
Soil type	**Sandy loam**	**Sandy loam**	**Loamy sand**
texture	Clay: 10.5 Silt: 21 Sand: 68.5	Clay: 9.5 Silt: 19 Sand: 71.5	Clay: 3 Silt: 26.5 Sand: 70.5
pH	5.35±0.23	4.5±0.45	6.29±0.46
Moisture (%)	23.22±2.56	22.49±1.87	30.39±13.57
N (%)	0.37±0.09	0.215±0.014	0.20±0.0034
P (%)	0.23±0.021	0.21±0.014	0.19±0.02
Organic carbon (%)	2.69±1	1.98±0.14	1.11±0.11
Na (ppm)	0.016±0.001	0.016±0.001	0.012±0.001
K (ppm)	0.07±0.003	0.065±0.002	0.072±0.0032

comprising path 115 and row 52 of IRS-1D, LISS III for February 2003 procured from the National Remote Sensing Agency (NRSA). The area was verified through Survey of India topo sheets (1:50,000 scale) No. 92 A/6, 7, 10 and 11. Using the software ERDAS Imagine (8.6 Version), the LISS-III data were geo-referenced for which feature controls obtained from the Survey of India topo sheets. The forest types categorized as per available classifications of Champion and Seth (1968), and Kaul and Haridasan (1987).

Vegetation stratification and structural analysis of the stands

For detailed structural analysis of the buffer zone, three major canopy compositions (namely, Altingia-mixed species (Stand I), Shorea-Dipterocarp (Stand II), and Albizia forest (Stand III) were considered. The Altingia-mixed species and Shorea-Dipterocarp stands fall under lowland tropical evergreen rainforest category while Albizia stands is composed of riverine semi-evergreen forest type. Five strata, that is, top canopy or emergent (A- stratum), mid canopy (B-stratum), lower canopy (C-stratum), shrub layer (D-stratum) and herb layer (E-stratum) were considered (Magurran, 2004). The structural analysis was done for three stands using the transect method that were placed randomly to cover stand diversity and maximum possible species (Muller-Dombois and Ellenberg, 1974). Transects (50×20 m size) were laid 50 m inside the forest trail to avoid any outside interference. A total of 12, 10 and 18 transects were identified for Stands I, II and III, respectively. Each transect was further sub-divided into 10×10 m quadrat size, and ten 100 m^2 quadrats were studied randomly from each transect. Care was taken to cover slopes, altitude, and most possible species in view of rapid changes in vegetation (Kent and Coker, 1994). Altogether 120, 100 and 180 quadrats were studied for the three stands, respectively. The total number of quadrats sampled was based on number of species and tree individuals repeated, (Sundriyal and Sharma, 1996; Magurran, 2004).

All tree individuals falling in each quadrat were marked with a permanent aluminum tag. Each individual tree was measured for its circumference at breast height (CBH), that is, at 1.3 m above ground level. Individuals with buttresses or other deformities at breast height were measured for its girth just above buttresses. Samples of all plant individuals were collected and later identified to family, genera and species, level using existing flora (Haridasan and Rao, 1985; Chauhan et al., 1996) and with the help of the experts of State Forest Research Institute (SFRI) and Botanical Survey of India (BSI), Itanagar, Arunachal Pradesh. The data gathered for each stand from all transect were subsequently pooled together for further analysis. All woody species ≥31.5 cm CBH (circumference at breast height) that is, ≥10 cm DBH (diameter at breast height) were considered as trees, and <31.5 cm CBH (<10 cm DBH) were considered as regenerating individuals (Sundriyal and Sharma, 1996; Uma, 2001). Regenerating individuals were further categorized into sapling (DBH <10 cm and height >20 cm) and seedling (individuals <20 cm in height). The vegetation data subsequently analyzed for each species for its frequency, density, and total basal area (Kent and Coker, 1994; Maguran, 2004). To provide comparative data among species, we analyzed the relative density, relative frequency and relative cover for each species in a stand (Philips, 1959). To assess dominance of species in stand Importance Value Index (IVI) was computed by summing up the relative density, relative frequency and relative dominance for each tree species for all individuals ≥31.5 cm CBH (Philips, 1959). Species represented with an IVI <1 were considered rare (Maguran, 2004).

All the identified species were grouped into genera and family for presenting floristic composition for each stand, thus plant families contributing maximum number of species were assessed for each stand (Deb, 2006).

Tree diameter class structure

The diameter class structure of a forest is considered an important criterion for assessing maturity of the stand. The DBH of all trees was measured for interpreting population structure (Sundriyal et al., 1994; Sundriyal and Sharma, 1996). Based on tree diameter at breast height, individual tree species were separated into 8 classes (10-20, 21-30, upto >150 cm). To reflect the status of individual species in different forest stands, the DBH classes were further pooled into small (DBH 10 to 30 cm), medium (DBH 31 to 60 cm), adult (DBH 61to 90 cm), and mature (DBH >91 cm) trees as such data would represent evolving state of a given stand.

Shrub and herbaceous flora

The density of shrub and herbs was estimated for Altingia-mixed

Table 2. Landuse/ cover statistics for buffer zone area of Namdapha National Park, Arunachal Pradesh.

Landuse categories	Area (km^2)	Percentage of total area
*!South Bank tropical wet evergreen (*Dipterocarpus*) Forests (dense)	141.642	79.832
*!South Bank tropical wet evergreen (*Dipterocarpus*) forests (open)	13.100	7.383
**Riverine semi-evergreen forests	0.640	0.360
Grasslands	13.110	7.389
Water body	1.400	0.789
Built-up land	0.013	0.007
Landslide area	0.490	0.276
River sand	7.030	3.962
Total land area	177.425	100

Forest classification (Champion and Seth 1968, Kaul and Haridasan 1987).

species and Albizia stands only. We used all 10×10 m size quadrat that were placed for tree structure analysis. Each 10×10 m^2 quadrate was sub-divided into four 5×5 m quadrats, of which one quadrat was randomly selected for assessing shrub and sapling species composition. A total of 120 and 180 quadrats were analyzed for two stands, respectively. The herbaceous flora was also assessed from same quadrate by placing at least two 2×2 m size quadrats. A total of 240 and 360 quadrats of 2×2 m^2 were analyzed for assessing density of herbaceous species in two stands, respectively. The herbarium of all shrub and herbaceous species were prepared for identification.

Stand heterogeneity

The heterogeneity among different stands was assessed by estimating species richness, diversity, concentration of dominance, and evenness in each stand by calculating Menhinick index, Shannon-Weiner information index, Simpson's concentration index, and Pielou's measure of evenness, respectively (Maguran, 2004).
The index of species richness was calculated following Menhinick (1964) as:

$$D=S/\sqrt{n}$$

Where S is the number of species and n is the number of individuals.
The diversity index was calculated using the following formula (Shannon and Wiener 1963):

$$H = -\sum_{i=1}^{s} pi \ln pi$$

Where, pi is the proportion of the ith species and the number of individuals of all the species (ni/N), and ln is the log $base_n$
The concentration of dominance (CD) was measured using the following formula (Maguran 2004):

$$CD = \sum pi^2$$

Where, pi is same as in Shannon-Weiner information function.

Dominance-diversity curve

The dominance-diversity curve was plotted to assess species niche separation and partitioning of community resources (Whittaker, 1975; Kent and Coker, 1994). The Dominance-diversity curves for each stand were prepared based on the IVI for trees and density for the saplings and seedlings. Since the shape and slope of dominance-diversity curve vary, it was interpreted with the help of at least four different distribution functions, that is, the geometric series and niche pre-emption hypothesis, the log series, the log normal distribution, and broken stick series (Kent and Coker, 1994).

Similarity indices

Similarity was measured amongst three stands to assess the degree to which the species composition (tree, shrubs and herbs) of different forests matches together. The similarity index was calculated by the formula given by Sorensen (1948):

$$=S_S = 2C/ 100(A+B)$$

Where C is the number of common species in two stands being compared; and A and B denote the total number of species at stand I and stand II, respectively. The coefficient is multiplied by 100 to give a percentage similarity value.

RESULTS

Landuse and covers of buffer zone

Analysis of remote sensing image data for buffer zone area of Namdapha National Park delineated eight cover classes of which four were vegetation types and remaining others non-vegetation categories (Table 2). Of the three major forest categories the tropical wet evergreen (*Dipterocarpus*) forest in dense (79.8%) and open (7.4%) categories covered the major part of the buffer zone area (Table 2). A small area of buffer zone (0.36%) was occupied by the riverine semi-evergreen forest type. The grasslands also comprised sufficient area (7.4%), while water body and sand and landslide area were other main landcover categories (Table 2). A small part was also categorized as built-up land that comprised houses and huts (temporary structures) created by the Forest Department inside the Park.

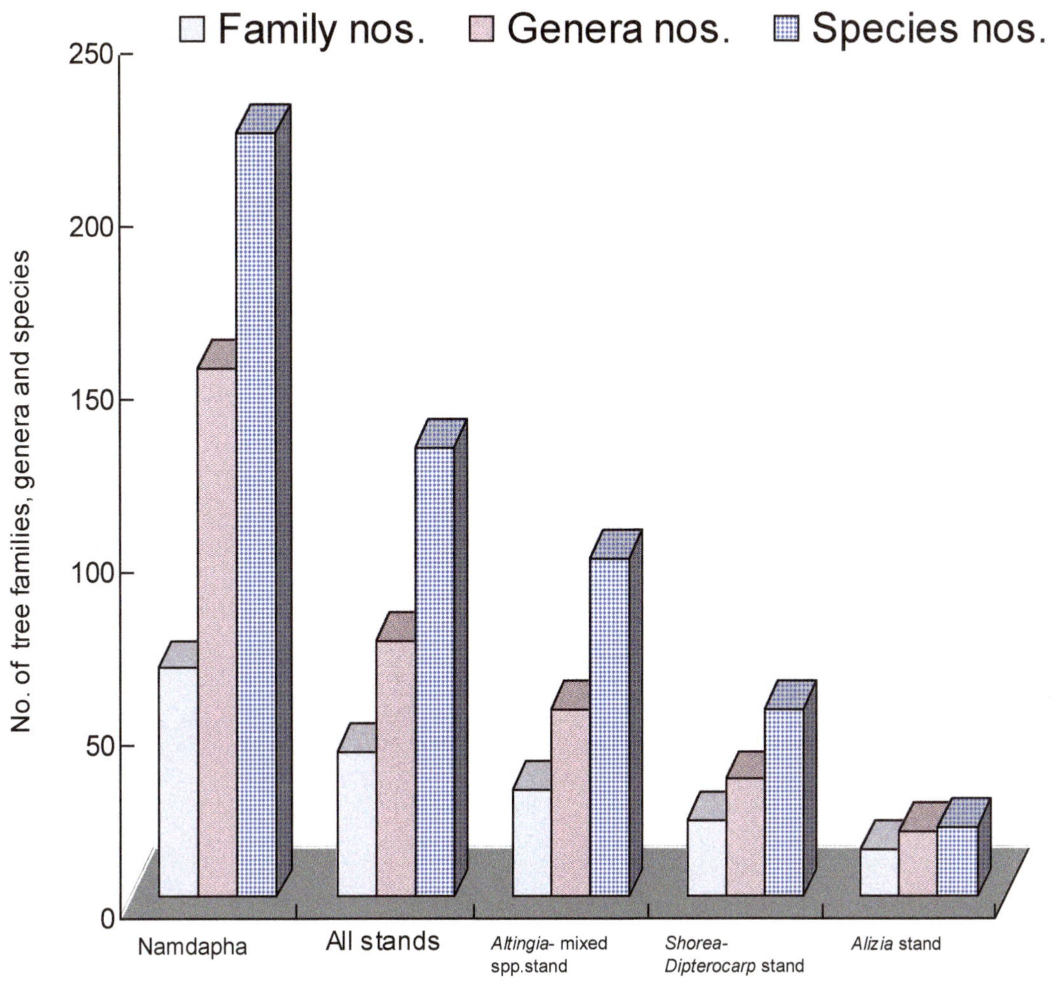

Figure 2. Status of tree species, genera and families at Namdapha National Park (all classes) and in three studied forest (for trees >31.4 cm CBH) stands in the buffer zone area.

Species number and vegetation stratification

A comparison of three different stands at Namdapha National Park in terms of plant family, genera and species is presented in Figure 2. The cumulative number of species, genera and families for three stand reached 130 species in 74 genera and 42 families (>10 cm DBH). The lowland tropical evergreen forest types exhibited five distinct strata. The A-stratum comprised emergent or top canopy species, mainly occupied by *Altingia excelsa*, *Dipterocarpus macrocarpus*, *Shorea assamica*, *Cinnamomum bejolghota*, *Ailanthes grandis*, *Michelia oblonga* and *Terminalia myriocarpa*, however they do not form a close canopy. Such trees had a straight bole reaching to 40 to 50 m height and a DBH of 60 to 120 cm. The B-stratum or mid-canopy comprised multitude of tree species, formed a close canopy, attained 20 to 30 m height, and mainly dominated with *Elaeocarpus aristatus*, *Ostodes paniculata*, *Mesua ferrea*, *Dysoxylum procerum*, *Talauma hodgsonii*, *Syzygium macrocarpum*, *Chisocheton paniculatus*, *Dysoxylum binecteriferum*, *Castanopsis indica*, *Castanopsis tribuloides* and *Beilschmiedia assamica*. The C-stratum or lower canopy had small trees with 10 to 15 m height, *Saprosma ternatum*, *Baccaurea ramiflora*, *Ardisia sp*., and Leea species being the major elements of this layer. The D-stratum composed of shrubs and lianas, while the E-stratum comprised herbs and newly germinated seedlings of tree species. The semi-evergreen forest (Albizia stand) that occurred along the river banks, riverine plains and swamps forming a narrow belt, showed emergent or top canopy (A-stratum) trees with a height of 25-40 m, comprising *Albizia procera*, *Bombax ceiba*, *A. grandis*, *Duabanga grandiflora* and *T. myriocarpa*. Generally the stand was devoid of a clear second storey, though tree species, such as *Dalbergia pinnata*, *Dalbergia sericia*, *Melia dubia*, *Erythrina stricta*, *Ehretia acuminate* and *Cordia dichotoma* recorded up to a height of 10 to 20 m. Shrubs showed scattered distribution in this stand. This stand receives frequent

Table 3. Comparative account of tree species, density, total basal area and other parameters for three-studied forest stands in the buffer zone of Namdapha National Park, Arunachal Pradesh.

Parameter	Forest stands		
	***Altingia*-mixed species**	***Shorea-Dipterocarp***	***Albizia***
Total number of species			
Tree*	98	54	20
Sapling	72	36	20
Seedling	36	20	3
Herb	29	17	16
Density (individuals ha^{-1})			
Trees	418	390	245
Saplings	4464	3860	345
Seedlings	13184	12250	423
Herb	43281	17500	25461
Total tree basal area (m^2 ha^{-1})	45.47	49.68	18.29
Trees with maximum total basal area	*Altingia excelsa*	*Terminalia myriocarpa*	*Albizia procera*
Their contribution to total basal area (m^2 ha^{-1})	17.58	8.09	13.03
No. of trees with buttressed	69	40	2
No. of deciduous	26	20	196

*The number of unidentified tree species was 31, 16 and 3 at *Altingia*-mixed species, *Shorea-Dipterocarp*, and *Albizia* stands, respectively.

flood during rains that adversely affect the ground flora. Saccharum grass was major E-stratum that dominates in riparian alluvial patches, which was categorized as grassland area.

Tree structure

The tree density and basal cover was recorded highest for Altingia-mixed species (Stand I), followed by Shorea-Dipterocarp (Stand II), and minimum for Albizia (Stand III) forests. When all the life forms, namely, trees, sapling, seedling, lianas, shrubs and herbs considered together, 198 species varying from 151 genera and 74 families were recorded from three studied stands (Figure 2). A total of 31, 16 and 3 tree species at Altingia-mixed species, *Shorea-Dipterocarp*, and Albizia stands, respectively, could not identified though definitely were new genera. The highest number of tree, sapling, seedling and herb species were recorded for Stand I, followed by Stand II and Stand III (Table 3). Similar was the trend for stand tree, sapling and seedling density. The total tree basal cover was recorded highest at Stand II, followed by Stand I and minimum at Stand III.

At least 14 to 21% of tree stems were recorded having buttresses at the two stands of lowland tropical evergreen forest. *A. excelsa* contributed 55% of total buttressed stems at Altingia-mixed species stand. *S. assamica* (8%) and *D. macrocarpus* (5%) also had buttressed stems at *Shorea-Dipterocarp* stand. Only two individuals of *B. ceiba* were recorded having buttresses at Albizia stand (Table 3). The percentage of deciduous stems was estimated 6.27 and 5.13% in Altingia-mixed species and *Shorea-Dipterocarp* stands though it was as high as 80% in Albizia stand, and 5, 6, and 14 tree species showed deciduous habit in respective stands.

The data for tree structure with reference to relative density and relative basal cover for different stands are given in Table 4. Of the total 98 tree species at Altingia-mixed species stand maximum relative density was recorded for *O. paniculata*, followed by *A. excelsa*, *D. macrocarpus*, and *T. hodgsonii*. It was interesting to note that 82 tree species recorded <1% relative density though collectively they contributed 42.47% of total stand density (Table 4). The species contribution to total basal cover (that is relative cover) was estimated highest for *A. excelsa*, followed by *D. macrocarpus*, *Schima wallichii* and *O. paniculata*. 84 tree species recorded a relative cover of <1% each, however when pooled together they contributed 31.33% of total relative cover at same stand (Table 4). The data showed that rare species contribute significantly to the stand density and total basal cover.

At Shorea-Dipterocarp stand, of the total 54 tree species, the maximum relative density was recorded for *D. macrocarpus*, followed by *S. assamica*, *O. paniculata*, *B. ramiflora*, and *Leea indica*. Nearly 29 tree species recorded <1% relative density though their total contribution to the stand density was 23.75% (Table 4). A total of 16 tree species of this stand contributed nearly 82% of total basal cover, while remaining 38 species,

Table 4. Relative density (%) and relative basal cover (%) of important tree species at *Altingia*-mixed species, *Shorea-Dipterocarpus*, and *Albizia* stand in the buffer zone area of Namdapha National Park (only tree individuals with ≥31.5 cm CBH are considered; species with <1% contribution for relative density and cover are pooled together).

Species	*Altingia*-mixed species stand		*Shorea-Dipterocarpus* stand		*Albizia* stand	
	Relative density (%)	Relative cover (%)	Relative density (%)	Relative cover (%)	Relative density (%)	Relative cover (%)
Actinodaphne obovata			1.54	0.82		
Ailanthus grandis			0.51	1.36		
Albizia procera					51.429	71.086
Altingia excelsa	8.37	38.66	1.54	10.93		
Aporusa dioca	0.24	0.02	2.05	0.29		
Baccaurea ramiflora	1.44	0.31	6.67	0.91		
Beilschmiedia sp.	0.24	0.02				
Bischofia javanica			1.03	2.94		
Bombax ceiba					7.347	6.601
Calicarpa arborea					1.024	0.927
Canarium strictum	0.24	0.07	3.59	3.54		
Castanopsis indica	0.72	0.24	0.51	2.79		
Castanopsis sp. 3	0.48	0.57	1.03	3.68		
Cedrella toona	0.48	2.31				
Chisocheton paniculatus	2.87	1.63	0.51	0.13		
Cinnamomum bejolghota	1.20	1.58				
Cinnamomum glaucescens	0.60	1.14	1.54	7.19		
Cinnamomum sp.	0.24	0.13	1.03	0.41		
Croton roxburghii	0.24	0.02	2.56	2.40		
Dalbergia pinnata					12.653	3.655
Dalbergia sericia					9.796	3.219
Dipterocarpus macrocarpus	4.31	5.08	10.26	10.06		
Dysoxylum binecteriferum	2.39	1.17				
Dysoxylum procerum	3.83	1.14				
Elaeocarpus aristatus	0.60	0.31	3.08	1.17		
Eriobotrya bengalensis	1.03	0.31				
Glochidion lanceolarium					3.673	0.873
Gmelina arborea					7.347	2.073
Griffithianthus fuscus	0.60	0.15	2.56	0.58		
Helicia robusta			1.03	0.26		

Table 4. Contd.

Leea indica			4.62	0.48		
Leea indica	2.15	0.29				
Lindera sp.			2.05	0.37		
Magnolia griffithii	0.24	0.02	1.54	2.16		
Ostodes paniculata	15.07	3.58	7.69	1.41		
Quercus lamellosa	0.24	0.02	1.03	0.35		
Saprosma ternatum	2.39	0.20	3.08	0.27		
Schima wallichii	1.79	3.91	0.51	3.34		
Shorea assamica	0.48	0.53	8.21	11.51		
Syzygium cumini	1.67	2.11	1.03	0.71		
Syzygium macrocarpum	3.11	1.17	2.05	0.35		
Talauma hodgsonii	4.19	2.55	1.54	0.95		
Talauma hodgsonii					0.816	0.382
Terminalia chebula			0.51	1.62		
Terminalia myriocarpa	0.48	2.64	3.59	16.28	1.633	5.619
Other species*	37.83	28.12	21.20	10.74	4.29	5.56

*No. of other species comprised 68, 24, and 10 species in *Altingia*-mixed species, *Shorea-Dipterocarp*, and *Albizia* stands, respectively.

which are rare in distribution, comprised just 18% of total basal cover. *S. assamica, T. myriocarpa, D. macrocarpus, A. excelsa, Cinnamomum glaucescens, S. wallichii* contributed most to the total cover of the stand (Table 4).

In contrary to the tropical wet evergreen stands, a few species showed clear dominance at semi-evergreen stand, of which *A. procera* comprised more than half of the relative density and two-third of the relative basal cover (Table 4). Other species contribution was much lower, that is, *D. pinnata* (31 trees ha^{-1}), *D. sericia* (24 trees ha^{-1}), *G. arborea and B. ceiba* (each having 18 trees ha^{-1}). Besides *A. procera*, other major contributors to the total basal area were *B. ceiba, T. myriocarpa, D. pinnata, Dalbergia sericia* and *G. arborea* (Table 4). Ten other species individually contributed <1% to total stand density and basal cover.

Plant family-density ranking of stands

Taxonomically the number of plant families representing different tree species were more in Altingia-mixed species stand (32 families), followed by Shorea-Dipterocarp stand (23 families) and minimum in Albizia stand (16 families) (Table 5). For tree stratum six families, namely, Lauraceae, Euphorbiaceae, Meliaceae, Magnoliaceae, Combretaceae and Verbenaceae were common among the three stands. Nearly 17 families were common between Altingia-mixed species and *Shorea-Dipterocarp* stands, and 8 families each between Altingia-mixed species and Albizia stands, and *Shorea-Dipterocarp* and Albizia stands (Table 5).

The density ranking of plant families also varied within and amongst different stands (Table 5). In Altingia-mixed species stand, Euphorbiaceae comprised 20% of total stand density, while Meliaceae, Magnoliaceae, Dipterocarpaceae and Lauraceae ranked subsequently with 10, 5, 5, and 4% contribution at Altingia-mixed species stand (Table 5). In Shorea-Dipterocarp stand, Dipterocarpaceae and Euphorbiaceae contributed 18 and 17% of the total stand density, respectively, and Fagaceae, Leeaceae, Lauraceae and Magnoliaceae contributed 7, 5, 4 and 3.5% to total stand density (Table 5). In Albizia stand, Leguminosae contributed 75% to

Table 5. Important plant families and no. of tree species, density (ha^{-1}, CBH >31.5 cm) and total basal area (TBA, $m^2 ha^{-1}$) in different forest stands in the buffer zone area.

Family	*Altingia*-mixed species stand			*Shorea-Dipterocarp* stand			*Albizia* stand		
	Species	**Density**	**TBA**	**Species**	**Density**	**TBA**	**Species**	**Density**	**TBA**
Alangiaceae	-	-	-	-	-	-	1	1	0.01
Annonaceae	2	6	0.17	1	10	0.29	-	-	-
Araliaceae	1	2	0.11	-	-	-	-	-	-
Bignoniaceae	-	-	-	-	-	-	1	1	0.02
Bombaceae	-	-	-	-	-	-	1	18	1.21
Boraginaceae	-	-	-	-	-	-	1	0	0.01
Buseraceae	1	1	0.03	-	-	-	-	-	-
Capparadiaceae				2	2	0.28	-	-	-
Celestraceae	2	6	0.13	-	-	-	-	-	-
Clusiaceae	1	3	0.06	-	-	-	-	-	-
Combretaceae	1	2	1.2	1	16	8.89	1	4	1.03
Dilleniaceae	-	-	-	2	2	0.21	-	-	-
Dipterocarpaceae	2	20	2.55	1	72	10.71	-	-	-
Ebenaceae	1	2	0.05				-	-	-
Elaeocarpaceae	2	3.5	0.16	4	12	0.58	-	-	-
Euphorbiaceae	6	77.5	2.00	5	68	2.76	1	9	0.17
Fagaceae	5	9.5	0.48	1	26	5.17			
Hammemelidaceae	1	35	17.58	3	6	5.43			
Juglandaceae	-	-	-	-	-	-	1	1	0.06
Lauraceae	8	17	1.54	1	16	1.00	1	1	0.14
Leeaceae	1	9	0.13	3	18	0.24			
Leguminosae	1	2	0.09	-	-	-	4	183	14.45
Magnoliaceae	4	22	2.22	3	14	1.83	1	2	0.07
Meliaceae	5	41	2.86	2	16	1.46	1	1	0.12
Moraceae	3	4.5	1.00	-	-	-	-	-	-
Myrsinaceae	-	-	-	-	-	-	1	1	0.01
Myristicaceae	1	2	0.02	-	-	-	-	-	-
Myrtaceae	2	20	1.49	1	12	0.53	-	-	-
Oleaceae	2	2	0.06	-	-	-	-	-	-
Proteaceae	-	-	-	1	4	0.13	-	-	-
Rosaceae	-	-	-	1	4	0.15	-	-	-
Rubiaceae	2	12	0.11	1	12	0.13	-	-	-

Table 5. Contd.

Rutaceae	2	6	0.12	-	-	-	-	-	-
Sapindaceae	1	2.5	0.1	-	-	-	-	-	-
Saurauceae	1	1	0.01	-	-	-	-	-	-
Simorubiaceae	-	-	-	1	2	0.68	1	0	0.01
Sonneritiaceae	1	2	1.41	1	6	3.57	-	-	-
Sterculiaceae	2	2.5	0.04	1	2	0.02	-	-	-
Styraceae	2	3.5	0.13				-	-	-
Theaceae	1	7.5	1.78	1	2	1.66	-	-	-
Urticaceae	1	2.5	0.32	-	-	-	-	-	-
Verbenaceae	2	2	0.09	1	2	0.17	1	20.5	0.55
Unidentified	1	89	7.42	1	66	3.78	2	2	0.47
Total	70	417.5	45.46	39	390	49.67	1	244	18.33

total stand density while Verbenaceae and Bombaceae had much lower contribution (9 and 7%). With respect to contribution for total basal area, Hammemelidaceae, Dipterocarpaceae and Leguminoceae contributed maximum at three stands, respectively (Table 5).

Tree diameter class distribution

Distribution of tree species and their individuals in different DBH classes varied among stands (Figure 3, Table 6). The relationship between tree girth sizes and number of individuals varied among species in forest stands. Both, the number of tree individuals and number of species in different girth classes decreased with increase in girth sizes in all the studied stands. Of the total 418 tree individuals ha^{-1} in ≥10 cm DBH in Altingia-mixed species stand, 49% tree individuals were recorded in DBH class 10 to 20 cm, 23% in 21 to 30 cm, 18% in 31 to 60 cm, and 10% in DBH class >60 cm (Figure 3). The data revealed that individuals in middle to lower DBH classes showed an overall dominance and only *A. excelsa*, *D. macrocarpus* and few other species reached >60 cm DBH. *O. paniculata*, a middle canopy species of this stand, was mostly represented in the <30 cm DBH class (Figure 3A). Similarly in Shorea-Dipterocarp stand, of the total of 390 tree individual ha^{-1}, nearly 45, 21, 22 and 13% tree individual were recorded in DBH classes 10 to 20 cm, 21 to 30 cm, 31 to 60 cm, and >60 cm, respectively (Table 6). The stand showed that tree individuals of *S. assamica*, *D. macrocarpus* and *T. myriocarpa* could reach DBH class >60 cm (Figure 3B). *T. myriocarpa* was not represented in DBH <30 cm. With a total of 245 tree individuals ha^{-1} in Albizia stand, 46, 27, 25 and 2% individuals were recorded in DBH classes 10 to 30, 31 to 60 and >61 cm classes, respectively (Table 6). Tree individual distribution showed the dominance of individuals in DBH between 10 to 30 cm (Figure 3C). Only *A. procera*, *B. ceiba* and *T. myriocarpa* represented the DBH class >61 cm (Figure 3C).

The representation of number of species in different DBH class structure showed an inverse relationship (Table 6). A total of 81, 60, 47, 37 and 18% species represented in DBH class <10, 11-20, 21-30, 31-60, and >61 cm in Altingia-mixed species stand, while 67, 52, 44, 41 and 26% species in *Shorea-Dipterocarp* stand, and 100, 80, 60, 40 and 20% species in Albizia stand represented similar girth class distribution (Figure 4). The individual: species ratio revealed higher values for lower girth sizes and it decreased as girth size increases (Table 6). The ratio showed a gradual lapse rate at Altingia-mixed species stand, and inconsistence trend in *Shorea-Dipterocarp* stand. For Albizia stand the ratio was higher than other stands showing lower number of species at this stand (Table 6). *A. excelsa* recorded a stem diameter up to 72 cm in Altingia-mixed species stand, and 457 cm in *Shorea-Dipterocarp* stand. Other large girth size species in tropical wet evergreen stands were

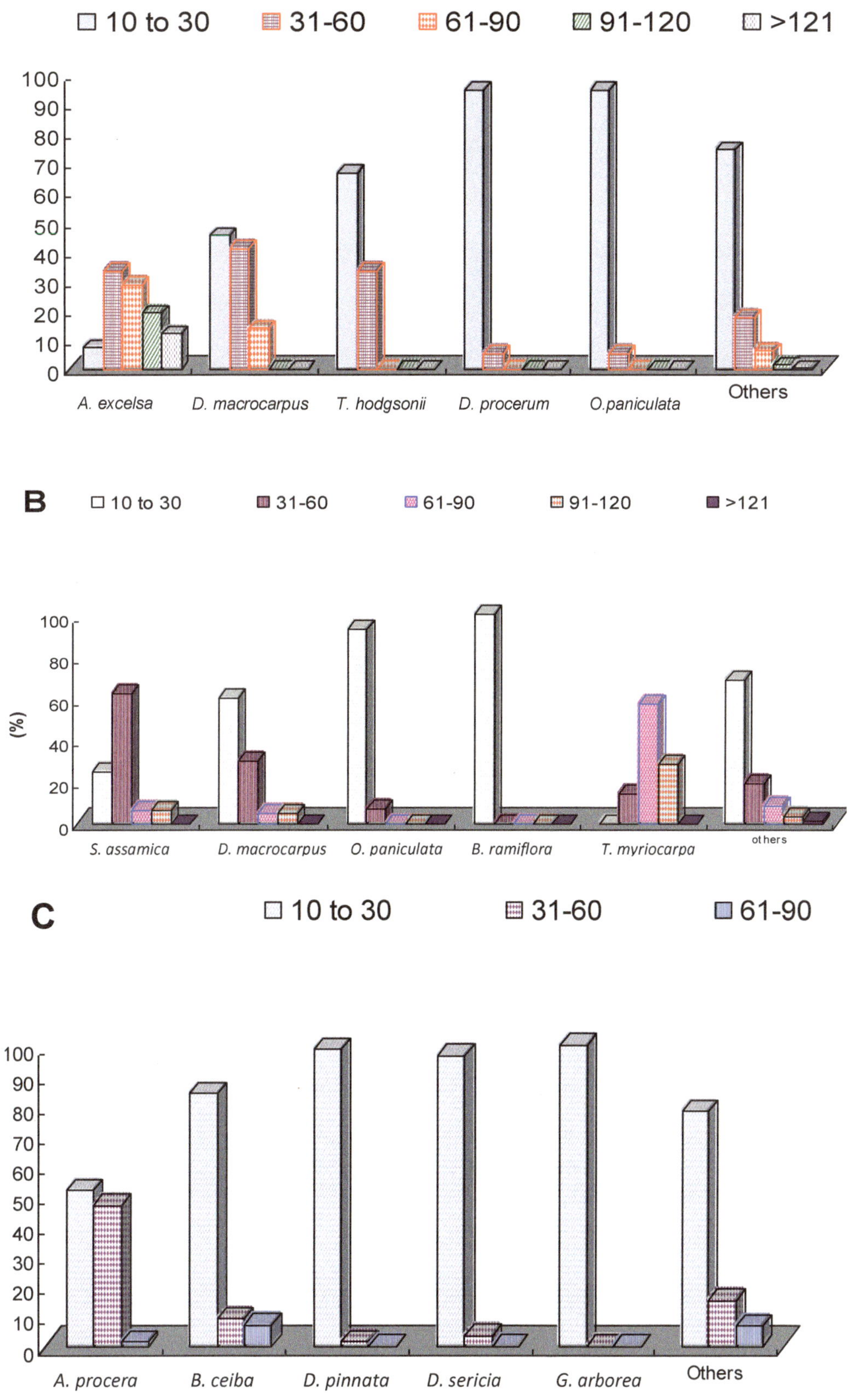

Figure 3. Diameter (DBH) classes of selected important trees species in different studied stands in the buffer zone of Namdapha National Park (A-*Altingia*-mixed spp., B-*Shorea-Dipterocarp*, and C- *Albizia* stands).

Table 6. Comparison of structural composition of three studied stands in the buffer zone area of Namdapha National Park, Arunachal Pradesh, India.

Parameter	**Forest stand**		
	***Altingia*-mixed species**	***Shorea-Dipterocarp* stand**	***Albizia* stand**
No. of tree individuals/ha (≥10 cm DBH)	418	390	245
dbh<10 cm	17791	16500	854
10-20 cm	202	176	114
21-30 cm	93	80	66
31-60 cm	74	84	61
>61 cm	43	50	6
No. of tree species	98	54	20
dbh <10 cm	79	36	20
10-20 cm	59	28	16
21-30 cm	43	24	12
31-60 cm	36	22	8
>61 cm	18	14	4
Ratio (individual : species*)			
dbh <10 cm			
10-20 cm	4.11	3.26	7.89
21-30 cm	2.60	2.86	9.9
31-60 cm	2.47	3.82	13.73
>61 cm	2.87	3.57	2.7
Percentage of most abundant species			
Species	*Altingia excelsa*	*D. macrocarpus*	*Albizia procera*
10-20 cm	0	30	13.73
21-30 cm	7.14	30	37.25
31-60 cm	33.33	30	47.06
>61 cm	59.52	10	1.96

*individuals ha^{-1} × total area sampled / No. of species.

Castanopsis sp. (DBH 50 cm), *D. macrocarpus* (DBH 55 cm), *D. grandiflora* (DBH 60 cm), Ficus sp. (DBH 55 cm), *S. wallichii* (DBH 55 cm), *S. assamica* (DBH 43 cm), *T. myriocarpa* (DBH 47 cm), and *T. myriocarpa* (DBH 57 cm). At Albizia stand, *T. myriocarpa* reached to a maximum diameter size (DBH 42 cm), while *C. dichotoma* (DBH 42 cm), *A. procera* (DBH 41 cm), and *B. ceiba* (DBH 38 cm) also attained higher diameter sizes. The percentage distribution of individuals from 10 to 20 cm to >61 cm DBH classes showed inverted pyramidical structure for A. excelsa in Stand I with increasing number of individuals from lower to higher girth classes (Figure 4). For *D. macrocarpus* in Stand II, it was fairly constant except for DBH >61 cm where it decreased. *A. procera* showed upright pyramidical structure (Figure 4).

The family wise distribution of tree individuals in different girth classes showed that Euphorbiaceae represented highest number of tree individuals (19.54%) in DBH >10 cm, followed by Meliaceae (9.72%) and Hamamelidaceae (8.50%) at Altingia-mixed species stand (Figure 5a). Dipterocarpaceae (18.46%), Euphorbiaceae (17.44%) and Fagaceae (6.67%) dominated the Shorea-Dipterocarp stand. Leguminosae was the dominant family at Albizia stand, followed by Verbenaceae (8.54%) and Bombacaceae (7.49%) (Figure 5b). By adding seedlings and saplings with mature tree individuals, Lauraceae and Euphorbiaceae showed dominance at Altingia-mixed stand, Dipterocarpaceae and Euphorbiaceae at Shorea-Dipterocarp stand, and Leguminosae at Albizia stand (Figure 5c).

Species dominance and rarity at stands

The dominance of each species in a stand was calculated by assessing their importance value index (IVI) (Appendix I). The IVI of different species varied greatly in

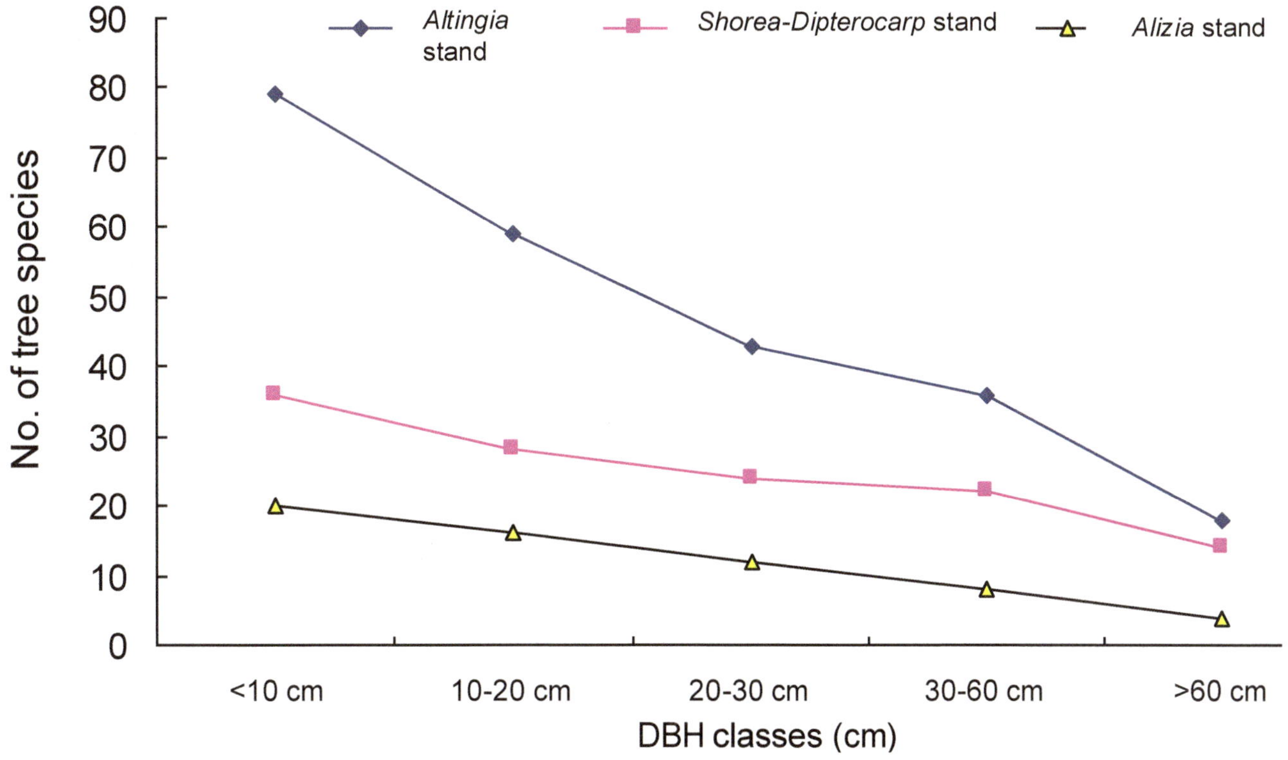

Figure 4. No. of species in different DBH classes in *Altingia*-mixed species stand, *Shorea-Dipterocarp* stand, and *Albizia* stand at the buffer zone area of Namdapha National Park, Arunachal Pradesh.

different stands. A. excelsa (IVI 51.72) was the most dominant species of the Altingia-mixed species stand. *O. paniculata, D. macrocarpus* and *T. hodgsonii* were the co-dominating species of the stand (Appendix I). Nearly 10 tree species had an IVI value between 5 to 8, while for another 20 species it varied between 2 to 5. The remaining 62 tree species recorded an IVI <2 which expressed their rarity in the stand (Appendix I).

In Shorea-Dipterocarp stand, the maximum IVI was recorded for *D. macrocarpus*, which was closely followed by *T. myriocarpa* and *S. assamica* (Appendix I). For *A. excelsa, C. indica, O. paniculata, B. ramiflora* and *D. grandiflora* the IVI value varied from 15 to 10. Other 46 tree species had an IVI value <10, which included 35 species with an IVI<5, and 19 species with an IVI value <2 (rare species) (Appendix I).

A. procera showed a clear dominance at riverine stand (Appendix I). *D. pinnata, B. ceiba, D. sericia, G. arborea, Glochidion lanceolarium* and *T. myriocarpa* were the other co-dominant species at this stand. Remaining 11 tree species had an IVI value <3, which also include 9 species with an IVI value <2 (Appendix I).

Shrub and herbaceous layer

The density of shrub and liana was estimated for Altingia-mixed species and Albizia stands only and it was higher in the former stand than the latter stand (Table 7). The most common liana at Altingia-mixed species stand was *Byttnera aspera*, *Fissistigma bicolor*, and *Milletia sp.* Tree clasping climber *Pothos scandens* and *Roydsia suaveolens* were also recorded (Table 7). It was interesting to note that 8% of total trees sampled were having lianas. Strobilanthes and Boehmeria were the most common shrub species at this stand. Strobilanthes sp. particularly dominated under the canopy openings and forest fringes, attains a height of 4 m and is supported by stilt roots. The most common shrub in the Albizia stand was Elaeagnus, while *Citrus medica* was another sturdy shrub at this site. *Phrynium sp.*, *Alpinia sp.* and *Phragmites karka* showed a gregarious presence at this site. An observation of the epiphytic loads on the trees showed that *Asplenium nidus*, *Ryncostylus retusus*, *Dendrobium sp.*, and *Aerides odorata* grow profusely on trees. A terrestrial orchid namely, *Calanthe sp.* was also recorded from the Albizia stand.

The occurrence and floristic composition of ground herbs varied among the three studied stands (Table 8).

A total of 46 species varying from 38 genera and 22 families were recorded in all the three stands. Piperaceae with 5 species was the most diverse herbaceous family followed by Arecaceae and Asteraceae (4 species each) and Zingiberaceae (3 species). In terms of number of

Table 7. Climber and shrub species (density/ha) at selected forest stand in the buffer zone area of Namdapha National Park.

Species	*Altingia*-mixed species stand (density ha^{-1})	*Albizia* stand (density ha^{-1})
Shrubs		
Citrus medica	-	25
Cudrania cochinchinensis	-	1
Desmodium sp.	-	1
Elaeagnus sp.	-	10
Leea edgeworthii	-	1
Solanum torvum	-	31
Strobilanthus sp.	1752	0
Boehmeria sp.	40	12
Amblyanthus sp.	113	12
Climber		
Byttneria aspera	17	-
Combretum sp.	1	-
Fissistigma bicolor	1	-
Milletia sp.	6	-
Pothos scandens	2	-
Premna sp.	-	1
Roydsia suaveolens	1	-
Unidentified climbers	25	-

individuals per family, Commelinaceae followed by Bolibitidae and Chloranthaceae dominated the Altingia-mixed species stand (Table 8). In the Shorea-Dipterocarp stand, Musaceae and Chloranthaceae were dominating families. In the riverine Albizia stand Asteraceae dominated all other families in terms of number of individuals, followed by Thelypteridaceae and Athyriaceae. The species richness and density for herbs recorded maximum at Altingia-mixed species stand (29 species), followed by Shorea-Dipterocarp stand (17 species), and Albizia stand (16 species). The Albizia stand remained covered by *Mikenia mikrantha* during the monsoon months (July to September). The herbaceous species differed in their density at different stands (Table 8). Some epiphytes, terrestrial orchids (*Calanthe sp.*), and ferns (*Trigonospora cilliata*, *Athyrium asperum*, *Diplazium esculentum* etc) were also recorded at this stand. During rainy season the ground and major shrubs were covered by a thick blanket of a weed called *M. mikrantha*. The riverine Albizia stand despite being a open forest, was having low herbaceous population than other stands because the area was frequently under floods during monsoons that removed the ground flora.

Stand heterogeneity

The Menhinick's species richness index for trees, saplings, seedlings and herbaceous layers was estimated highest for Altingia-mixed species stand, followed by *Shorea-Dipterocarp* stand and minimum for Albizia stand (Table 9). The Shannon-Weiner diversity index (H) also recorded similar trend for different vegetation layers (Table 9). The concentration of dominance value was estimated low for Altingia-mixed species and *S. Dipterpcarp* stands than the Albizia stand, which showed that many species shared the dominance in former stands. In Albizia stand, however, only a few species showed their dominance.

Dominance-diversity curve

The dominance-diversity curves for the tree layer (on the basis of IVI) were drawn for all the three studied stands for eliciting the community organization in terms of resource share and niche space (Figure 6). The dominance-diversity curve for tree layer in Altingia-mixed species and the *Shorea-Dipterocarp* stands fits the log normal situation, whereas for Albizia stand it fits the geometric series. For the sapling layer as well the curves followed same patterns as for trees. However, the riverine Albizia forest stand showed a poor status of sapling and seedling (Figure 6).

Similarity Index

The similarity of trees, saplings and seedlings layers

Table 8. Herbaceous density (individuals/ha) of the three stands in the lowland tropical forests of Namdapha National Park.

Species	Family	*Altingia*-mixed species stand	*Shorea-Dipterocarp* stand	*Albizia* stand
Acanthus leucostachys Nees	Acanthaceae	260.42	0	0
Ageratum conyzoides L.	Asteraceae	0	0	1538.46
Alpinia sp.	-	0	0	576.92
Althia sp.	Malvceae	0	0	115.38
Amaranthus sp.	Amaranthaceae	0	0	38.46
Amphineuron immersum (Bl.) Holtt	Thelypteridaceae	0	0	2961.54
Athyrium asperum (Blume) Milde.	Athyriaceae	0	0	1115.38
Bidens pilosa	Asteraceae	0	0	115.38
Bolbitis heteroclite (Presl) Ching.	Bolibitidaceae	5000.00	1625.00	0
Calamus sp. (Hukabeth)	Arecaceae	104.17	125.00	0
Calamus sp. (Lejai beth)	Arecaceae	52.08	0	0
Calamus sp.(Beth)	Arecaceae	52.08	0	0
Chasalia sp.	-	156.25	0	0
Chloranthus elatior R.	Chloranthaceae	6458.33	1375.00	0
Cissus sp.	Vitaceae	2239.58	625.00	0
Collocasia sp.1	Araceae	52.08	0	0
Collocasia sp.2	Araceae	0	0	115.38
Crassocephalum crepidioides	Asteraceae	0	0	500.00
Diplazium esculentum (Retz.) Sw. Schrad	Athyriaceae	0	0	192.31
Elatastemma sessile Forst.	Urticaceae	52.08	375.00	0
Eugenofolia binectifida, Egenolfia bipinnatifida J. Sm.	Bolibitidaceae	1718.75	625.00	0
Forrestia sp.	Commelinaceae	16979.17	8125.00	0
Globba sp.	Zinziberaceae	312.50	0	0
Kharkaria loapani (local name)	-	677.08	0	0
Mikenia mikrantha	Asteraceae	0	0	14999.4
Musa sp.	Musaceae	156.25	1500.00	0
Peripinnate	-	520.83	125.00	0
Phrynium sp. (Koapat)	Marantaceae	104.17	0	0
Piper betleoides	Piperaceae	364.58	250.00	0
Piper rytidocarpum	Piperaceae	156.25	0	0
Piper sp.	Piperaceae	364.58	0	0
Piper sp.1	Piperaceae	0	0	576.92

Table 8. Contd.

Piper sylvaticum Roxb.	Piperaceae	1354.17	250.00	0
Ploygonum sp.	Polygonaceae	1562.50	125.00	0
Poaceae	Poaceae	0	0	269.23
Pteris biaurita L.	Pteridiaceae	364.58	250.00	0
Pteris semipinnata L.	Pteridiaceae	1718.75	0	0
Pteris sp.	Pteridiaceae	1822.92	500.00	0
Selaginella sp.	Selaginellaceae	364.58	1250.00	0
Trigonospora cilliata (Wall. ex Benth.)	Thelypteridaceae	0	0	2115.38
Urticia dioca	Urticaceae	0	0	192.31
Zalacca secunda	Araceae	0	250.00	0
Zinziberaceae	Zinziberaceae	104.17	0	0
Unidentified climber	-	156.25	125.00	0
Unidentified sp. 1	-	0	0	38.46
Unidentified Spiny sp	-	52.08	0	0
Total		43281.13	17500.00	10461.51

between Altingia-mixed species stand andShorea-Dipterocarp stand was estimated 28.19, 33.33 and 14%, respectively (Table 10). Between Altingia-mixed species and Albizia stands it was recorded 4.35, 4.17 and 0%, respectively, for trees, samplings and seedling layers; while between Shorea-Dipterocarp and Albizia stand, the same value was 11.94, 6.67 and 0% for respective layers (Table 10). The overall similarity showed that all the three stands were quite distinct, and it was more marked between tropical evergreen and semi-evergreen forests.

DISCUSSION

Baseline information on forest inventories, resource assessment and status are prerequisites for long-term planning and strategy development for management of any forest stand (Magurran, 2004). Arunachal Pradesh has nearly 11.44% of its total area under protected area network (Ringu, 2002). The state also has other legal forest categories, such as Reserve Forests with nearly 11.61% area and Unclassed State Forests (USF) with 36.91% area. Namdapha National Park with an area of 1985 km^2 forms 20.71% of the total protected area of the state. It comprised 1119 species of plants of which 674 were angiosperms with 60 to 70% of the state's total floral wealth (Chauhan et al., 1996). The park has all major forest types of the region and also had floral and faunal affinities with other northeastern states, Indo-Malayan, Indo-Chinese and oriental regions (Chauhan et al., 1996; Deb and Sundriyal, 2005). Therefore, the park deserves significant conservation measures, which is often jeopardized due to lack of information on floristic composition, species structure, and dominance-diversity relationship. The RS/GIS study was used for the first time to map buffer zone of the park. The RS/GIS reconnaissance of the buffer zone area (177.425 km^2) of Namdapha National Park revealed species rich and diverse tropical wet evergreen forests extended over 80% land area while open forest on 7% area. It also has a small area (<1%) under semi-evergreen riverine forest. Besides, considerable area also falls under grasslands (7%) and river sand (4%). Of three studied stands in this investigation, two comprised tropical wet evergreen forests, which are reported of restricted distribution in revised classification (Kaul and Haridasan, 1987). The landuse map of the buffer zone could also be used for future monitoring of the area, thus may play a leading role for management and conservation of this highly diverse park. The Tropical wet evergreen forest

Table 9. Stand heterogeneity (species richness, species diversity and concentration of dominance) of three studied forest stands in the buffer zone of Namdapha National Park, Arunachal Pradesh.

Parameter	Forest stands		
	***Altingia*-mixed species**	***Shorea-Dipterocarp***	***Albizia***
Species richness (Menhinnick index)			
Tree	4.79	2.73	1.27
Sapling	1.17	0.55	0.81
Seedling	0.36	0.17	0.15
Herb	0.14	0.13	0.10
Species diversity (Shannon-Weiner index)			
Tree	3.85	3.55	1.71
Sapling	3.29	2.69	2.35
Seedling	2.54	2.36	1.04
Herb	2.18	1.98	1.54
Concentration of dominance (Simpson index)			
Tree	0.044	0.042	0.31
Sapling	0.08	0.11	0.09
Seedling	0.17	0.13	0.37
Herb	0.20	0.25	0.37

showed high species richness and also a tiered forest structure. The emergent top canopy (A stratum) included *D. macrocarpus*, *S. assamica*, *A. excelsa*, *A. grandis* and *T. myriocarpa*. The Riverine semi-evergreen forests occur along river banks, riverine plains and swamps forming a narrow belt, the trees were mostly deciduous, lack a dense canopy, and categorized as 3C/C3b East Himalayan moist deciduous forests and 3/1S2 Eastern Hollock forests (Champion and Seth, 1968).

The tropical evergreen forests under this investigation exhibited similar feature as those of 'Lowland evergreen tropical rainforest formation' sensu Whitmore (1998) and Proctor et al. (1998). The top canopy species were 25 to 40 m high consisting *B. ceiba*, *A. procera*, *A. grandis*, and *T. myriocarpa*. There were distinct middle and lower canopy species. Besides, the woody climbers were abundant and trees comprise considerable epiphytic loads. Such distinct stratification of the vegetation is well reported for other tropical evergreen rainforests as well (Whitmore, 1998; Richards, 1996; Sist and Saridan, 1988; Swamy et al., 2000; Wilkie et al., 2004). Buttresses are common features in Dipterocarp rainforests. In this study A. excelsa contributed 55% of total buttressed stems, which was similar to other Dipterocarp lowland evergreen rainforests (Whitmore, 1998).

For the three sampled plots altogether 198 species (trees, shrubs and herbs) in 151 genera and 74 families were recorded. Such high diversity of tree and other species within the sampled area represent about 34% of total floral species and 50% of total tree species of Namdapha National Park, which is significant. The tree species richness of 98, 54 and 20 species in the three studied stands was close to the richness recorded in various parts of western Ghats (Ganesh et al., 1996; Elouard et al., 1997; Swamy et al., 2000), Amazonian forest (De Oliveira and Mori, 1999), neotropical forests (Zent and Zent, 2004), and Peninsular (Kochummen et al., 1990). 31 species were not identified to genera or species levels though they were clearly different specimens, such species categorized as morphospecies (Proctor et al., 1983; Wilkie et al., 2004). There is inherent problem of identification of the tropical trees because of sporadic flowering and fruiting, which take several years for many species as well as difficulty of collecting material from some tall trees (Whitmore, 1998; Wilkie et al., 2004). Altogether, the floristic richness of Namdapha National Park reveals 674 species of flowering plants of which 266 were reported as tree species (Chauhan et al., 1996).

All enumerated tree species in the present study belonged to 44 families (range 16 to 32), which falls well within the range of 16 to 58 families found in other tropical forests (Campbell et al., 1992; Pascal and Pelissier, 1996; Parthasarathy and Karthikeyan, 1997; Swamy et al., 2000). Our values are higher than 36 families recorded for Haldibari area (Proctor et al., 1998) and for core and other neighboring areas (27 families) of Namdapha National Park (Nath et al., 2005). The difference in family wise distribution may be attributed to heterogeneity in the habitats of the park. When all individuals (seedling, sapling and adult trees) of tree species of the families Lauraceae, Dipterocarpaceae,

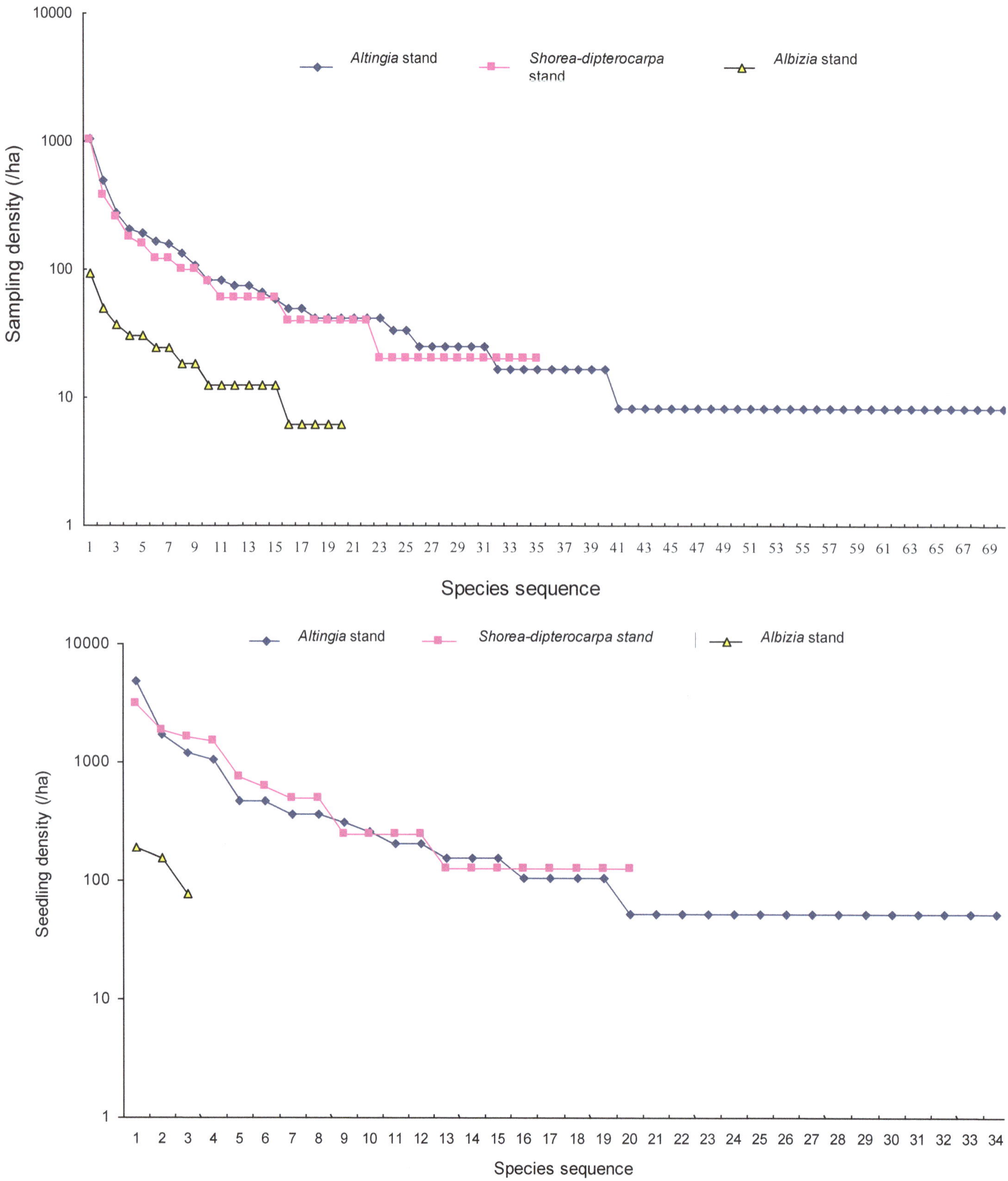

Figure 6. Density-Dominance curve for the saplings (<31.4 cm CBH and height > 20 cm) (top graph) and seedlings (< 20 cm) (lower graph) for *Altingia* mixed, *Shorea-Dipterocarp* and *Albizia* stands.

Table 10. Comparison of Sorensen's similarity index of tree, sapling, seedling and herb components in the three forest stands.

Stands	Trees	Saplings	Seedlings	Herbs
*Altingia-m*ixed species stand and *Shorea-Dipterocarp* stand	28.19	33.33	14	96.96
Altingia Mixed stand and *Albizia* stand	4.35	4.17	0	0
Shorea-Dipterocarp stand and *Albizia* stand	11.94	6.67	0	0

Euphorbiaceae, Fagaceae and Leguminosae are combined together, it exhibited a high number of species and reveals that the forest comprises the attributes of lowland evergreen tropical rainforests (Whitmore, 1998; Proctor et al., 1998).

The tree density of *Shorea-Dipterocarp* and Altingia-mixed stand (390 and 418 tree ha^{-1} respectively) is within the range of 300 to 700 tree ha^{-1} for tropical rainforests (Richards, 1996). The total tree basal area (45 to 49 m^2 ha^{-1}) was also within the range recorded for tropical forests of south-east Asia (25.2 to 67.4 m^2 ha^{-1}) (Newbery et al., 1992; Swamy et al., 2000), however our values were lower than forests of Silent Valley (Singh et al., 1981), Southeast Asian rainforests (Proctor et al., 1998; Manokaran and LaFrankie, 1990; Newbery et al., 1992), Peninsular forests (Manokaran and LaFrankie, 1990), Malaysian rainforest (Proctor et al., 1983; Sist and Saridan, 1998) and neotropical forests (Knab-Vispo et al., 1999; Zent and Zent, 2004).

An analysis of the IVI of species can be used to recognize the dominance pattern of different species in a community (Parthasarathy and Karthikeyan, 1997). A. excelsa dominated the Altingia-mixed species stand, while *O. paniculata*, *D. macrocarpus*, *T. hodgsonii*, *Dysoxylum procerum* were co-dominant species in this stand. In *Shorea-Dipterocarp* stand, *Dipterocarpus macrocarpus* was the most dominant followed by *Shorea assamica*, *Terminalia myriocarpa*, *A. excelsa*, *D. grandiflora*, *C. tribuloides*. At Albizia stand, *A. procera* was the dominant species all through the riverine area.

As many as 91, 83 and 85% species in three studies stands, respectively, had fewer than 10 individuals, while 35, 22 and 50% species had ≤ 2 individual ha^{-1} thus showing high species rarity. For tropical rainforests tree species rarity is a common feature (Whitmore, 1998). Species rarity also underlines the need to conserve such forests as reserves. At Western Ghats nearly 60-63% species showed rarity at local level (Swamy et al., 2000). Our values of species rarity are higher than those of tropical forests of Malaysia (8%) (Poore, 1968), and similar to those of Peninsular region (40%) (Manokaran and Kochumnen, 1987) and South India (26 to 61%) (Parthsarathy and Karthikeyan, 1997).

With increasing DBH size the stem density decreased, which is perhaps an important characteristics of the primary forests where large trees, though very few in density, comprised considerable share of total basal cover. The differences in mean tree density and basal area were low between two stands of tropical evergreen forest. However, it varied markedly high between the evergreen and semi-evergreen stands.

The structural pattern of the different stands showed a heterogeneous distribution of trees and can be considered as one of the highly diverse forests in the Eastern Himalaya (Bhuyan, et al., 2003). The Shannon-Weiner diversity index is generally high for tropical forests, ranging from 0.81 to 4.1 for the Indian subcontinent (Bhuyan et al., 2003; Singh et al., 1984; Swamy et al., 2000). The diversity index values of the present study (1.71 to 3.85) are within the earlier reported range. Within studied stands, it was recorded higher for tropical evergreen stands than semi-evergreen stand. Our values are lower than those reported for young (H=5.06) and old (H=5.4) tropical forest stands of Barro Colorado Island, Panama (Knight, 1975).

The overall liana density was 53 individuals ha^{-1}, and nearly 8% trees carried them. *Buettneria aspera* was major contributor to the density and a few individuals could reach to >30 cm CBH at Altingia-mixed species stand. At Sabah, Malaysia, large numbers of lianas (range 833 to 929 individuals ha^{-1}) have been reported that are carried by 54 to 61% trees (Campbell et al., 1992), 391 to 440 lianas ha^{-1} are reported from Sarawak, Malaysia (Proctor et al., 1983; Newbery et al., 1992). For humid tropical forests of south India, 273 to 1015 lianas and climbers ha^{-1} were recorded and carried by 1 to 14% trees (Swamy et al., 2000). The high number of lianas in a forest stand is a reflection of the opening of canopy in the past (Newbery et al., 1992). In Namdapha National Park, the forest gaps were few and small which can explain the low density of liana at this site (Deb and Sundriyal, 2007). Climbers and lianas are conspicuous feature of almost all rainforest and compete actively with the trees for light and space.

Dominance-diversity has inverse relationship and if the dominance increases the diversity decreases (Maguran, 2004). The dominance-diversity figure for the tree layer in the Altingia-mixed species and the Shorea-Dipterocarp stand fits the Preston's log normal distribution, which reveals that the community niche space was substantially partitioned by large number of species (Sugihara, 1980). For Albizia stand figure represents the geometric series which corroborates the niche preemption hypothesis that is supposed to be commonly applicable for species poor stands with a dominance of one or few species only (Whittaker, 1975). The lognormal forms are common in mixed type of communities (Whittaker, 1975; Upreti et al., 1985). The concept of dominance plays an important part in theories on plant strategies and competition/ co-existence mechanisms.

The similarity index (Sorensen, 1948) was also assessed to understand the commonality of species occurrence between different stands. The difference in species similarity among different forest stands may be due to the varied microclimate, habitat conditions and soil factors which may support different species composition in studied forests

Implications for management

The tropical forests of the world are increasingly being affected by human activities (Proctor et al., 1983; Newbery et al., 1992; Swamy et al., 2000). Therefore, an investigation of such forests for analysis of different components of biodiversity and their quantification has high implications for forest management. This study presented data from Namdapha National Park, Northeast India and revealed that the lowland tropical evergreen forests were highly diverse and well comparable to other similar forests of the globe. These forests comprised a large number of species that were rare at local level, however they contributed significantly to the stand diversity. On the contrary, the diversity of lowland tropical semi-evergreen forest was low, most species had successional status. All forests, however, showed an evolving population. The information on these forests has high implications for conservation, As most of the forests surrounding Namdapha National park are depleting fast because of community pressure, logging, and habitat alteration (Deb and Sundriyal, 2005). the management planning for the park must take into account the rich diversity of the buffer zone comprising best timber and other utility tree species including dipterocarp species. Availability of such species may attract community attention in near future; therefore maximum vigil is required at this site so as to avoid any outside disturbance. This highlights the need to investigate community resource dependency and their linkages with the park so as to improve the management of such primary forest stands everlastingly. It is also reported that a few weed species are invading at the park boundary, particularly nearby the Albizia stand. It is expected that such proliferation of alien weed species may extend to the other parts of the park in near future when resources in the nearby forests are depleted. Such kind of perturbation in the ecosystem may result in alteration of the structure of the biological community and proliferation of noxious weeds (Ramakrishnan, 1992). Therefore, assessing patterns of vegetation cover, species accumulation, richness and diversity, commonness and rarity, plant family expression, forests diameter class structure, and tree, shrub and herb composition are key parameters to determine ecosystem stability, and such information is useful in management as well as restore the health of bioresources in forest stands. This calls for further analysis of broad patterns of species distribution across different vegetation covers and landscapes types in biodiversity rich areas. Since the information on biodiversity characterization is meager in northeast India, particularly in protected areas (Proctor et al., 1998; Bhuyan et al., 2003; Nath et al., 2005), it is suggested that more and more area be investigated as it will open now intricacies of biodiversity in the area.

ACKNOWLEDGEMENTS

We thank the Director, G.B. Pant Institute of Himalayan Environment and Development for extending facilities for the work. The Department of Environment and Forests, Government of India provided financial assistance (F.10/35/CS/BR). The SRSAC, Itanagar is thanked for the help in analyzing RS/GIs data. We sincerely acknowledge help of the Arunachal Pradesh Government for permission to work in Namdapha National Park, and Field Director, Research Officer and the administrative/ field staff of the Park to help in collection of information.

REFERENCES

Barik SK, Rao P, Tripathi RS, Pandey HN (1996). Dynamics of tree seedling population in a humid subtropical forest of northeast India as related to disturbances. Can. J. For. Res., 26: 584-589.

Bhat DM, Naik MB, Patagar SG, Hedge GT, Kanade YG, Hedge GN, Shastri CM, Shetti DM, Furtado RN (2000). Forest dynamics in the tropical rainforest of Uttara Kanada district in Western Ghat, India. Curr. Sci., 79(7): 975-985.

Bhuyan P, Khan ML, Tripathi RS (2003). Tree diversity and population structure in undisturbed and human-impacted stands of tropical wet evergreen forests in Arunachal Pradesh, Eastern Himalayas, India. Biodivers. Conserv., 12: 1753-1773.

Campbell DG, Stone JL, Rosa Jr A (1992). A comparison of the phytosociology and dynamics of three floodplain (varzea. Forest of known ages, Rio Jurua, Western Brazilian Amazon. Bot. J. Linn. Soc., 108: 213-237.

Champion HG, Seth SK (1968). A revised survey of the forest types of India. Govt. of India press, Nasik, India.

Chauhan AS, Singh KP, Singh DK (1996). A contribution to the flora of Namdapha, Arunachal Pradesh. *In*: Hajra PK (eds.). Botanical Survey of India, Calcutta, p. 422.

Deb P (2006). Tree structure, regeneration, stand biomass and community dependence in the buffer zone area of Namdapha National Park, Arunachal Pradesh. Unpublished Doctoral thesis, Kumaun University, Nainital, Uttarakahand, India.

Deb P, Sundriyal RC (2005). Status review of Namdapha National Park: Need for community conservation linkages. Ind. J. For., 28(1): 85-96.

Deb P, Sundriyal RC (2007). Tree species gap phase performance in the buffer zone area of Namdapha National Park, Eastern Himalaya, India. Trop. Ecol., 48(2): 209-225.

Deb P, Sundriyal RC (2008). Tree regeneration and seedling survival patterns in old-growth lowland tropical rainforest in Namdapha National Park, North-East India. For. Ecol. Manag., 255: 3995-4006.

FRA: Forest Resource Assessment (2005). Global forest resource assessment: Progress towards sustainable forest management. FAO Forestry Paper Italy. 147.

Ganesh T, Ganesan R, Devy MS, Davidar P, Bawa KS (1996).

Assessment of plant biodiversity at a mid elevation evergreen forest of Kalakad-Mundanthurai Tiger Reserve, Western Ghats, India. Curr. Sci., 71: 379-392.

Ghosh AK (1987). Qualitative analysis of faunal resources of proposed Namdapha Biosphere Reserve. Zoological Survey of India, Calcutta, India.

Haridasan K, Rao RR (1985). Forest flora of Meghalaya. Bishen Singh Mahendra Pal Singh, Dehradun. (1&2).

IIRS (2002). Biodiversity Characterization at Landscape Level in North East India Using Satellite Remote Sensing and Geographic Information System. Indian Institute of Remote Sensing (IIRS), Dehradun.

Kaul RN, Haridasan K (1987). Forest types of Arunachal Pradesh, a preliminary study. J. Ecol. Tax. Bot., 9: 379-389.

Kent M, Coker P (1994). Vegetation description and analysis, A practical approach. International Book Distributor, Dehradun, India.

Khan ML, Rai JPN, Tripathi RS (1986). Regeneration and survival of tree seedlings and sprouts in tropical deciduous and subtropical forests of Meghalaya, India. For. Ecol. Manage., 14: 293-304.

Khan ML, Rai JPN, Tripathi RS (1987) Population structure of some tree species in disturbed and protected subtropical forests of northeast India. Acta Oecol., 8: 247-255.

Knab VC, Berry P, Rodriguez G (1999). Floristic and structural characterization of a lowland rain forest in the lower Caura watershed, Venezuelan Guayana. Acta Bot. Venez., 22: 325-359.

Knight DH (1975). A phytosociological analysis of species rich tropical forest on Barro Colorado Island, Panama. Ecol. Monogr., 45: 259-284.

Kochummen KM, LaFrankie JV, Manokaran N (1990). Floristic composition of Pasoh Forest Reserve, a lowland rainforest in Peninsular Malaysia. J. Trop. For. Sci., 3: 1-13.

Kothari A., Pande P, Singh S, Variava D (1989). Management of National Parks and Sanctuaries in India: A status Report. Indian Institute of Public Administration, New Delhi.

Maguran AE (2004). Measuring Biological Diversity assessment. Blackwell Publishing Company, p. 256.

Manokaran N, Kochummen KM (1987). Recruitment, growth and mortality of tree species in a lowland dipterocarp forest in Peninsular Malaysia. J. Trop. Ecol., 3: 315-330.

Manokaran N, LaFrankie Jr JV (1990). Stand structure of Pasoh forest reserve, a lowland rainforest in peninsular Malaysia. J. Trop. For. Sci., 3: 14-24.

McNeely JA, Nees G (1996). In Human population, biodiversity and protected areas: Science and Policy Issues 9th ed. Dompka, Victoria. American Association for the Advancement of Science, pp. 19-70, Washington DC.

Menhinick EF (1964). A comparison of some species-individual diversity indices applied to samples of field insects. Ecology, 45: 858-862.

Menon S, Ponttius RGJr, Rose J, Khan ML, Bawa KS (2001). Identifying conservation-priority areas in the tropics: a land-use change modeling approach. Conserv. Biol., 15: 501-512.

Mueller-Dombois D, Ellenberg H (1974). Aims and Methods of Vegetation Ecology. John Wiley and Sons, New York.

Myers N, Mittermeier RA, Mittermeier CG, da Fonseca GAB, Kent J (2000). Biodiversity hotspots for conservation priorities. Nature, 403: 853-858.

Nath PC, Arunachalam A, Khan ML, Arunachalam K, Barbhuiya AR (2005). Vegetation analysis and tree population structure of tropical wet evergreen forests in and around Namdapha National Park, Northeast India. Biol. Conserv., 14: 2109-2136.

Newbery DM, Campbell EJF, Lee YE, Ridsdale CE, Still MJ (1992). Primary lowland dipterocarp forest at Danum Valley, Sabah, Malayasia: structure, relative abundance and family composition. Phil. Trans. R. Soc. Lond., B 335: 341-356.

Oliver CD, Larson BC (1990). Forest stand dynamics. McGraw Hill Inc. New York.

Parthasarathy N, Karthikeyan R (1997). Biodiversity and population density of woody species in a tropical evergreen forest in Courtallum reserve, Western Ghats, India. Trop. Ecol., 38(2): 297-306.

Pascal JP, Pelissier R (1996). Structure and floristic composition of a tropical evergreen forest in southwest India. J. Trop. Ecol., 12: 191-214.

Philips EA (1959). Method of vegetation study. Henry Holt and Co. Inc. New York.

Philips MS (1994). Measuring trees and forests. 2nd Ed. CAB International, New York. P. 310.

Poore MED (1968). Studies in Malayasain forests 1. The forest on Triassic sediments in Jengka forest reserve. J. Ecol., 56: 143-196.

Proctor J, Anderson JM, Chai P, Vallack HW (1983). Ecological studies of four contrasting lowland rainforest types of Gunung Mulu National Park, Sarawak.I. Forest environment, structure and floristics. J. Ecol., 76: 320-340.

Proctor J, Haridasan K, Smith GW (1998). How far does lowland evergreen tropical rainforest go? Global Ecology and Biogeography Letters, 7: 141-146.

Ramakrishnan PS (1992). Shifting Agriculure and Sustainable Development of North-Eastern India. UNESCO MAB Series, Paris, and Parthenon Publishing, UK. (Reprinted by Oxford University Press, New Delhi, 1993).

Rao P, Barik SK, Pandey HN, Tripathi RS (1997). Tree seed germination and seedling establishment in tree fall gaps and understory in a subtropical forest of north-east India. Australian J. Ecol., 22: 136-145.

Richards PW (1996). The tropical rainforest (2nd edn). Cambridge University press.

Ringu P (2002). An action plan for the strengthening protected area network and eco-development in Arunachal Pradesh. In: Arunachal Pradesh: Environmental Planning and Sustainable Development, (eds. Sundriyal RC, Singh T and Sinha GN), Himavikash. 16: 249-254.

Rodgers WA, Panwar HS, Mathur VB (2000). Wildlife Protected Area Network in India: A Review Executive Summary. Wildlife Institute of India, Dehradun, India. pp. 44.

Romero-Duque LP, Jaramillo VJ, Perez-Jimenez A (2007). Structure and diversity of secondary tropical dry forest in Mexico, differing in their prior land-use history. For. Ecol. Manag., 253: 38-47.

Shannon CE, Wiener W (1963). The mathematical theory of communities. University of Illinois press, Urbana, Illinois, p.117.

Singh JS, Pandey U, Tiwari AK (1984). Man and forests: a Central Himalayan case study. Ambio., 13: 80-87.

Singh JS, Singh SP, Saxena AK, Rawat YS (1981). The Silent Valley forest ecosystem and possible impact of proposed hydroelectric project. Report on the Silent valley study. Ecology Research Circle, Kumaun University, Nainital, India.

Sist P, Saridan A (1988). Description of the primary lowland forest of Berau. In: Silvicultural Research in a lowland mixed Dipterocarp Forest of East Kalimantan. (eds. Bertault J, Kadir K), Contribution of STREK project, CIRAD-forest, Indonesia. pp. 435-457.

Sorensen T (1948). A method of establishing groups equal amplitude in plant sociology based on similarity of species content. Det. Kong. Danske Vidensk, Selsk Biology Skr (Copenhagen) 5: 1-34.

Sugihara G (1980). Minimal community structure: an explanation of species abundance patterns. Am. Nat., 116: 770-787.

Sundriyal RC, Sharma E (1996). Anthropogenic pressure on tree structure and biomass in temperate forest of Mamlay watershed in Sikkim. For. Ecol. Manage., 81: 113-134.

Sundriyal RC, Sharma E, Rai LK, Rai SC (1994). Tree structure, regeneration and woody biomass removal in a subtropical forest of Mamlay watershed in the Sikkim Himalaya. Vegetation, 113: 53-63.

Swamy PS, Sundarapandian SM, Chandrasekaran S, Chandrasekar P (2000). Plant species diversity and tree population structure in a humid tropical forests of Tamil Nadu, India. Biodivers. Conserv., 9: 1-27.

Uma S (2001). A case of high tree diversity in a sal (*Shorea robusta*)-dominated lowland forest of Eastern Himalaya: floristic composition, regeneration and conservation. Curr. Sci., 81(7): 776-786.

Upreti N, Tewari JC, Singh SP (1985). The oak forests of Kumaun Himalaya (India): Composition, Diversity and regeneration. Mount. Res. Dev., 5(2): 163-174.

Whitmore TC (1998). An introduction to tropical rain forests (2nd Edition). Clarendon Press, Oxford.

Whittaker RH (1975). Communities and ecosystems. 2nd ed. Macmillan

Publi. Co., New York.
Wilkie P, Argent G, Campbell E, Saridan A (2004). The diversity of 15 ha of lowland mixed dipterocarp forest, Central Kalimantan. Biodivers. Conserv., 13: 695-708.
Zent EL, Zent S (2004). Floristic composition, structure and diversity of four forests in the Siera Maigualida, Venezuelan Guayana. Biodivers. Conserv. 13(13): 2453-2483.

Appendix I. Importance Value Index (IVI) of tree species in the three studies forest stands at the buffer zone area of Namdapha National Park, Arunachal Pradesh.

Species	*Altingia*-mixed species stand	*Shorea-Dipterocarp* stand	*Albizia* stand
Altingia excelsa Noronha	51.72	15.10	-
Ostodes paniculata Bl. Bigdr	23.55	13.05	-
Talauma hodgsonii Hk.f & Th.	10.60	3.80	2.25
Dipterocarpus macrocarpus Vesque	10.30	24.26	-
Dysoxylum procerum Hiern	8.04	4.07	-
Schima wallichii (DC) Korth	7.24	5.17	-
Chisocheton paniculatus (Roxb.) Hiern.	6.74	1.96	-
Leea indica (Burn.f. &) Merr	6.29	9.04	-
Saprosma ternatum Hk.f	6.02	5.97	-
Syzygium cumini (Linn.) Skeels	6.01	3.06	-
Dysoxylum binectariferum Hk.f. et. Bedd.	5.89	-	-
Syzygium macrocarpum (Roxb.) Balak.	5.18	-	-
Cinnamomum bejolghota (Buch-Ham) Sweet	5.08	-	-
Duabanga grandiflora (Roxb ex DC.) Walp.	5.00	10.04	-
Terminalia myriocarpa Heurck & Muell-Arg	4.55	23.81	9.64
Ficus sp. 2	4.18	-	-
Baccaurea ramiflora Lour	4.00	12.84	-
Bridelia assamica Hk. f.	3.63	-	-
Michelia oblonga Wall. ex. Hk.f &Th.	3.51	2.40	-
Cedrella toona Roxb. ex. Roth	3.49	-	-
Micromellum sp.	3.46	-	-
Litsea salicifolia (Roxb. ex. Nees) Hk. f.	3.30	-	-
Cinnamomum glaucescens (Nees) Meissn	3.27	-	-
Elaeocarpus aristatus Roxb	3.19	8.19	-
Euonymus sp. 2	2.67	-	-
Castanopsis indica (Roxb.) A. DC.	2.55	13.71	-
Miliusa roxburghiana (Wall) Hk.f & Th.	2.51	-	-
Mesua ferrea Linn.	2.44	-	-
Aphania rubra (Roxb.) Radlk.	2.34	-	-
Castanopsis tribuloides (Smith) A. DC.	2.28	1.87	-
Griffithianthus fuscus Merr.	2.27	5.78	-
Antidesma acuminatum Linn.	2.25	-	-
Beilschmiedia sp.	2.14	-	-
Milletia sp.	2.10	-	-
Laportea pterostigma Wedd.	2.08	-	-
Castanopsis sp. 3	1.96	4.62	-
Shorea assamica Dyer	1.70	23.66	-
Styrax serruletum Roxb	1.66	-	-
Pterygota alata (Roxb.) R. br.	1.48	-	-
Macropanax dispermus (Bl.) O. Ktze	1.43	-	-
Euonymus sp. 1	1.32	-	-
Diospyros sp.	1.28	-	-
Murraya paniculata (Linn.) Jack	1.26	-	-
Knema angustifolia (Roxb.) Warb.	1.23	-	-
Lasianthus longicauda Hk. F.	1.23	-	-
Persea sp.	1.16	-	-
Magnolia sp. A	1.15	-	-
Premna sp.2	1.12	-	-

Appendix I. Contd.

Cinnamomum sp.	1.11	-	-
Ficus sp. 1	1.09	-	-
Olea dentata Wall ex DC	1.08		-
Beilschmiedia assamica Meissn.	1.07	-	-
Quercus Semiserrata Roxb.	1.04	-	-
Canarium strictum Roxb.	1.03	-	-
Elaeocarpus sp.	1.03	-	-
Linociera macrophylla Wall. ex. DC	1.03	-	-
Meliacecae	1.03	-	-
Premna sp.	1.02	-	-
Aporosa dioica (Roxb.) Muell-Arg	1.01	4.97	-
Croton roxburghii Balakr.	1.01	-	-
Ficus sp. 3	1.01	-	-
Lindera latifolia Hk. f.	1.01	-	-
Magnolia griffithii Hook.f.& Thomson	1.01	-	-
Pterospermum lancifolium	1.01	-	-
Quercus lamellosa Smith	1.01	-	-
Saurauia cerea Griff. ex Dyer	1.01	-	-
Styrax sp.	1.00	-	-
Actinodaphne obovata (Nees) Bl.	-	6.30	-
Ailanthus grandis Prain	-	3.19	0.97
Alangium chinense (Lour) Harms	-	-	1.17
Albizia procera (Roxb.) Benth	-	-	155.30
Bischofia javanica Bl.	-	6.60	-
Bombax ceiba avet. non Linn	-	-	25.05
Callicarpa arborea Roxb.	-	2.18	2.74
Capparis acutifolia (Hook. f. & Thomson) Jacobs.	-	2.40	-
Castanopsis sp. 4	-	7.33	-
Cordia dichotoma Forst. f.	-	-	0.97
Dalbergia pinnata (Lour.)Prain	-	-	36.67
Dalbergia sericia G. Don	-	-	24.19
Dillenia indica Linn.	-	2.25	-
Dysoxylum sp.	-	7.60	-
Ehretia acuminata R. Br.	-	-	1.36
Engelhardtia spicata Leschen ex Bl. Bigdr	-	-	1.43
Eriobotrya bengalensis Hk. f.	-	3.97	-
Erythrina stricta Roxb.	-	-	2.71
Glochidion lanceolarium (Roxb.) Voigt	-	-	10.28
Gmelina arborea Roxb.		-	15.06
Grewia disperma. Roxb.	-	-	1.00
Helicia robusta Wall ex Benn	-	2.60	-
Lindera sp.	-	6.37	-
Litsea monopetala (Roxb.) pers	-	-	1.89
Litsea sp.	-	2.65	-
Maesa indica (Roxb.) Wall	-	-	1.16
Magnolia sp. B	-	6.33	-
Melia dubia Cavv.	-	-	1.76
Pterospermum acerifolium Willd.	-	1.86	-
Quercus sp.	-	2.69	-
Syzygium sp.	-	3.72	-

Appendix I. Contd.

Terminalia chebula Retz.	-	3.45	-
Unidentified*	42.53	31.12	4.40

Impact of interspecific competition by compatriot aquatic weeds on water hyacinth *Eichhornia crassipes* (Martius) Solms growth and development in the Kagera River

Katagira, F.[1]*, Kyamanywa, S.[2], Tenywa, J. S.[2], Rajabu, C. A.[3], Sombe, D.[4], and Ndunguru, J.[3]

[1]Ministry of Agriculture, Food Security and Cooperatives, P. O. Box 9071, Dar-es-Salaam, Tanzania.
[2]Makerere University, Faculty of Agriculture, P. O. Box 7062, Kampala, Uganda.
[3]Mikocheni Agricultural Research Institute, Box 6226, Dar es Salaam, Tanzania.
[4]Lake Victoria Environment Management Project, Box Bukoba, Kagera, Tanzania.

The phenotype of a plant is strongly influenced by the presence of neighbouring plants often resulting into change in growth and development. We have demonstrated to our knowledge for the first time, the influence of three aquatic weed species (*Commelina* sp., *Justicia* sp. and *Vossia cupsidata*) on growth and development of water hyacinth (*Eichhornia crassipes*). We found that *Commelina* sp., *Justicia* sp. and *V. cupsidata* significantly (P < 0.001) suppressed water hyacinth total fresh weight when growing together both in the Kagera River and greenhouse. *Justicia* sp. gave the highest (51%) water hyacinth fresh weight reduction when compared to fresh weight in pure stand probably due to competition effect. In the Kagera River but not greenhouse, water hyacinth plants were significantly taller ($F_{(4,1371)}$ = 150.53, P < 0.001) in pure stands than when growing in mixtures. *V. cupsidata* strongly suppressed water hyacinth plant height although water hyacinth compensated by producing elevated number of ramets (daughter plants). In greenhouse environment, the number of ramets per plant was higher in mixtures than in pure water hyacinth treatment. Correlation analysis revealed a strong significant negative relationship between plant height and number of ramets. Competition for space and resources was suggested to be the major influential factor as water hyacinth in pure stands tended to have higher density and total biomass per unit area than when grown in mixtures. Analysis of water hyacinth ratios to other aquatic weeds showed a significant effect on water hyacinth growth and development in terms of fresh weight, plant height, leaves per plant and ramets. These data show that neighbouring aquatic weeds are important component in the regulation of water hyacinth growth and development in the aquatic ecosystem.

Key words: *Commelina* sp., *Justicia* sp., Kagera river, water hyacinth.

INTRODUCTION

Eichhornia crassipes (Martius) Solms is often regarded as the most troublesome aquatic weed in the world (Holm et al., 1991; Lu et al., 2007). *E. crassipes* possesses specialized growth habits, physiological characteristics, and reproductive strategies that allow for rapid growth and expansion in freshwater environments and has spread rapidly throughout the tropics and subtropics (Pushpa and John, 2010). It has become a serious weed in freshwater habitats in rivers, lakes and reservoirs in tropical and warm temperate areas worldwide, where it displaces native aquatic plant and animal communities (Navarro and Phiri, 2001). *E. crassipes* forms large, free-floating, monospecific mats that compete with other aquatic species for light, nutrients and oxygen (Pushpa and John, 2010). Water hyacinth floating mats reduce dissolved oxygen levels and light and significantly alter invertebrate and vertebrate communities. As biomass from mats decomposes, organic input to sediments

*Corresponding author. E-mail: jndunguru2003@yahoo.co.uk

increases dramatically (Gopal, 1987; Jafari, 2010). *E. crassipes* grows in shallow temporary ponds, wetlands and marshes, sluggish flowing waters and large lakes, reservoirs, and rivers (Center et al., 2005). Plants can tolerate extremes of water level fluctuation and seasonal variations in flow velocity, and extremes of nutrient availability, pH, temperature and toxic substances (Gopal, 1987; Pushpa and John 2010). *E. crassipes* forms dense, monospecific, free-floating mats in still to slow moving waters. Winds or currents may disperse these mats. During mat development, plants allocate most production to root biomass with little increase in average plant size. As plants mature, they increase in average biomass and production of daughter plants with reduced allocation to roots (Center et al., 2005). At peak density, daughter-plant production is reduced but average plant size continues to increase, resulting in plant mortality (Madsen, 1993; Center et al., 2005). Young plants in low-density mats form a great deal of float tissue. In higher density mats, the proportion of float tissue decreases as surrounding plants support each other (Sculthorpe, 1985; Center et al., 2005). In existing mats, plants reallocate biomass to the emergent shoots following winter. In early spring, ramet production increase, resulting in high leaf densities and high foliar height diversity (Center et al., 2005). Some smaller plants are lost, resulting in lower absolute density (Center and Spencer 1981; Center et al., 2005).

The spatial arrangements of the plants in herbaceous communities are seldom random on the scales at which competitive interactions are likely to occur (Cain et al., 1995; Keddy, 2001). Neighbourhood models relate the performance of individual plants to the competitors surrounding them, by relating the growth, seed-production, or survival of target plant to the number, proximity, and sometimes size, of competitors (Mack and Harper, 1977; Weiner, 1984; Center et al., 2005). Kagera River, which originates in the highlands of northern Rwanda and discharge into the Lake Victoria in Northwestern Tanzania, is densely colonized by aquatic water hyacinth. Water hyacinth may be found in association with a variety of other deep water or free-floating aquatic plants. In the this river, water hyacinth is commonly found growing with macrophytes, especially hippograss *Vossia cupsidata*, *Justicia* sp., *Cemmelina* sp. and papyrus, *Cyperus papyrus* (Ndunguru et al., 2000). This study was initiated with an objective of finding out if at all there was an influence of the neighbouring aquatic plants on the growth and development of water hyacinth in the Kagera River.

MATERIALS AND METHODS

The effect of compatriot aquatic weed species, their ratios and their interaction on the growth performance of water hyacinth was conducted in the Kagera River (natural habitat) and in the greenhouse (at Kyaka station) in Tanzania between 2005 and 2006.

Study species

Three aquatic weed species commonly growing and fringing the open water of the Kagera River in mixture with water hyacinth (*E. crassipes*) was chosen for this study. Hippograss, *V. cupsidata* (Roxb.) Griff. is a characteristic grass of the stump areas African flood plains and forms dense, semi floating, beds and established mats persist for an indefinite period under less than optimum period (Narror and Phiri, 2001). Plants usually grow in water up to 1 m deep at the margins of water bodies and fringes the open waters, develop large number of roots and rootless from the nodes of the spongy stem (Bor, 1960; Pushpa and John, 2010). *Justicia* sp. grows in shallow waters, margins and beds of streams, marshy shores and ponds, lakes and ditches. It can grow to a depth of 1.3 m (Carter and Grace, 1986). Once established, plants spread by rhizomes and often form large colonies and form roots at the nodes and plants can ascend above to about 1 m tall (Penfound, 1940; Carter and Grace, 1986).

Description of study site

Kagera River that originates from the highlands of Burundi and Rwanda flows into the Lake Victoria on the northwestern side of Tanzania. The average width of Kagera River is between 40 and 70 m with a total depth of 3 to 6 m on average. Water temperature is between 20 to 22°C (Rutagemwa, 2001). Large part of Kagera River consists of scattered clumps of papyrus and *Vossia* fringing both banks of the river. Water hyacinth infestation in the Kagera River is heavy usually interspersed with clumps of waterweeds such as *Vossia* sp., *Commelina* sp., *Justicia* sp., papyrus and others fringing the river banks. Occasionally, pure water hyacinth 2 to 5 m wide fringes several other locations of the river and there is active flow of water hyacinth down the river. Water hyacinth in the Kagera River was reported for the first time in the 1980s probably originating in the upper reaches of the river probably in Rwanda (Labrada, 1996). Despite introduction of *Neochetina* weevil for water hyacinth control in the Kagera River, several impact assessment surveys revealed virtually no weevil establishment. It was hypothesized that the only stress-induced on fringing water hyacinth is that resulting from competition by other neighbouring aquatic weed species for resources like food, light, water, space for growth and reproduction.

Field site sampling

Wooden frame measuring (1 × 1 m) was used for sampling by placing it on the weed mats. To avoid confounding edge effects, sampling was done by throwing the frame into weed mats at the centre to define 1 m^2 sampling units in July 2005. Only a weed mixture containing at least 20% of the competing species (*V. cupsidata*, *Justicia* sp. and *Commelina* sp) was chosen for sampling. The weed mixtures sampled were: (1) Water hyacinth + *Vossia cupsidata*; (2) Water hyacinth + *Justicia* sp.; (3) Water hyacinth + *Commelina* sp., (4) Water hyacinth + *V. cupsidata* + *Justicia* sp. + *Commelina* sp., and (v) Water hyacinth alone (Figure 1). For each weed mixture, a total number of water hyacinth plants and the other species in the frame were counted. Then 10 water hyacinth plants were randomly removed from each frame and individual plants weighed after draining the water briefly to obtained total plant fresh weight using a spring balance. Plant height of water hyacinth was measured using a special calibrated ruler from the base of the plant to the apical tip of the tallest leaf. In addition, number of leaves and ramets (stoloniferous daughter plant) per plant was counted and recorded. For each weed mixture, a total of 32 sampling points (320 plants) selected randomly on both sides of the Kagera River were sampled using a motorized 15´ fibre grass

Figure 1. Different water hyacinth/aquatic weed mixtures sampled in the Kagera River (A to E): (A) Pure water hyacinth stand; (B) Water hyacinth + *Commelina* sp.; (C) Water hyacinth + *Justicia* sp.; (D) Water hyacinth + *V. cupsidata*; (E) Water hyacinth + *Commelina* sp. +*Justicia* sp. + *V. cupsidata,* and (F) Elevated number of water hyacinth ramets in response to light competition with *V. cupsidata*.

boat following a zigzag pattern. Distance between sampling point was between 100 to 150 m. A similar sampling was conducted in 32 points comprising of pure water hyacinth stand to serve as control treatment.

Greenhouse experiment

Young water hyacinth (*c.* 15 cm long) collected from the Kagera River were planted in plastic containers (53 cm diameter) filled with water 50 L from the same Kagera River. Water hyacinth was grown in monoculture, and in combination (1) Water hyacinth + *V. cupsidata*, (ii) Water hyacinth + *Justicia* sp. (iii) Water hyacinth + *Commelina* sp. and (iv) Water hyacinth + *V. cupsidata* + *Justicia* sp. + *Commelina* sp. Both *Commelina* and *Justicia* sp. plants used for this experiment were young with at least 5 nodes long. The size of *V. cupsidata* ranged from 15 to 20 cm long. To avoid intra species competition for water hyacinth, only a single plant was placed in each container. The number of other weed species introduced per container was five and there were four replicates for each

Table 1. Water hyacinth growth characteristics sampled when growing with different aquatic weed combinations in the Kagera River.

Weed combination	Fresh weight (g)	Plant height(cm)	Leaves/plant	Plant density/m^2	Total biomass/ m^2	Ramets/plant
Water hyacinth	652.7 ± 56.9 ^{c}	38.4 ± 0.6 ^{e}	9.3 ± 7 ^{b}	51.7 ±3.4 ^{c}	30.98 ± 4.21 ^{c}	1.3 ± 6 ^{b}
Water hyacinth + *Commelina* sp	452.2 ± 12.7 ^{b}	33.7 ± 0.6 ^{d}	9 ± 7 ^{ab}	38.9 ± 2.7 ^{b}	16.09 ± 1.27 ^{b}	0.8 ± 5 ^{a}
Water hyacinth + *Justicia* sp	320.2 ± 9.6 ^{a}	26.5 ± 0.9 ^{b}	8.6 ± 0.27 ^{a}	37.9 ± 2.3 ^{b}	12.3 ± 1.01 ^{a}	0.8 ± 5.5 ^{a}
Water hyacinth + Hippograss	332.7 ± 27 ^{a}	21.8 ± 0.5 ^{a}	8.6 ± 7 ^{a}	26 ± 2.5 ^{a}	7.77 ± 0.82 ^{a}	1.2 ± 7.5 ^{b}
Water hyacinth + *Commelina* sp + *Justicia* sp + Hippograss (*Vossia cupsidata*)	342 ± 11.7 ^{a}	29.2 ± 0.6 ^{c}	9 ± 0.28 ^{ab}	28.3 ±2.7 ^{a}	9.50 ± 1.21 ^{a}	0.73 ± 5.4

[a]Means (± SE) followed by the same letters are not statistically significant different at 0.05.

Table 2. Variations of water hyacinth growth parameters at different aquatic weed combination in controlled on-station experiment (N = 24).

Weed combination	Fresh weight (g)	Plant height(cm)	Leaves/plant	Plants/container	Ramets/plant
Water hyacinth	180.8 ± 8.52^c	8.69 ± 0.42^a	10.54 ± 0.38^a	1.2 ± 8.46^a	0.21 ± 8.46^a
Water hyacinth + *Commelina* sp.	129.08 ± 9.26^{bc}	8.8 ± 0.6^a	10.7 ± 0.46^a	1.5 ± 0.15^{ab}	0.5 ± 0.15^a
Water hyacinth + *Justicia* sp.	151.66 ± 9.56^c	8.88 ± 0.43^a	9.9 ± 0.38^a	1.87 ± 0.14^{bc}	0.87 ± 0.13^{bc}
Water hyacinth + Hippograss	98.19 ± 4.57^a	8.24 ± 0.29^a	9.75 ± 0.46^a	1.37 ± 0.1^a	0.37 ± 0.1^a

treatment. Plants were left to grow in the pots for a period of 4 months (July to October 2005). The containers were randomly arranged on four lines on the floor in the greenhouse and re-randomized twice three times in the course of the experiment. Water hyacinth growth parameters similar to that collected in a typical natural population (in the Kagera River) were taken in the date of experimental initiation and onward every after 2 weeks for a period of three months. Plants were not fertilized.

Data analysis

Averages taken from all plants examined in each sampling unit were used in all analysis (n = 32 units per weed mixture). Per unit biomass of water hyacinth in the field samples was calculated by multiplying mean fresh weight by its density per unit area. Pearson correlations were used to test for associations between fresh weight and plant height, life leaves per plant, and ramets. The effect of weed mixture on water hyacinth growth and development and their mixture ratios for field samples were assessed with weed mixture as a fixed effect in analysis of variance (ANOVA) and Duncan mean comparison ($P < 0.05$) using SPSS 10.0 statistical package. For greenhouse experiment, effect of weed combination, time and its interaction were assessed with weed combination as main factor using a multivariate analysis of variance GLM (SPSS 10.0) and Duncan mean comparisons ($P < 0.05$). Effect of weed mixtures on per unit area biomass of water hyacinth was assessed using one-way ANOVA.

RESULTS

Total plant fresh weight

Total water hyacinth fresh weight was significantly higher ($F_{(4,1371)} = 23.50$, $P < 0.001$) in water hyacinth growing alone than when growing with other aquatic weeds in the natural habitat of the Kagera River (Table 1). The lowest fresh weight (320.2 ± 9.6 g (g) was recorded for water hyacinth growing with *Justicia* sp. *Commelina* sp. had less negative impact on water hyacinth fresh weight than other aquatic weed species (Table 1). As observed in the natural aquatic habitat in the Kagera River, water hyacinth grown alone in green house displayed significantly ($F_{(4,1371)}$, $P < 0.001$) more fresh weight than when grown in mixture (Table 2). Fresh weight was the lowest (98.19 ± 4.57 g) for water hyacinth plants grown with *V.cupsidata* compared to control (water hyacinth grown alone) (Table 2).

Water hyacinth plant height

For the study in the Kagera River, statistical

Table 3. Multivariate analysis of variance describing significance of factors weed combination (WC), and time (T) on water, hyacinth growth parameters.

Source	Dependent variable	df	Mean square	F	Sig.
	Fresh weight (g)	4	26510.554	13.899	0.000 S
	Plant height (cm)	4	2.993	0.707	0.591 NS
	Number of leaves per plant	4	4.096	0.8	0.528 NS
WC	Ramets per plant	4	4.221	8.268	0.000 S
	Number of plants/container	4	4.221	8.268	0.000 S
	Error		115		
	Total		119		
	Fresh weight (g)	5	3586.212	1.330	0.257 NS
	Plant height (cm)	5	24.774	7.479	0.000 S
	Number of leaves/plant	5	80.073	44.507	0.000 S
T	Ramets/plant	5	2.088	3.654	0.004 S
	Number of plants/container	5	2.088	3.654	0.004 S
	Error		114		
	Total		119		
	Fresh weight (g)	20	743.352	0.359	0.994 NS
	Plant height (cm)	20	4.167	1.328	0.182 NS
	Number of leaves/plant	20	1.888	1.124	0.341 NS
T × WC	Ramets/plant	20	0.276	0.581	0.917 NS
	Number of plants/container	20	0.276	0.581	0.917 NS
	Error		90		
	Corrected total		119		

analysis showed that water hyacinth plants growing alone (without competition) were significantly ($F_{(4,1371)} = 150.53$ P 0.001) taller than those growing in association with other aquatic weeds (Table 1). Hippograss (*V. cupsidata*) strongly suppressed growth of water hyacinth as indicated by a reduction in plant height from 39.4 ± 0.6 cm in pure stand to 21.8 ± 0.5 cm in mixture with *V. cupsidata* (Table 1). In controlled environment in the green house experiment, none of the aquatic weed species had a significant competitive effect on the plant height of water hyacinth (Table 3). Similarly, there was no significant effect of other aquatic weeds on water hyacinth number of leaves per plant in both greenhouse and the River Kagera (Table 3).

Impact of neighbouring aquatic weeds on water hyacinth plant density and biomass

In the analysis of the full data set obtained from the Kagera River, water hyacinth plant density was significantly higher ($F_{(4,115)} = 13.32$, $P < 0.001$) when growing alone than in mixture(Table 1). Compared to the control treatment (pure water hyacinth stand), water hyacinth growing together with hipppograss (*V. cupsidata*) showed the least plant density (26.03 ± 2.5 plants/m^2) about 49.6% reduction (Table 1). For water hyacinth in pure stand, plant density ranged from 25 to 102 plants/m^2 while those in mixture with *V. cupsidata*, *Justicia* sp. or *Commelina* sp. ranged only from 5 to 75 plants/m^2. Mean per unit area water hyacinth biomass varied significantly between weed mixtures ($F_{(4,154)}$ = 18.587, $P < 0.001$). The highest biomass was recorded for pure water hyacinth (30.98 ± 4.21 Kg/m^2) on average and the lowest (7.7 ± 0.82 Kg/m^2) for water hyacinth + *Vossia cupsidata* mixture, an almost 75% reduction (Table 1). For greenhouse experiment, number of water hyacinth plants per container significantly changed with weed combination (Table 3). When water hyacinth was grown alone, the number of water hyacinth plants averaged 1.2 ± 8.46 per container. In the overall greenhouse experiment, the highest water hyacinth plant number per container was observed in water hyacinth + *Commelina* sp. + *Justicia* sp. + *V. cupsidata* combination and ranged from one to four (Table 1).

Impact of neighbouring aquatic weeds on water hyacinth ramet production

Water hyacinth reproduces mainly asexually by means of stoloniferous ramets produced. To investigate the

Table 4. Pearson correlation analysis of water hyacinth growth parameters for on-station experiment after growing with other aquatic weeds.

Field study	FW	Ph	Lp	R[a]
FW	1.000			
Ph	0.261**	1.000		
Lp	0.065*	0.101**	1.000	
RM	0.178**	-0.053*	0.059*	1.000
Greenhouse study				
FW	1.000			
Ph	0.161	1.000		
LP	-0.047	-0.120	1.000	
RM	0.191*	-0.229*	0.255**	1.000

** Correlation is significant at the 0.01 level (2-tailed); *Correlation is significant at the 0.05 level (2-tailed); [a]FW = fresh weight (gm); Ph = plant height (cm); LP = Leaves/plant; RM = ramets/plant.

interspecific competition effect of neighbouring aquatic weeds on water hyacinth's ramet production in the Kagera River, the number of ramets for individual water hyacinth plant in each weed combination was counted. The number of water hyacinth ramets was significantly lower ($P < 0.001$) in weed mixtures than in pure stand (Table 3) except in water hyacinth + *V. cupsidata* weed combination. In this treatment, number of ramet was higher (1.2 ± 7.5 ramets/plant) by 37% than in other mixtures. The maximum number of ramets per plant in this weed mixture reached four per plant. The least number of ramets per plant on average was recorded for water hyacinth + *Commelina* sp. + *Justicia* sp. + *V. cupsidata* combination (Table 1). In the greenhouse experiment, number of ramets significantly varied between treatments (Table 3). Number of ramets per plant was significantly higher in mixtures than in water hyacinth monoculture. Water hyacinth + *Commelina* sp. + *Justicia* sp. + *V. cupsidata* combination gave the highest number of ramets (1.25 ± 0.2 ramets/plant) on average followed by water hyacinth + *Justicia* sp. combination (Table 2).

Relationship between water hyacinth growth components

When Pearson correlation analysis was performed on Kagera River sample data, there was a negative significant relationship between plant height and number of ramets (Table 4). A similar relationship was observed for the greenhouse experiment. Plants with high fresh weight tended to have significantly large number of ramets both in the field and greenhouse (Table 4). In addition, they appeared to be taller in the natural habitat than in greenhouse condition. There was no correlation between mean water hyacinth fresh weight and plant height in the greenhouse experiment.

Changes in water hyacinth plant growth parameters over time in the greenhouse

The water hyacinth plant growth parameters (plant height, number of leaves per plant, ramets per plant, and number of plants per container) significantly changed with time ($P < 0.001$) (Table 3). Water hyacinth plant fresh weight did not significantly change with time ($P > 0.05$) and there was no interaction between time and weed combination factors. After a period, water hyacinth plant height significantly decreased from 10.7 ± 0.46 cm at T0 to 8.17 ± 0.46 cm at T5 (Figure 2). However, mean number of water hyacinth plants in the containers significantly increased from one plant at T0 to 1.85 ± 0.19 plants at T5 (Figure 3). Number of water hyacinth plant ramets also significantly increased over time in such that by T5 (3 months), there were 0.85 ± 0.19 ramets/plant on average as compared to zero at T0 (beginning of the experiment).

Effect of ratios of water hyacinth to competitors on water hyacinth growth performance

To avoid genetic influence of different neighbour plant species on the results, the impact of water hyacinth to other aquatic weed ratio on water hyacinth growth and development was examined separately. The ratio was calculated by dividing the total number of water hyacinth plants per square meter by total number of other aquatic weed in the mixture. High value indicates that the number of the other species exceeds that of water hyacinth in the mixture and vice versa when the number is low. For Water hyacinth + *Commelina* sp. weed mixture, all the water hyacinth plant growth parameters examined (fresh weight, plant height, number of leaves and ramets) effect of water hyacinth to *Commelina* sp. ratio on water hyacinth growth and development was significant, $F_{(31, 281)}$

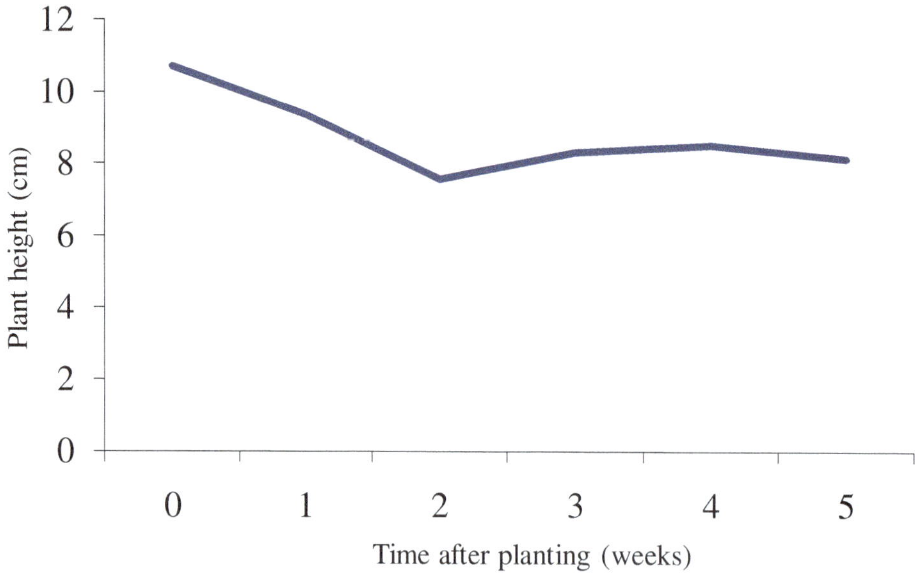

Figure 2. Changes in water hyacinth plant height with time in greenhouse experiment grown with different aquatic weed combination.

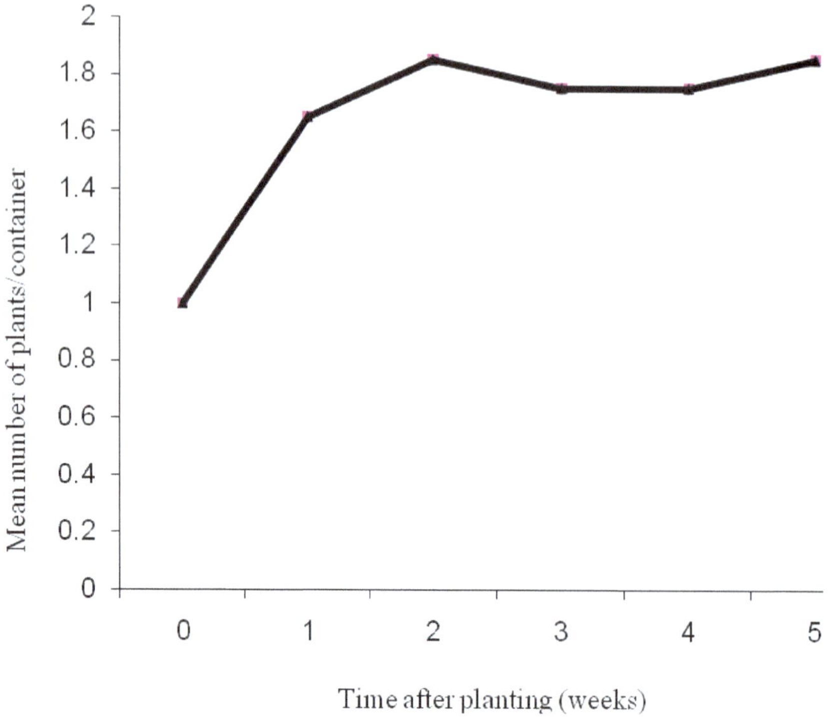

Figure 3. Changes in water hyacinth number of plants per container during the period of the experiment in the greenhouse.

= 4.00, $P < 0.001$ for fresh weight, $F_{(31, 281)} = 12.776$, $P < 0.001$) for plant height, $F_{(31, 281)} = 2.646$, $P < 0.001$ for number of leaves per plant and $F_{(31, 281)} = 3.385$, $P < 0.001$ for number of ramets per plant. Changes in water hyacinth total plant fresh weight in response to water hyacinth:*Commelina* sp ratio is presented in Figure 4a. With a few exceptions, as expected, low ratio values (when *Commelina* sp. density exceeded that of water hyacinth in the mixture) were associated with declining fresh weight. Similarly, high ratio values (example 1.45)

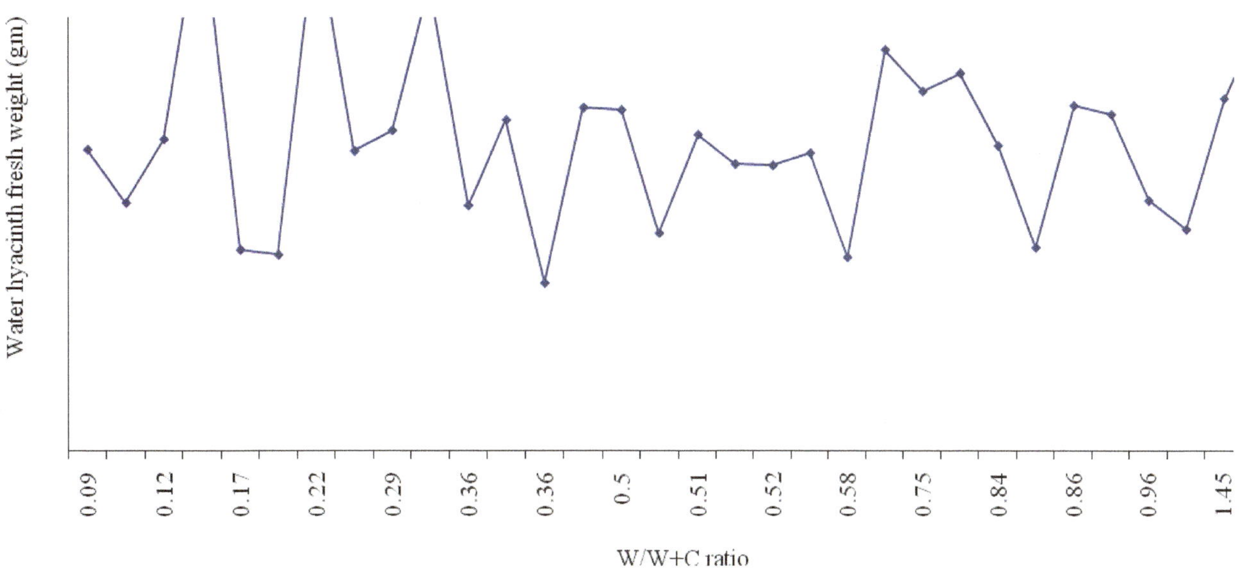

Figure 4a. Changes in water hyacinth fresh weight under water hyacinth (W) and *Commelina* sp. (C) weed mixture at different w/c ratios.

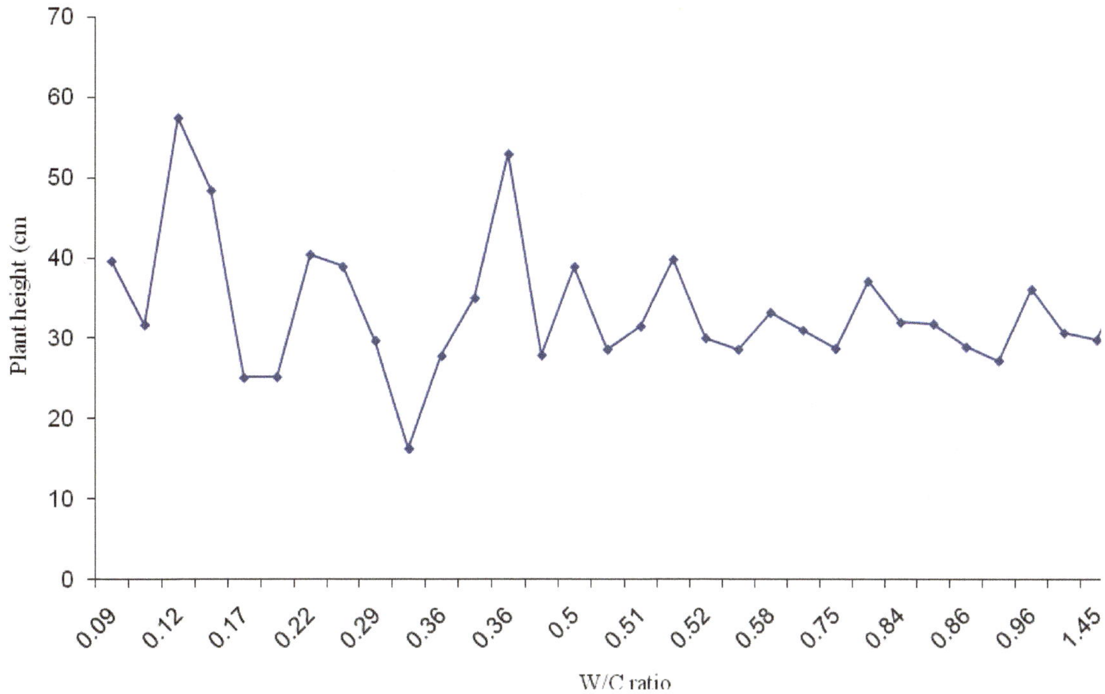

Figure 4b. Changes in water hyacinth plant height under water hyacinth (W) and *Commelina* sp. (C) weed mixture at different W/C ratio.

were associated with high fresh weight because of decreased density of *Commelina* sp. in the mixture, which imparted minimum competition effect on water hyacinth fresh weight (Figure 4a). In terms of water hyacinth plant height (Figure 4b) at low ratios (0.09 to 0.5), there was high water hyacinth mean plant height while at higher ratios (0.51 to 1.45), there was a relatively low water hyacinth plant height uniformly.

For water hyacinth + *Justicia* sp mixture, the ratio of water hyacinth (W) to *Justicia* sp. (J) had a significant effect on water hyacinth growth performance. Water hyacinth fresh weight significantly ($F_{(31, 277)}$ = 4.258, P <

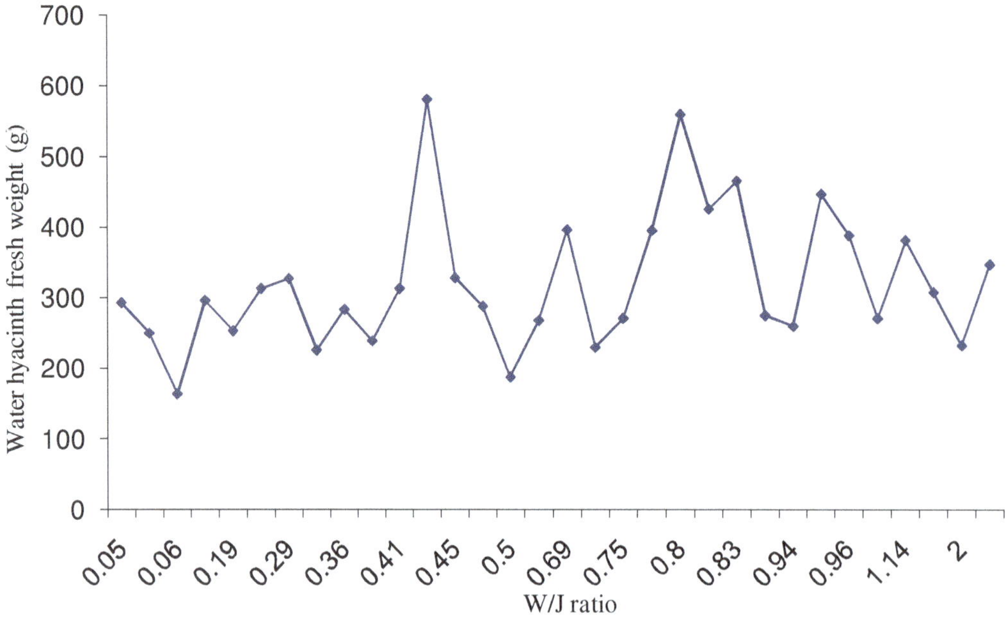

Figure 5a. Changes in water hyacinth fresh weight under water hyacinth (W) and *Justicia* sp. (J) weed mixture at different W/J ratios in the Kagera River.

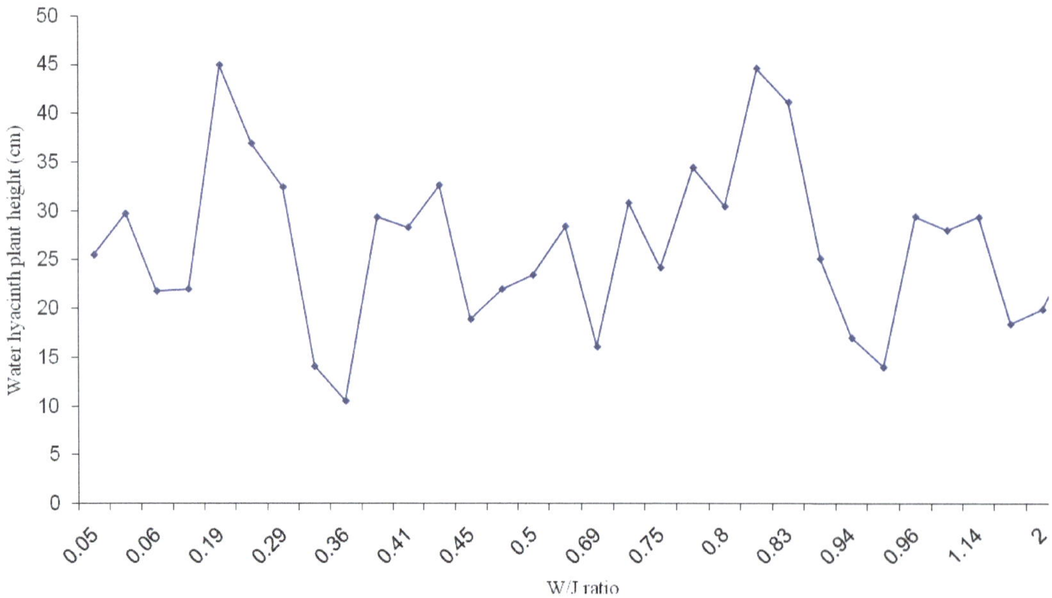

Figure 5b. Changes in water hyacinth plant height under water hyacinth (W) and *Justicia* sp. (J) weed mixture at different W/J ratios in the Kagera River.

0.001) changed with ratio. This pattern was also shown for water hyacinth plant height, mean number of leaves and ramets per plant. The W/J ratio ranged from 0.05 to 2.0. Figure 5 presents changes in water hyacinth fresh weight in response to W/J ratio. At low values (0.05 to 0.41) fresh weight was significantly low (Figure 5a) probably because of increased competition effect of *Justicia* as a result of elevated density in the weed mixture. At higher ratios, there was a significantly high fresh weight resulting from decreased competition effect from *Justicia* due to low density in the mixture. However, the trend was different for water hyacinth plant height (Figure 5b).

There was no significant effect of water hyacinth (W) to *V. cupsidata* (V) ratio on water hyacinth plant fresh weight (P = 0.291) and mean number of leaves per plant

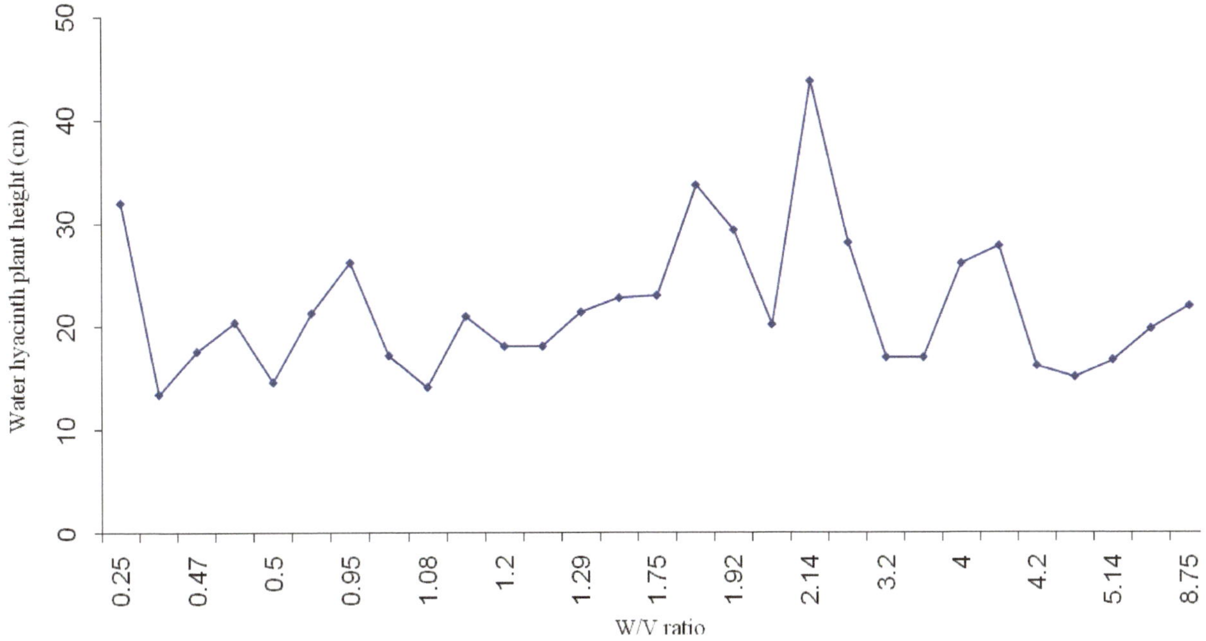

Figure 6a. Changes in water hyacinth plant height under water hyacinth (W) and *V. cupsidata* (V) weed mixture at different W/V ratios in the Kagera River.

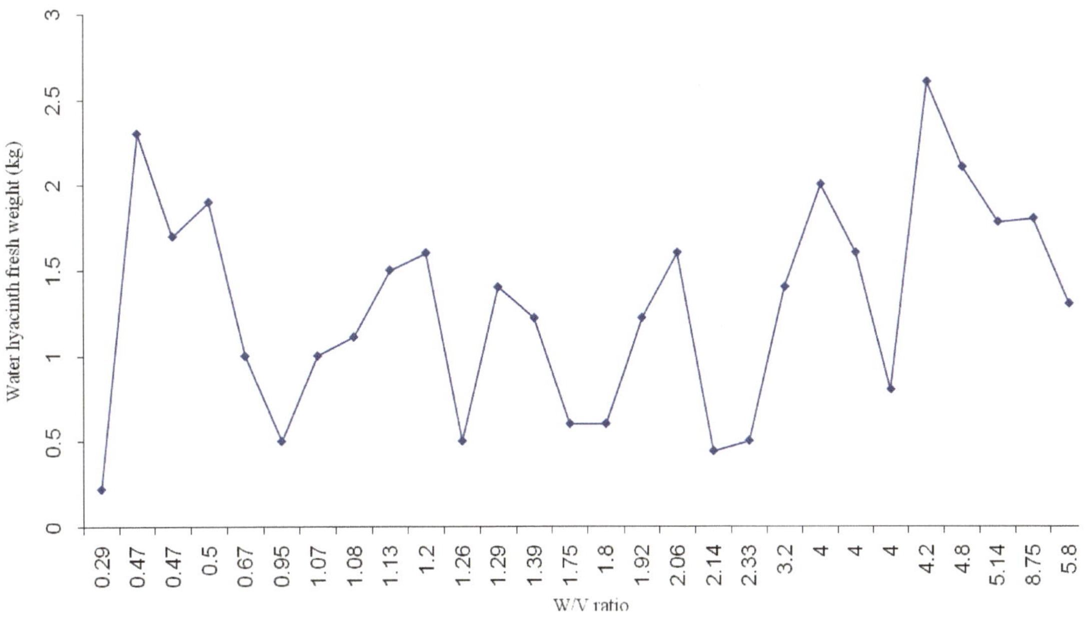

Figure 6b. Changes in water hyacinth fresh weight under water hyacinth (W) and *V. cupsidata* (V) weed mixture at different W/V ratios in the Kagera River.

($P = 0.094$). However, W/V ratio significantly influenced water hyacinth plant height ($F_{(28, 254)} = 11.298$, $P < 0.001$) and mean number of ramets per plant ($F_{(28, 254)} = 2.754$, $P < 0.001$). At low W/V ratios (Figure 6a and 6b), there was a reduced water hyacinth plant height as compared to that observed in high ratios. The effect of W/C+J+V ratio had significant effect on water hyacinth fresh weight, plant height as well as mean ramets per plant ($P < 0.001$). However, the ratio had no significant impact on number of life leaves per plant ($P = 0.399$). At elevated density of C+J+V in the mixture (low ratios), fresh weight was significantly reduced than in high ratios (0.56 to 1.19)

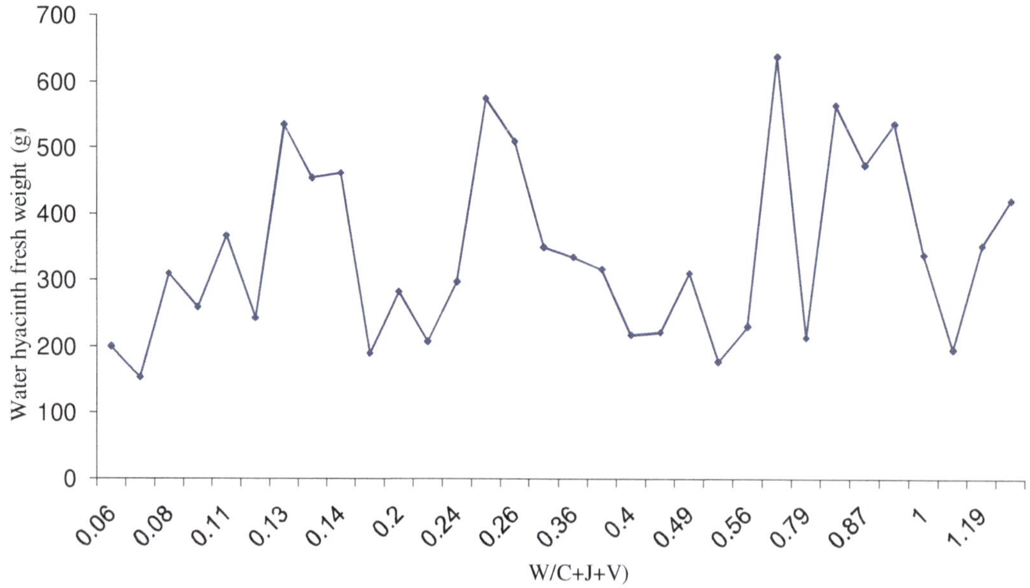

Figure 7a. Changes in water hyacinth fresh weight under water hyacinth (W+C+J+W) weed mixture at different W/C+J+V ratios in the Kagera River

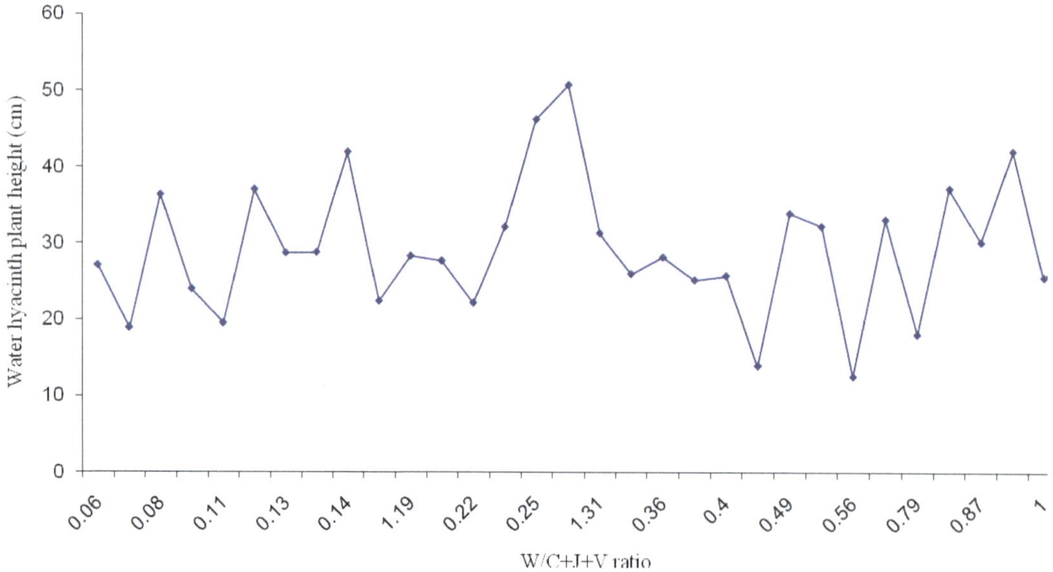

Figure 7b. Changes in water hyacinth plant height under water hyacinth (W+C+J+W) weed mixture at different W/C+J+V ratios in the Kagera River.

(Figure 7a). The pattern was not consistence probably because of inter species competition ability differences. The relationship between the ratios to water hyacinth plant height was therefore not clear (Figure 7b).

The interaction between weed mixtures and water hyacinth/other aquatic weed ration on growth and development of water hyacinth was significant in terms of all parameters measured (Table 5). Similarly water hyacinth to other aquatic weed ratios produced no significant effect on water hyacinth density across all the sampling units (Table 5).

DISCUSSION

The phenotype of a plant is strongly influenced by the presence of neighbouring plants. The present study demonstrated significant impact of neighbour plants on

Table 5. Summary of analysis of variance performed on the effect of weed mixtures, their combination ratios and interactions on water hyacinth growth performance in the Kagera River.

		Fresh weight		Plant height		Ramets		Plant density	
Factor	**df**	**F ratio**	**P**	**F ratio**	**P**	**F ratio**	**P**	**F ratio**	**P**
Weed combination (A)	4	23.506	<0.001	150.537	<0.001	25.964	<0.001	13.324	<0.001
Water hyacinth/other aquatic weed ratio (B)	31	1.729	<0.01	8.543	<0.001	2.506	<0.01	0.599	0.947
A × B	121	1.903	<0.001	8.401	<0.001	3.264	<0.001	ND	ND

growth and development of water hyacinth growing in mixture with other aquatic weed species. There was a strong suppression of hyacinth growth in terms of fresh weight, plant height, ramets and plant density probably due, to interspecific competition. In some occasions, however, water hyacinth plants responded to the competitive stress by feedback mechanisms, which involved production of elevated number of ramets, as was the case of water hyacinth growing in mixture with hippograss (*V. cupsidata*) in the Kagera River. This observation could be interpreted as shade avoidance reaction when growing with *V. cupsidata* which compete for light and water hyacinth plants could attempt to allocate functional leaves and roots in the resource (light) rich zones of their surrounding environment as this response increase the amount of light that can be captured, leading to increased plant fitness in crowded condition. This phenomenon has been recently demonstrated in other plant species by Pierik et al. (2004). Of the three aquatic weed species used in the present study, *Justicia* sp. when growing in high proportion with water hyacinth displayed a strong negative impact on water hyacinth fresh weight in the Kagera River suggesting that it exerted a high competitive stress on water hyacinth. This could be so because *Justicia* sp. produces a massive root system to help in competing for water resources, particularly nutrient. In the greenhouse experiment however, the highest reduction in water hyacinth fresh weight occurred in plants grown with *V. cupsidata*.

Observation from the study shows that water hyacinth plants growing under competition-free conditions (pure stand) had significantly higher fresh weight than when growing with other aquatic weeds both in the field and greenhouse. This is likely due to negative competition effect from other aquatic weeds in the mixtures for resources in the crowding environment resulting to reduction in fresh weight. Since the magnitude and extend of the reduction differed between plant species involved in the mixture clearly suggest that competition intensity also differed between individual plant species as demonstrated earlier (Grace, 1995; Center et al., 2005). Spencer and Ksander (2000) found a reduction in individual plant weights of hydrilla aquatic weed as a result of increased density of American pondweed in a mixture. Water hyacinth total biomass per unit area was also reduced in the presence of other aquatic species with *V. cupsidata* giving the highest reduction of 75% suggesting that it imparted a strong negative competition effect in the mixture, an indication of its dominance.

Interference between individual plant species in an ecological niche can affect individual plant survival, growth, and reproduction (Grace and Tilman, 1990; Keddy, 1989, Lu et al., 2007). Water hyacinth plant height in the Kagera River was affected by the presence of other aquatic weeds in the mixture. There was a reduction in plant height by other aquatic weeds with the highest reduction recorded when it was growing with *V. cupsidata*. The large size nature of *V. cupsidata* could have caused light shading effect probably by altering amount of light available to water hyacinth, which is a creeping plant and poor at invading tall grass. Spencer and Ksander (2000) using light measurement experiment reported that taller American pondweed plants clearly altered suppressed height of hydrilla in the mixtures by reducing amount of light that could be captured by it. Reduction in water hyacinth plant height by other aquatic weeds was not reproduced in greenhouse because three months was not long enough to allow for effective weed density of the competitor weed to build up large enough to bring about light reduction effect on water hyacinth. This observation is supported by a decrease in plant height with time in the greenhouse experiment. Rees (1995) large-seeded annual plants have greater competitive ability and that a competition-colonization trade off promotes co-existences, in which the smaller-seeded annuals are weak competitors that co-exist with more competitive species by virtue of their greater seed production. In the present study, the small sized water hyacinth plants survive by producing large number of ramets.

In the Kagera River, it was apparently that *V. cupsidata* significantly suppressed water hyacinth plant density and total biomass per unit area. This result may be a consequence of high interspecific competition for space and resources by *V. cupsidata* leading to a reduced biomass production and density particularly under elevated densities of *V. Cupsidata*. Decreased growth of individual submerged aquatic plant becuase of increasing plant density of neighbours has been reported for *Eleocharis acicularis* and *Juncus pelacarpus* forma submerses (MacGreary and Carpenter, 1987, Ndunguru et al., 2001). Correlation analysis revealed a significant negative relationship between plant height and ramets both in the field and greenhouse suggesting that decreased plant height was associated with elevated number of ramets per plant. This was observed when water hyacinth was growing with *V. cupsidata* and could be attributed to compensation effect as a light shade avoidance response. Center et al. (2005) observed nutrient availability as a factor that can influence interspecific weed competition. The influence of nutrients on water hyacinth competitional outcomes observed in the present study cannot be totally excluded.

Across all treatments in the Kagera River, water hyacinth growth parameters varied significantly with water hyacinth, other aquatic weed ratios such that at elevated density of the competitor (low ratios) resulted to a significant suppression plant height, and fresh weight reduction except for water hyacinth + *V. cupdiata* where effect of ratios were not significant. Water hyacinth plant density was not significantly affected by ratios. The effect of W/C+J+V ratio on water hyacinth fresh weight was significant but the pattern was not consistence probably because three different species with different competing abilities were involved in the crowding environment.

Although it had been shown that density-dependence only weakly regulates plant population sizes; this investigation in contrast, highlights the fact that high density of competing neighbour species in the mixture resulted to suppressed growth of water hyacinth at least in the Kagera River. The fact that some positive effect of competition from neighbourhood species was observed in the Kagera River suggests that other aquatic weed have both positive and negative effect on water hyacinth proliferation in the river. Interaction between water hyacinth and other aquatic weeds suggest that efforts to manage the weed should not only be focused on water hyacinth but rather on other aquatic weeds growing with it (ecosystem approach of weed management). Inability to reproduce the field data in the greenhouse needs further investigation. However, the competitive ability of a species depends on the environment.There is no 'super species' that are competitively superior in all environments; rather, there are some trade offs among the traits that are beneficial in some environments, but which cause plants to be poor competitors in other environments. For a plant to compete successfully in particular environment, it must have specific eco-physiological traits that allow effective growth in that environment. It would be unrealistic to expect parameters measured in the field to produce the same data in greenhouse, but the greenhouse allowed us to remove the variability of the real world, thus allowing the impact of some factors to be measured in isolation, and then use the results to generate hypothesis than can be further tested in the field.

REFERENCES

Cain ML, Pacala SW, Silander Jr JA, Fortin MJ. (1995). Neighborhood models of clonal growth in the white clover, Trifolium repens. Amer. Natur., 145: 888-917.

Carter MF, Grace JB. (1986). Relative effects of *Justicia americana* litter on germination, seedlings and established plantsof polygonum lapathifolium. Aquat. Bot., 23: 341-349.

Center TD, Spencer NR (1981). The phenology and growth of water hyacinth. (*Eichhornia crassipes* (Mart. Solms) Solns) in a eutrophic North-Central Florida lake. Aquat. Bot., 10:1-32.

Center TD, Van TK, Dray FA, Franks SJ, Rebelo MT, Pratt PD, Rayamajhi MB. (2005). Herbivory alters competitive interactions between two invasive aquatic plants. Biol. Control, 33: 173-185.

Gopal B (1987). Aquatic Plant Studies 1. Water Hyacinth. Elsevier Publishing, New

Grace JB (1995). On the measurement of plant competition intensity. Ecol., 76: 305–308.

Holm LG, Plucknnet DL, Pancho JV, Herberger JP. (1991). The world's worst weeds: Distribution and biology. Kreiger Publishing Co., Malabar, F.L.

Jafari N (2010). Ecological and socio-economic utilization of water hyacinth (*Eichhornia crassipes* Mart Solms). J. Appl.ied Sci. Environ.mental Manag., 14: 43-49.

Keddy PA (2001). Competition (2nd ed.). Kluwer Academic Publishers, Dordrecht, The Netherlands.

Labrada R (1996). Status of water hyacinth in developing countries. In: Charudattan R, Labrada R, Center TD, Kelly-Begazo C (Eds.), Strategies for Water Hyacinth Control. Report of a Panel of Experts Meeting. September 11–14, 1995, Fort Lauderdale, Florida. FAO, Rome, Italy, pp. 3–11.

Lu JB, Wu JG, Fu ZH, Zhu L (2007). Water hyacinth in China: A sustainability science based Management framework. Environ. Manag., 40: 823-830.

MacGreary NJ, Carpenter SR (1987). Density development growth interaction *Eleocharis acicularis* (L) R & S and Juncus pelocarpus forma submerses Fassett. Aquat. Bot., 27: 229-241.

Mack RN, Harper L (1977). Interference in dune annuals: spatial pattern and neighbourhood effects. J. Ecol., 65: 345-363.

Madsen JD (1993). Growth and biomass allocation pattern during water hyacinth mat development. J. Aquat. Plant Manag., 31: 134-137.

Navarro LA, Phiri G (2001). Water hyacinth in Africa survey of problems and solutions. International Development Research Center. Pp. 140-140.

Ndunguru J, Katagira F, Mjema P, Rajabu CA (2001). Aerial survey of Aerial survey of water hyacinth infestation and its control impact in Lake Victoria in Tanzania. Office of the Vice President. Lake Victoria Environmental Management Project (LVEMP). Technical Report, March 2001. Water Hyacinth Control Component, Mwanza, Tanzania.

Ndunguru J, Mjema P, Rajabu CA, Katagira F (2000). Assessment of water hyacinth infestation, impact and agricultural activities in the river Kagera, Tanzania. Office of the Vice President. Lake Victoria Environmental Management Project (LVEMP). Technical Report, November 2001.Water Hyacinth Control Component, Mwanza, Tanzania.

Pierik R, Cuppens MLC, Voesenek LACJ, Visser EJW (2004). Interactions between ethylene and gibberellins in phytochrome-mediated shade avoidance responses in tobacco. Plant Physiol.,136: 2928-2936.

Pushpa GS, John CV (2010). Does water hyacinth (*Eichhornia crassipes*) compensate for simulated defoliation? Implications for effective biocontrol. Biocontrol, 54: 35-40.

Rees M (1995). Community structure in sand dune annuals: is seed weight a key quantity? J. Ecol., 83: 857-653.

Rutagemwa DK (2001). Water quality monitoring: The LVEMP (Tanzania) experience. Proceedings of the LVEMP-Tanzania 2001 Scientific Conference, Mwanza, Tanzania. Pp. 145-157.

Sculthorpe CD (1985). The biology of aquatic vascular plants. Koeltz Scientific Books, Konigstein, West Germany.

Spencer DF, Ksander GG (2000). Interactions between American Pondweed and Monoecious Hydrilla Grown in Mixtures. New York,USA. J. Aquat. Plant Manag., 38: 5-13.

Weiner J (1984). Neighborhood interference amongst *Pinus rigida* individuals. J. Ecol., 72: 183-195.

25

Distribution patterns of cultivated watermelon forms in Zimbabwe using DIVA-GIS

Claid Mujaju[1,2]* and Moneim Fatih[3]

[1]Department of Plant Breeding and Biotechnology, Swedish University of Agricultural Sciences, Balsgård, Fjälkestadsvägen 459, SE-291 95 Kristianstad, Sweden.
[2]Seed Services, Department of Research and Specialist Services, Ministry of Agriculture, Mechanization and Irrigation Development, Box CY550, Harare, Zimbabwe.
[3]Box 91, The Swedish Biodiversity Center (CBM), Swedish University of Agricultural Sciences, SE-230 53 Alnarp, Sweden.

Identification of potential areas and regions for collection of watermelon germplasm is pivotal for better utilization of the available diversity. DIVA-GIS were used on all watermelon collections in the National Genebank of Zimbabwe to explore the pattern of distribution of two forms of watermelons; sweet watermelon (*Citrullus lanatus* var. *lanatus*) and cow-melon (*Citrullus lanatus* var. *citroides*). DIVA-GIS analysis revealed that most watermelon accessions have been collected at altitudes ranging from 160 to 1550 m above sea level, and that sandy loam and sand soils apparently are preferred. Distribution of the two forms of watermelons across soil types, altitudinal ranges and provinces is similar. In order to maintain watermelon diversity, promotion of *in-situ* or on-farm conservation of watermelon genetic resources using best practices in traditional farming system is recommended. Furthermore, this should be complemented by *ex-situ* conservation through gap collection missions in order to identify and document additional diversity hotspots.

Key words: DIVA-GIS, sweet watermelon, cow-melon, *Citrullus lanatus*, *in-situ*, on-farm conservation.

INTRODUCTION

Watermelon (*Citrullus lanatus*), plays an important role as a food security crop in the semi-arid areas of Zimbabwe, where droughts and erratic rainfalls affect agricultural production. The semi-arid areas of the country are host to over 75% of the Zimbabwean populace, wholly depending on subsistence farming for their livelihoods and survival. The semi-arid areas cover agro-ecological regions III, IV and V, as described by Rukuni and Eicher (1994), with an estimated 88.9% (32,510,000 ha) of the area of Zimbabwe. In these areas, access to inter- and intra-specific watermelon diversity is a survival strategy used by farmers to ensure food security. Farmers have thus developed landraces to meet specific purposes for human use (dessert, oil, seed and porridge) and animal feed. By contrast, they have not adopted modern varieties despite being exposed to these. Two botanical varieties of watermelon such as *C. lanatus* var. *lanatus* the sweet watermelons and *C. lanatus* var. *citroides* (citron melons or cow-melons) are found in cultivation. The ancestral forms of watermelon (also known as the Tsamma melons belonging to var. *citroides*) are also found in Southern Africa, in the wild. Southern Africa is generally regarded as the centre of origin for watermelon (Dane and Liu, 2007).

The conservation of biodiversity is critical for sustaining a growing population and enhancing global nutrition (Frankel, 1977). Genetic resources of indigenous plant species are important for crop and dietary diversity. Efforts to sustain population demands on agricultural productivity will necessitate an increase in the effective use of germplasm during plant improvement (Pistorius, 1997). Under the regional programme of Southern Africa Development Cooperation Genetic Resources Centre and National Plant Genetic Resources Centre Network, watermelon has been considered an underutilized crop, implying that its full potential has not been explored. It is

*Corresponding author. E-mail: mujajuclaid@yahoo.com.

therefore considered a mandate crop for conservation in the whole of the Southern African region.

Traditional local cultivars embody substantial diversity, and continue to provide an essential component of sustainable crop production for many of Africa's poor. The diversity of watermelon uses in Africa as shown by various studies for instance in Nigeria (Badifu, 1993); Namibia (Maggs-Kölling et al., 2000); Cote d'Ivoire (Loukou et al., 2007); Benin (Achigan-Dako et al., 2008); Mozambique (Munisse et al., 2011) and Mali (Nantoumé, 2011) is a clear indication of its importance in human diets. The link between local watermelon diversity and diverse uses, provide the rationale for enhancing the availability and use of local crop varieties in the arid and semi-arid regions of sub-Saharan Africa. Often, due to the drought-tolerant nature of watermelon, its importance as a food source is linked to agro-ecological conditions. Apart from diverse uses, farmers also need watermelon diversity for security against unpredictable conditions and to enhance optimum use of their resource base.

Farmers in traditional agriculture depend on informal systems to supply local watermelon seeds for cultivation and they maintain much of this diversity within their agroecosystems (on-farm). However, adequate access to diverse varieties is limited by lack of information on these varieties, placing traditional varieties at a disadvantage relative to varieties from the formal sector. In order to support and strengthen the conservation and use of watermelon, there is a need to gain insight into the genetic, ecological and geographic patterns associated with its distribution. The status of traditional watermelon cultivation and on-farm conservation in Zimbabwe is insufficiently documented. Limited information also exists on the diversity of the genus *Citrullus* and the extent of its distribution in Zimbabwe. With increasing commercial production of sweet watermelons from neighbouring countries (South Africa in particular), there is a potential risk that cultivation of indigenous, well-adapted landraces may be abandoned as farmers wish to try sweeter, modern cultivars through trade.

Proper identification and collection of plant material is an important step for effective conservation and utilization of the available germplasm. The use of DIVA-GIS has been designed to assist the plant genetic resources and biodiversity communities to map the range of distribution of species in which they are interested (Hijmans et al., 2002). Such analyses are essential for identification of potential areas of diversity and collection gaps for consideration when designing future collection missions. In the present study, DIVA-GIS was used to analyse and map cultivated watermelon forms in Zimbabwe.

MATERIALS AND METHODS

Data for DIVA-GIS use (Table 1 and 2) was gathered from passport information recorded on a total of 89 watermelon accessions stored in the National Genebank of Zimbabwe; collector's number (CollNo), type of watermelon (Type), province (Prov), name of farmer, longitude (Long), latitude (Lat), altitudinal range (m) and soil type. Taxon information recorded on accessions collected was mainly watermelon types (Type), aimed at separating watermelons into two major forms (SWM sweet watermelons, and CWM cow-melons) based on indigenous knowledge derived from local farmers. The separation of sweet watermelons (*C. lanatus* var. *lanatus*) and cow-melons (*C. lanatus* var. *citroides*) were further confirmed by molecular studies (Mujaju et al., 2010; Mujaju and Nybom, submitted).

The sampling of watermelons was carried out around the whole country as part of multi-species collection trips. These collections mostly followed the main roads, with stops every 5 kilometres for collecting new material. An additional targeted collection mission was performed in Masvingo province (Chitanga village) for the purposes of assessing the watermelon diversity within a single community (Mujaju and Nybom, submitted). Analysis of the spatial distribution of watermelon accessions collected throughout Zimbabwe was done using DIVA-GIS version 5.2 (Hijmans et al., 2002). The analysis involved generating maps of the distribution pattern of watermelon in relation to soil type and altitudinal range with the help of point-to-grid option applying simple method. Simple method assign watermelon collection points to the grid cell they fall in.

RESULTS AND DISCUSSION

Diversity analysis of the existing collections using DIVA-GIS (Figure 1), demonstrated that Masvingo province had the highest number of watermelon accessions. This province is situated in a drought-prone area, with average rainfall of 600 mm per annum. However, the high number of observations in Masvingo province was mostly due to the targeted collection in Chitanga village (21°17´S, 30°45´E), where 29 accessions were collected. While arguments about assessing genetic diversity by the number of accessions exists, previous studies on the same accessions (Mujaju et al., 2010; Mujaju and Nybom, submitted; Mujaju et al., submitted) using dominant molecular markers have revealed individual accession profiles. Consequently, the distribution of watermelon reflects the existence of diverse forms of watermelon in cultivation. Watermelon cultivation has been reported to be prevalent in drought-prone, semi-arid areas with an annual rainfall below 650 mm (Mujaju et al., 2010). If therefore further targeted collections are done in other drier provinces of the country, they are expected to have similar numbers of watermelon farming communities as Masvingo province. Subjecting all provinces to the same sampling regime (applied during multiple species germplasm collection) by excluding watermelon collections from Chitanga village, provides a fairly similar number of watermelon observations across the drier provinces of Matabeleland South, Matabeleland North, Midlands, Mashonaland West and Masvingo. A multi-species germplasm collection mission focuses on collecting diverse types of crops which limits the quantity of each crop collected. Other provinces like Harare, Bulawayo, Mashonaland Central, Mashonaland East and Manicaland, which are found within the high-rainfall belt of the country, only exhibited a few collections each

Table 1. List of watermelon material from the National Genebank of Zimbabwe.

CollNo	Type	Prov	Long	Lat
MMM014	CWM	Mashonaland Central	31.34885	-16.3828
MMM036	CWM	Mashonaland Central	31.34885	-16.3828
MMM049	CWM	Mashonaland West	29.49436	-16.5986
MMM062	CWM	Mashonaland West	30.09509	-17.4604
MMM063	CWM	Mashonaland West	30.09509	-17.4604
MMM064	CWM	Mashonaland West	30.09509	-17.4604
MMM068	SWM	Mashonaland West	29.53672	-17.3921
MMM076	SWM	Midlands	28.59085	-17.4275
MMM077	SWM	Midlands	28.59085	-17.4275
MMM084	CWM	Midlands	28.58981	-17.5676
MMM101	CWM	Midlands	28.52666	-18.1119
MMM102	CWM	Midlands	28.52666	-18.1119
MMM103	CWM	Midlands	28.52666	-18.1119
MMM112	CWM	Matabeleland North	28.06491	-17.2781
MMM123	CWM	Matabeleland North	27.41267	-17.5787
MMM127	CWM	Matabeleland North	27.41706	-17.5787
MMM133	CWM	Matabeleland North	27.0647	-18.1108
MMM156	SWM	Midlands	29.119	-19.0319
MMM162	SWM	Midlands	29.57793	-19.0771
MMM170	CWM	Midlands	29.17218	-19.0814
MMM177	SWM	Matabeleland South	28.58826	-19.5857
MMM179	CWM	Matabeleland South	28.58826	-19.5857
MMM180	CWM	Matabeleland South	28.58826	-19.5857
MMM181	CWM	Matabeleland South	28.58826	-19.5857
MMM186	CWM	Matabeleland North	27.32282	-19.2885
MMM190	CWM	Matabeleland South	27.50096	-20.3042
MMM191	SWM	Matabeleland South	27.50096	-20.3042
MMM194	SWM	Matabeleland South	27.39455	-20.2605
MMM202	CWM	Matabeleland South	27.39455	-20.2605
MMM203	CWM	Matabeleland South	27.39455	-20.2605
MMM210	SWM	Midlands	29.00014	-20.2806
MMM211	CWM	Midlands	30.10352	-20.2806
MMM216	CWM	Midlands	30.27996	-20.4482
MMM218	CWM	Masvingo	30.42494	-20.4453
MMM219	CWM	Masvingo	30.42494	-20.4453
MMM227	CWM	Mashonaland East	31.3834	-18.2301
MMM228	SWM	Mashonaland East	31.3834	-18.2301
MMM253	SWM	Masvingo	31.30661	-21.1896
MMM265	SWM	Masvingo	32.06837	-20.5994
MMM274	CWM	Manicaland	32.35514	-18.4688
MMM314	CWM	Mashonaland Central	31.54382	-17.1409
MMM321	SWM	Matabeleland North	27.32282	-19.2885
MMM322	SWM	Matabeleland North	27.32282	-19.2885
MMM323	SWM	Matabeleland North	27.32282	-19.2885
MMM324	SWM	Matabeleland North	27.32282	-19.2885
MMM325	SWM	Matabeleland North	27.32282	-19.2885
MMM326	SWM	Matabeleland North	27.32282	-19.2885
MMM328	SWM	Mashonaland West	30.09509	-17.4604
MMM329	SWM	Mashonaland West	30.09509	-17.4604
MMM330	SWM	Mashonaland West	30.09509	-17.4604
MMM331	SWM	Mashonaland West	30.09509	-17.4604

Table 1. Contd.

MMM332	SWM	Matabeleland South	27.39455	-20.2605
MMM333	SWM	Matabeleland South	27.39455	-20.2605
MMM334	SWM	Matabeleland South	27.39455	-20.2605
MMM338	CWM	Matabeleland South	27.39455	-20.2605
MMM339	CWM	Matabeleland South	27.39455	-20.2605
CM1	SWM	Masvingo	30.7542	-21.2943
CM2	SWM	Masvingo	30.7542	-21.2943
CM3	CWM	Masvingo	30.7542	-21.2943
CM4	CWM	Masvingo	30.7542	-21.2943
CM9	SWM	Masvingo	30.75538	-21.2921
CM10	SWM	Masvingo	30.75538	-21.2921
CM11	SWM	Masvingo	30.75538	-21.2921
CM12	SWM	Masvingo	30.75538	-21.2921
CM13	CWM	Masvingo	30.75538	-21.2921
CM14	CWM	Masvingo	30.75538	-21.2921
CM15	SWM	Masvingo	30.75538	-21.2921
CM16	SWM	Masvingo	30.75538	-21.2921
CM17	SWM	Masvingo	30.75538	-21.2921
CM18	SWM	Masvingo	30.75538	-21.2921
CM19	SWM	Masvingo	30.75538	-21.2921
CM20	SWM	Masvingo	30.75538	-21.2921
CM21	SWM	Masvingo	30.75538	-21.2921
CM22	SWM	Masvingo	30.75538	-21.2921
CM29	CWM	Masvingo	30.75417	-21.2885
CM30	CWM	Masvingo	30.75417	-21.2885
CM31	CWM	Masvingo	30.75417	-21.2885
CM32	CWM	Masvingo	30.75417	-21.2885
CM33	CWM	Masvingo	30.75417	-21.2885
CM38	SWM	Masvingo	30.75417	-21.2885
CM39	SWM	Masvingo	30.76422	-21.2914
CM40	SWM	Masvingo	30.76422	-21.2914
CM41	SWM	Masvingo	30.76422	-21.2914
CM42	SWM	Masvingo	30.76422	-21.2914
CM43	SWM	Masvingo	30.76422	-21.2914
CM44	CWM	Mashonaland East	32.41667	-17.25
CM45	SWM	Masvingo	30.75	-21
CM46	CWM	Masvingo	30.75	-21
CM47	CWM	Masvingo	30.75	-21

SWM: Sweet watermelon; CWM: Cow-melons.

suggesting that watermelon is grown mainly in a drier habitat. GIS has been successfully used in identifying areas of high diversity, e.g., for phaseolus bean (Jones et al., 1997), wild potatoes (Hijmans et al., 2000) and black pepper (Parthasarathy et al., 2006). The watermelon collection mission undertaken in Masvingo province in 2009 revealed a potential risk faced by farmers in maintenance of diversity on-farm. Most of the active farmers involved were elderly, whereas most of the young family members were reported to be working in towns (Mujaju and Nybom, submitted). This skewed age distribution implies a risk concerning the continuity of traditional knowledge about crop cultivation (Montes-Hernandez et al., 2005) and may advance the existence of superstitious myths that hinder seed exchange (Mujaju and Nybom, submitted). Until now, the common practice of exchanging seeds, and the different methods for watermelon planting that can promote gene flow across fields, may have helped to decrease the potential loss of heterozygosity within watermelon forms. In the closely related crop squash, the exchange of seeds for planting among farmers was considered useful as a method to

Table 2. The number of sweet watermelon and cow-melon accessions found within a particular soil type, altitudinal range and province in Zimbabwe. Numbers within parentheses indicate accessions collected in Chitanga village in Masvingo province.

Parameter	Sweet watermelon	Cow-melon
Soil type		
Clay	1	3
Clay loam	2	5
Loam	4	4
Sandy loam	33 (20)	25 (9)
Sand	6	6
Altitudinal range (m)		
158 – 620	22 (20)	9 (9)
620 – 1081	3	13
1081 – 1543	20	19
1543 – 2004	1	2
2004 – 2466	0	0
Province		
Mashonaland Central	0	3
Mashonaland West	5	4
Mashonaland East	1	2
Masvingo	23 (20)	13 (9)
Manicaland	0	1
Midlands	5	7
Matabeleland North	6	5
Matabeleland South	6	8
Harare	0	0
Bulawayo	0	0

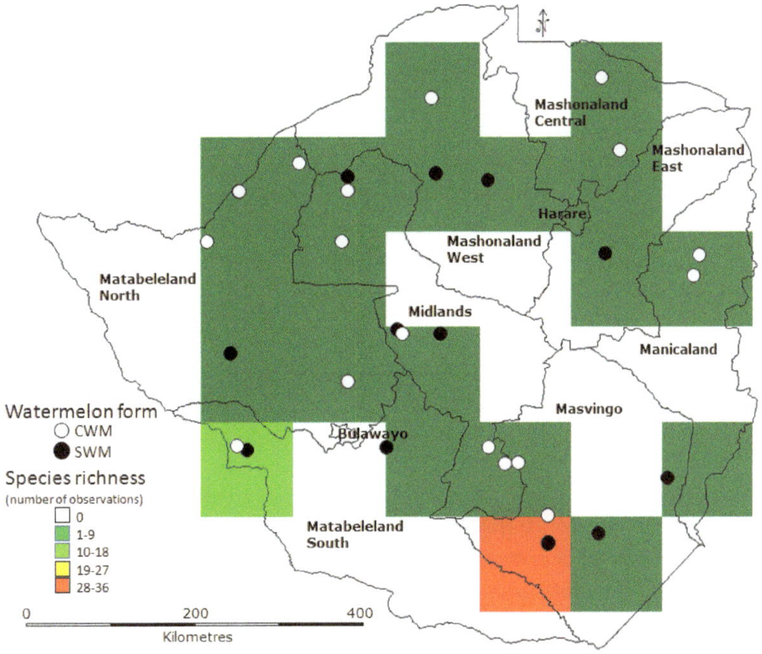

Figure 1. Distribution and number of accessions in the Zimbabwe genebank for the two watermelon forms using DIVA-GIS. CWM cow-melon, SWM sweet watermelon. Names of provinces are inserted.

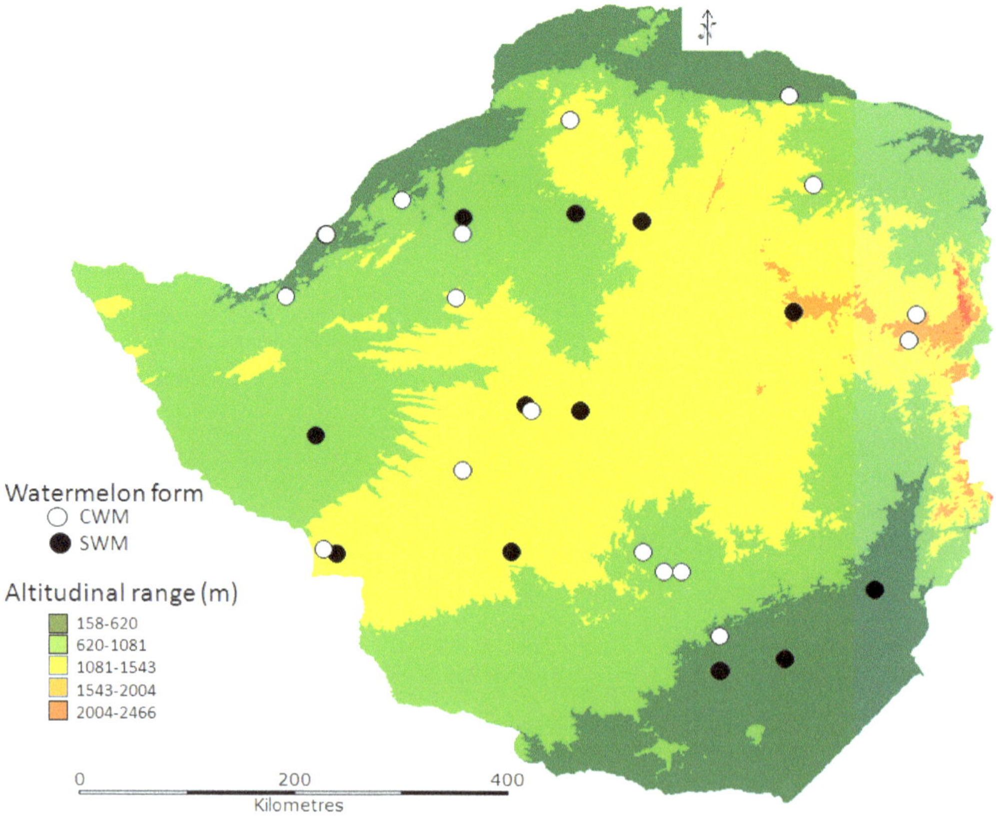

Figure 2. Distribution of collections in the Zimbabwe genebank for the two watermelon forms according to altitude using DIVA-GIS. CWM cow-melon, SWM sweet watermelon.

increase the genetic basis of squash germplasm (Montes-Hernandez et al., 2005). Use and benefits of increasing diversity by exchanging seeds for planting hasbeen reported also in local varieties of cowpea in Africa (Uguru, 1998). Distribution patterns of the two watermelon forms across soil types, relative to altitudinal ranges and provinces in Zimbabwe were assessed (Table 2, Figure 2 and 3). The maps for the distribution of watermelons according to soil type (Figure 3) and altitudinal range (Figure 2) showed that most accessions had been collected at 158 to 620 m and 1081 to 1543 m above sea level. The latter range harbored 22 sweet watermelons and 9 cow-melons whereas the former was host to 20 sweet watermelons and 19 cow-melons but most of these came from one village, Chitanga in Masvingo province (Table 2). The altitudinal range 620 to 1081 m harbored 3 sweet watermelons and 13 cow-melons. Very few accessions were obtained for the range 1543 to 2004 m (1 sweet watermelon and 2 cow-melons) and none at even higher altitudes. This is most likely due to the increase in rainfall above 1543 m, which is unfavorable for the growing of watermelons. Apart from Chitanga village, where watermelon cultivation was initiated only about ten years ago as a result of the land redistribution, almost all accessions were thus collected at medium high altitudes from 600 to 1600 m. Sandy loam soils followed by sandy soils are predominant in the areas where watermelon was collected (Figure 3). Watermelon has been reported to perform especially well under savannah environments in well-drained sandy loam soil (Van der Vossen et al., 2004).

The distribution map also demonstrates areas where very little explorations have been undertaken, suggesting gaps in collections. This has an implication on conservation which should also focus on the need for comprehensive collection of watermelon germplasm in order to identify and document additional diversity hotspots. Any conservation strategy that seeks to promote on farm conservation, should also consider long-term sustainability, where security of greater diversity results from the complementary application of *in-situ* and *ex-situ* techniques (Maxted et al., 2002).

Conclusion

The two cultivated forms of watermelon are apparently grown in the same habitats in Zimbabwe. These habitats are characterized by sandy soils and low rainfall, while altitudes vary substantially; from just above sea level to at

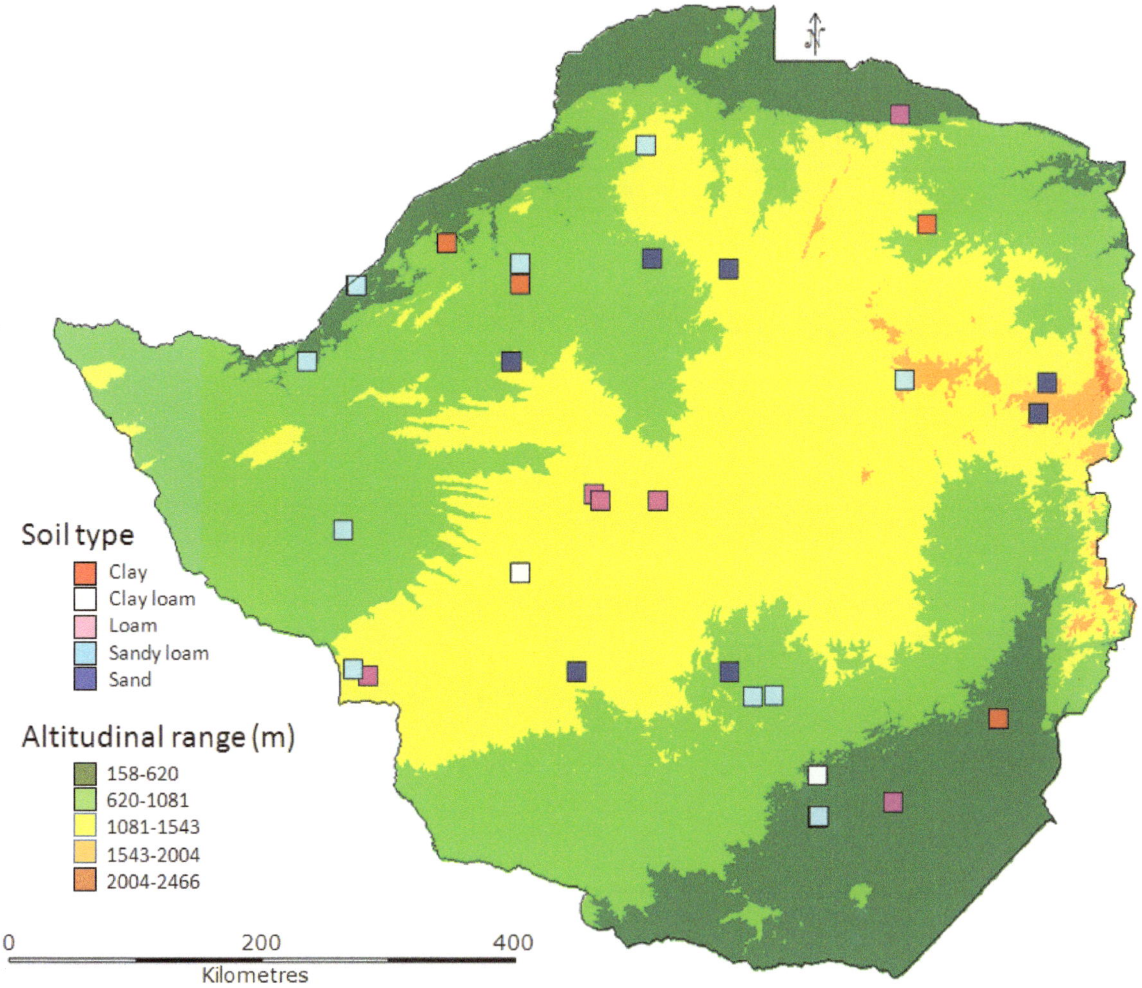

Figure 3. Distribution of watermelon collections in the Zimbabwe genebank according to soil types and relative altitudinal range using DIVA-GIS.

least 1500 m. Future collection missions should focus on those areas where the distribution map shows major gaps. A detailed study based on climatic data is warranted to predict the climatic and ecogeographic conditions required for watermelon and also to forecast its distribution in other parts of Zimbabwe.

REFERENCES

Achigan-Dako EG, Fagbemissi R, Avohou HT, Vodouhe RS, Coulibaly O, Ahanchede A (2008). Importance and practices of Egusi crops (*Citrullus lanatus* (Thunb.) Matsum. & Nakai, *Cucumeropsis mannii* Naudin and *Lagenaria siceraria* (Molina) Standl. cv. 'Aklamkpa') in sociolinguistic areas in Benin. Biotechnol. Agron. Soc. Environ., 12: 393-403.

Badifu GIO (1993). Food potentials of some unconventional oilseeds grown in Nigeria - A brief review. Plant Food. Hum. Nutr., 43: 211-224.

Dane F, Liu JR (2007). Diversity and origin of cultivated and citron type watermelon (*Citrullus lanatus*). Genet. Resour. Crop Evol., 54: 1255-1265.

Frankel OH (1977). Natural variation and its conservation. In: Muhammed et al. (eds) Genetic diversity in plants, Plenum Press, New York, pp. 21-44.

Hijmans RJ, Garrett KA, Huaman Z, Zhang DP, Schreuder M, Bonierbale M (2000). Assessing the geographic representativeness of genebank collections: the case of Bolivian wild potatoes. Conserv. Biol., 14: 1755-1765.

Hijmans RJ, Guarino L, Rojas E (2002). DIVA-GIS a geographic information system for the analysis of biodiversity data. Manual, International Potato Center, Lima, Peru.

Jones PG, Beebe SE, Tohme J, Galway NW (1997). The use of geographical information systems in biodiversity exploration and conservation. Biodivers. Conserv., 6: 947-958.

Loukou AL, Gnakri D, Dje Y, Kippre AV, Malice M, Baudoin JP, Bi IAZ (2007). Macronutrient composition of three cucurbit species cultivated for seed consumption in Cote d'Ivoire. Afr. J. Biotechnol., 6: 529-533.

Maggs-Kölling GL, Madsen S, Christiansen JL (2000). A phenetic analysis of morphological variation in *Citrullus lanatus* in Namibia. Genet. Resour. Crop Evol., 47: 385-393.

Maxted N, Guarino L, Myer L, Chiwona EA (2002). Towards a methodology for on-farm conservation of plant genetic resources. Genet. Resour. Crop Evol., 49: 31-46.

Montes-Hernandez S, Merrick LC, Eguiarte LE (2005). Maintenance of squash (*Cucurbita* spp.) landrace diversity by farmers' activities in Mexico. Genet. Resour. Crop Evol., 52: 697-707.

Mujaju C., Sehic J., Werlemark G., Garkava-Gustavsson L., Fatih M.,

Nybom H. (2010). Genetic diversity in watermelon (*Citrullus lanatus*) landraces from Zimbabwe revealed by RAPD and SSR markers. Hereditas, 147(4): 142-153.
Munisse P, Andersen SB, Jensen BD, Christiansen JL (2011). Diversity of landraces, agricultural practises and traditional uses of watermelon (*Citrullus lanatus*) in Mozambique. Afr. J. Plant Sci., 5: 75-86.
Nantoumé AD (2011). Cultivated indigenous watermelons (*Citrullus lanatus*) in Mali: Diversity, cultivation and use. PhD thesis, Faculty of Life Sciences, University of Copenhagen, Copenhagen, Denmark.
Parthasarathy U, Saji KV, Jayarajan K, Parthasarathy VA (2006). Biodiversity of *Piper* in South India - application of GIS and cluster analysis. Curr. Sci., 91: 652-658.
Pistorius R (1997). Scientists, Plants and Politics: A History of the Plant Genetic Resources Movement. International Plant Genetic Resources Institute, Rome, pp. 1-134.
Rukuni M., Eicker C.K. (1994). Zimbabwe Agricultural Revolution. University of Zimbabwe Publishers, Jongwe Printing and Publishing, pp. 1-418.
Uguru MI (1998). Traditional conservation of vegetable cowpea in Nigeria. Genet. Resour. Crop Evol., 45: 135-138.
Van der Vossen H.A.M., Denton O.A., El Tahir I.M. (2004). *Citrullus lanatus* (Thunb.) Matsum. and Nakai. In: Grubben et al. (eds) Plant Resources of Tropical Africa 2 Vegetables, Wageningen, Netherlands: PROTA Foundation, pp. 185-191.

Distribution and abundance of terrestrial orchids of the genus *Bletia* in sites with different degrees of disturbance, in the Cupatitzio Natural Reserve, México

María de los Ángeles Beltrán-Nambo[1], Pilar Ortega Larrocea[2], Rafael Salgado Garciglia[3], Joel Tupac Otero Ospina[4], Miguel Martínez Trujillo[1] and Yazmín Carreón-Abud[1]*

[1]Laboratory of Genetics and Microbiology, Faculty of Biology, Universidad Michoacana de San Nicolás de Hidalgo, México.
[2]Bio-edaphic Microcosm Laboratory, Department of Edaphology, Institute of Geology, Universidad Nacional Autónoma de México.
[3]Plant Biotechnology Laboratory, Institute of Chemical Biologics, Universidad Michoacana de San Nicolás de Hidalgo, México.
[4]National University of Colombia, Sede Palmira, Colombia.

In this work, we determined the distribution and relative abundance of terrestrial orchids of the genus *Bletia* in the Cupatitzio Natural Reserve, Michoacán, México, during two flowering seasons. Four different sampling sites were identified according to degree of disturbance and site quality. Relative abundance was established with an arbitrary scale of four values: abundant (>40 individuals), medium (20 to 40 individuals), low (<20 individuals) and null. Three species of orchids were found: *Bletia roezlii* was the most abundant, with populations of over 40 specimens per 1000 m^2 in some sites (zones 1 and 3). It was present in all the reserve's sites, although it showed better development in areas near roads and/or in open sites. *Bletia purpurata* had medium distribution and abundance, with more of 30 plants per site, having more presence in conserves sites (zone 3). *Bletia punctata* had the lowest abundance and distribution, with populations of 10 or less individuals, presenting a mean abundance of low disturbance sites. The distribution of the specimens of *Bletia* spp. is determined by factors such as the degree of disturbance of the sites where they develop, the adaptability of the species, the amount and morphotype of the mycorrhizal fungus, which is associated with their dispersibility of them, among other factors.

Key words: Orchid distribution, relative abundance, disturbance.

INTRODUCTION

Orchidaceae is the most diverse of all angiosperm families, with estimates of <25000 species (Dressler, 1993; Mabberley, 1997; Cribb et al., 2003); and more than any other plant family they have a high proportion of threatened genera, with most containing threatened species. Two-thirds of orchid species are epiphytes and lithophytes, with terrestrial species comprising the remaining third, yet almost half of the extinct species according to The World Conservation Union (IUCN, 1999) are terrestrial herbaceous perennials. Terrestrial orchids thus represent a life form class likely to experience a greater extinction risk as a result of the multiplicity of threatening processes, particularly under current climatic change scenario (Swartz and Dixon, 2009).

An observation of ecological significance is that organisms differ greatly in distribution and abundance; consequently, rare species may be recognized as those

*Corresponding author. Email: ycabud@umich.mx, ycabud@gmail.com.

of low numerical abundance compared with others (Pate and Hopper, 1993; Harper, 1981) classified rare species according to space, time or group relatedness. A space-dependent species may be locally abundant, but only occur in a limited number of sites, restricted due to high niche specificity or barriers reducing dispersal potential. These species are often local endemics, particularly vulnerable to threatening processes. A time-dependent rare species results from fluctuations in population numbers following adverse sporadic or cyclical events, such as drought or fire (Koopowitz et al., 2003). Populations of a rare species, occupying a specialized niche with a limited distribution, represent group-dependent rarity associated with certain ecotypes often at ecological frontiers for species. Orchids are found in all these classes. Although a significant literature exists on the many causes of rarity in plants, drivers of rarity in orchids are more often than not linked to their unique habitats and pollinator requirements. Ecological specialization has not only contributed to the great species diversity in Orchidaceae, but has also resulted in the high level of threat in this family (Cribb et al., 2003). However, it is the complexity of ecological specialization that makes orchids ideal model species for developing and testing conservation strategies.

In Mexico, terrestrial orchids are represented by genera such as *Spiranthes*, *Bletia*, *Govenia*, *Habenaria* and *Malaxis* (Soto, 1988). Among these, one of the most representative neotropical genera is *Bletia*, since its centre of diversification is in Mexico, with approximately 50 known species, out of which 13 have been described in the State of Michoacan, distributed throughout 19 municipalities (Hágsater et al., 2005; Sosa, 2007; data from Herbarium of Ecological Institute, Pátzcuaro, Michoacán, 2008).

Human activities such as logging and agriculture can severely damage forest ecosystems by changing forest structure, ecosystem function, and biodiversity. These changes may have long-lasting consequences, which influence forest recovery. The fact that loss of natural habitats due to changes in land use and other factors such as tree felling, contamination, introduction of exotic species, and collection of wild specimens, the populations of species of this genus, as of the rest of the orchids, have been reduced through time, and so their study is important for conservation purposes. Conservation through reserves alone is now considered unlikely to achieve protection of plant species necessary to mitigate direct losses of habitat and the pervasive impact of global climate change. Some of these species, like *Bletia campanulata* (La Llave are Lex), are able to grow in areas with scarce, disturbed soils, in reforested areas, or in areas with eroded soils. Others, like *Bletia reflexa* Lind., show very limited distribution within the Michoacán State, México (data from Herbarium of Ecological Institute, Pátzcuaro, Michoacán, 2008). Records of species found in different ecosystems are limited to taxonomic descriptions, and only a few studies conducted in Mexico offer information on abundance and distribution. This makes it hard to verify whether there has been an alarming reduction in their population through time, so as to be able carry out acts of conservation or legislation. In addition, it is important to determine the proficiency of different species to adapt the habitat transformation, in order to conserve those inhabited by orchids with the least ability to live in perturbed niches and which are therefore in greatest risk of being lost.

In spite of being a protected area, the Cupatitzio Natural Reserve, located in the municipality of Uruapan, Michoacán state, México (Figure 1), is affected by its proximity of the urban zone of Uruapan. Its neighbouring areas have undergone changes in vegetation cover due to being constantly stepped on by humans, to introduction of exotic species, to changes in land use from forests to horticultural orchards, mainly avocado and peach, and to the presence of plagues that force cutting down diseased trees, giving rise to areas that are more open and allow more light penetration. Whereas areas farthest from the locality and which have more restricted or difficult access, show vegetation that is better conserved and less disturbed.

Identification of sites with different characteristics within the reserve allowed studying the distribution and relative abundance of some species of the genus *Bletia*, with the objective of determining whether they showed changes in these parameters, related to different site quality and disturbance conditions. This will enable us to generate a first diagnosis, which will serve as a basis for conducting subsequent demographic studies that will help conserve the most threatened species and habitats in this reserve.

The study was based on the hypothesis that the species that were found would show broader distribution and relative abundance in less-disturbed sites, but they were also represented in disturbed sites within the same habitat, which reflects their capacity to adapt. Currently, only a few species of this genus are considered threatened or endangered according to Official Mexican Standard (NOM-ECOL-059-2001). Nevertheless, even if they are not officially considered to be threatened, some species show small populations and are scarce within Michoacán State (data from Herbarium of Ecological Institute, Pátzcuaro, Michoacán, 2008).

MATERIALS AND METHODS

Four zones with different degrees of disturbance within the reserve were established (zones 1, 2, 3 and 4) based on 6 recognition trips, marking the boundaries with 50 cm wooden stakes and by taking photographs (Sony DSC-W270 digital camera, 7.2 megapixels). The trips were made on July 29, August 29, September 6, October 10 and 24, and November 18, in 2008 and June16, September 9 and November 28, in 2009. The sampling sites were delimited considering previously described characteristics such as primary and secondary vegetation type (Bello and Madrigal, 1996; Gómez, 2005; Zavala, 2006), soil (Gómez, 1985), depth of humus, light penetration, altitude (GARMIN GPS unit), presence of exotic species, road routes and other infrastructure building sites.

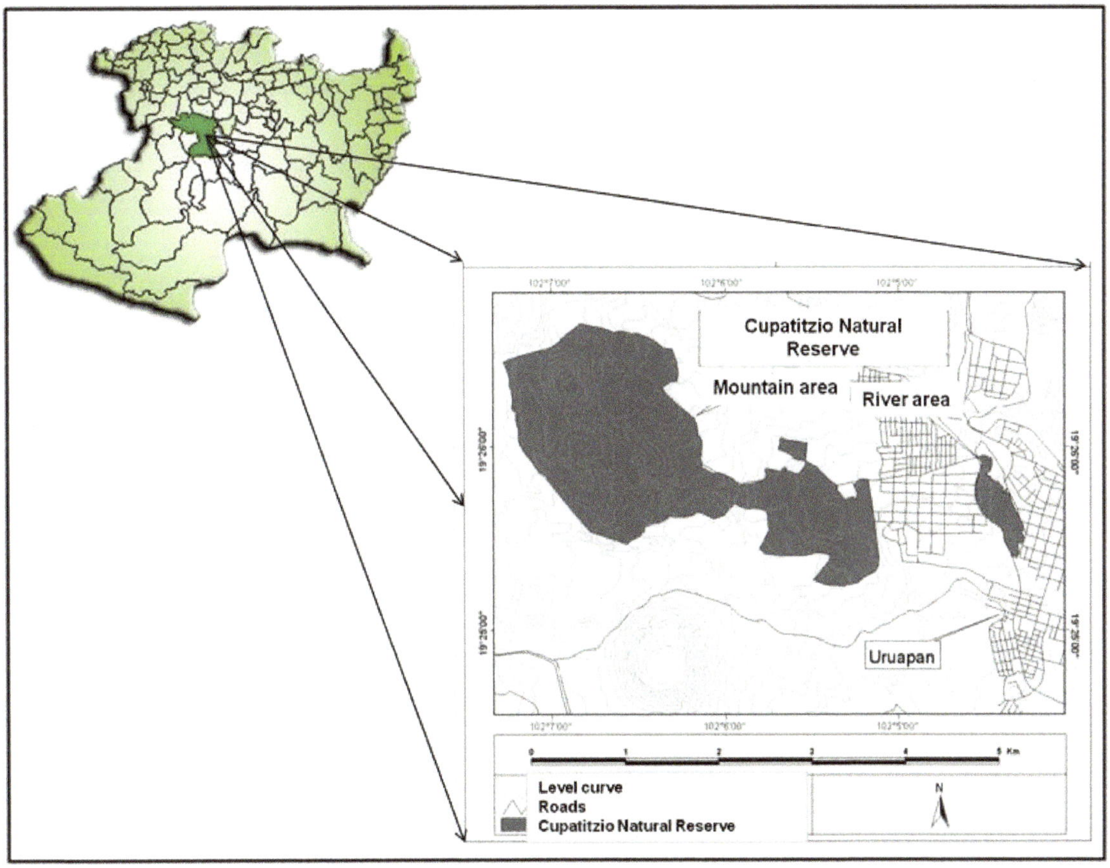

Figure 1. Location of the Cupatitzio Natural Reserve. Dark green represents the municipality of the Reserve. The gray areas represent all the reserve. (Source: Topographic map 1: 50 000 and map of the state of Michoacán 1: 75 000, INEGI).

Boundaries between sites were established considering the main and secondary roads that exist within the reserve (Figure 2).

Sites identified as 1 and 4 are adjacent to main roads that serve as transit route for people visiting the reserve and for the city's inhabitants, in addition to being close to the main access and the urban area. These sites include the orchards and the cabins, which accommodate tourists and researchers (Figures 2 and 3). This has resulted in changes in secondary vegetation, increased light penetration, soil erosion and therefore a greater degree of disturbance.

The sites identified as 2 and 3 are located in the portion of reserve area, near the main road, which is less travelled due to being far and hard to access, where vegetation is denser and less disturbed (Figures 2 to 4).

The four sites show the presence of land formed by volcanic rock types, however this is more abundant in site 4. These soils are defined as ground generated by the rapid cooling and solidification of volcanic lava and one for this characteristics is that they are considered as eroded soils with little water content, and their surface is very rough and it is difficult to navigate through them (Table 1).

Data cards and herbarium specimens of the genus *Bletia* were consulted in the Institute of Ecology of the city of Patzcuaro, Michoacan and the Faculty of Biology of the Michoacana University, with the objective of determining the species that have been recorded near the reserve. Later, during the flowering season (June to November, 2008 and 2009), specimen marking was performed to identify the species and thus be able to determine which locations have populations of these orchids. Marking was conducted using aluminum plates containing number of specimen and species, tied to the plant's pseudobulb with a nylon string.

Species identification was conducted in the Orchid Garden of the of Morelia city (Michoacán, México). Plant geo-referencing was performed with GPS (Garmin® 3 m) during the trips for later positioning in digital images provided by Mexico's National Institute of Statistics and Geography (INEGI) using Autocad 2007 software, where the reserve borders and main topographic features were specified.

Relative abundance was determined quantitatively based on an arbitrary scale of four values conducted in four quadrat 20 × 50 m (1000 m^2) per site, selected on topographic maps based on physical features, where populations of over 40 specimens were considered abundant; 20 to 40 specimens, medium; less than 20 specimens, low; and null when no specimen. The results are reported in average values for each species per site and we performed statistical analyses of variance and Tukey test when the values showed significant differences, using the program JMP V.8.0s could be identified.

RESULTS

According to Gomez-Reyes (2005), four vegetation units were observed: a) arboretum, corresponding to a plantation zone with species of the genera *Eucalyptus, Cupressus* and *Pinus,* located at 2000 m altitude; b) avocado (*Persea americana* L.) and peach (*Prunus*

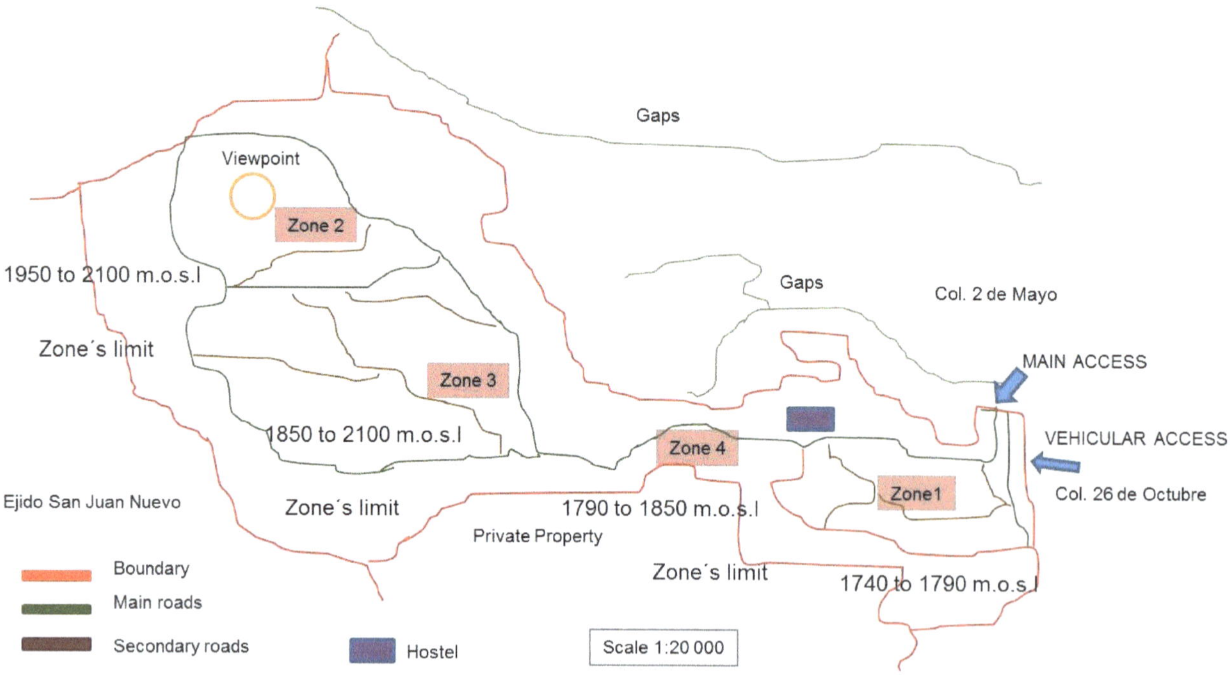

Figure 2. Delimitation of sampling areas in the Cupatitzio Natural Reserve (elaborated based on a Reserve Route Map).

Figure 3. Appearance of sites designated as disturbed (sites 1 and 4). Where roads are observed, the so-called secondary vegetation and badlands (bottom left).

Figure 4. Appearance of sites designated as preserved areas (sites 2 and 3). Where there is dense vegetation, less illuminated and secondary vegetation.

persica L. Stokes) orchards near the main access and the facilities; c) pine forests (*Pinus lawsonii, Pinus douglasiana, Pinus michoacana* and *Pinus leiophylla*), which are the dominant vegetation in the reserve; and d) pine-oak forests, considered to be the original vegetation throughout the reserve, dominated by the previously described pine species and by *P. michoacana* var. *cornuta* and *Quercus obtusata, Quercus castanea*, *Quercus candicans*, *Quercus magnoliifolia* and *Quercus resinosa*.

During the previously mentioned flowering season, only three species were found in the reserve areas: *Bletia purpurata* (A. Rich and Galleoti), *Bletia punctata* (Llave and Lex) and *Bletia roezlii* (Reichb. f. Linnaean) (Figure 5). 1360 specimens that had inflorescences, therefore allowing taxonomic identification, were marked. Plants that did not flower during the trips were not considered for abundance, and so the populations might be

Table 1. Characterization of sampling sites of the reserve area, according to site quality and degree of disturbance.

Zone	Vegetation	Soil (FAO)	Depth of humus	Altitude (m.a.s.l.)	Light penetration	Disturbance	Malpais presence
1	Pine, pine- oak, orchards, exotic vegetation	Lithosol and Andosol in equal proportions	From 1 to 15 cm in some areas outside the roads	1740 to 1770	High	Change of land use, nearby urbanization, introduction of exotic species, people passing through, and looting.	Yes
2	Pine, pine-oak and arboretum	Lithosol and Andosol	From 5 to 15 cm	1950 to 2050	Lower due to denser arboreal vegetation	Low disturbance due to its remoteness and hard access.	Yes
3	Pine-oak and pine	Lithosol	From 5 to 20 cm	1850 to 2100	Lower due to denser arboreal vegetation	Low disturbance, mostly preserves its original vegetation	Yes
4	Pine	Lithosol	From 1 to 10 cm	1790 to 1840	High	Moderate disturbance due to its proximity to the urban zone, little organic matter in soil, and abundant malpais.	Yes

Figure 5. Species found: a) *B. purpurata* (B. Rich and Galeotti), b) *B. roezlii* (B. Rchb. f. Linnaean) and c) *B. punctata* (La Llave and Lex).

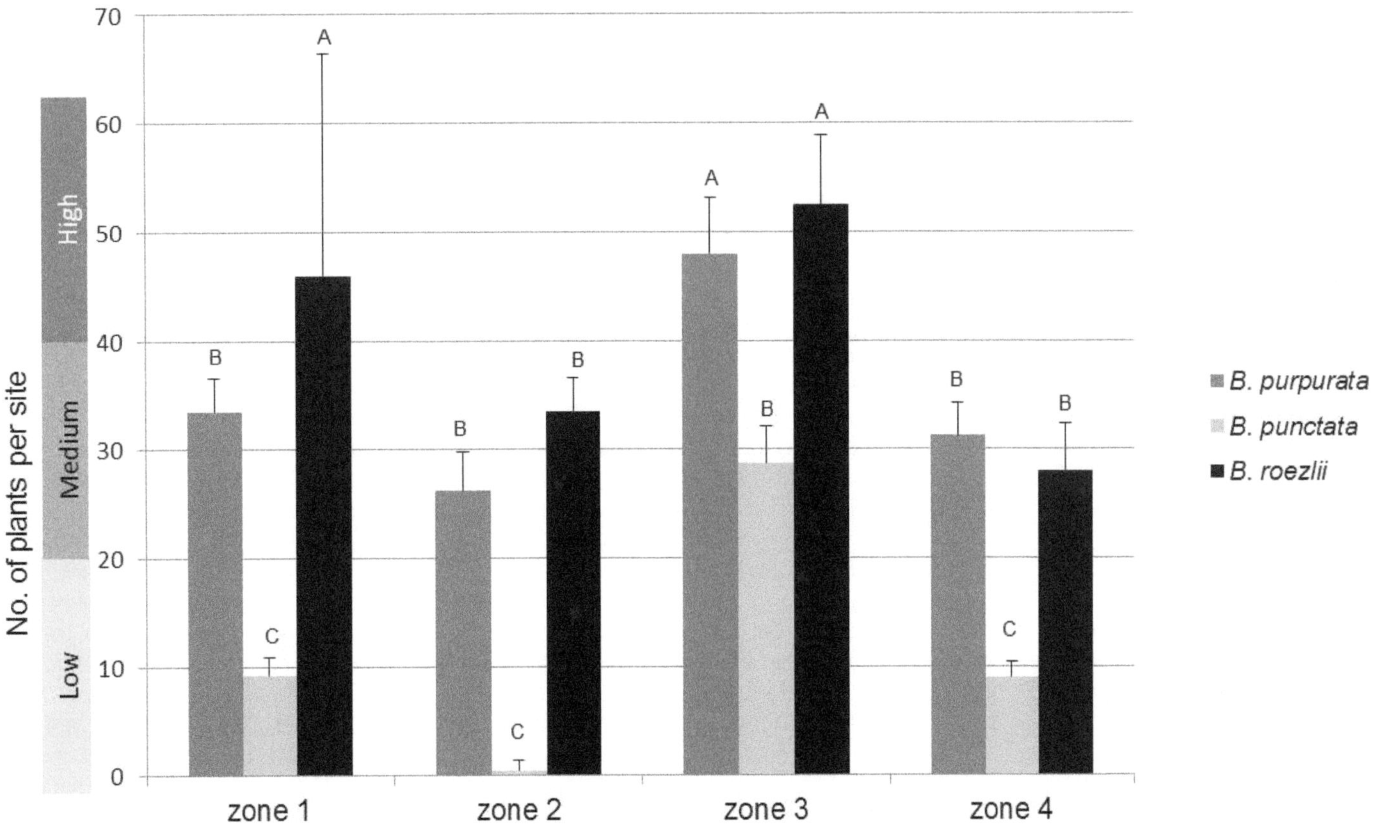

Figure 6. Relative abundance (abscissa) of the three species of the genus *Bletia* in the four study areas, in Cupatitzio Natural Reserve. We present mean values with standard deviation, different letters indicate significant differences (Tukey, P ≤ 0.05).

underestimated.

The most widely distributed and abundant species in the reserve area for the year of sampling was *B. roezlii* (B. Rchb. f. Linnaean), with populations of more than 35 plants per site in the four areas. This species developed well in both very disturbed areas, in sites near of roads or in open areas (zone 1) and of low disturbance areas (zone 3), with 46 plants and 53 plants, respectively (Figure 6). Espejo et al. (2002) mention it in pine-oak forests and in stony soils, as found by Hágsater et al. (2005) in temperate forests.

B. purpurata (B. Rich and Galeotti) was also well represented in the four areas of the reserve, developing better in preserves areas (zone 3), with 48 plants. However, this species presented a medium abundance with 33 and 31 specimen in disturbed sites, zones 1 and 4, respectively (Figure 6). The species seems to adapt well to places with low soil development (Ortega-Larrocea and Rangel-Villafranco, 2007) and/or disturbance (data from the Herbarium of the Ecological Institute, Pátzcuaro).

B. punctata (La Llave and Lex) was hard to locate in most of the study sites and was not found in some of them. In zone 3, site of low disturbance, the abundance was medium with 29 plants. Nevertheless, in zone 2, as well considered a site of low disturbance, the abundance was low with only one plant. Less than 10 plants were found in sites with greater disturbance (zones 1 and 4) (Figure 6).

DISCUSSION

The reasons for the absence or reduced abundance of some species of *Bletia* in the reserve, like *B. punctata* in disturbed areas are not known.

Bletia establishment in these areas may be affected by a number of different factors. These factors can be intrinsic and extrinsic. The intrinsic factors related to limits on abundance and distribution as a result of natural factors. In the case of terrestrial orchids, for example, range and abundance may be driven by factors pertaining to the underground and above-ground life history phases of species (Woolcock and Woolcock, 1984; Clements, 1988; Dixon, 1989). The first need, represented in the underground phase, is a mycorrhizal association with a specific fungal endophyte (Warcup, 1971; Ramsay et al., 1986; Rasmussen, 2002).

A population decrease in *B. punctata* has been described in other habitats in the country where it was believed extinct (Hagsater et al., 2005), and so studies of its populations should be considered in samplings that are conducted over longer time periods. Other species such as *B. purpurata and B. roezlli* are more abundant, and it is possible that lower mycorrhizal specificity could favor their distribution over a wide range of habitats where different mycorrhizal partners are available.

Previous works report that the orchid *B. roezlii* was significantly more colonized than the other two species. There were no significant differences in any mycorrhizal colonization parameter among sites (Beltrán-Nambo et al., 2010). Additionally, this orchid was the most abundant at all sites. It seemed quite possible that the mycotrophic status of this species was not influenced by habitat transformation, resulting in a better adaptation of this species to a larger heterogeneity in soil, light and altitude, or possibly a lower specificity (McCormick et al., 2004).

Conversely, the orchid's *B. purpurata* and *B. punctata* were more affected in their mycotrophic as a consequence of habitat transformation, such as an increase in light resources when the forest was replaced by exotic species (Beltrán-Nambo et al., 2010). Habitat degradation is comprised not only of changes in plant coverage, but species composition and soil erosion processes. Indirect changes in light availability for orchids that grow in shade forest and their dependence on mycotrophic could be indirectly affected. The response of orchids to such transformations is not known or extensively documented.

The fact that some species were more abundant at all sites and that others seemed sensitive to habitat transformation could be due in part to changes in fungal partners or the status of mycorrhizal colonization. The abundance or distribution of orchids could be influenced by mycorrhizal dynamics and root colonization levels in response to several factors.

The second need is an effective pollination/fertilization in the above-ground phase (Stoutamire, 1983; Roberts, 2003). The great taxonomic diversity of Orchidaceae is often attributed to specialization of these two requirements, either independently or in combination, the effects of which place species on a theoretical risk continuum from low to high in the event of environmental and habitat change.

Orchids produce vast numbers of minute seeds lacking storage reserves, such as an endosperm, found in many other angiosperms (Arditti and Ghani, 2000; Batty et al., 2000). Seed recruitment success varies from species to species, with some temperate terrestrial species possessing a requirement for stratification or ageing in soil to release dormancy (Stoutamire, 1974).

Production of large numbers of small seeds favours high dispersal rates, plant fecundity and expression of genetic variability across geographical and ecological boundaries while minimizing parental investment per seed (Batty et al., 2002; Zettler et al., 2003). Although abundant seed is released, few germinate, and even fewer develop into mature plants. Under natural conditions, seeds of most terrestrial orchid species will germinate only in association with a compatible mycorrhizal fungus (Warcup, 1981; Ramsay et al., 1986; Arditti et al., 1990). Due to limited food reserves, orchid seeds have a complete dependency on nutrients supplied by the mycorrhizal association during early germination and seedling establishment phases (Rasmussen, 1995), although some species may substitute or acquire new mycorrhizal associates depending upon plant maturity (Bidartondo and Read, 2008).

Rangel-Villafranco and Ortega-Larrocea (2007) in a preliminary study for the mycorrhizal fungi diversity associated to germination seed in terrestrial orchids in southern Mexico City, found that it looks like some genera are highly specific for their mycobiont as the couples *Bletia - Epulorhiza, Dichromanthus – Ceratorhiza, Habenaria – Epulorhiza,* and *Malaxis – Epulorhiza*. More evidence is required to confirm whether this specificity always occurs in nature, due to the fact that some symbiotic cultures can be developed *in vitro* with different isolates. Bioassays confirm specificity at some levels, they noticed that *Bletia* species are less specific for both genus isolates than *Dichromanthus*. However, results for the *in vitro* propagation must be interpreted carefully and do not reflect the specificity in nature. *In situ* bioassays demonstrate that specificity can be developed through the life history of the plant. The *Bletia* spp. can be probably less dependent on fungi for *in vitro* germination and less specific because they photosynthesize rapidly; but in nature, populations are more endangered than *Dichromantthus* and seedlings are very difficult to observe where asexual corm propagation is common (Ortega and Rangel, 2007).

This agrees with the fact that of the three species found in *Bletia* in the sampling years (2008, 2009), *B. punctata* (Llave and Lex) was the one with lower distribution and abundance. It is possible that this species has a lower capacity to adapt to the transformation of their habitat because in sites that were identified with some degree of disturbance, its presence was little or none.

This coincides with that is the specie of lower distribution in the state. It has been described in only two municipalities. *B. roezlii* has been recorded in four municipalities and *B. purpurata* in eight municipalities (data from the Herbarium of Ecology Institute, Pátzcuaro, 2008; Herbarium of Biology Faculty, Michoacana University, 2008). All these factors may influence the distribution of the three species of *Bletia* in the reserve, and these results can be compared with the data obtained from Bergman et al. (2006), who described the analysis of the *W. calcarata's* abundance in the different historic canopy cover class areas that showed that the orchid was most abundant in areas which had been minimally impacted by human activity like the result in the present study.

Conclusions

This study allows concluding that disturbance affects the distribution of three *Bletia* species (*B. punctata, B, purpurata* and *B. roezlli*). They showed the highest distribution and abundance in sites with a low disturbance and a low number of plants in sites with higher degree of disturbance. *B. punctata* presented the lower distribution and abundance. It is possible that this species is less able to adapt to habitat transformation, since its presence was very low or null in areas with some degree of disturbance. An understanding of ecosystem dynamics and the role of interactions between plants and their environment are essential to plant conservation (Schemske et al., 1994). In the case of *Bletia*, we have found that land-use history dramatically affects the suitability of sites for the occurrence of the orchid.

Anthropogenic processes often also accelerate environmental and habitat change (kick-on effects), adversely impacting environmental conditions necessary for sustaining orchid populations. These include such factors as spread of disease and pests, changed fire regimes, salinization and desertification (Sahagian, 2000). Extrinsic factors with knock-on effects pose some of the most significant and pervasive of all threats to orchid conservation, particularly in the face of climate change (Dixon et al., 2003).

Prospective studies such as this one are important because they provide basic information on the relative abundance and adaptation degree exhibited by some species of orchids. However, population studies and interactions with other organism must be designed through the time, so as to generate useful data for future conservation and/or restoration practices.

ACKNOWLEDGEMENTS

We dedicate our cordial thanks to Biol. Nuria Gómez Dorantes and Q.F.B Yolanda Mendez Sántiz for their support during field trips.

REFERENCES

Arditti J, Ernst R, Wing YT, Glabe C (1990). The contribution of orchid mycorrhizal fungi to seed germination: A speculative review. Lindleyana, 5: 249–255.

Arditti J, Ghani AKA (2000). Tansley review no. 110. Numerical and physical properties of orchid seeds and their biological implications. New Phytol., 145: 367-421.

Batty AL, Dixon KW, Sivasithamparam K (2000). Soil seed bank dynamics of terrestrial orchids. Lindleyana, 15: 227-236.

Batty AL, Dixon KW, Brundrett MC, Sivasithamparam K (2002). Orchid conservation and mycorrhizal associations. Microorganisms in plant conservation and biodiversity. Dordrecht,The Netherlands: Kluwer Academic Publishers.

Bello GMA, Madrigal SX (1996). Estudio florístico del campo experimental "Barranca del Cupatitzio", Uruapan, Michoacán. INIFAP. *Scientific brochure* Uruapan, Mich. México., 2: 47.

Beltrán NMA, Carreón AY, Ortega LMP (2010). Mycorrhizal Colonization Patterns of Three *Bletia* Species in a Natural Forest Reserve in Michoacán, Mexico. Acta Horticulturae, 878: 43-52.

Bergman E, Ackerman JD, Thompson J, Zimmerman JK (2006). Land-use history affects the distribution of the saprophytic orchid *Wullschlaegelia calcarata* in Puerto Rico's tabonuco forest. Biotropica, 38: 492–499.

Bidartondo MI, Read DJ (2008). Fungal specificity bottlenecks during orchid germination and development. Mol. Ecol., 17: 3707-3716.

Clements MA (1988). Orchid mycorrhizal associations. Lindleyana, 3: 73-86.

Cribb PJ, Kell SP, Dixon KW, Barrett RL (2003). Orchid conservation: a global perspective. In: Dixon KW, Kell SP, Barrett RL, Cribb PJ, eds. *Orchid conservation*. Kota Kinabalu, Sabah, Natural History Publications, pp.1–24.

Dixon KW (1989). Seed propagation of ground orchids. In: Dixon KW, Buirchell BJ, Collins MJ, eds. *Orchids of Western Australia,* 2nd edn. Victoria Park: Native Orchid Study and Conservation Group Inc, pp. 18-26.

Dixon KW, Cribb PJ, Kell SP, Barrett RL eds. (2003). Orchid conservation. Kota Kinabalu, Sabah: Natural History Publications.

Dressler RL (1993). Phylogeny and classification of the orchid family. Portland, OR: *Dioscorides* Press.

Espejo SA, García CJ, López FA, Jiménez MR, Sánchez LS (2002). Orquídeas del Estado de Morelos. Herbario AMO, Mexico City, p.332.

Gómez RV (2005). Diversidad de hongos ectomicorrízicos y su relación con diferentes unidades ambientales en el Parque Nacional Barranca del Cupatítzio, Uruapan, Mich. Master's Thesis. Natural Resource Conservation and Management. Faculty of Biology, UMSNH, p.67.

Gómez TA (1985). Levantamiento de suelos del Campo Experimental Forestal Barranca del Cupatítzio y sus relaciones con la vegetación de coníferas. Master's Thesis. Faculty of Sciences, UNAM. Mexico, p. 212.

Hágsater E, Soto AA, Salazar CHA, Jiménez-Machorro R, López-Rosas MA, Dressler RL (2005). Las orquídeas de México. Instituto Chinoín, Mexico, p.304.

Harper JL (1981). The meaning of rarity. In: Synge H, ed. The biological aspects of rare plant conservation. New York: J. Wiley and Sons Ltd, pp.189–203.

Instituto de Ecología, AC (2008). Centro Regional del Bajío. Av. Lázaro Cárdenas No. 253 A.P. 386 C.P. 61600 Pátzcuaro, Michoacán, Mexico.

Instituto Nacional de Estadística, Geografía e Informática (INEGI) (2008). Carta topográfica 000 E13B39. Morelia, Michoacán. México, pp.1-50

IUCN (1999). IUCN guidelines for the prevention of biodiversity loss due to biological invasion. Species, pp.31–42.

Koopowitz H, Lavarack PS, Dixon KW (2003). The nature of threats to orchid conservation. In: Dixon KW, Kell SP, Barrett RL, Cribb PJ, eds. *Orchid conservation*. Kota Kinabalu, Sabah: Natural History Publications, pp.25–42.

Mabberley DJ (1997). The plant book. Cambridge: Cambridge University Press.

McCormick MK, Whigham DF, O'Neill JP (2004). Mycorrhizal diversity in photosynthetic terrestrial orchids, New Phytol., 163: 425-438.
NORMA Oficial Mexicana NOM-059-ECOL-2001.

Ortega-LMP, Rangel VM (2007). Fungus-assisted reintroduction and long-term survival of two Mexican terrestrial orchids in the natural habitat. Lankesteriana, 7: 317-321.

Pate JS, Hopper SD (1993). Rare and common plants in ecosystems, with special reference to the south Western Australian flora. In: Schulze ED, Mooney HA, eds. *Ecosystem function of biodiversity*. New York: Springer Verlag, pp. 293–325.

Ramsay RR, Sivasithamparam K, Dixon KW (1986). Patterns of infection and endophytes associated with Western Australian orchids. Lindleyana, 1: 203-214.

Rangel VM, Ortega LMP (2007). Efforts to conserve endangered terrestrial orchids *in situ* and *ex situ* at two natural reserves within Central Mexico. Lankesteriana, 7: 326-333.

Roberts DL (2003). Pollination biology: the role of sexual reproduction in orchid conservation. In: Dixon KW, Kell SP, Barrett RL, Cribb PJ, eds. *Orchid conservation*. Kota Kinabalu, Sabah: Natural History

Publications, pp.113–136.
Rasmussen HN (1995). Terrestrial orchids, from seed to mycotrophic plant. Cambridge: Cambridge University Press.
Rasmussen HN (2002). Recent developments in the study of orchid mycorrhiza, Plant. Soil, 244: 149-163.
Sahagian D (2000). Global physical effects of anthropogenic hydrological alterations: sea level and water redistribution. Glob. Planet. Change, 25: 39-48.
Schemske DW, Husband BC, Ruckelshaus MH, Goodwillie C, Parker IM, Bishop JG (1994). Evaluating approaches to the conservation of rare and endangered plants. Ecol., 75: 584-606.
Sosa V (2007). Molecular and Morphological Phylogenetic Study of Subtribe Bletiinae (Epidendreae, Orchidaceae). Syst. Bot., 32: 34-42.
Soto MA (1988). Listado actualizado de las orquídeas de México. Orquídea (Mex.), 11: 233-277.
Stoutamire WP (1974). Terrestrial orchid seedlings. In: Withner CL (ed.) The orchids: scientific studies. New York: John Wiley & Sons, pp.101–108.
Stoutamire WP (1983). Wasp-pollinated species of Caladenia (Orchidaceae) in south-western Australia, Aust. J. Bot., 31: 383-394.
Swartz ND, Dixon KW (2009) Terrestrial orchid conservation in the age of extinction. Ann. Bot., 104: 543-556.
Warcup JH (1971). Specificity of mycorrhizal association in some Australian terrestrial orchids. New Phytol., 70: 41-46.
Warcup JH (1981). The mycorrhizal relationships of Australian orchids, New Phytol., 87: 371-381.
Woolcock CE, Woolcock DT (1984). Australian terrestrial orchids. Melbourne: Thomas Nelson.
Zavala AC (2006). Pteridoflora del Parque Nacional Barranca del Cupatitzio, Uruapan, Michoacán, Mexico. Thesis. UMSNH, p. 87.
Zettler LW, Sharma J, Rasmussen FN (2003). Mycorrhizal diversity. In: Dixon KW, Kell SP, Barrett RL, Cribb PJ, eds. *Orchid conservation*. Kota Kinabalu, Sabah: Natural History Publications, pp.205–226.

Permissions

All chapters in this book were first published in IJBC, by Academic Journals; hereby published with permission under the Creative Commons Attribution License or equivalent. Every chapter published in this book has been scrutinized by our experts. Their significance has been extensively debated. The topics covered herein carry significant findings which will fuel the growth of the discipline. They may even be implemented as practical applications or may be referred to as a beginning point for another development.

The contributors of this book come from diverse backgrounds, making this book a truly international effort. This book will bring forth new frontiers with its revolutionizing research information and detailed analysis of the nascent developments around the world.

We would like to thank all the contributing authors for lending their expertise to make the book truly unique. They have played a crucial role in the development of this book. Without their invaluable contributions this book wouldn't have been possible. They have made vital efforts to compile up to date information on the varied aspects of this subject to make this book a valuable addition to the collection of many professionals and students.

This book was conceptualized with the vision of imparting up-to-date information and advanced data in this field. To ensure the same, a matchless editorial board was set up. Every individual on the board went through rigorous rounds of assessment to prove their worth. After which they invested a large part of their time researching and compiling the most relevant data for our readers.

The editorial board has been involved in producing this book since its inception. They have spent rigorous hours researching and exploring the diverse topics which have resulted in the successful publishing of this book. They have passed on their knowledge of decades through this book. To expedite this challenging task, the publisher supported the team at every step. A small team of assistant editors was also appointed to further simplify the editing procedure and attain best results for the readers.

Apart from the editorial board, the designing team has also invested a significant amount of their time in understanding the subject and creating the most relevant covers. They scrutinized every image to scout for the most suitable representation of the subject and create an appropriate cover for the book.

The publishing team has been an ardent support to the editorial, designing and production team. Their endless efforts to recruit the best for this project, has resulted in the accomplishment of this book. They are a veteran in the field of academics and their pool of knowledge is as vast as their experience in printing. Their expertise and guidance has proved useful at every step. Their uncompromising quality standards have made this book an exceptional effort. Their encouragement from time to time has been an inspiration for everyone.

The publisher and the editorial board hope that this book will prove to be a valuable piece of knowledge for researchers, students, practitioners and scholars across the globe.

List of Contributors

Simone Alves de Oliveira
Federal University of Juiz de Fora, MG, Brazil

Alexander Machado Auad
Embrapa Dairy Cattle Research Center, Cep. 36038-330. Juiz de Fora - MG. Brazil

Brígida Souza
Federal University of Lavras, Brazil, Caixa Postal 3037, CEP 37200-000, Brazil

Marcy das Graças Fonseca
Embrapa Dairy Cattle Research Center, Cep. 36038-330. Juiz de Fora - MG. Brazil

Tiago Teixeira Resende
Embrapa Dairy Cattle Research Center, Cep. 36038-330. Juiz de Fora - MG. Brazil

Rama Shankar
Ayurveda Regional Research Institute, Itanagar, India-791111

G. S. Lavekar
Central Council for Research in Ayurvedic Science, New Delhi- 110058, India

Sourabh Deb
Department of Forestry and Biodiversity, Tripura University, Suryamaninagar, Tripura, India

B. K. Sharma
Department of Microbiology, Tripura University, Suryamaninagar, Tripura, India

M. S. Rawat
National Medicinal Plants Board, Chandralok Building, Janpath, New Delhi, 110001, India

M. Latha
National Bureau of Plant Genetic Resources Regional station, KAU P.O., Thrissur, Kerala 680 656 India

M. Abdul Nizar
National Bureau of Plant Genetic Resources Regional station, PKV Campus, Akola, Maharashtra 444 104, India

Z. Abraham
National Bureau of Plant Genetic Resources Regional station, KAU P.O., Thrissur, Kerala 680 656 India

K. Joseph John
National Bureau of Plant Genetic Resources Regional station, KAU P.O., Thrissur, Kerala 680 656 India

R. Asokan Nair
National Bureau of Plant Genetic Resources Regional station, KAU P.O., Thrissur, Kerala 680 656 India

S. Mani
National Bureau of Plant Genetic Resources Regional station, KAU P.O., Thrissur, Kerala 680 656 India

M. Dutta
National Bureau of Plant Genetic Resources, Pusa Campus, New Delhi 110012, India

E. P. S Assédé
Laboratory of Applied Ecology, Faculty of Agronomic Sciences, University of Abomey-Calavi, 03 BP 1974, Cotonou, Benin

E. P. S Adomou
Laboratory of Applied Ecology, Faculty of Agronomic Sciences, University of Abomey-Calavi, 03 BP 1974, Cotonou, Benin

B. Sinsin
Laboratory of Applied Ecology, Faculty of Agronomic Sciences, University of Abomey-Calavi, 03 BP 1974, Cotonou, Benin

Achyut Aryal
Ecology and Conservation Group, Institute of Nature Sciences, Massey University, New Zealand
The Biodiversity Research and Training Forum (BRTF), Nepal

Sven Gastaur
The Biodiversity Research and Training Forum (BRTF), Nepal

Steffen Menzel
The Biodiversity Research and Training Forum (BRTF), Nepal

Til Bahadur Chhetri
The Biodiversity Research and Training Forum (BRTF), Nepal

Jack B. Hopkins
Department of Ecology, Montana State University, Bozeman, MT 59717-3460, USA

S. H. Sharaf El Din
Oceanography Department, Faculty of Science, Alexandria University, Egypt

F. M. Eid
Oceanography Department, Faculty of Science, Alexandria University, Egypt

N. N. Saad
National Institute of Oceanography and Fisheries, Alexandria, Egypt

K. A. Alam El Din
Oceanography Department, Faculty of Science, Alexandria University, Egypt

M. El Sharkawy
National Institute of Oceanography and Fisheries, Alexandria, Egypt

T. Ramesh
Loyola Institute of Frontier Energy (LIFE), Loyola College, Chennai- 600 034, India

K. Jahir Hussain
Environmental and Industrial Safety Section, Safety Group, Indira Gandhi, Centre for Atomic Research, Kalpakkam- 603102, India

M. Selvanayagam
Loyola Institute of Frontier Energy (LIFE), Loyola College, Chennai- 600 034, India

K. K. Satpathy
Environmental and Industrial Safety Section, Safety Group, Indira Gandhi, Centre for Atomic Research, Kalpakkam- 603102, India

M. V. R. Prasad
Environmental and Industrial Safety Section, Safety Group, Indira Gandhi, Centre for Atomic Research, Kalpakkam- 603102, India

Tsewang Rinchen
Conservation Ecology Lab, Centre for Biodiversity Studies, School of Biosciences and Biotechnology, BGSBU University, Rajouri-185 131, J and K, India

Shreekar Pant
Conservation Ecology Lab, Centre for Biodiversity Studies, School of Biosciences and Biotechnology, BGSBU University, Rajouri-185 131, J and K, India

Mohmad Anwar
Conservation Ecology Lab, Centre for Biodiversity Studies, School of Biosciences and Biotechnology, BGSBU University, Rajouri-185 131, J and K, India

Wafaa. M. Amer
Department of Botany, Faculty of Science, University of Cairo, Giza 12613- Egypt

Ahmad. K. Hegazy
Department of Botany, Faculty of Science, University of Cairo, Giza 12613- Egypt

Safwat. A. Azer
Flora and Phytotaxonomy Researches Department, Agriculture Research Center, Dokki 12618, Giza- Egypt

K. K. Sharma
Department of Zoology, University of Jammu, Jammu-180006, Jammu and Kashmir, India

Komal Bangotra
Department of Zoology, University of Jammu, Jammu-180006, Jammu and Kashmir, India

Minakshi Saini
Department of Zoology, University of Jammu, Jammu-180006, Jammu and Kashmir, India

Yosef Mamo
Hawassa University, P.O.box 5, Hawassa, Ethiopia

Girma Mengesha
College of Forestry and Natural Resources, Hawassa University, Wondo Genet, P.O. Box 128, Ethiopia

Aramede Fetene
Debere Markos University, P. O. box 269, Debre Markos, Ethiopia

Kefyalew Shale
College of Forestry and Natural Resources, Hawassa University, Wondo Genet, P.O. Box 128, Ethiopia

Mezemir Girma
Mezemir Girma, Southern Nation Nationalities People Region, P.O. Box 69, Hawassa, Ethiopia

H. Badsi
Laboratory of Oceanography and Limnology, Faculty of Sciences, Department of Biology, Ibn Zohr University, P.O. Box. 8106, Agadir 80060, Morocco

H. Oulad Ali
Laboratory of Oceanography and Limnology, Faculty of Sciences, Department of Biology, Ibn Zohr University, P.O. Box. 8106, Agadir 80060, Morocco

M. Loudiki
Laboratory of Biology and Biotechnology of Microorganisms Faculty of Sciences Semlalia Department of Biology, Cadi Ayyad University, Marrakech, Morocco

M. El Hafa
Laboratory of Oceanography and Limnology, Faculty of Sciences, Department of Biology, Ibn Zohr University, P.O. Box. 8106, Agadir 80060, Morocco

R. Chakli
Laboratory of Oceanography and Limnology, Faculty of Sciences, Department of Biology, Ibn Zohr University, P.O. Box. 8106, Agadir 80060, Morocco

A. Aamiri
Laboratory of Oceanography and Limnology, Faculty of Sciences, Department of Biology, Ibn Zohr University, P.O. Box. 8106, Agadir 80060, Morocco

Rahatullah
Department of Zoology, Hazara University, Mansehra, Pakistan

Faizul Haq
Department of Botany, Degree College, Battagram, Pakistan

Sardar Azhar Mehmood
Department of Zoology, Hazara University, Mansehra, Pakistan

Kausar Saeed
Department of Zoology, Hazara University, Mansehra, Pakistan

Shamsur Rehman
Department of Science, High School Battamori, Battagram, Pakistan

B. P. Harini
Department of Zoology, Drosophila Culture Laboratory, Bangalore University, Jnana Bharathi Campus, Bangalore 560 056, India

D. S. Pranesh Sujaymeendra
Department of Zoology, Drosophila Culture Laboratory, Bangalore University, Jnana Bharathi Campus, Bangalore 560 056, India

R. Brilliant
Department of Environmental Science, St. John's College, Anchal, India

Vincy Mary Varghese
School of Environmental Sciences, Mahatma Gandhi University, Kottayam, India

Joby Paul
School of Environmental Sciences, Mahatma Gandhi University, Kottayam, India

A. P. Pradeepkumar
Department of Geology, University of Kerala, Kariavattom, India

S. A. Shameem
Division of Environmental Sciences, SKUAST-Kashmir, Srinagar (J and K), Forest Research Institute University, Dehradun, Uttarakhand, India

P. Soni
Ecology and Environment Division, Forest Research Institute Dehradun, Uttarakhand, India

G. A. Bhat
Postgraduate Department of Environmental Sciences/ CORD, University of Kashmir, Srinagar (J and K), India

Kabi Prasad Pokhrel
Geography, Research Division, TU, Kirtipur, Tribhuvan University, Nepal

Peerzada Arshid Shabir
Economic Botany and Reproductive Biology Laboratory, Department of Botany, University of Kashmir, Srinagar-190006. J&K, India

Irshad Ahmad Nawchoo
Economic Botany and Reproductive Biology Laboratory, Department of Botany, University of Kashmir, Srinagar-190006. J&K, India

Aijaz Ahmad Wani
Economic Botany and Reproductive Biology Laboratory, Department of Botany, University of Kashmir, Srinagar-190006. J&K, India

F. A. Bhat
Faculty of Fisheries, Sher-e-Kashmir University of Agricultural Sciences and Technology of Kashmir, Srinagar, India -190 006

A. R. Yousuf
Centre of Research for Development (CORD), University of Kashmir, Srinagar, India- 190 006

A. Aftab
Centre of Research for Development (CORD), University of Kashmir, Srinagar, India- 190 006

J. Arshid
Centre of Research for Development (CORD), University of Kashmir, Srinagar, India- 190 006

M. D. Mahdi
Centre of Research for Development (CORD), University of Kashmir, Srinagar, India- 190 006

M. H. Balkhi
Faculty of Fisheries, Sher-e-Kashmir University of Agricultural Sciences and Technology of Kashmir, Srinagar, India -190 006

Blandine Marie Ivette Nacoulma
Department of Plant Biology and Physiology, Laboratory of Plant Biology and Ecology, University of Ouagadougou, UFR-SVT, 09 BP 848, Ouagadougou 09, Burkina Faso

Salifou Traoré
Department of Plant Biology and Physiology, Laboratory of Plant Biology and Ecology, University of Ouagadougou, UFR-SVT, 09 BP 848, Ouagadougou 09, Burkina Faso

Karen Hahn
Department of Ecology and Geobotany, Institute for Ecology, Evolution and Diversity, J. W. Goethe University, Siesmayerstraße 70, 60323 Frankfurt am Main, Germany

Adjima Thiombiano
Department of Plant Biology and Physiology, Laboratory of Plant Biology and Ecology, University of Ouagadougou, UFR-SVT, 09 BP 848, Ouagadougou 09, Burkina Faso

K. A. Monney
Department of Entomology and Wildlife, School of Biological Sciences, University of Cape Coast, Cape Coast, Ghana

M. L. Darkey
Department of Entomology and Wildlife, School of Biological Sciences, University of Cape Coast, Cape Coast, Ghana

K. B. Dakwa
Department of Entomology and Wildlife, School of Biological Sciences, University of Cape Coast, Cape Coast, Ghana

R. Shyam Kumar
Department of Biotechnology, Kamaraj College of Engineering and Technology, Virudhunagar, 626 001, India
Department of Botany and Microbiology, College of Science, King Saud University, Riyadh, KSA, Saudi Arabia

N.Thajuddin
Department of Microbiology, Bharathidasan University, Trichirapalli, 620 024, India

D. K. Upreti
Lichenology Laboratory, National Botanical Research Institute, Lucknow, 226 001, India

P. Deb
G.B. Pant Institute of Himalayan Environment and Development, Kosi-Katarmal, Almora, Uttaranchal, 263643 India
Centre for Biodiversity and Natural Resources Conservation, Assam University, Silchar, India

R. C. Sundriyal
G.B. Pant Institute of Himalayan Environment and Development, Kosi-Katarmal Almora, Uttaranchal, 263643 India
Herbal Research and Development Institute, Gopeshwar (Chamoli) - 246401, Uttarakhand, India

F. Katagira
Ministry of Agriculture, Food Security and Cooperatives, P. O. Box 9071, Dar-es-Salaam, Tanzania

S. Kyamanywa
Makerere University, Faculty of Agriculture, P. O. Box 7062, Kampala, Uganda

J. S.Tenywa
Makerere University, Faculty of Agriculture, P. O. Box 7062, Kampala, Uganda

C. A. Rajabu
Mikocheni Agricultural Research Institute, Box 6226, Dar es Salaam, Tanzania

D. Sombe
Lake Victoria Environment Management Project, Box Bukoba, Kagera, Tanzania

J. Ndunguru
Mikocheni Agricultural Research Institute, Box 6226, Dar es Salaam, Tanzania

Claid Mujaju
Department of Plant Breeding and Biotechnology, Swedish University of Agricultural Sciences, Balsgård, Fjälkestadsvägen 459, SE-291 95 Kristianstad, Sweden
Seed Services, Department of Research and Specialist Services, Ministry of Agriculture, Mechanization and Irrigation Development, Box CY550, Harare, Zimbabwe

Moneim Fatih
Box 91, The Swedish Biodiversity Center (CBM), Swedish University of Agricultural Sciences, SE-230 53 Alnarp, Sweden

María de los Ángeles Beltrán-Nambo
Laboratory of Genetics and Microbiology, Faculty of Biology, Universidad Michoacana de San Nicolás de Hidalgo, México

Pilar Ortega Larrocea
Bio-edaphic Microcosm Laboratory, Department of Edaphology, Institute of Geology, Universidad Nacional Autónoma de México

Rafael Salgado Garciglia
Plant Biotechnology Laboratory, Institute of Chemical Biologics, Universidad Michoacana de San Nicolás de Hidalgo, México

Joel Tupac Otero Ospina
National University of Colombia, Sede Palmira, Colombia

Miguel Martínez Trujillo
Laboratory of Genetics and Microbiology, Faculty of Biology, Universidad Michoacana de San Nicolás de Hidalgo, México

Yazmín Carreón-Abud
Laboratory of Genetics and Microbiology, Faculty of Biology, Universidad Michoacana de San Nicolás de Hidalgo, México